Catalogue of Indian Insects

THE EDITOR

Thomas Bainbrigge Fletcher (1878-1950) was an English entomologist. He worked in the Royal Navy; he became an expert on the "microlepidoptera" and was appointed as the second Imperial Entomologist in India, succeeding Harold Maxwell-Lefroy at the Imperial Agricultural Research Institute at Pusa. He was a meticulous naturalist and very careful on matters of systematics and taxonomic nomenclature.

He produced a list of publications on India Entomology and Catalogue of Indian Insects. He also wrote general works on entomology including Some South Indian Insects, and a Veterinary Entomology for India and hints on collecting and preserving insects. He was Fellow of the Linnean Society, the Royal Entomological Society, Zoological Society of London and a President of the Cotteswold naturalists field club.

Catalogue of Indian Insects

Volume 1

–Editor–

Thomas Bainbrigge Fletcher

2016

Daya Publishing House®

A Division of

Astral International Pvt. Ltd.

New Delhi – 110 002

Catalogue of Indian Insects (5 Volumes Set)

Publisher's note:

The book "***Catalogue of Indian Insects***" by Thomas Bainbrigge Fletcher was published in 1921. But presently complete set not available anywhere. We have tried to reproduced the complete valuable set from different sources in a form as close to the original possible. This historical reproduction fulfill this long felt needs of students, teachers and researchers engaged in the field of entomology, zoology and other related discipline.

Cataloging in Publication Data—DK
Courtesy: D.K. Agencies (P) Ltd. <docinfo@dkagencies.com>

Catalogue of Indian insects / editor, Thomas Bainbrigge Fletcher.

volumes cm
First published in 1921.
ISBN: 978-93-86071-06-4 (Vol. 1)
ISBN: 978-93-5130-922-2 (Set) (International Edition)

1. Insects—Catalogs and collections—India. I. Fletcher, T. B. (Thomas Bainbrigge), 1878-1950, editor.

QL468.2.C38 2016 DDC 595.70954 23

Published by : **Daya Publishing House®**
A Division of
Astral International Pvt. Ltd.
– ISO 9001:2008 Certified Company –
4760-61/23, Ansari Road, Darya Ganj
New Delhi-110 002
Ph. 011-43549197, 23278134
E-mail: info@astralint.com
Website: www.astralint.com

Printed at : **Replika Press Pvt. Ltd.**

डा. एस. अय्यप्पन
सचिव एवं महानिदेशक
Dr. S. AYYAPPAN
SECRETARY & DIRECTOR GENERAL

भारत सरकार
कृषि अनुसंधान और शिक्षा विभाग एवं
भारतीय कृषि अनुसंधान परिषद
कृषि एवं किसान कल्याण मंत्रालय, कृषि भवन, नई दिल्ली 110 001

GOVERNMENT OF INDIA
DEPARTMENT OF AGRICULTURAL RESEARCH & EDUCATION
AND
INDIAN COUNCIL OF AGRICULTURAL RESEARCH
MINISTRY OF AGRICULTURE AND FARMERS WELFARE
KRISHI BHAVAN, NEW DELHI 110 001
Tel.: 23382629; 23386711 Fax: 91-11-23384773
E-mail: dg.icar@nic.in

Foreword

With rapidly increasing human population that hardly matched by the rate of increase in agriculture production, the per capita availability of food and other products is bound to fall short. Therefore, increasing production per unit area is the most plausible alternative to sustain self sufficiency in food, fodder and fibre. Insects have got both positive and negative impacts in agriculture. Insects become pests when they interfere with our welfare, aesthetics or profits. A reduction or loss in yield and quality of crops is partly due to result of insect feeding activities and as vector/host to pathogen transmissions. On the other hand these small creatures serve us a lot as pollinating agents of many important agricultural crops, as biological controlling agents of pests and weeds as well as by producing honey, silk and many other materials essential to us.

This book 'Catalogue of Indian Insects' is divided into 5 volumes and covers 29 parts. This historical compilation is a dedicated effort to present the different aspects of the insects in a very concise and relevant manner. I believe that the students and scientist working in the area of entomology, agriculture, fisheries, zoology, botany and veterinary science will find this book as a source of useful and authentic information.

I appreciate the efforts of Daya Publishing House, A Division of Astral International (P) Ltd. New Delhi.

(Dr. S. Ayyappan)
Secretary (DARE) & Director General (ICAR)
Krishi Bhawan, New Delhi – 110001

Preface

There is a immense number of insect species on the planet. The class Insecta is a large and diverse as well as important ecologically. When compared to other groups, the number of insect species is much larger and represents more than three quarters of all known animal species. Most insects are either beneficial or harmless. Yet there are large numbers which are economically important as pests, pollinators, natural enemies and other beneficial insects like honey bees, silkworms, etc. Insects are fascinating to investigate, their classification is complex. It important to learn the main differences among insects so you can distinguish one group from another. Knowledge of insect classification is of primary importance in carrying out proper control procedures and quality pest management. Proper insect classification and identification is extremely important. If a beneficial or harmless insect is improperly identified, the resultant unnecessary and undesirable pesticide application usually disrupts natural control agents.

Insects deliver a host of ecological services fundamental to the survival of humankind. Insects play an important role in plant reproduction. Thousands of pollinator species have been identified and almost all of these are insects. Over 90 percent of the flowering plant species depend on pollinators. Insects play an equally vital role in waste biodegradation. They clean up dead plant matter, breaking down organic matter until it is fit to be consumed by fungi and bacteria. Insects also play a significant role in decomposing animal carcasses and manure. Insects have a large array of natural enemies, predators and parasitoids, keeping them under economic threshold levels. Due to large number of insects species it is necessary that these catalogued so as to enable locating the species into correct classification. One shall also know what are the species knows from a specific group. Especially in case of Indian Insects where large number of species are known. These are important Catalogue of Indian Insects as a compilation serves this purpose.

"Catalogue of Indian Insects" is divided into 5 volumes and covers 29 parts. Volume 1 deals with Part I to Part IX. It covers Acrydidae, Culicidae, Bombyliidae, Trypetidae, Nitidulidae, Stapylinidae, Lasiocampidae, Amatidae and Zygaenidae. Volume 2 gives a detailed account of Part X to Part XVII. This volume covers Stephanidae, Brenthidae, Tabanidae,

Cicindelidae, Palpicornia, Cecidomyidae, Cosmopterygidae and Yponomeutidae. Volume 3 is devoted to Part XVIII i.e. Carabidae. Volume 4 covers in detail of Part XIX to Part XXV. This volume deals with Gyrinoidea, Alucitidae, Lycidae, Phalonidae and Chlidanotidae, Chalcidoidea, Evanidae and Thysanoptera. Volume 5 of the book describes the Part XXVI to Part XXIX. This volume deals with Serphoidea, Isoptera, Anthribidae and Asilidae.

This historical reproduction fulfill the long felt needs of students, teachers, researchers engaged in the field of entomology, agriculture, ecology, fisheries, forestry, paleontology, zoology, botany, plant science, medical and veterinary science, towards knowing insects.

We are extremely grateful to Dr. S. Ayyapppan, Secretary and Director General, Indian Council of Agricultural Reserarch, New Delhi for providing a foreword. The cooperation and support of Dr. V.V. Ramamurthy, Principal Scientist (Retd.), Division of Entomology, IARI, New Delhi is duly acknowledged.

All suggestions for future improvement for this book will be highly solicited.

Publisher

Contents

VOLUME 3

VOLUME 4

VOLUME 5

Part I
ACRYDIDAE (TETTIGIDÆ)

TRIPETALOCERINÆ

Tripetalocerini, Kirby, F.I. Orth. p. 11 (1914) [key to genera]

Tripetalocerinae, Hancock, Rec. Ind. Mus. XI 55-56 (1915) [key to genera].

1. Tripetalocera, Westwood (1834).

Type : *ferruginea,* Westw.

Westw., Zool. Journ. V 444 (1834); Bol., Ann. S. E. Belg. XXXI 190, 196, 310 (1887); Hanc., Gen. Ins. Tetr. p. 4 (1906); Kirby, Cat. Orth. III, 1 (1910), F.I. Orth. p. 12 (1914) [redescr.]

ferruginea, Westw., Zool. Journ. V 444, t. 22 f. 3 (1834)[1]; de Haan, Temm. Verh. Orth. pp. 166, 168, t. 22 f. 11 (1842) [*Acridium (Tetrix) ferrugineum*][2]; Bol., Ann. S. E. Belg. XXXI 310, t. 5 f. 28 (1887)[3]; Kirby, Cat. Orth. III 1 (1910)[4], F. I. Orth. pp. 12-13, ff. 15, 16 (1914)[5].	Travancore[5] Malabar[4] ——— Penang[4] Borneo[4] (eadem?)[4]

2. Birmana, Brunner (1893).

Type : *gracilis,* Brunn.

Brunn., Ann. Mus. Genova XXXIII 113-114 (1893); Hanc., Gen. Ins. Tetr. p. 4 (1906); Kirby, Cat. Orth. III 1 (1910), F.I. Orth. pp. 13-14 (1914); Hanc., Rec. Ind. Mus. XV 61 (1915).

gracilis, Brunn., An Mus. Genova XXXIII 114, t. 5f. 47, (1893)[1]; Kirby, Cat. Orth III[1] (1910),[2] F. I. Orth p. 14 (1914)[3]; Hanc, Rec. Ind. Mus. XI 61 (1915)[4].	Bhamo[1, 4]

CLADONOTINÆ

Cladonotini, Kirby, F. I. Orth. p. 14 (1914) [key to genera]

Cladonotinæ, Hanc., Rec. Ind. Mus. XI 56, 61 (1915) [key to genera].

3. Fieberiana, Kirby (1914).

Type : *pachymerus,* Fieb.

Plagiocephalus [*nec* Macq.], Fieber, Abh. Königl.-böhm. Ges. Wiss. (5) III 407 (1845).
Type: *pachymerus,* Fieb.

Fieberiana, Kirby, F. I. Orth. p. 14 (1914).
Type: *pachymerus* Fieb.

pachymerus, Fieber, Abh. Königl.-böhm. Ges. Wiss. (5) III 407 (1845) [*Plagiocephalus*][1]; Kirby, Cat. Orth. III 7 (1910) [*Piezotettix*][2], F. I. Orth. p. 15 (1914) [redescr.][3]
India[3]

4. Oxyphyllum, Hancock (1909).

Type : *pennatum,* Hanc.

Hanc., T. E. S. 1908, 393 (1909), Rec. Ind. Mus. XI 56, 61 (1915).

pennatum, Hanc., T. E. S. 1908, 393-394, t. 22 f. 3 (1909)[1], Rec. Ind. Mus. XI 61 (1915)[2]
Darjiling[1, 2]

5. Deltonotus, Hancock (1904).

Type : *subcucullatus,* Wlk.

Pœcilotettix [*nec* Scudd.], Bol., Ann. S. E. Fr. LXX 580 (1902).
Type : *gibbiceps,* Bol.

Deltonotus, Hanc., Spol. Zeyl. II 107, 111 (1904), Gen. Ins. Tetr. pp. 9, 14 (1906); Kirby, Cat. Orth. III 7 (1910), F. I. Orth. p. 15 (1914) [redescr.]; Hanc., Rec. Ind. Mus. XI 56, 61 (1915).
Type : *subcucullatus,* Wlk.

gibbiceps, Bol., Ann. S. E. Fr. LXX 580 (1902) [*Pœcilotettix*][1]; Kirby, Cat. Orth. III 7 (1910)[2], F. I. Orth. pp. 16-17 (1914)[3]; Hanc., Gen. Ins. Tetr. p. 14 (1906)[4], Rec. Ind. Mus. XI 61 (1915)[5].
Madura[1, 2, 3, 5]

subcucullatus, Wlk., Cat. Derm. Salt. V 830 (1871) [*Tettix subcucullata*][1].
Ceylon[1, 5]
Pundaluoya[2, 4, 6]

Deltonotus tectiformis, Hanc., Spol. Zeyl. II 107, 111, 154, t. 1 ff. 2, 2*a* (1904)[2], Gen. Ins. Tetr. p. 14, t. 1 f. 1 (1906)[3], T. E. S. 1907, 215 (1907)[4].
Hantane[2]
Kandy[7]

Deltonotus subcullatus, Kirby, Cat. Orth. III 7 (1910)[5], F. I. Orth. pp. 15-16, f. 17 (1914)[6].

Deltonotus subcullatus, Hanc., Rec. Ind. Mus. XI 61 (1915)[7].

6. Cladonotus, Saussure (1861).

Type : *humbertianus,* Sauss.

Sauss., Ann. S. E. Fr. (4) I 478 (1861); Bolivar, Ann. S. E. Belg. XXXI 184, 192, 208 (1887); Hancock, Spol. Zeyl. II 107, 112-113 (1904) [redescr.], Gen. Ins. Tetrig. pp. 10, 16 (1906) [charact.], T. E. S. 1908 394-395 (1909) [revised]; Kirby, Cat. Orth. III 9 (1910), F. I. Orth. p. 17 (1914) [redescr.]; Hanc., Rec. Ind. Mus. XI 56, 61-62 (1915) [key to spp.].

humbertianus, Sauss., Ann. S. E. Fr. (4) I 478 (1861)[1]; Bol., Ann. S. E. Belg. XXXI 209, t. 4. ff. 10, 10*a* (1887)[2]; Hanc., Spol. Zeyl. II 107, 113-114 (1904)[3]; Kirby, Cat. Orth. III 9 (1910)[4], F. I. Orth. pp. 17-18, f. 18 (1914)[5]; Hanc., Rec. Ind. Mus. XI 62 (1915)[6]

Ceylon[4, 6]
Peradeniya[2, 5]
Trincomali[2, 5]

latiramus, Hanc., Spol. Zeyl. II 107, 113, 114-116, t. 1 ff. 1, 1*a* (1904)[1], Gen. Ins. Tetrig. p. 16, t. 1 f. 3 (1906)[2]; Kirby, Cat. Orth. III 9 (1910)[3], F. I. Orth. p. 19 (1914)[4]; Hanc., Rec. Ind. Mus. XI 62 (1915)[5]

Ceylon[3, 5]
Kandy[1, 4]

turrifer, Wlk., Cat. Derm. Salt. V 843 (1871)[1]; Kirby, Cat. Orth. III 9 (1910)[2], F. I. Orth., pp. 18-19, f. 19 (1914)[3]; Hanc., Rec. Ind. Mus. XI 62 (1915)[4].

Ceylon[1-4]

7. Tettilobus, Hancock (1909).

Type : *spinifrons,* Hanc.

Tettilobus, Hanc., T. E. S. 1908 396 (1909), Rec. Ind. Mus. XI 56, 62 (1915).

Type : *spinifrons,* Hanc

Mnema, Kirby, F. I. Orth. p. 19 (1914).

Type : *pelops,* Wlk.

pelops, Wlk., Cat. Derm. Salt. V 843 (1871) [*Cladonotus*][1]; Kirby, Cat. Orth. III 9 (1910) [*Cladonotus*][2]; Kirby, F. I. Orth. p. 20, ff. 20, 21 (1914) [*Mnema*][3]; Hanc., Rec. Ind. Mus. XI 62 (1915)[4].

Ceylon[1-4]

spinifrons, Hanc., T. E. S. 1908 396-397, t. 22 f. 4 (1909)[1], Rec. Ind. Mus. XI 62 (1915)[2].

Ceylon[1, 2]

8. Gignotettix, Hancock (1909).

Type : *burri,* Hanc.

Hanc., T. E. S. 1908 397-398 (1909), Rec. Ind. Mus. XI 56, 62 (1915).

burri, Hanc., T. E. S. 1908 398, t. 22 f. 5 (1909)[1], Rec. Ind. Mus. XI 62 (1915)[2].

Pundaluoya[1]

9. Potua, Bolivar (1887).

Type : *coronata,* Bol. [Borneo; Malacca].

Bol., Ann. S.E. Belg. XXXI 183, 192, 208 (1887); Hanc., Gen. Ins. Tetrig. pp. 10, 19 (1906) [charact.]; Kirby, Cat. Orth. III 11 (1910); Hanc., Rec. Ind. Mus. XI 56, 62 (1915) [charact.]

sabulosa, Hanc., Rec. Ind. Mus. XI 62-63 (1915)[1]. — Satara distr.[1]

SCELIMENINÆ.

Scelimenini, Kirby, F. I. Orth. pp. 20-21 (1914) [key to genera].

Scelimeninæ, Hanc., Rec. Ind. Mus. XI 56-58 [key to genera], 63 (1915).

10. Amphibotettix, Hancock (1906).

Type : *longipes,* Hanc. [Borneo].

Hanc., Entl. News XVIII 86 (1906), Gen. Ins. Tetrig. p. 22 (1906); Kirby, Cat. Orth. III 12 (1910); Hanc., Rec. Ind. Mus. XI 57, 63 (1915).

rosaceus, Hanc., Rec. Ind. Mus. XI 63-64 (1915)[1]. — Dawna Hills[1]

11. Scelimena, Serville (1839).

Type : *producta,* Serv. [Java].

Scelimena, Serv., Ins. Orth. p. 762 (1839); Bol., Ann. S. E. Belg. XXXI 184, 193, 215 (1887); Hanc., Spol. Zeylan. II 107, 116, 154 (1904) (redescr.], Gen. Ins. Tetrig. pp. 21, 23 (1906) [charact.]; Kirby, Cat. Orth. III 12 (1910), F. I. Orth. p. 21 (1914) [redescr.]; Hanc., Rec. Ind. Mus. XI 57, 64-65 (1915) [charact.; key to Indian spp.] — Type : *producta,* Serv.

Scelymena, Sauss., Ann. S. E. Fr. (4) I 484 (1861). — Type : *producta,* Serv.

Scelhymena, Bol. Ann. S. E. Fr. LXX 581 (1902). — Type : *harpago,* Serv.

contracta, Wlk., Cat. Derm. Salt. V 841 (1871)[1]; Kirby, Cat. Orth. III 13 (1910) [*Scelimena ?*][2]. — India[2]

gavialis, Sauss., Ann. S. E. Fr. (4) I 485 (1861) [*Scelymena*][1].

Scelymena nodosa, Wlk., Cat. Derm. Salt. V 840 (1871)[2].

Ceylon[5]
Peradeniya[4, 7]
Pundaluoya[4, 7, 9]
Dambulla[4, 7]

Scelimena harpago [*nec* Serv.], Green, Ent. Mo. Mag. XXXVIII 215 (1902)[3].

Scelimena gavialis, Hanc., Spol. Zeyl. II 107, 117-120, 154, t. 1 ff. 4, 4 *a-c* (1904)[4]; Kirby, Cat. Orth. III 13 (1910)[5]; Hanc., Spol. Zeyl. VI 141-143 (1910) [habits][6]; Kirby, F. I. Orth. pp. 24-25, f. 24 (1914)[7]; Hanc., Rec. Ind. Mus. XI 66 (1915)[8], T. E. S. 1907, 218-219 (1907)[9].

Maskeliya[4, 6, 7] Madulsima[6, 8] Haputale[6] Labugama [6, 8] Kandy[9] Pollibetta (Coorg).

harpago, Serv., Ins. Orthopt. p. 763 (1839) [*Tetrix*][1].

Tetrix harpago, de Haan, Temm., Verh., Orth. p. 166 (1842)[2].

Scelimena harpago, Bol., Ann. S. E. Belg. XXXI 216-217, t. 4 f. 13 (1887)[3].

Scelhymena harpago, Bol., Ann. S. E. Fr. LXX 581 (1902)[4].

Scelimena harpago, Kirby, Cat. Orth. III 13 (1910)[5]; Hanc., Ind. Agric. Ent. Mem. IV 132 (1912)[6]; Kirby, F. I. Orth. pp. 23-24, f. 23 (1914)[7]; Hanc., Rec. Ind. Mus. XI 65-66 (1915) [variation][8].

Bombay[1, 4, 5, 7] Matheran[6] Bandra[7] Igatpuri[8] Satara distr.[8] Kasara[6] Almora[7] Coimbatore[7] S. India[4] Ceylon[4] Ceylon[3]

logani, Hanc., Spol. Zeyl. II 107, 120-122, t. 1 ff. 5, 5 *a-c* (1910)[1], Gen. Ins. Tetrig. p. 24, t. 2 f. 15 (1906)[2]; Kirby, Cat. Orth. III 13 (1910)[3]; Hanc., Spol. Zeyl. VI. 143 (1910) [habits][4]; Kirby, F. I. Orth. pp. 25-26, f. 25 (1914)[5]; Hanc., Rec. Ind. Mus. XI 65 (1915) [charact.][6], T. E. S. 1907 218 (1907)[7]; Lefroy, Ind. Ins. Life, f. 21 on p. 76 (1909)[8].

Kandy[1, 5] Haragama[1, 5] Pundaluoya[5] Madulsima[4] Kelawewa (N.C.P.)[7]

spinata, Hanc.

Scelimena gavialis [*nec* Sauss.], Kirby, F. I. Orth. pp. 24-25 [part.] (1914)[1].

Scelimena spinata, Hanc., Rec. Ind. Mus. XI 66-67 (1915)[2].

Trivandrum[1, 2]

uncinata, Serv., Orth. pp. 763-764 (1839) [*Tetrix*][1].

Scelymena contracta, Wlk., Cat. Derm. Salt. V 851 (1871)[2].

Scelimena uncinata, Bol., Ann. S. E. Belg. XXXI 216, 218 (1887)[3].

Scelimena ? uncinata, Kirby, Cat. Orth. III 13 (1910)[4].

Scelimena uncinata, Kirby, F. I. Orth. p. 26, f. 26 (1914)[5]; Hanc., Rec. Ind. Mus. XI 65-66 (1915) [doubtful sp.][6].

Bombay[1, 5]

12. Eugavialidium, Hancock (1906).

Type : *dentiumeris*, Hanc. [Borneo].

Hanc., Gen. Ins. Tetrig., pp. 21, 24 (1906); Kirby, Cat. Orth. III 14 (1910), F. I. Orth. p. 30 (1914) [redescr.]; Hanc., Rec. Ind. Mus. XI 57, 67-70 (1915) [key to spp.].

angulatum, Hanc., Rec. Ind. Mus. XI 73 (1915)[1]. — Calcutta[1]

birmanicum, Brunn., Ann. Mus. Genova XXXIII 104, t. 5 f. 37 (1893) [*Gavialidium*][1]. — Burma[3, 5]; Karen Hills[1, 4]

Eugavialidium birmani, Hanc., Gen. Ins. Tetrig. p. 25 (1996)[2].

Eugavialidium birmanicum, Kirby, Cat. Orth. III 14 (1910)[3].

Scelimena ? birmanica, Kirby, F. I. Orth. p. 27 (1914)[4].

Eugavialidium birmanicum, Hanc., Rec. Ind. Mus. XI 73 (1915)[5].

discale, Hanc.

Scelimena producta [*nec* Serv.], Brunn., Ann. Mus. Genova XXXIII 103 [1893)[1]; Kirby, F. I. Orth. p. 22 [part.] (1914)[2]. — Karen Hills[1, 2]; Sibsagar[3]

Eugavialidum discalis, Hanc., Rec. Ind. Mus. XI 71 (1915)[3].

feæ, Bol., R. Soc. Españ. Nat. Hist. IX 396 (1909)[1]; Hanc., Rec. Ind. Mus. XI 73 (1915)[2]. — Karen Hills[1, 2]

flavopictum, Bol., Real. Soc. Españ. Nat. Hist. IX 394-395 (1909)[1]; Hanc., Rec. Ind. Mus. XI 73 (1915)[2]. — Calcutta[1, 2]

india, Hanc.

Scelimena india, Hanc., T. E. S. 1907 219 (1907) [male][1]; Kirby, Cat. Orth. III 576 (1910)[2], F. I. Orth. p. 23 (1914) [orig. descr. reprod.][3] — Cherrapunji[1, 3]; Assam[1, 2, 4]; Shillong

Eugavialidium indicum, Hanc., Rec. Ind. Muc. XI 72 (1915) [charact.][4]

kempi, Hanc.

Scelimena india [*nec* Hanc. 1907], Hanc., Rec. Ind. Mus. VIII 311 (1913)[1]. — Aborland[1, 2]

Eugavialidium kempi, Hanc., Rec. Ind. Mus. XI 72-73 (1915)[2].

multidentatum, Hanc.

Scelimena producta [*nec* Serv.], Kirby, F. I. Orth., p. 22 [part.] (1914)[1]. — Dawna Hills[1, 2]

Eugavialidium multidentatum, Hanc., Rec. Ind. Mus. XI 70-71 (1915)[2].

saussurei, Hanc., Rec. Ind. Mus. XI 74 (1915)[1]. — Calcutta[1]

13. Abbasia, Kirby (1914).

Type : *subserrata*, Kirby.

Kirby, F. I. Orth., pp. 27-28 (1914).

subserrata, Kirby, F. I. Orth. p. 28, f. 27 (1914)[1]. — Tenmalai[1] [Travancore]

14. Gavialidium, Saussure (1861).

Type : *crocodilus*, Sauss.

Sauss., Ann. S. E. Fr. (4) I 481 (1861); Bol., Ann. S. E. Belg. XXXI 184, 193, 218 (1887); Hanc., Spol. Zeyl. II 107, 122 (1904) [redescr.], Gen. Ins. Tetrig. pp. 22, 25 (1906) [charact.]; Kirby, Cat. Orth. III 14 (1910), F. I. Orth. p. 28 (1914) [redescr.]; Hanc., Rec. Ind. Mus. XI 57 [charact.], 74 (1915).

alligator, Sauss., Ann. S. E. Fr. (4) I 483 (1861) [*Scelymena*][1]; Bol., Ann. S. E. Belg. XXXI 218, 219 (1887)[2]; Hanc., Spol. Zeyl. II 107, 122, 125 (1904)[3]; Kirby, Cat. Orth. III 15 (1910)[4], F. I. Orth. pp. 29-30 (1914)[5]; Hanc., Rec. Ind. Mus. XI 74 (1915) [probably immature *crocodilus*][6]. — Ceylon[1, 3, 4]; Kadugannawa[5]

crocodilus, Sauss., Ann S. E. Fr. (4) I 481 (1861) [*Scelymena*][1]; Bol., Ann. S. E. Belg. XXXI 218, 219 (1887)[2]; Hanc., Spol. Zeyl. II 107, 122, 123-124, t. 2 *ff.* 11, 11 *a,b.* (1904)[3], Gen. Ins. Tetrig. p. 25, t. 2 ff. 16, 16*a* (1906)[4]; Kirby, Cat. Orth. III 14 (1910)[5]; Hanc., Spol. Zeyl. VI 143-144 (1910) [habits][6]; Kirby, F. I. Orth. p. 29, f. 28 (1914)[7]; Hanc., Rec. Ind. Mus. XI 74 (1915)[8], T. E. S. 1907 220 (1907)[9]. — Ceylon[5]; Kadugannawa[1, 3, 7]; Peradeniya[8]; Pundaluoya[3, 7, 8, 9]; Haputale[6]; Madulsima[6]

15. Bolotettix, Hancock (1907).

Type : *validispinus*, Hanc. [N. W. Borneo].

Hanc., T. E. S. 1907 223 (1907); Kirby, Cat. Orth. III 577 (1910); Hanc., Rec. Ind. Mus. XI 57 [charact.], 75 [key to Indian spp.] (1915).

anomalus, Hanc., Spol. Zeyl. VI 146 (1910) [*Systolederus*][1]; Kirby, F. I. Orth. p. 46 (1914) [*Systolederus*][2]; Hanc., Rec. Ind. Mus. XI 76 (1915)[3]. — Madulsima[1-3]

armatus, Hanc., Rec. Ind. Mus. XI 76-77 (1915)[1]. — Dawna Hills[1]

exsertus, Bol.

Criotettix exsertus, Bol., Ann. S. E. Fr. LXX 583 (1902)[1]; Kirby, Cat. Orth. III 16 (1910)[2], F. I. Orth. pp. 34-35, f. 33 (1914)[3]. — Kodaikanal[1-4]

Bolotettix exsertus, Hanc., Rec. Ind. Mus. XI 79 (1915)[4].

inermis, Hanc., Rec. Ind. Mus. XI 77-78 (1915)[1]. — Darjiling distr.[1] (4,000 feet)

lobatus, Hanc., — Shillong, Lebong[1], Ghumti (Darjiling distr.; 4,000 feet)[3], Kurseong[3]

Systolederus labatus, Hanc., Ind. Agric. Ent. Mem. IV 143-144 (1912)[1].

Mazarredia lugubris, Kirby, F. I. Orth. p. 55, f. 51 (1914)[2].

Bolotettix lobatus, Hanc., Rec. Ind. Mus. XI 78-79 [redescr.] (1915)[3].

oculatus, Bol.,

Criotettix oculatus, Bol., Ann. Mus. Genova XXXIX 71 (1898)[1], Ann. S. E. Fr. LXX 584 (1902)[2]; Kirby, Cat. Orth. III 16 (1910)[3], F. I. Orth. p. 34, f. 34 on p. 37 (1914)[4]; Bol., Rev. R. Acad. Cienc. Madrid (2) XVI 283 (1918)[5]. — Kodaikanal[2, 4, 6], Coorg[5] ——— Java[3, 4, 6], Sumatra[3, 4, 6]

Bolotettix oculatus, Hanc., Rec. Ind. Mus. XI 76 (1915)[6].

pictipes, Hanc., Rec. Ind. Mus. XI 77 (1915). — Shevaroys[1]

quadratus, Hanc., Rec. Ind. Mus. XI 79-80 (1915)[1]. — Singla (Darjiling distr.; 1,500 feet)[1]

triangularis, Hanc., Rec. Ind. Mus. XI 80 (1915)[1]. — Sikkim[1], Sibsagar[1]

16. Thoradonta, Hancock (1909).

Type : *dentata,* Hanc. [Penang].

Hanc., T. E. S. 1908 407 (1909), Rec. Ind. Mus. XI 57 [charact.], 80 (1915), T. E. S. 1910 358-359 (1910).

apiculata, Hanc., Rec. Ind. Mus. XI 81-82 (1915)[1]. — Dawna Hills[1], Tenasserim[1], Sibsagar[1], Singla (Darjiling distr.; 1,500 feet)[1]

sinuata, Hanc., Rec. Ind. Mus. XI 81 (1915)[1]. — Satara distr.[1]

spiculoba, Hanc., Ind. Agric. Ent. Mem. IV 138 (1912)[1], Rec. Ind. Mus. VIII 312-13 (1913) [variation][2], Rec. Ind. Mus. XI 80-81 [1915)[3]. — Bihar[1], Dibrugarh[2], Calcutta[3,], Rangoon[3], Kandy[3]

17. Criotettix, Bolivar (1887).

Type : *tricarinatus*, Bol.

Bol., Ann. S. E. Belg. XXXI 184, 193, 222 (1887); Hanc., Spol. Zeyl. II 108, 128 (1904) [redescr.], Gen. Ins. Tetrig. pp. 22, 27, (1906) [charact.]; Kirby, Cat. Orth. III 16 (1910), F. I. Orth. p. 31 (1914) [redescr.]; Hanc., Rec. Ind. Mus. XI 58 [charact.], 82-84 [key to spp.] (1915).

æqualis, Hanc., Ind. Agric. Ent. Mem. IV 135-136 (1912)[1], Rec. Ind. Mus. XI 87, t. 14 f. 3 (1915)[2].	Lebong[1, 2]
annandalei, Hanc., Rec. Ind. Mus. XI 87, t. 14 f. 6 (1915)[1].	Paresnath (4,300 ft.)[1]
dohertyi, Hanc., Rec. Ind. Mus. XI 86-87, t. 14 f. 4 (1915)[1].	Upper Assam[1]
extremus, Hanc., Ind. Agric. Ent. Mem. IV 132-133 (1912)[1], Rec. Ind. Mus. XI 89, t. 14 f. 11 (1915). [possibly a form of *maximus*][2].	Shevaroys[1, 2] Nilgiris (5,000 ft.)
fastiditus, Bol., Rev. R. Acad. Cienc. Madrid (2) XVI 284 (1918)[1].	Taliparamba[1]
flavopictus, Bol., Ann. S. E. Fr. LXX 582 (1902)[1]; Kirby, Cat. Orth. III 18 (1910) [*Acanthalobus*][2], F. I. Orth. p. 41 [1914] [*Acanthalobus*][2]; Hanc., Rec. Ind. Mus. XI 88, t. 14 f. 10 (1915)[4], T.E.S. 1907 220 (1907)[5].	Kodaikanal[1, 3] Anamalais[4] Dawna Hills[4] Cherrapunji[5]
grandis, Hanc., Ind. Agric. Ent. Mem. IV 134-135 (1912)[1], Rec. Ind. Mus. XI 89, t. 14 f. 8 (1915)[2].	Cherrapunji[1,2] Shillong Khasis (3100 ft.) Darjiling distr. (1,500 ft.)[2] , Shan Hills[2]
gravelyi, Hanc., Rec. Ind. Mus. XI 88, t. 14 f. 9 (1915)[1].	Ghumti (Darjiling distr.; 4,000 ft.)[1]
indicus, Bol., Ann. S. E. Fr. LXX 581-582 (1902)[1]; Kirby, Cat. Orth. III 16 (1910)[2], F. I. Orth. pp. 33-34 (1914)[3]; Hanc., Rec. Ind. Mus. XI 85 (1915)[4]; Bol., Rev. R. Acad. Cienc. Madrid (2) XVI 283 (1918)[5]. *Criotettix maculatus*, Kirby, F. I. Orth. p. 32, f. 30 (1914)[6].	Sikkim[1] South India[1-3] Burma[6]
maximus, Hanc., Rec. Ind. Mus. VIII 311-12, t. 15 ff. 1, 1[a] (1913)[1], Rec. Ind. Mus. XI 89, t. 14 f. 7 (1915)[2].	Aborland[1, 2] Janakmukh[1, 2] Ghumti (Darjiling distr.; 4,000 ft.)[2]
montanus, Hanc., Ind. Agric. Ent. Mem. IV 133-134 (1912)[1], Rec. Ind. Mus. XI 87, t. 14 f. 1 (1915)[2].	Simla[1, 2]

orientalis, Hanc., Rec. Ind. Mus. VIII 312, t. 15 ff. 4, 4^a (1913)[1], Rec. Ind. Mus. XI 85, t. 14 f. 12 (1915)[2].

Dibrugarh[1, 2]

pallidus, Hanc., Rec. Ind. Mus. XI 85-86, t. 14 f. 5 (1915)[1].

Tenasserim[1]

rugosus, Bol., Ann. S. E. Belg. XXXI 228 (1887)[1]; Brunn., Ann. Mus. Genova XXXIII 105 (1893)[2]; Hanc., Rec. Ind. Mus. XI 84-85, t. 14 f. 2 (1915)[3].

Pegu[2], Rangoon[3]
———
Borneo[1]

spinilobus, Hanc., Spol. Zeyl. II 108, 129-131, t. 3, ff. 12, $12^{a, b}$ (1904)[1], Gen. Ins. Tetrig. p. 28, tab. f. 13 (1906)[2]; Kirby, Cat. Orth. III 16 (1910)[3]; Hanc., Spol. Zeyl. VI 145 (1910)[4]; Kirby, F. I. Orth. pp. 31-32 (1914)[5]; Hanc., Rec. Ind. Mus. XI 85, t. 14 f. 13 (1915)[6].

Pundaluoya[1, 5, 6]
Maskeliya[4]
Travancore Coast[6, 7]

Criotettix obscurus, Kirby, F. I. Orth. pp. 32-33, f. 31 (1914)[7].

tricarinatus, Bol., Ann. S. E. Belg. XXXI 184, 223, 224 (1887)[1], Ann. S. E. Fr. LXX 583 (1902)[2]; Hanc., Spol. Zeyl. II 108, 128-129, t. 3 ff. 15, $15^{a, b}$ (1904)[3]; Kirby, Cat. Orth. III 16 (1910)[4]; Hanc., Spol. Zeyl. VI 144 (1910) [habits][5], Ind. Agric. Ent. Mem. IV 135 (1912)[6]; Kirby, F. I. Orth. p. 33 f. 32 (1914)[7]; Hanc., Rec. Ind. Mus. XI 85, t. 14 f. 14 (1915)[8]; Bol., Rev. R. Acad. Cienc. Madrid (2) XVI 283 (1918)[9].

Ceylon[1,4],
Weligama[5]
Polgahawela[5]
Kadugannawa[3, 7]
Peradeniya[3, 7, 8]
Kandy[3, 7, 8], Sigiriya[8]
Pundaluoya[3, 7]
Madulsima[5], Kodaikanal[2],
Shevaroys[6], Coimbatore[9]
Virajpet (Coorg)

18. Loxilobus, Hancock (1904).

Type : *acutus,* Hanc.

Hanc., Spol. Zeyl. II 108, 134 (1904), Gen. Ins. Tetrig. pp. 22, 29 (1906) [charact.]; Kirby, Cat. Orth. III 18 (1910), F. I. Orth. p. 41 (1914) [redescr.]; Hanc., Rec. Ind. Mus. XI 58, 91, (1915) [charact.].

acutus, Hanc., Spol. Zeyl. II 108, 134-135, t. 1 f. 3, t. 3 ff. 16, $16^{a, b}$ (1904)[1], Gen. Ins. Tetrig. p. 29, f. 14, t. 2 f. 17 (1906)[2]; Kirby, Cat. Orth. III 18 (1910)[3], F. I. Orth. p. 42 (1914)[4]; Hanc., Rec. Ind. Mus. XI 90 (1915)[5], [*nec* Hanc., Ind. Agric. Ent. Mem. IV 137 (1912), male, = *parvispinus.*]

Pundaluoya[1, 4]
Ceylon[2, 3, 5]

assamus, Hanc., T. E. S. 1907 223 (1907)[1]; Kirby, Cat. Orth. III 577 (1910)[2]; Hanc., Ind. Agric. Ent. Mem. IV 136 (1912)[3]; Kirby, F. I. Orth. p. 43 (1914)[4]; Hanc., Rec. Ind. Mus. XI 90 (1915)[5].

Cherrapunji[1, 4]
Lebong[3]

hancocki, Kirby.

Loxilobus rugosus [*nec* Bol.], Hanc., Spol. Zeyl. II 108, 134, 135-136, t. 3 ff. 17, 17[a, b] (1904)[1], Gen. Ins. Tetrig. p. 30 (1906)[2], T. E. S. 1908 401 (1909)[3], Spol. Zeyl. VI 145 (1910)[4]. — Pundaluoya[1, 3, 7] Ceylon[2, 5] Weligama[4] Bassein Fort[6]

Loxilobus hancocki, Kirby, Cat. Orth. III 18 (1910)[5]; Hanc., Ind. Agric. Ent. Mem. IV 137 (1912)[6]; Kirby, F. I. Orth., pp. 42-43, f. 40 (1914)[7]; Hanc., Rec. Ind. Mus. XI 90 (1915)[8].

parvispinus, Hanc., Rec. Ind. Mus. XI 90-91 (1915)[1]. — Pusa[1, 2]

Loxilobus acutus [*nec* Hanc. 1904], Hanc., Ind. Agric. Eng. Mem. IV 137 (1912) [male][2].

striatus, Hanc., Rec. Ind. Mus. XI 91 (1915)[1]. — Calcutta[1] Dawna Hills[1]

subulatus, Bol.,

Criotettix subulatus, Bol., Ann. S. E. Belg. XXXI 185, 223, 227 (1887)[1]; Kirby, Cat. Orth. III 16 (1910)[2], F. I. Orth., p. 35 (1914)[3]. — E. Indies[1-3] Ceylon[5]

Loxilobus subulatus, Hanc., Rec. Ind. Mus. XI 90 (1915)[4]; Bol., Rev. R. Acad. Cienc. Madrid (2) XVI 284 (1918)[5].

vidali, Bol.,

Criotettix vidali, Bol., Ann. S. E. Belg. XXXI 185, 223, 227 (1887)[1]; Brunn., Ann. Mus. Genova XXXIII 105 (1893)[2]; Kirby, Cat. Orth. III 16 (1910)[3], F. I. Orth. pp. 35-36 (1914)[4]; Hanc., Rec. Ind. Mus. XI 89 (1915)[5]. — Karen Hills[2, 4, 5] ——— Philippines[1, 3]

Loxilobus vidali, Bol., Rev. R. Acad. Cienc. Madrid (2) XVI 284 (1918)[6].

19. Acanthalobus, Hancock (1904).

Type : *miliarius,* Bol.

Hanc., Spol. Zeyl. II 108, 131 (1904), Gen. Ins. Tetrig. pp. 22, 28 (1906); Kirby, Cat. Orth. III 17 (1910), F. I. Orth. p. 36 (1914) [redescr.]; Hanc., Rec. Ind. Mus. XI 58, [charact.], 91-92 [key to spp.] (1915).

bispinosus, Dalman,

Acrydium bispinosum, Dalm., Vet. Acad. Handl. 1818, p. 77[1]. — Burma[8] Karen Hills[8, 10]

Tettix bispinosa, Dalm., Anal. Entom. p. 86 (1823)[2]. — Sibsagar[11] Brahmaputra River

Tettix pallitarsis, Wlk., Cat. Derm. Salt. V. 822-823 (1871)[3].	———above Jorhat
Tettix armigera, Wlk., Cat. Derm. Salt. V. 826-827 (1871)[4].	Penang[12]
Tettix latispina, Wlk., Cat. Derm. Salt. V. 837 (1871)[5].	Java[6, 7], Sumatra[10]
Tettix bispinosus, Stål, Recens, Orth. I 150 (1873)[6].	Borneo[7, 10]
Criotettix bispinosus, Bol., Ann. S. E. Belg. XXXI 185, 223, 226 (1887)[7]; Brunn., Ann. Mus. Genova XXXIII 105 (1893)[8].	Hongkong[3, 4] China[3, 6, 7, 9, 10]
Acanthalobus bispinosus, Kirby, Cat. Orth. III 17-18 (1910)[9], F. I. Orth. pp. 37-38 (1914)[10]; Hanc., Rec. Ind. Mus. XI 93-94 (1915)[11], T. E. S. 1907, 222 (1907)[12].	
cuneatus, Hanc., Spol. Zeyl. II 108 (1904)[1], Spol. Zeyl. II 133-134 (1904) [*A. miliarius cuneatus*]; Kirby, Cat. Orth. III 17 (1910)[3], F. I. Orth. pp. 40-41, ff. 37-39 (1914)[4]; Hanc., Rec. Ind. Mus. XI 92 (1915) [apparently shortwing form of *miliarius*][5].	Colombo[2, 4]
curticornis, Hanc., Rec. Ind. Mus. XI 92-93 (1915)[1].	Satara Distr.[1]
inornatus, Wlk.,	
Tettix inornata, Wlk., Cat. Derm. Salt. V 834 (1871)[1].	Burma[1, 5]
Criotettix saginatus. Bol., Ann. S. E. Belg. XXXI 185, 223, 225-226 (1887)[2]; Brunn., Ann. Mus. Genova XXXIII 104, t. 5 f. 38 (1893)[3]; Bol., Ann. S. E. Fr. LXX 583 (1902)[4].	Rangoon[3, 6, 7] Madura[4]
Acanthalobus inornatus, Kirby, Cat. Orth. III 17 (1910)[5], F. I. Orth. pp. 39-40, f. 36 (1914)[6].	————
Acanthalobus saginatus, Hanc., Rec. Ind. Mus. XI 93 (1915)[7], T. E. S. 1907, 221 (1907)[7A].	Java[2, 5, 6, 7A] Sumatra[5], China[7A]
Acanthalobus inornatus, Bol., Rev. R. Acad. Cienc. Madrid (2) XVI 285 (1918) [doubts whether *inornatus* and *saginatus* are really only one sp.][8].	
miliarius, Bol., Ann. S. E. Belg. XXXI 185, 223, 226 (1887) [*Criotettix*][1]; Hanc., Spol. Zeyl. II 108, 131, 132-133, t. 2 ff. 8, 8[a] (1904)[2], Gen. Ins. Tetrig. p. 29, t. 2, ff. 19, 19[a] (1906)[3]; Kirby, Cat. Orth. III 17 (1910)[4]; Hanc., Spol. Zeyl. VI 145 (1910)[5]; Kirby, F. I. Orth. p. 37, [*nec* f. 34] (1914)[6]; Hanc., Rec. Ind. Mus. XI 92 (1915)[7], T. E. S. 1907 222 (1907)[8].	Ceylon[1, 4] Colombo[2, 5, 7] Galle[5] Trincomali[7] Kandy[2, 6] Peradeniya[2, 6] Pundaluoya[8]
rufescens, Kirby, F. I. Orth. pp. 38-39, f. 35 (1914)[1].	Gangtok[1], Kurseong[1] Calcutta[1]

20. Tettitelum, Hancock (1915).

Type : *hastulatum,* Kirby.

Tettitellum, Hanc., Rec. Ind. Mus. XI 94 (1915).

Tettitellum, Hanc., Rec. Ind. Mus. XI 58 (1915) [error typogr.].

hastulatum, Kirby, F. I. Orth., p. 30, f. 29 (1914) [*Eugavialidium*][1].
Tenasserim[1, 2]

Tettitelum hastatum, Hanc., Rec. Ind. Mus. XI 94-95 (1915)[2].

METRODORINÆ.

Metrodorini, Kirby, F. I. Orth. pp. 43-44 (1914) [key to genera].

Metrodorinœ, Hanc., Rec. Ind. Mus. XI 58-59 [charact.; key to genera], 95 (1915).

21. Systolederus, Bolivar (1887).

Type : *haani,* Bol. [Philippines].

Bol., Ann. S. E. Belg. XXXI 185, 194, 234 (1887); Hanc., Spol. Zeyl. II 108, 136 (1904) [redescr.], Gen. Ins. Tetrig. pp. 30, 33 (1906) [charact.]; Kirby, Cat. Orth. III 20 (1910), F. I. Orth. p. 44 (1914) [redescr.]; Hanc., Rec. Ind. Mus. XI 58 [charact.], 95 (1915).

cinereus, Brunner, Ann. Mus. Genova XXXII 105 (1893)[1]; Kirby, Cat. Orth. III 20 (1910)[2], F. I. Orth. pp. 44-45, f. 41 (1914)[3]; Hanc., Rec. Ind. Mus. XI 95 (1915)[4].
Karen Hills[1-4]

greeni, Bol., Ann. S. E. Fr. LXX 584 (1902)[1]; Hanc., Spol. Zeyl. II 108, 137-138, t. 2 ff. 9, 9[a, b] (1904)[2], Gen. Ins. Tetrig. p. 34, t. 2. f. 14 (1906)[3]; Kirby, Cat. Orth. III 20 (1910)[4]; Hanc., Spol. Zeyl. VI 145-146 (1910) [habits][5]; Kirby, F. I. Orth. p. 45, f. 42 (1914)[6]; Hanc., Rec. Ind. Mus.4 XI 95 (1915)[7].
Pundaluoya[1, 2, 6, 7] Madulsima[5, 7] Haputale[5], Maskeliya[5, 7] Kadugannawa[2, 6] Kodaikanal[1, 6] Nilgiris[7], Virajpet (Coorg), Pusa

22. Hancockia, Kirby (1914).

Type : *portentosa,* Kirby.

Kirby, F. I. Orth. p. 46 (1914).

portentosa, Kirby, F. I. Orth. pp. 46-47, f. 43 (1914)[1].
Travancore[1]

23. Eurymorphopus, Hancock (1906).

Type : *cunctatus,* Bol. [New Caledonia].

Hanc., Gen. Ins. Tetrig. pp. 30, 35 (1906); Kirby, Cat. Orth. III 21 (1910), F. I. Orth. pp. 48-49 (1914) [redescr.]; Hanc., Rec. Ind. Mus. XI 58 [charact.], 95 (1915).

latilobus, Hanc., Spol. Zeyl. V 113-114, f. 1 (1908)[1]; Kirby, F. I. Orth. p. 49, ff. 46, 47 (1910)[2], Hanc., Rec. Ind. Mus. XI 95 (1915)[3]. — Undugoda[1-3] Kandy[3]

24. Spadotettix, Hancock (1910).

Type : *fletcheri,* Hanc.

Hanc., Spol. Zeyl. VI 146-147 (1910), Ind. Agric. Ent. Mem. IV 141-142 (1912); Kirby, F. I. Orth. p. 70 (1914) [redescr.]; Hanc., Rec. Ind. Mus. XI 58 [charact.], 95 (1915).

fletcheri, Hanc., Spol. Zeyl. VI 147-148, figs. (1910)[1]; Kirby, F. I. Orth. p. 71 (1914)[2]; Hanc., Rec. Ind. Mus. XI 95 (1915)[2]. — Madulsima[1, 2]

provertex, Hanc., Ind. Agric. Ent. Mem. IV 142-143 (1912)[1], Rec. Ind. Mus. XI 95 (1915)[2]. — Shevaroys[1]

25. Apterotettix, Hancock (1904).

Type : *obtusus,* Hanc.

Hanc., Spol. Zeyl. II 108, 140 (1904), Gen. Ins. Tetrig. pp. 30, 35 (1906) [charact.]; Kirby, Cat. Orth. III 21 (1910), F. I. Orth. p. 47 (1914) [redescr.]; Hanc., Rec. Ind. Mus. XI 58 [charact.] 95 (1915).

obtusus, Hanc., Spol. Zeyl. II 108, 140-141, t. 3 ff. 13, 13[a,b] (1904)[1], Gen. Ins. Tetrig. p. 35, f. 16 (1906)[2]; Kirby, Cat. Orth. III 21 (1910)[3], F. I. Orth. p. 48, ff. 44, 45 (1914)[4]; Hanc., Rec. Ind. Mus. XI 96 (1915)[5]. — Pundaluoya[1, 4] Peradeniya[1, 4, 5]

26. Amphinotus, Hancock (1915).

Type : *pygmœus,* Hanc.

Hanc., Rec. Ind. Mus. XI 58-59, 96 (1915).

pygmœus, Hanc., Rec. Ind. Mus. XI 96-97 (1915)[1]. — Hakgala[1], Pundaluoya[1]

27. Mazarredia, Bolivar (1887).

Type : *gemella,* Bol. [Philippines].

Bol., Ann. S. E. Belg. XXXI 185, 194, 236-237 (1887); Brunn., Ann. Mus. Genova XXXIII 106 (1893); Hanc., Spol. Zeyl. II 108, 138 (1904) [redescr.], Gen. Ins. Tetrig. pp. 32, 42 (1906) [charact.]; Kirby, Cat. Orth. III 24 (1910), F. I. Orth. pp. 48-51 (1914) [redescr.]; Hanc., Rec. Ind. Mus. XI 59 [charact.], 97-99 [Key to spp.] (1915).

convergens, Brunn., Ann. Mus. Genova XXXIII 106, 107-108, t. 5 f. 40 (1893)[1]; Kirby, Cat. Orth. III 26 (1910)[2], F. I. Orth. p. 52 (1914)[3]; Hanc., Rec. Ind. Mus. XI 101 (1915)[4]. — Karen Hills[1, 3] Tenasserim[1, 3]

cristulata, Bol., Ann. S. E. Fr. LXX 584-585 (1902)[1]; Kirby, Cat. Orth. III 25 (1910)[2], F. I. Orth. pp. 52-53 (1914)[3]; Hanc., Rec. Ind. Mus. XI 102 (1915)[4]. — Madura[1, 3]

dubia, Hanc., Ind. Agric. Ent. Mem. IV 141 (1912), Rec. Ind. Mus. XI 101 (1915)[2]. — Lebong[1, 2] Cherrapunji

ghumtiana, Hanc., Rec. Ind. Mus. XI 100-101 (1915)[1]. — Ghumti (Darjiling distr.)[1]

inæqualis, Brunn., Ann. Mus. Genova XXXIII 106-107, t. 5 f. 39 (1893)[1]; Kirby, Cat. Orth. III 25 (1910)[2], F. I. Orth. p. 53, f. 49 (1914)[3]; Hanc., Rec. Ind. Mus. XI 99 (1915)[4]. — Karen Hills[1, 2]

insignis, Kirby, F. I. Orth. p. 54, f. 50 (1914)[1]. — Bhim Tal[1], Shillong, Coorg

insularis, Bol., Ann. S. E. Belg. XXXI 237, 239 (1887)[1]; Hanc., Spol. Zeyl. II 108, 138-140, t. 2 ff. 7, $7^{a,b}$ (1904)[2], T. E. S. 1907 230 (1907)[3]; Kirby, Cat. Orth. III 25 (1910)[4], F. I. Orth. p. 52 (1914)[5]; Hanc., Rec. Ind. Mus. XI 102 (1915)[6]. — Ceylon[1] Pundaluoya[2, 3, 5] Hantana[2, 5] Peradeniya[5] Kandy[3]

laticeps, Bol., Bol. Soc. Easpañ. IX 399 (1909)[1]; Hanc., Rec. Ind. Mus. XI 102 (1915)[2]. — Upper Assam[1, 2]

latifrons, Hanc., Ind. Agric. Ent. Mem. IV 139 (1912)[1]; Rec. Ind. Mus. XI 102 (1915)[2]. — Lebong[1, 3]

lativertex, Brunn., Ann. Mus. Genova XXXIII 106, 108, t. 5 f. 41 (1893)[1]; Kirby, Cat. Orth. III 26 (1910)[2], F. I. Orth. p. 54 (1914)[3]; Hanc., Rec. Ind. Mus. XI 100 (1915)[4]. — Karen Hills[1, 3]

ophthalmica, Bol., Bol. Soc. Españ. IX 399 (1909)[1]; Hanc., Rec. Ind. Mus. XI 102 (1915)[2]. — Sibsagar[1, 2]

perplexa, Hanc., Rec. Ind. Mus. XI 101 (1915)[1]. — Sikkim[1]

sculpta, Bol., Ann. S. E. Belg. XXXI 237, 238 (1887)[1]; Brunn., Ann. Mus. Genova XXXIII 107 (1893)[2]; Hanc., T. E. S. 1908 405 (1909)[3]; Kirby, Cat. Orth. III 25 (1910)[4], F. I. Orth. p. 51 [*nec* f. 48] (1914)[5]; Hanc., Rec. Ind. Mus. XI 100 (1915)[6]; Boliv., Rev. R. Acad. Cienc. Madrid (2) XVI 285 (1918)[7]. — "East Indies"[1]; Tenasserim[2, 4, 5]; Karen Hills[2, 5]; Pegu[2, 4, 5]; Assam[3]; Sukna[6]

sikkimensis, Bol., Bol. Soc. Españ. IX 398-399 (1909)[1]; Hanc., Rec. Ind. Mus. XI 100 (1915)[2]. — Sikkim[2]

singlaënsis, Hanc., Rec. Ind. Mus. XI 99-100 (1915)[1]. — Singla (Darjiling distr.; 1,500 feet)[1]

28. Xistra, Bolivar (1887).

Type : *gogorzæ,* Bol. [Philippines].

Bol., Ann. S. E. Belg. XXXI 186, 194, 242 (1887); Brunn., Ann. Mus. Genova XXXIII 108 (1893); Hanc., Gen. Ins. Tetrig. pp. 32, 45 (1906) [charact.]; Kirby, Cat. Orth. III 27 (1910), F. I. Orth. p. 55 (1914) [redescr.]; Hanc., Rec. Ind. Mus. XI 59 [charact.], 102 (1915).

dubia, Brunn.,

Xistra ? dubia, Brunn., Ann. Mus. Genova XXXIII 108-109, t. 5 f. 42 (1893)[1]; Kirby, Cat. Orth. III 27 (1910)[2], F. I. Orth. p. 56 (1914)[3]. — Karen Hills[1, 3]

Xistra dubia, Hanc., Rec. Ind. Mus. XI 102 (1915)[4].

sikkimensis, Hanc., Rec. Ind. Mus. XI 102-103 (1915)[1]. — Sikkim[1]

stylata, Hanc., T. E. S. 1907 231-232 (1907)[1], T. E. S. 1908 409 (1909)[2]; Kirby, Cat. Orth. III 577 (1910)[3], F. I. Orth. p. 56 (1914)[4]; Hanc., Rec. Ind. Mus. XI 102 (1915)[5]; Boliv., Rev. R. Acad. Cienc. Madrid (2) XVI 286 (1918)[6]. — Puttalam[1, 4]; Kandy[2]; Nagody (S. Kanara)[6]; Coorg

29. Xistrella, Bolivar (1909).

Type : *dromadoria,* Bol.

Bol., Bol. Soc. Españ. IX 400-401 (1909); Hanc., Rec. Ind. Mus. XI 59 [charact.], 103 (1915).

dromadaria, Bol., Bol. Soc. Españ. IX 401 (1909)[1]; Hanc., Rec. Ind. Mus. XI 103 (1915) [female descr.][2] — Sikkim[1, 2]

30. Lamellitettix, Hancock (1904).

Type : *acutus,* Hanc.

Hanc., Spol. Zeyl. II 107, 125-126 (1904), Gen. Ins. Tetrig. pp. 31, 41 (1906) [charact.]; Kirby, Cat. Orth. III 24 (1910), F. I. Orth. p. 49 (1914) [redescr.]; Hanc., Rec. Ind. Mus. XI 59 [charact.], 103 (1915).

acutus, Hanc., Spol. Zeyl. II 107, 126-128, t. 2 ff. 6,6[a,b] (1904)[1], Gen. Ins. Tetrig. p. 42, t. 2 ff. 21, 21[a] (1906)[2], T. E. S. 1908 404 (1909)[3]; Kirby, Cat. Orth. III 24 (1910)[4], F. I. Orth. p. 50 (1914)[5].

Maskeliya[1, 5]
Hewaheta[3]

fletcheri, Hanc., Rec. Ind. Mus. XI 103 (1915)[1].

Anamalais[1]

pluricarinatus, Hanc., T. E. S. 1908 404 (1909)[1], Spol., Zeyl. VI 148 (1910)[2], Rec. Ind. Mus. XI 103 (1915)[3].

Deltota[1, 3]
Maskeliya[2]

31. Hyboella, Hancock (1915).

Hanc., Rec. Ind. Mus. XI 59 [charact.], 104-109 [Key to spp.] (1915).

Type : *tentata,* Hanc.

acuteterminatus, Brunn., Ann. Mus. Genova XXXIII 112 (1893) [*Coptotettix*][1]; Kirby, Cat. Orth. III 55 (1910)[2], F. I. Orth. pp. 75-76, f. 70 (1914)[3] [*Coptotettix*]; Hanc., Rec. Ind. Mus. XI 106 (1915)[4].

Pegu[1-3]

angulifrons, Hanc., Rec. Ind. Mus. XI 108-109 (1915)[1].

Dawna Hills[1]

conioptica, Hanc., Rec. Ind. Mus. XI 109 (1915)[1].

Singla (Darjiling distr.; 1,500 feet)[1]
Assam[1]

dilatata, de Haan,;

Acridium (Tetrix) dilatatum, de Haan, Temm., Verh. Orth. pp. 167, 169, t. 22 f. 12 (1842)[1].

Tettix dilatatus, Bol., Ann. S., E. Belg. XXXI 187, 257, 260 (1887)[2], Brunn., Ann. Mus. Genova XXXIII 109 (1893)[3].

Acrydium dilatatum, Kirby, Cat. Orth. III 45 (1910)[4], F. I. Orth. p. 68 (1914)[5].

Hyboella dilatata, Hanc., Rec. Ind. Mus. XI 107 (1915)[6].

Karen Hills[3, 5]

Java[1, 2, 4, 5]

latifrons, Brunn.,

Coptotettix latifrons, Brunn, Ann. Mus. Genova XXXIII 112, t. 5 f. 44 (1893)[1]; Kirby, Cat. Orth. III 55 (1910)[2], F. I. Orth. pp. 74-75, ff. 67-69 (1914)[3]. — Pegu[1, 3] Pusa[3]

Hyboella latifrons, Hanc., Rec. Ind. Mus. XI 106 (1915)[4].

nullipennis, Hanc., Rec. Ind. Mus. VIII 314, t. 15 ff. 2, 2[a] (1913) [*Coptotettix*][1], Rec. Ind. Mus. XI 106 (1915)[2]. — Janakmukh[1, 2] (N.E. Assam)

obesa, Hanc., Rec. Ind. Mus. XI 107-108 (1915)[1]. — Ghumti (Darjiling distr.; 4,000 feet)[1]

problematica, Bol.,

Coptotettix problematicus, Bol., Real Soc. Españ. IX 401 (1909)[1]. — Upper Assam[1, 2]

Hyboella problematica, Hanc., Rec. Ind. Mus. XI 109 (1915)[2].

tentata, Hanc., Rec. Ind. Mus. XI 106-107 (1915)[1]. — Sibsagar[1]

tumida, Hanc.,

Coptotettix tumidus, Hanc., Rec. Ind. Mus. VIII 313-314, t. 15 ff. 3, 3[a] (1913)[1]. — Dibrugarh[1, 2]

Hyboella tumida, Hanc., Rec. Ind. Mus. XI 106 (1915)[2].

ACRYDINÆ.

Acrydiini, Kirby, F. I. Orth. p. 57 (1914) [key to genera].

Tettiginæ, Hanc., Rec. Ind. Mus. XI 59-60 [charact.; key to genera], 109 (1915).

32. Teredorus, Hancock (1906).

Type : *stenofrons,* Hanc. [Peru]

Teredorus, Hanc., Gen. Ins. Tetrig. pp. 51-52 (1906); Kirby, Cat. Orth. III 30 (1910); Hanc., Rec. Ind. Mus. XI 109-110 (1915).

Terederus [error typogr.], Hanc., Rec. Ind. Mus. XI 59 [charact.] (1915).

carmichaeli, Hanc., Rec. Ind. Mus. XI 110 (1915)[1]. — Singla (Darjiling distr.; 1,500 feet)[1]

frontalis, Hanc., Rec. Ind. Mus. XI 110 (1915)[1]. — Simla Hills (5,000 ft.)[1]

33. Paratettix, Bolivar (1887).

Type : *meridionalis,* Rambur [Mediterranean Region].

Batrachidia [*nec* Serv.], Sauss., Rev. Zool. (2) XIII 401 (1861).

Paratettix, Bol., Ann. S. E. Belg. XXXI 187, 195, 270 (1887); Finot, Faune France, Orth. pp. 100, 169 (1889); Lugger, Orthopt. Minnesota pp. 196, 199 (1897); Morse, Biol. Centr. Am., Orth. II 5, 11 (1900); Hanc., Tetrig. N. Amer. pp. 36, 39, 108 (1902); Jacobs. and Bianchi, Prem. i Lozhn. Ross. Imp. pp. 164, 175, 211 (1902); Blatchley, Rep. Indiana Dep. Geology XXVII 218, 227 (1903); Hanc., Spol. Zeyl. II 108, 110, 144 (1904) [redescr.], Gen. Ins. Tetrig. pp. 52, 55 (1906) [charact.]; Kirby, Cat. Orth. III 32 (1910); Hanc., Ind. Agric. Ent. Mem. IV 149 (1912); Kirby, F. I. Orth. pp. 60-61 (1914) [redescr.]; Hanc., Rec. Ind. Mus. XI 60 [charact.], 111-112 [key to spp.] (1915).

alatus, Hanc., Rec. Ind. Mus. XI 113-114 (1915)[1].	Ghumti (Darjiling distr.; 4,000 feet)[1]
curtipennis, Hanc., Ind. Agric. Ent. Mem. IV 146-147 (1912) [*Coptotettix*][1], Rec. Ind. Mus. XI 113 (1915)[2].	Lebong[1], Darjiling distr. (4,000 feet)[2] Upper Shillong (6,000 feet)
dorsifer, Wlk.	
Tettix dorsifera, Wlk., Cat. Derm. Salt. V 825 (1871)[1].	Bombay[1-3]
Acrydium dorsiferum, Kirby, Cat. Orth. III 45 (1910)[2].	
Paratettix ? dorsifer, Kirby, F. I. Orth. p. 63, f. 55 (1914)[3].	
hirsutus, Brunn., Ann. Mus. Genova XXXIII 110, t. 5 ff. 43[a, b] (1893)[1]; Kirby, Cat. Orth. III 33 (1910)[2], F. I. Orth. p. 65, f. 57 (1914)[3]; Hanc., Rec. Ind. Mus. XI 112 (1915)[4].	Teinzo[1, 3] Katha[1, 3] Bhamo[1, 3] Sibsagar[4]
latipennis, Hanc., Rec. Ind. Mus. XI 113 (1915)[1].	Moradabad[1]
rotundatus, Hanc., Rec. Ind. Mus. XI 112-113 (1915)[1].	Tezpur[1]
scaber, Thunberg	
Acrydium scabrum, Thnb., Nova Acta Upsal, VII 159 (1815)[1].	South India[5, 8] Coimbatore[10] Palamcottah[10]
Tettix subpustulata, Wlk., Cat. Derm. Salt. V 819 (1871)[2].	Kezanathum[10] Bangalore[10]
Tettix scaber, Stãl, Recens, Orth. I 149 (1873)[3].	Nandyal[10]

Paratettix scaber, Bol., Ann. S. E. Belg. XXXI 188, 279 (1887)[4], Ann. S. E. Fr. LXX 585-586 (1902)[5].

Paratettix scaber ugandensis, Karny, Sitz. Akad. Wiss. Wien. CXVI 26 (1907)[6].

Paratettix scaber, Kirby, Cat. Orth. III 32 (1910)[7], F. I. Orth. p. 62 (1914)[8]; Bol., Tr. Linn. Soc. (2) XV 267 (1912)[9], Rev. R. Acad. Cienc. Madrid (2) XVI 287 (1918)[10].

Annampallee[10]

South Africa[2, 3, 7]
Zanzibar[4, 8]
East Africa[8]
Gaboon[4, 8]
Aldabra[9]
Seychelles[9]

semihirsutus, Brunn., Ann. Mus. Genova XXXIII 110-111 (1893)[1]; Kirby, Cat. Orth. III 33 (1910)[2], F. I. Orth. p. 66, f. 58 (1914)[3]; Hanc., Rec. Ind. Mus. XI 113 (1915)[4].

Karen Hills[1, 3]

34. Acrydium, Geoffroy (1762).

Type : *subulatum,* Linn. [Europe, W. and N. Asia]

Type : *subulatum,* Linn.

Acrydium, Geoff., Hist. Ins. I 390 (1762); Fabr., Syst. Entom. p. 278 (1775); Rossi, Faun. Entrusc. p. 262 (1790); Oliv., Enc. Méthod. Ins. VI 209 (1791); Leach, Edinb. Encycl. IX 120 (1815); Curt., Brit. Ent. X t. 439 (1833); Steph., III, Brit. Ent., Mandib. VII 32 (1835); Kirby, Cat. Orth. III 35 (1910), F. I. Orth. pp. 66-67 (1914); Hanc., Rec. Ind. Mus. XI 60 [charact.], 114 (1915).

Bulla, Schrank, Enum. Ins. Austr. p. 242 (1781).

Type : ....?

Acridium, Schrank, Fauna Boica II 30 (1801); Zett., Orth. Suec. p. 104 (1821), Faun. Ins. Lapp. I 452 (1828); Phil., Orthopt. Berol. p. 40 (1830); Zett., Ins. Lapp. p. 250 (1840).

Type : ....?.........

Tetrix, Latr., Hist. Nat. Crust. Ins. III 284 (1802), XII 161 (1804); St. Farg. and Serv., Encycl. Meth. Ins. X 598 (1825); Serv., Ann. Sci. Nat. XXII 264, 291 (1831), Ins. Orthopt. p. 757 (1839); Wesm., Bull. Acad. Brux. V 596 (1838); Burm., Handb. Ent. II 657 (1838); Borck, Skand. Rätv. Ins. p. 138 (1848); Finot. Orth. France pp. 39, 82 (1883), Faune France Orth. pp. 99, 165 (1889); Jacobs. and Bianchi, Prem. i Lozhn. Ross. Imp. pp. 174, 208 (1902); Hanc., Tett. N. Amer. pp. 35, 38, 66, 68 (1902), Spol. Zeyl. II 108, 110, 141 (1904) [redescr.], Gen. Ins. Orth. pp. 52, 57 (1906) [charact.], Ind. Agric. Ent. Mem. IV 149 (1912); Lucas, Brit. Orth. p. 206 (1920).

Type : *subulata,* Linn.

Tettix, Charp. Germ. Zeitschr. Ent. III 315 (1841); Fieb., Abh. Böhn. Ges. (5) III 408 (1844); Fisch. Waldh., Orth. Russ. p. 347 (1846); Fisch., Orth. Eur. p. 215

Type : *subulatus,* Linn.

(1853); Fieber, Lotos III 139 (1853), Syn. Eur. Orth. p. 29 (1854); Türk, Wien. Ent. Mon. VI 206 (1862); Friv., Ertek. Termész. Köreb. I (12) p. 188 (1868); Stål, Recens. Orth. I 146 (1873); Bol., Orthopt. Esp. p. 161 (1876), An. Soc. Esp. V 367 (1876); Brunn., Prodr. Eur. Orth. pp. 87, 234 (1882); Bol., Ann. S. E. Belg. XXXI 187, 195, 257 (1887), Ann. Sci. Nat. Porto IV 221, V 35 (1897-98); Burr, Brit. Orth. pp. 33, 46 (1897); Auriv., Ent. Tidskr. XXI 241, 247 (1900); Tümpel, Geradfl. Mitteleuropas pp. 220, 253 (1900); Hanc., Spol. Zeyl. II 141-142 (1904) [redescr.].

bipunctatum, Linn.

Gryllus Bulla bipunctatus, Linn., Syst. Nat. (ed. X) p. 427 (1758)[1], Sulz., Gesch. Ins. p. 82, t. 8 f. 6 (1776)[2].

Acrydium scutellatum, de Geer, Mém. Ins. III 483, t. 23, f. 15 (1773)[3].

Acrydium bipunctatum, Herbst, Fuess. Arch. p. 189, t. 52 f. 1 (1786)[4].

Acrydium opacum, Herbst., Fuess. Arch. p. 190, t. 52 f. 2 (1786)[5].

Acrydium gibbum, Olive., Enc. Meth. Ins. VI 233 (1791)[6].

Tetrix bipunctata, Serv., Ins. Orthopt. p. 765 (1839)[7].

Tettix schrankii, Fieb., Abhandl. Böhm. Ges. (5) III 412, t. 10 ff. 17-19 (1843)[8].

Tettix bipunctatus, Brunn., Prodr. Eur. Orth. pp. 234, 235 (1882)[9]; Bol., Ann. S. E. Belg. XXXI 263 (1887)[10]; Burr. Brit. Orth. p. 46, t. 4 f. 2 (1897)[11]; Tümpel, Geradfl. Mitteleuropas pp. 220, 253, t. 20 (1900)[12].

Tetrix bipundata, Jacobs. and Bianchi, Prem. i Lozhn. Ross. Imp. pp. 174, 209, t. 7 (1902)[13].

Tettix bipunctatus, Bol., Ann. S. E. Fr. LXX 585 (1902)[14].

Tetrix bipunctata, Hanc., T. E. S. 1910 359 (1910)[15].

Acrydium bipunctatum, Kirby, Cat. Orth. III 39-42 (1910) [numerous other refs. and synonyms][16], F. I. Orth. pp. 68-69, ff. 59, 60 (1914)[17]; Hanc., Rec. Ind. Mus. XI 114 (1915)[18].

Tetrix bipunctatus, Lucas Brit. Orth. pp. 212-217, t. 21 f. 1 (1920)[19].

Kodaikanal[14, 17]
Sikkim[18]

Persian Gulf[18]
N. & W. Asia[16]
N. Africa[16]
Europe[16, 18, 19]

ceylonum, Hanc.

Tettix atypicalis ceylonus, Hanc. Spol. Zeyl. II 108, 142, 143-144 (1904)[1].

Tetrix atypicalis ceylonus, Hanc., T. E. S. 1907 239 (1907)[2].

Acrydium ceylonicum, Kirby, Cat. Orth. III 45 (1910)[3].

Acrydium atypicale ceylonum, Kirby, F. I. Orth. pp. 67-68 (1914)[4].

Acrydium ceylonicum, Hanc., Red. Ind. Mus. XI 115 (1915) [short-wing form of *variegatum,* Bol.][5].

Pundaluoya[1, 2, 4]
Maskeliya[1, 4]
Peradeniya[5]

cingalense, Wlk.

Tettix cingalensis, Wlk., Cat. Derm. Salt. V 827 (1871)[1].

Paratettix variegatus, Bol., Ann. S. E. Belg. XXXI 188, 272, 280 (1887)[2]; Hanc., Spol. Zeyl. II 108, 144-145 (1904)[3].

Tettix atypicalis, Hanc., Spol. Seyl. II 108, 142-143, t. 3 ff. 14, 14[a, b] (1904)[4].

Tetrix atypicalis, Hanc., T. E. S. 1907 239 (1907)[5].

Tetrix variegatus, Hanc., T. E. S. 1908 413 (1909)[6].

Euparatettix cingalensis, Kirby, Cat. Orth. III 31 (1910)[7].

Paratettix variegatus, Kirby, Cat. Orth. III 33 (1910)[8].

Acrydium atypicalis, Kirby, Cat. Orth. III 45 (1910)[9].

Euparatettix variegatus, Hanc., Spol. Zeyl. VI 148 (1910)[10].

Tetrix variegatus, Hanc., Ind. Agric. Ent. Mem. IV. 149 (1912)[11].

Paratettix cingalensis, Kirby, F. I. Orth. pp. 63-64, f. 56 (1914)[12].

Acrydium atypicale, Kirby, F. I. Orth. p. 67 (1914)[13].

Acrydium variegatum, Hanc., Rec. Ind. Mus. XI 114-115 (1915)[14].

Paratettix cingalensis, Bol., Rev. R. Acad. Cienc. Madrid (2) XVI 287 (1918)[15].

Ceylon[1, 2, 8, 12, 11]
Madulsima[10]
Kandy[4, 13]
Peradeniya[4, 13, 19]
Pundaluoya[5]
Dikoya[4, 13]
Ratnapura[14]
Pondichery[11]
Taliparamba[15] (Malabar)
Singla (Darjiling distr.; 1,500 feet)[14]
Moulmein[14]

Java[6]

discale, Wlk.

Tettix discalis, Wlk., Cat. Derm. Salt. V 826 (1871)[1].

Acrydium discalis, Kirby, Cat. Orth. III 45 (1910)[2].

Ceylon[1, 2]

indicum, Bol.

Paratettix indicus, Bol., Ann. S. E. Belg. XXXI 188, 272, 281 (1887)[1], Ann. S. E. Fr. LXX 585 (1902)[2]; Kirby, Cat. Orth. III 33 (1910)[3]; Hanc., Ind. Agric. Ent. Mem. IV 150 (1912)[4]; Kirby, F. I. Orth. pp. 64-65 (1914)[5]; Bol., Rev. R. Acad. Cienc. Madrid (2) XVI 288 (1916)[6]. — "East Indies"[1] South India[2, 5] Coimbatore[6] Bihar[4, 7] ——— China[1-2, 5]

Acrydium indicum, Hanc., Rec. Ind. Mus. XI 115 (1915)[7].

indum, Olivier.

Acrydium indum, Oliv., Enc. Méth. Ins. VI 234 (1791)[1]. — "East Indies[1, 2]"

Acrydium ? indum, Kirby, Cat. Orth. III 49 (1910)[2].

Acrydium indicum, Kirby, F. I. Orth. p. 77 (1914)[3]; Bol., Rev. R. Acad. Cienc. Madrid (2) XVI 288 (1918)[4].

lineosum, Wlk.

Tettix lineosa, Wlk., Cat. Derm. Salt. V, Suppl. p. 90 (1871)[1]. — Bombay[1, 2]

Acrydium lineosum, Kirby, Cat. Orth. III 45 (1910)[2].

mundum, Wlk., Cat. Derm. Salt. V 823-824 (1871) [*Tettix munda*][1]; Kirby, Cat. Orth. III 45 (1910)[2]; Hanc., Rec. Ind. Mus. XI 115 (1915)[3]. — North India[1, 2]

nigricolle, Wlk., Cat. Derm. Salt. V 826 (1871) [*Tettix nigricollis*][1]; Kirby, Cat. Orth. III 45 (1910)[2]. — Bombay[1, 2]

obliquiferum, Wlk., Cat. Derm. Salt. V 825 (1871) [*Tettix obliquifera*][1]; Kirby, Cat. Orth. III 45 (1910)[2]. — Bombay[1, 2]

quadriplagiatum, Wlk.,

Tettix quadriplagiata, Wlk., Cat. Derm. Salt. V 828 (1871)[1]. — North India[1-3]

Hedotettix quadriplagiata, Kirby, Cat. Orth. III 49 (1910)[2].

Acrydium quadriplagiatum, Kirby, F. I. Orth. p. 69, f. 61 (1914)[3].

tectitergum, Hanc., Rec. Ind. Mus. XI 115-116 (1915)[1]. — Hoshangabad[1], Surat[1]

umbriferum, Wlk., Cat. Derm. Salt. V 824 (1871) [*Tettix umbrifera*][1]; Kurby, Cat. Orth. III 45 (1910)[2]. — Bombay[1, 2]

vittiferum, Wlk., Cat. Derm. Salt. V 824-825 (1871) [*Tettix vittifera*][1]; Kirby Cat. Orth. III 45 (1910)[2]. — Bombay[1, 2]

35. Coptotettix, Bolivar (1887).

Type : *asperatus*, Bol. [Cochin China].

Bol., Ann. S. E. Belg. XXXI 188, 195, 287 (1887); Hanc., Spol. Zeyl. II 108, 111, 152 (1904) [redescr.], Gen. Ins. Tetrig. pp. 52, 65 (1906) [charact.]; Kirby, Cat. Orth. III 55 (1910), F. I. Orth. p. 74 (1914) [redescr.]; Hanc., Rec. Ind. Mus. XI 60 [charact.], 116-118 [key to spp.] (1915).

annandalei, Hanc., Rec. Ind. Mus. XI 119 (1915)[1].	Singla (Darjiling distr.; 1,500 feet)[1] Calcutta[1] N. Shan Hills[1]
artolobus, Hanc., Rec. Ind. Mus. XI 121 (1915)[1].	Ceylon[1]
capitatus, Bol., Ann. S. E. Belg. XXXI 188, 287, 289 (1887)[1]; Brunna., Ann. Mus. Genova XXXIII 111 (1893)[2]; Kirby, Cat. Orth. III 55 (1910)[3], F. I. Orth. pp. 76-77 (1914)[4]; Hanc., Rec. Ind. Mus. XI 118 (1915)[5]; Bol., Rev. R. Acad. Cienc. Madrid (2) XVI 288 (1918)[6].	Karen Hills[2-4] Somwarpet (Coorg)[6] ———— Java[1, 3]
conspersus, Hanc., Rec. Ind. Mus. XI 119-120 (1915)[1].	Sibsagar[1] Siliguri[1] Bhim Tal[1]
fossulatus, Bol., Ann. S. E. Belg. XXXI 188, 287, 288-289 (1887)[1]; Hanc., Spol. Zeyl. II 108, 153 (1904)[2]; Kirby, Cat. Orth. III 55 (1910)[3], F. I. Orth. p. 76-(1914)[4]; Hanc., Rec. Ind. Mus. XI 118 (1915)[5].	Ceylon[1-4] Puri distr.[5]
indicus, Hanc., Ind. Agric. Ent. Mem. IV 144-145 (1912)[1], Rec. Ind. Mus. VIII 313 (1913) [female descr.][2], Rec. Ind. Mus. XI 120 (1915)[3].	Kalyan[1] Satara distr.[3] Aborland[3]
interruptus, Bol., Ann. S. E. Belg. XXXI 189, 288, 291-292 (1887)[1]; Brunn., Ann. Mus. Genova XXXIII 113 (1893)[2]; Kirby, Cat. Orth. III 56 (1910)[3], F. I. Orth. p. 78 (1914)[4]; Hanc., Rec. Ind. Mus. XI 118 (1915)[5].	Rangoon[2, 4] Bhamo[2, 4] Karen Hills[2, 4] ______ Java[1, 3, 4]
parvulus, Hanc., Ind. Agric. Ent. Mem. IV 145-146 (1912)[1], Rec. Ind. Mus. VIII 314 (1913)[2], Rec. Ind. Mus. XI 118 (1915)[3].	Chapra[1, 3], Calcutta[3] Kushtea (Bengal)[3] Singla (Darjiling distr.; 1,500 feet)[3] Sikkim[3], Janakmukh[2] Kobo[2], Dibrugarh[2]
retractus, Hanc., Rec. Ind. Mus. XI 120-121 (1915)[1].	Pusa[1]

testaceus, Bol., Ann. S. E. Belg. XXXI 189, 288, 291 (1887)[1]; Hanc., Spol. Zeyl. II 108, 153-154 (1904)[2]; Kirby, Cat. Orth. III 56 (1910)[3], F. I. Orth. p. 77 (1914)[4]; Hanc., Rec. Ind. Mus. XI 121 (1915)[5].

Ceylon[1-4]
Chota Nagpur[5]

36. Hedotettix, Bolivar (1887).

Type : *gracilis,* de Haan.

Bol., Ann. S. E. Belg. XXXI 188, 195, 283 (1887); Hanc., Spol. Zeyl. II 108, 111, 148 (1904) [redescr.], Gen. Ins. Tetrig. pp. 52, 60 (1906) [charact.]; Kirby, Cat. Orth. III 49 (1910), F. I. Orth. p. 71 (1914) [redescr.]; Hanc., Rec. Ind. Mus. XI 60 [charact.], 121-122 [key to spp.] (1915).

attenuatus, Hanc., Spol. Zeyl. II 108, 149, 151-152, t. 3 ff. 18, 18[a, b] (1904)[1], Gen. Ins. Tetrig. p. 60, f. 23 (1906)[2]; Kirby, Cat. Orth. III 50 (1910)[3], F. I. Orth. p. 73 (1914)[4]; Hanc., Rec. Ind. Mus. XI 123 (1915)[5].

Colombo[1, 4, 5]
Trincomali[5], Puri[5]
Sibsagar[5], Assam-Bhutan Frontier[5]

cristitergus, Hanc., Rec. Ind. Mus. XI 122 (1915)[1].

Hoshangabad[1]

costatus, Hanc., Ind. Agric. Ent. Mem. IV 147-148 (1912)[1], Rec. Ind. Mus. XI 123 (1915)[2].

Bengal[1, 2], Bihar[1, 2]
Nepal[2],
Allahabad[2]

gracilis, de Haan.

Acridium (Tetrix) gracile, de Haan, Temm., Verh., Orth. pp. 167, 169 (1842)[1].

Hedotettix gracilis, Bol., Ann. S. E. Belg. XXXI 188, 283, 284 (1887)[2].

Heditettix festivus, Bol., Ann. S. E. Belg. XXXI 188, 284, 286, t. 5 ff. 24, 24[a] (1887)[3].

Hedotettix gracilis, Brunn., Ann. Mus. Genova XXXIII 111 (1893)[4]; Bol., Ann. S. E. Fr. LXX 586 (1902) [variation][5]; Hanc., Spol. Zeyl. II 108, 149-151, t. 3 ff. 19, 19[a, b.] (1904)[6].

Hedotettix gracilis abortus, Hanc., Spol. Zeyl. II 108, 149, 151 (1904)[7].

Hedotettix gracilis, Hanc., T. E. S. 1907 240 (1907)[8], T. E. S. 1908 422 (1909)[9].

Hedotettix gracilis abortus, Hanc., T. E. S. 1908 422 (1909)[10].

Hedotettix gracilis, Kirby, Cat. Orth. III 49-50 (1910)[11]; Hanc., Spol. Zeyl. VI 149 (1910) [habits][12], Ind. Agric. Ent. Mem. IV 147 (1912)[13].

Throughout India (to Peshawar), Burma & Ceylon.

Siam[11], Sumatra[11]
Java[2, 4, 8, 11, 15]
Celebes[11, 15]

Hedotettix diminutus, Hanc., Ind. Agric. Ent. Mem. IV 148-149 (1912)[14].

Hedotetti gracilis, Kirby, F. I. Orth. p. 72, ff. 64, 65 (1914)[15]; Hanc., Rec. Ind. Mus. XI 123 (1915)[16]; Bol., Rey. R. Acad. Cienc. Madrid (2) XVI 288 (1918)[17].

grossus, Hanc., Rec. Ind. Mus. XI 124 (1915)[1]. — Singla (Darjiling distr.; 1,500 feet)[1]

lineifer, Wlk.

Tettix lineifera, Wlk., Cat. Derm. Salt. V 824 (1871)[1]. — Bombay[1-3]

Acrydium linceiferum, Kirby, Cat. Orth. III 45 (1910)[2].

Hedotettix ? lineiferum, Kirby, F. I. Orth. p. 73, f. 66 (1914)[3].

37. Euparatettix, Hancock (1904).

Type : *personatus,* Bol.

Euparatettix, Hanc., Spol. Zeyl. II 108, 111, 145 (1904), Gen. Ins. Tetrig. pp. 51, 55 (1906) [charact.]; Kirby, Cat. Orth. III 31 (1910); Hanc., Ind. Aric. Ent. Mem. IV 150-151 (1912) [key to spp.]; Kirby, F. I. Orth. p. 57 (1914) [redescr.]; Hanc., Rec. Ind. Mus. XI 60 [charact.], 124-125 [key to spp.] (1915). — Type : *personatus,* Bol.

Ergatettix, Kirby, F. I. Orth. pp. 69-70 (1914). — Type : *scabripes,* Bol. (=*tarsalis,* Kby.)

balteatus, Wlk.

Tettix balteata, Wlk., Cat. Derm. Salt. V 828-829 (1871)[1]. — South India[1-3] Salem distr.[4]

Acrydium balteatum, Kirby, Cat. Orth. III 45 (1910)[2]. — Palur[4]

Paratettix balteatus, Kirby, F. I. Orth. pp. 61-62, f. 54 (1914)[3].

Euparatettix balteatus, Bol., Rev. R. Acad. Cienc. Madrid (2) XVI 286 (1918)[4].

corpulentus, Hanc., Ind. Agric. Ent. Mem. IV 153 (1912), Rec. Ind. Mus. XI 126 (1915)[2]. — Throughout India (to Peshawar), Burma, and Ceylon.

personatus, Bol.

Paratettix personatus, Bol., Ann. S. E. Belg. XXXI 188, 278 (1887)[1]; Brunn., Ann. Mus. Genova XXXIII 109 (1893)[2]. — Throughout India (to Peshawar), Burma & Ceylon.

Euparatettix personatus, Hanc., Spol. Zeyl. II 108, 145, 146-148, t. 2 ff. 10, 10[a, b], t. 3 ff. 20, 20[a, b] (1904)[3], Gen. Ins. Tetrig. p. 55, t. 3 f. 32 (1906)[4], T. E. S. 1907 238 — Java[1, 3, 5, 6, 10] Philippines[1, 3, 10]

(1907)[5], T. E. S. 1908 409 (1909)[6], Spol. Zeyl. VI 148-149 (1910) [habits][7]; Kirby, Cat. Orth. III 31 (1910)[8]; Hanc., Ind. Agric. Ent. Mem. IV 152 (1912)[9]; Kirby, F. I. Orth. p. 58, ff. 52, 53 (1914)[10]; Hanc., Rec. Ind. Mus. XI 125 (1915)[11].	
personatus birmanicus, Hanc., Rec. Ind. Mus. XI 125-126 (1915)[1].	Rangoon[1], Assam-Bhutan Frontier.[1]
scabripes, Bol.	
Paratettix scabripes, Bol., Ann. Mus. Genova XXXIX 76 (1898)[1], Ann. S. E. Fr. LXX 585 (1902)[2]; Kirby, Cat. Orth. III 35 (1910)[3].	South India[2] Tinnevelly[6] Salem distr.[6] Coimbatore[6]
Ergatettix tarsalis, Kirby, F. I. Orth. p. 70, ff. 62, 63 (1914)[4].	Taliparamba (Malabar)[6], Calcutta[4]
Euparatettix scabripes, Kirby, F. I. Orth. p. 59 (1914)[5]; Bol., Rev. R. Acad. Cienc. Madrid (2) XVI 286 (1918)[6].	Rajshashi (Bengal)[5] ——— Java[2,3], Sumatra[2, 5]
tenuis, Hanc., Ind. Agric. Ent. Mem. IV 151-152 (1912)[1], Rec. Ind. Mus. VIII 315 (1913)[2], Rec. Ind. Mus. XI 126 (1915)[3].	Throughout United Prov., Bihar, Bengal and Assam. Dawna Hills[3]
variabilis, Bol.	
Paratettix variabilis, Bol., Ann. S. E. Belg. XXXI 188, 271, 276-277 (1887)[1], Ann. S. E. Fr. LXX 585 (1902); Hanc., T. E. S. 1907 236 (1907)[3]; Kirby, Cat. Orth. III 33 (1910)[4], F. I. Orth. pp. 62-63 (1914)[5].	Anuradhapura[7] Madura[1, 4, 5], Puri[7] Asansol[7], Berhampur[7] Rajshashi[7]
Euparatettix variabilis, Hanc., Ind. Agric. Ent. Mem. IV 150 (1912)[6], Rec. Ind. Mus. XI 126 (1915)[7].	Rajmahal[7] Pusa, Shillong Mandalay[7] ——— Borneo[1, 3, 5] Philippines[1, 4, 5]

38. Indatettix, Hancock (1915).

Type : *nodulosus,* Hanc.

Hanc., Rec. Ind. Mus. XI 60 [charact.], 127-129 [key to spp.] (1915).

callosus, Hanc., Rec. Ind. Mus. XI 130 (1915)[1].	Singla (Darjiling distr.; 1,500 feet)[1]

crassipes, Hanc., Ind. Agric. Ent. Mem. IV 153-164 (1912) [*Euparatettix*][1], Rec. Ind. Mus. XI 130-131 (1915)[2].	Throughout United Prov., Bihar, Bengal, Gauhati.
crassipes bengalensis, Hanc., Ind. Agric. Ent. Mem. IV 154-155 (1912) [*Euparatettix bengalensis*][1], Rec. Ind. Mus. XI 131 (1915)[2].	Pusa[1] Muzaffarpur[1] Dacca[1]
crassipes hybridus, Hanc., Rec. Ind. Mus. XI 131 (1915)[1].	Dibrugarh[1] Madhupur[1], Nepal[1]
crassipes var. A, Hanc., Rec. Ind. Mus. XI 131 (1915)[1].	Chapra[1], Pusa[1]
interruptus, Brunn.,	
Paratettix interruptus, Brunn., Ann. Mus. Genova XXXIII 109-110 (1893)[1].	Karen Hills[1, 4] Bhamo[1, 4]
Euparatettix interruptus, Hanc., T. E. S. 1910 359-360 (1910)[2]; Kirby, Cat. Orth. III 31 (1910)[3], F. I. Orth. p. 59 [? f. 48 on p. 51] (1914)[4]; Bol., Rev. R. Acad. Cienc. Madrid (2) XVI 59 (1918)[5].	Teinzo[1, 4], Damukdia[6] Madhurpur[6] Calcutta[6], Puri[6] Coimbatore[5]
Indatettix interruptus, Hanc., Rec. Ind. Mus. XI 130 (1915)[6].	Marikuppam[6] Travancore[6] Batticaloa (Ceylon)[6] ——— Johore[2]
interruptus var. A, Hanc., Rec. Ind. Mus. XI 130 (1915)[1].	Satara distr.[1]
interruptus lobulatus, Hanc., Rec. Ind. Mus. XI 130 (1915)[1].	Hoshangabad[1]
nodulosus, Hanc., Ind. Agric. Ent. Mem. IV 155-156 (1912) [*Euparatettix*][1], Rec. Ind. Mus. XI 129 (1915)[2].	Throughout United Prov., Bihar and Bengal. Garhwal[2] Nagpur, Satara[2]
parvus, Hanc.	
Euparatettix parvus, Hanc., Spol. Zeyl. II 108, 145-146 (1914)[1].	Elephant Pass (Ceylon)[1, 3], Weligama (Ceylon)[3]
Euparatettix pilosus, Hanc., T. E. S. 1908 409-410 (1909)[2], Spol. Zeyl. VI 149 (1910)[3].	S. Malabar Mysore[2], Central India[2]
Euparatettix parvus, Kirby, Cat. Orth. III 31 (1910)[4]; Hanc., Ind. Agric. Ent. Mem. IV 156 (1912)[5], Rec.	Puri[8] Calcutta[8]

Ind. Mus. VIII 315 (1913)[6]; Kirby, F. I. Orth. p. 60 (1914)[7]. — Chaumahani (Barisal) Igatpuri[8], Amangarh (UP)[8]

Indatettix parvus, Hanc., Rec. Ind. Mus. XI 129 (1915)[8]. — Dhampur (UP)[8], Bhim Tal[8], Peshawar, Dibrugarh[6], Aborland[6], Chapra[5], Pusa[5]

BATRACHIDIANÆ

Batrachidiini, Kirby, F. I. Orth. p. 78 (1914).

Batrachidinœ, Hanc., Rec. Ind. Mus. XI 60 [charact.], 131 (1915).

39. Saussurella, Bolivar (1887).

Type : *cucullifera,* Wlk. (*cornuta,* Bol., *nec* de Haan.)

Bol., Ann. S. E. Belg., XXXI 189, 196, 303 (1887); Hanc., Gen. Ins. Tetrig. pp. 67, 71 (1906); Kirby, Cat. Orth. III 61 (1910); Hanc., Ind. Agric. Ent. Mem. IV 156-157 [key to supp.] (1912); Kirby, F. I. Orth. pp. 78-79 (1914) [redescr.]; Hanc., Rec. Ind. Mus. XI 60 [charact.], 131 (1915).

cucullifera, Wlk. — Pegu[3, 4, 6, 7] Sibsagar[8]

Tettix cucullifera, Wlk. Cat. Derm. Salt. V 829 (1871)[1].

———

Saussurella cornuta [*nec* de Haan], Bol., Ann. S. E. Belg. XXXI 189, 303 (1887)[2]; Brunn., Ann. Mus. Genova XXXIII 113, t. 5 f. 45 (1893)[3]; Kirby, Cat. Orth. III 61 (1910) [part.][4] — Amoy[1, 5, 6] Hongkong[6]

Saussurella cucullifera, Kirby, Cat. Orth. III 61 (1910)[5], F. I. Orth. p. 79, f. 71 (1914)[6].

Saussurella brunneri, Hanc., Ind. Agric. Ent. Mem. IV 156 (1912)[7], Rec. Ind. Mus. XI 131-132 (1915)[8].

curticornu, Hanc., Ind. Agric. Ent. Mem. IV 158-159 (1912)[1], Rec. Ind. Mus. XI 131 (1915)[2]. — Pusa[1, 2]

decurva, Brunn., Ann. Mus. Genova XXXIII 113, t. 5 f. 46 (1893)[1]; Kirby, Cat. Orth. III 61 (1910)[2], F. I. Orth. pp. 79-80 (1914)[3]; Hanc., Rec. Ind. Mus. XI 131 (1915)[4]. — Pegu[1, 3] Tenasserim[4] N. Lakhimpur[4]

indica, Hanc., Ind. Agric. Ent. Mem. IV 157-158 (1912)[1], Rec. Ind. Mus. XI 131 (1915)[2]. — Lebong[1, 2]

TOTAL 169 species.

PART II
CULICIDÆ.

Giles, Bom. Jo. XIII, 603-610, [Prodromus]; Brun. R. I. M., X 1-73 [Review of genera].

GENERAL ACCOUNT OF INDIAN. — Hawlett, Ind. Ins. L. 564-575; Fletcher, S. Ind. Ins. 177-183, figs., [List of S. Ind. spp.]; Patton and Cragg, T. B. Med. Ent. 187.

ANATOMY OF. — Patton and Cragg, T. B. Med. Ent. 29, tab VII, (mouh parts); 74, [thorax]; 81, [Venation]; 109, [oesophagus]; 120, [salivary glands]; 133, [male genitalia]; 137, [female genitalia]; 195, [eggs]; 419, [dissection]; Prashad, I. J. Med. R. III. 503, Pl., [halteres]; *id. loc. cit.;* III. 497. 2 Pl., [male genitalia]; Ralph., Proc. Linn. Soc. I (1) ii, (1856) [mouth parts]; Edw., Ann. Trop. Med. Paras. XIV. 23-40, 2ff., [male hypopygium nomenclature]; Wesché, B. Ent. R. I. 7, [diagnostic points in larvæ and pupæ].

BIONOMICS OF. — Patton and Cragg, T. B. Med. Ent. 266, [collection, identification, feeding]; Sen I. J. Med. R., II, 681, [respiration]; *id. loc. cit.*: IV. 279 and Pusa Sci. Rpts. (1919-20) 103, [blood in ovulation]; Blacklock and Carter, Ann. Trop. Med. Paras. XIV 115-126, figs. and Pl., [tree-hole breeders and descript. of collecting apparatus for such]; Balfour, B. Ent. R. XII, 29, [Saline water breeders, bibliography]; Martini, Arch. Schiffs. u. Trop. Hyg. XXV. 120-121, [artificial breeding]; Davis, Ent. News, XII, 185 (1901), [distance travelled]; Macfie, B. Ent. R., VII 277-295, [larval respiration].

ENEMIES, PARASITES AND CONTROL. — Wilson, I. J. Med. R., I 691, [fish]; Hodgson, *loc. cit;* II. 405, [fish of Delhi]; Green, Trop., Ag. XXVIII. 297, [Ceylon duckweeds]; Nicholls, B. Ent. R., I. 213, [natural enemies]; Graham, *loc. cit.,* II 137, [fish of Nigeria]; Macfie, *loc. cit.,* VII, 277-295 [limits of kerosine]; Howlett, Pusa Sci. Rpt. (1912-13) 81, [calomel as larvicide]; Sharma, *loc. cit.,* (1919-20), 105, [repellents]; Campbell, Trop. Ag. XLII, 508-512 [eradication by bats]; Stevens & Wenyon, T. Vet. B. III, 129, [*Lankesteria culicis*]; Novy, McNeal, and Torry, Jl. Infec. Dis., IV. 223-276, (abst. J. Trop. Vet. Sci. III 97, 99, [trypanosomas of –].

SPECIES LISTS. — Bentley, Malaria in Bombay, 62, [Bombay]; Iyengar, I. J. Med. R., 1920 Sci. Cong. No., 8-17, (abst. R. A. E. [B], IX, 124-125, [Calcutta]; Senior White, *loc. cit.,* VIII, 304, [Ceylon Rubber Estate]; Cruickshank & Wright, *loc. cit.,* I. 741, [Cochin]; James, *loc. cit.,* II. 227, Pl. [Colombo]; Macfarlane, B. Ent. R., VI. 67, [Hong Kong]; Iyenger, *loc. cit.,* VII, Spl. Sci. Cong. No., 26, [Indore];

Barraud, B. Ent. R., X 323, [Lower Mesopotamia]; Edw., *loc. cit.*, XI, 133, [Mauritius, etc.]; Barraud, *loc. cit.*, XI, 387, [Palestine]; Dyar, Ins. Ins. Mens., VIII. 175-186, [Philippines]; Stanton, B. Ent. R., X, 333, [Ports of Far East]; Fletcher, Pusa Sci. Rpt. (1915-16), 83 [Pusa]; Green, Spol. Zeyl. IV, 183, [Trincomali].

CHAOBORINÆ (Corethrinæ).

Giles, Handbook, 2nd edn., 500; Brun. R.I. M., IV. 317, 504; id. *loc. cit.*, VI. 227; Patton and Cragg. T. B. Med. Ent., 190, [Key to genera].

1. **Chaoborus,** Lichtenstein.

Chaoborus, Licht., Arch. Zool. I, (1800). [Genotype, *Tipula plumicornis,* F. as *antisepticus,* Licht.]. Brun. R. I. M. XVII, 184.

Corethra, Meig., Illig. Mag. II, 260, (1803).

Sayomyia, Coq., Can. Ent. XXXV, 189, (1903). [Genotype, *C. punctipennis,* Say.]. Theo., Gen. Ins. fasc. XXVI, 42; Brun., R. I. M. I. 368; *id. loc. cit.* X. 71; *id. loc. cit.* XVII, 184.

manilensis, Schiner.

Corethra manilensis, Sch., Novara Reise, 30. 1. (1868)[1]; Wulp, Cat. Dipt. S. Asia, 34[2]; Giles, Handbook, 360; *id. loc. cit.* (2nd. edn.) 504.

[3]Bengal.
[3]Shahjahanpur.
[3]Satara Dist.

Sayomyia manilensis, Brun. R. I. M., I. 368.

Chaoborus manilensis, Edw., B. Ent. R., IV, 242; Prashad, R. I. M., XV, 154, Pl. XIX, ff. 1-7 [early stages]; Brun. R. I. M., XVII, 184[3].

[1, 2, 3]Philippines.

Corethra asiatica, Giles, Bom. Jo., XIII, 610, (1901); Brun. R. I. M., I. 368; Theo., Monog. Culic. II. 294, fig.; Giles, Handbook, (2nd. edn.), 506.

Chaoborus asiatica, Brun., R. I. M., IV, 505.

2. **Corethrella,** Coquillet.

Corethrella, Coq., Jo. N. Y. Ent. Soc., X, 191, (1902); [Genotype, *C. brakeleyi,* Coq.]. Brun., R. I. M., XVII, 185.

Ramcia, Anand., Spol. Zeyl., VII, 187, (1911), [Genotype, *R. inepta,* Ann.].

inepta, Annandale.

Ramcia inepta, Ann., Spol. Zeyl., VII, 189, figs. and Pl. [larva, etc.], (1911);[1] Macdougall, *loc. cit.*, VIII, 71, [notes]. [1,2] Diyatalawa

Corethrella inepta, Brun., R. I. M., XVII, 185[2].

CULICINÆ.

Theo., Gen. Ins. Fasc., XXVI, 26, 30; *id.*, Monog. Culic., IV, 19, 147, [tab. genera]; *id., loc. cit.*, V, 112 [tab. genera]; Blanch., Moustiques, 152, 231, 390, [tab. genera]; Giles, Handbook, (2nd. edn.) 334, [tab. genera]; Leic., Cublic. Malaya, 64; Alcock, Ent. Med. Off., 94; Brun., R. I. M., I., 326; *id., loc. cit.*, IV, 349; Edw., B. Ent. R., II, 241, [tab. African genera]; Patton and Cragg, T. B. Med. Ent., 192, 205, [classification].

ANOPHELINI.

Anophelini, Brun., R. I. M., X, 32, [notes, tab. genera]; Chris., I. J. Med. R., III, 454-488, [Key to Indian spp. synonomy].

Anophelinœ, Theo., Gen. Ins. Fasc., XXVI, 5; *id.*, Monog. Culic., IV, 18-22, (tab. genera]; *id. loc. cit.*: V, 1, [tab. genera]; Blanch., Moustiques, 157, 159, [tab. genera]; Leic., Culic. Malaya, 18, [tab. genera]; Brun., R. I. M., I, 302; *id., loc. cit.*, IV, 414; Chris., Ann. Trop. Med. Paras. VII, 1, (1913) [divides into three groups]; *id.,* Proc. Ind. Sanit. Conf., V, 1-5.

Anophelina, Theo., Monog, Culic. I, 97; *id., loc. cit.*, III, 9; Giles, Handbook, (2nd edn.], 280; *id.*, Revision of Anophelinæ, (London, 1904); Patton and Cragg, T. B. Med. Ent. 220 [Key to sub-genera], 256 [bionomics].

Anophelines, How., Dyar and Knab, Mosq. N. Amer, III, Pt. I, 193; James and Liston, Mon. Anph. Mosq. of India, (two editions).

Epialurgi, Alcock, Ent. Med. Off. 64.

3. Anopheles, Meigen.

SYSTEMATICS.

Anopheles, Mg., Syst., Besch., I, 10, (1818); [Genotype, *A. moculipennis,* Mg.]. R. D., Mem. Soc. H. Nat., III, 410; Mq., H. N. Dipt., I, 32; Zett., Dipt. Scand, IX, 3466; Sch., F. Aust., II, 624; Kert., Cat. Dipt., I, 250; Brun., R. I. M., I, 302; Edw., B. Ent. R., II, 141, [nomenclature, sub-genera]; Steph. and Chris., Rpt., Malaria. Commn., VII, 3-14, 4 Pl. [classification]; Chris., I. J. Med. R., III, 454-489, (1916) [Revision; Key to spp.]; Brun., R. I. M., XVII, 92-95, [cat. Oriental spp.]

Aldrichia, Theo., Monog. Culc. III App. 353, (1903); [Genotype, *A. error,* Theo.]; *id.,* Gen. Ins., Fasc., XXVI, 11; Blanch., Moustiques, 624; Brun., R. I. M., I. 322; *id., loc., cit.,* X, 55.

Aldrichinella, Theo., Monog. Culic. V, 77, (1910) [*nom. nov.* for *Aldrichia,* praeocc.]; Brun., R. I. M., IV, 433; *id. loc. cit.,* X, 55.

Arribalzagia, Theo., Monog. Culic. III, 81, (1903); [Genotype, *A. maculipes,* Theo.]; Alcock, Ent. Med. Off., 177.

Cellia, Theo., Monog. Culc. III, 107, (1903); [Genotype, *A. pharoensis,* Theo.]; *id., loc., cit.,* IV, 105, [list and tab. spp.]; *id., loc., cit.,* V, 67, [tab. spp.]; *id.,* Gen. Ins. Fasc. XXVI, 11; Blanch., Moustiques, 214; James, R. I. M., IV, 102; Leic., Culic. Malaya, 46; Alcock Ent. Med. Off., 90; Brun., R. I. M., I, 321; *id., loc., cit.,* IV, 431 *id., loc., cit.,* X. 57.

Christophersia, James, R. I. M., IV, 103, (1910); [Genotype, *C. halli,* J.]; *id.,* Paludism, I. 33, (July, 1910, *nom. nud.*); Brun., R. I. M., IV, 431; *id., loc., cit.,* X. 57.

Christya, Theo., Rpt. R. S. SI. Sick. Comn., VII, 34, (1903); [Genotype, *C. implexa,* Theo.]; Giles, Revis., Anoph., 40; Alcock, Ent. Med. Off., 77.

Coelodiazesis, Dyar and Knab, Jo. N.Y. Ent. Soc., XIV, 77, (1906); [Genotype, *A. barberi,* Coq. [Erected on larval chrs. only].

Cycloleppteron, Theo., Monog. Culic. II, 312, (1901); [Genotype, *A. grabhamii,* Theo.]; *id.,* J. Trop. Med. IV, 234, (1901, *nom. nud.*).

Cyclolepidopteron, Blanch., Moustiques, 185 (1905) [*Cycloleppteron,* emend.].

Dactylomyia, News, and Cart., Ann. Trop. Med. Paras., IV. 377, (1910); [Genotype, *D. ceylonica,* N. and C.].

Feltinella, Theo., Monog, Culic. IV, 56, (1907); [Genotype, *F. pallidopalpis,* Theo.].

Grassia, Theo., Jo. Trop. Med. V, 181, (1902); [Genotype, *A. rossii,* Giles.].

Howardia, Theo., Jo. Trop. Med. V, 181, (1902); [Genotype, *A. costalis,* Lw.].

Kerteszia, Theo., Ann. Mus. Hung., III, 66, (1905); [Genotype, *K. boliviensis,* Theo.]; *id.,* Monog. Culic. IV, 117; Brun., R. I. M., IV. 431; *id., loc., cit.,* X, 63.

Laverania, Theo., Jo. Trop. Med. V, 181, (1902); [Genotype, *A. argyrotarsis,* R.D.].

Lophomyia, Giles, Jo. Trop. Med. VII, 366, (1904); [Genotype, *Lophoscelomyia asiatica,* Leic.].

Lophoscelomyia, Theo., Entom., XXXVII, 12, (1904); [Genotype, *Lophoscelomyia asiatica,* Leic.]; Blanch., Moustiques, 635; Theo. Monog. Culic., IV, 91; Leic., Culic. Malaya, 21; Brun., R. I. M., IV, 427; *id., loc. cit.,* X, 64.

Lophocelomyia, Theo., Gen. Ins. fasc. XXVI, 10, [*laps.*]; Brun., R. I. M., I, 316.

Manguinhosia, Cruz in Peryassu, Os. Culic. Brazil, 112, (1908); [Genotype, *M. lutzi,* Cruz.].

Memnemyia, Strick., I. J. Med. R. III, 201, (1905); [Genotype, *A. brevipalpis,* Rop.].

Myzomyia, Blanch., C. R. Soc. Biol., LIV, 795, (1902); [nom. nov. for *Grassia,* præocc.]; Theo. Monog. Culic., III, 12, 24; *id., loc. cit.,* IV, 41, 42, [tab. spp.]; *id., loc. cit.,* V, 16, [tab. spp.]; *id.,* Gen. Ins. Fasc. XXVI, 7; Blanch., Moustiques, 177; James and List., Mon. Anoph. Ind. (2nd. edn.) 40; James, R. I. M., IV, 98; Leic., Culic. Malaya, 23; Alcock, Ent. Med. Off. 79; Brun., R. I. M., I, 305; *id., loc. cit.,* IV, 417; *id., loc. cit.,* X, 66.

Myzorhynchella. Theo., Monog. Culic., IV, 78, (1907); [Genotype, *M. nigra,* Theo.].

Myzorhynchus, Blanch., C.R. Soc. Biol., LIV. 795, (1902); [*nom. nov.* for *Rossia,* præocc.]; Theo., Monog. Culic., III, 84, Pl. V.; *id., loc. cit.,* IV, 81, [list and tab.]; *id., loc. cit.,* V, 49, [list and tab.]; *id.,* Gen. Ins. fasc. XXVI, 9; Blanch., Moustiques, 190; James, R. I. M., IV, 102; Leic., Culic. Malaya, 29, (*Myzorhyncus*); Struck., B. Ent. R. IV., 135-142; [M. group in Malaya]; Alcock, Ent. Med. Off. 74; Brun., R. I. M., I, 313; *id., loc. cit.,* IV, 424; *id., loc. cit.,* X, 66.

Neocellia, Theo., Monog. Culic. IV, 111, (1907); [Genotype, *N. indica,* Theo.], *id., loc. cit.,* V, 73, [tab. spp.]; James, R. I. M., IV, 101; Brun., *loc. cit.,* IV, 432; *id., loc., cit.,* X, 66; Alcock, Ent. Med. Off. 89.

Neomyzomyia, Theo., Monog. Culic, V, 29, (1910); [Genotype, *A. elegans,* James]; James, R. I. M., IV, 109; Brun., *loc., cit.,* IV, 417; *id., loc. cit.,* X, 68.

Neostethopheles, James., R. I. M., IV, 98, (1910) [Genotype, *A. aitkeni,* James]. James and List., Mon. Anoph. Mosq. Ind., 2nd edn. 40; Brun., R. I. M., IV, 422; *id., loc., cit.,* X, 67.

Nototricha, Coq., Class. Mosq. N. Amer., 12, (1906) [Genotype, *Cycloleppteron mediopunctatus,* Theo.].

Notonotricha, Theo., Monog. Culic., V., 33, (1910); [Emend].

Nyssomyzomyia, James, R. I. M., IV, 101, (1901) [Genotype, *N. rossii,* Giles, præocc. in *Myzomyia*]; James and List., Mon. Anoph. Mosq. Ind. (2nd edn.) 43; Brun., R. I. M., IV, 422; *id., loc. cit.,* X, 67.

Nyssorhynchus, Blanch., C. R. Soc. Biol. LIV, 795, (1902) [*nom. nov.* for *Laverania,* præocc.]; Theo., Monog. Culic., III, 92, Pl. V; *id., loc. cit.,* IV, 95, 96, [list and table spp.]; *id.,* Gen. Ins. fasc., XXVI, 10; Blanch, Moust., 202; James and List., Monog. Anoph. Mosq. Ind., (2nd. edn.) 43; James R. I. M., IV, 100; Leic., Culic. Malaya, 39; Brun., R. I. M., I, 317; *id., loc. cit.,* IV, 428; *id., loc. cit.,* X, 67; Alcock, Ent. Med. Off. 85.

Patagiamyia, James, R. I. M., IV, 98, (1910) [Genotype, *A. gigas,* Giles.]; James and List., Mon. Anoph,

Mosq. Ind., (2nd. edn.) 41; Brun., R. I. M., IV, 417; *id., loc., cit.,* X, 68.

Pyretophorus, Blanch., C. R. Soc. Biol., LIV, 795, (1902) [*nom. nov.* for *Howardia,* præocc.]; Theo., Monog. Culic., III 66; *id., loc. cit.,* IV, 63-64, [List and tab. spp.]; *id., loc., cit.,* V, 36, [tab. spp.]; *id.,* Gen. Ins. fasc. XXVI, 8; Blanch, Moustiques, 186; James and List. Mon. Anoph. Mosq. Ind., (2nd edn.), 41; James, R. I. M., IV, 99; Brun., *loc. cit.,* I, 312; *id., loc. cit.,* IV, 423; *id., loc. cit.,* X, 70; Leic., Culic. Malaya, 37.

Rossia, Theo., Jo. Trop. Med., V, 181, (1902) [Genotype, *A. sinensis,* Wd.].

Stethomyia, Theo., Jo. Trop. Med., V, 181, (1902) [Genotype, *S. nimba,* Theo.]; *id.,* Monog. Culic, III, 62, Pl. VIII; *id., loc. cit.,* IV, 59, [tab. spp.]; *id., loc. cit.,* V, 35, [tab. spp.]; *id.,* Gen. Ins. fasc. XXVI, 8; Blanch, Moustiques, 186; Brun., R. I. M., I, 312; *id., loc. cit.,* IV 423; *id., loc. cit.,* X, 72.

ANATOMY.

Prashad, I. J. Med. R., V, 610, pl. LVII-LXI, [Thorax and Wing]; *id., loc. cit.,* VI, 641, [imaginal buds of thorax]; Patton and Cragg, T. B. Med. Ent., 120, [salivary glands]; 148, [nervous system]; 195, [egg]; Chris, I. J. Med. R., III, 371, 6 pl., [male genitalia]; *id., loc., cit.,* III, 362, pl., [pilotaxy]; Steph., B. Ent. R., II, 1-8, 5 figs., [dis-section for parasites].

LARVÆ.

Steph. and Chris., Rpt. Malaria Comn. VII, 3-14, 4 pl., [notes]; Stanton, B. Ent. R., III, 387, [changes during growth]; *id., loc. cit.,* VI, 159-172, 15 figs. [descriptions, Key to Malayan spp.]; Patton and Cragg, T. B. Med. Ent., 204, [Key to Indian spp.]; James and List., Monog. Anoph. Mosq. Ind., (2nd. edn.), 57-58 [do]; Zetek, B. Ent. R. XI, 73, [breeding among water-lettuce]; Lamborn, *loc., cit.,* XII, 91 figs., [caudal tufts of Malayan]; Hacker, F. M. S. Mal. Bur. II, Rpt., 25-28, [in artificial coll. of water]; *id., loc. cit.,* [in delta]; id., loc. cit., II, 45-47, [among mangroves]; Ghosh in Fry, Malaria in Bengal, app. ii, [in rice fields].

BIONOMICS.

Steph. and Chris., Rpt. Mal. Comn. VI, 11-120, [Bengal spp.]; Editorial, I. J. Med. R., III, 559, [effect of jungle clearing]; Strickland, *loc. cit.*, IV, 256, [favourite breeding grounds]; Hacker, F. M. S. Mal. Bur. II Rpt., 7-9, 11, 12, 13, 23, [group associations with various kinds of water]; Geiger, etc. J. Amer. Med. Ass., LXXII, 844-847, [flight experiments];

PREDATORS, ETC.

Sinton, I. J. Med. R. V, 192, pl. XXX, [Trematode parasite]; Soparkar, *loc. cit.*, V, 512, [*do*]; Southwell, Ann. Trop. Med. Paras. XIV 181-186, [fish].

DISEASE.

Neumann, Trans. Rev. Vet., (1911), (abst. J. Trop. Vet Sci., VII, 62-64), [*Filaria immitis* transmitted]; Donovan, I. J. Med. R., I, 181 [failed transmit Kala azar]; Gunasekera, Ind. Med. Gaz., (abst.), LIV, 471-2, [Kurunegalla compaign]: Wright, *loc., cit.*, LIV, 361, maps [Mercara campaign]; Editorial, I. J. Med. R., III, 559, [effect on malaria of jungle clearing]; Strick., *loc. cit.*, IV, 256, [epidemiology in Malaya]; Geiger, etc., J. Amer. Med. Ass., LXII., 844-847, [control in rice field district of U. S. A.]; Purdy, Rpt., Cal. St. Bd. of Health, Nov. 1920, [prohibition of rice growing near towns]; Geiger and Purdy, So. Med. Jo., Aug. 1920, [malaria and rice fields in U. S. A.]; *id.* Jo. Amer. Med. Ass. LXXII, 774-779, [experimental mosquito control in rice fields]; Geiger, etc., *loc. cit.*, LXXI, 1283-1285, [malarial endemicity in rice field area]; Swell. and de Graaf, Parasitology, XII, 180-198, [Malaria Survey in Malay Archip.]; Chris. & Short, I. J. Med. R. VIII 29-32, 2 maps, [anti-malaria operations at Busra].

SPECIES LISTS.

Chris., Sci. Mem. M. O. Ind., LVI, 6-7, [Andamans]; Chris. & Chand, I. J. Med. R., III, 180 [Arabia; with synonomy of Patton's spp.]; Fry, Malaria in Bengal, 11, [Bengal]; Roper, I. J. Med., R. V., 137-147, 2 maps, [Br. N. Borneo]; Evans & Rennie, Jo. Trop. Vet., Sci., V, 240-256, [Burma]; Chalmers, Spol. Zeyl., II, 165-178, 10 figs. [Ceylon]; Fry,

Malaria in Bengal, 29, 30, [Chota Nagpur]; Acton, etc., I. J. Med. R., VIII, 751-752, [Dagshai]; Hodgson, *loc. cit.*, II, 405, 6 pl., map and figs., [Delhi]; Ghosh in Fry, Malaria in Bengal, app. i, [Dum Dum]; Swell, and de Graaf, Tijd. V. Ent., LXIII., 96, *id.*, B. Ent. R. XI, 77, [Dutch East Indies]; de Sá, Bol. Geral. Med. Farm., V, (8), 290-294, (abst. R. A. E. (B.), IX, 6), [Goa]; Cogill, Bom. J., XV, 327-336, [India]; Perry, I. J. Med. R., II, 456, pl. and figs., [Jeypore Hill Tracts]; Gill, *loc. cit.*, VII, 610-617, [Kashmir]; Sinton, *loc. cit.*, V, 195, [Kohat]; Hacker, F. M. S. Mal. Bur., II Rpt., 5, [Kuala Lumpur]; Gunasekera, Ceylon S. P. 37 of 1913, 7, [Kurunegalla]; Hodgson, I. J. Med. R., I. 702, [Madras city]; Stanton, Jo. Lon. Sch. T. Med., ii, (1913), [Malaya]; *id.*, I. J. Med. R., I. 203, [Malaya]; Hacker, F. M. S. Mal. Bur. Rpt. I, [Malaya; many bionomic notes]; Chris. & Chand, I. J. Med. R., III, 180, 2 pl., [Mesopotamia]; Chris. & Short, *loc. cit.*, VIII, 516-529. [Mesopotamia]; Chris., *loc. cit.*, VII, 710, 3 maps [Middle East; Synonomy]; Aitken, Jo. Trop. Med., V, 325-6, 341-2, [North Kanara]; Sinton I. J. Med. R. V, 204, [Parachinar]; de Mello and de Sá, Arquiv. Ind. Port. Med. Hist. Nat., I, 1-10, pl. 1-4, [Portuguese India]; Chris., Sci. Mem. M. O. Ind., XLVI, 75-89, [Punjab]; Chris. & Chand, I. J. Med. R., III, 638 [South India]; Kenkenschrijven, Gens. Tijd. Ned. Ind. LXI, 202, [abst. R. A. E. (B) IX, 135] [Sumatra]; Sinton, I. J. Med. R. V, 204, [Thal]; Gill & Singh, Supt. Govt. Print. Lahore, 1920; [abst. R. A. E. (B), IX, 130], [Thanesar Town].

aitkeni, James.

Anopheles aitkeni, James in Theo. Monog. Culic., III, 22, (1903)[1]; James & List., Monog. Anoph. Mosq. Ind., 119, pl. IX, 3, pl. XIII; Blanch. Moustiques, 620; Brun. R. I. M. I. 303[2]; *id. loc. cit.*, IV, 415; *id., loc. cit.*, XVII, 110[3]; Chris., I. J. Med. R., III, 461[4]; Stant., Jo. Lon. Sch. Trop. Med., II (1) 4; Alcock, *loc. cit.*, II (3) 139; Theo., R. I. M., II, 287[5]; Stanton. B. Ent. R., VI, 163, [larva]; Patton and Cragg, T. B. Med. Ent. 228; Strick., I. J. Med. R., IV, 259, [not biting man]; Hacker, F. M. S. Mal. Bur. I Rpt. 73, [index to numerous refs.]; *id., loc. cit.*, II Rpt. 21, [faunal associations]; Swell., B. Ent. R., XI, 91[6], [distrib. in D. E. Ind.].

[1, 2, 3]Karwar.
[2, 3]Goa Frontier.
[3, 5]Bengal Duars.
[3, 4]Nilgiris.
[3]Bengal.
[3, 4]Assam.
[3, 4]Burma.
[4]W. Coast of S. India, Matale-Kurunegalla (Ceylon) White).

Anopheles aitkeni var. insulaeflorum., Swell, B. Ent. R., XI, 81, (1920), fig. [larval variety only].

Anopheles aitkeni var. papuœ, Swell., B. Ent. R., XI, 81-82 (1920), [larval variety only].

Neostethopheles aitkeni, James & List., Monog. Anoph. Mosq. Ind. (2nd. edn.), 59, figs.

Stethomyia fragilis, Theo. Entom., XXXVI, 257 (1903); Brun., R. I. M., I, 312[8]; *id., loc. cit.*, IV, 423; Patton & Cragg, T. B. Med. Ent. 237.

Stethomyia pallida, Ludl., Can. Ent., XXXVII, 129, (1905); Brun., R. I. M. I, 312[9]; *id., loc. cit.*, IV, 423.

Anopheles treacheri, Leic., Culic, Malaya, 19, (1908); Brun., R. I. M., IV, 417[7]; *id., loc. cit.*, XVII, 113.

[3, 9]Philippines.
[7, 8]F. M. S. widely.
[6]Malayan Archip.

annandalei, Prashad.

Anopheles annandalei, Prashad, R. I. M., XV, 123, pl. XVII, (1918)[1]; Brun., *loc. cit.*, XVII, 114[2].

[1, 2]Darjiling.

asiaticus, Leicester.

Lophoscelomyia asiatica, Leic., Entom., XXXVII, 13, (1904)[1]; Theo., Monog. Culic., IV, 92; Brun., R. I. M., I, 317; *id., loc. cit.*, IV, 427; Leic. Culic. Malaya, 21.

Lophomyia asiatica, Giles, Jo. Trop. Med., VII, 366.

Myzorhynchus asiaticus, Strick., B. Ent. R., IV, 141.

Anopheles asiatica, Chris. I. J. Med. R., III, 462; Brun., R. I. M., XVII, 113; James, I. J. Med. R., II, 263.[2]

Anopheles asiaticus, Strick. Parasitology, VII, 12, [larva] Stanton, B. Ent. R., VI, 164, [larva]; Patton and Cragg, T. B. Med. Ent., 237; Lamborn, B. Ent. R., XII, 93, figs. [larva and pupa]; Hacker, F. M. S. Mal. Bur., II. Rpt., 23, [faunal associations].

[2]Ceylon.

---

[1]Kuala—Lumpur.

barbirostris, van der Wulp.

Anopheles barbirostris, Wulp, Notes Leyd. Mus., VI, 248,[19] (1884); *id.*, Tijd. V. Ent., XXVII, 79, pl. IV, 1; Theo. Monog. Culic. I, 146, 151, ff. 33, 34, pl. A; Giles, Handbook, 2nd edn., 308, pl. VIII, 13*a*; James and List., Anoph. Mosq. Ind. 77 pl. X, 1 pl. V, pl. II; Roper, B. Ent., R. V, 146; Stanton, I. J. Med. R. III, 253[13]; Chris., *loc. cit.*, III, 462; Brun., R. I. M. XVII,

[1, 6, 7, 8]Calcutta dist.
[2]Madras.
[3, 15]Malabar Coast.
[6]Kanara.
[6, 9]Assam.
[6, 9]Upper Burma.
[6]Lahore.
[4]Delhi.

97[9]; Chatterjee, R. I. M. I, 82[1]; Wilson, I. J. Med. R. I, 691[2], [breeding places]; Hodgson, *loc. cit.*, II, 411[4]; James, *loc. cit.*, II, 263[5]; Perry, *loc. cit.*, II, 469[10]; Iyengar, *loc., cit.*, VII, Sci., Cong. No. 26[11]; Stanton B. Ent. R., VI, 165, [larva]; *id., loc. cit.*, X, 333, [distribution in far Eastern ports]; Patton and Cragg, T. B. Med. Ent. 235; Ross, Prevention of Malaria, 632, [doubtful carrier]; Iyengar, I. J. Med. R., 1920 Sci. Cong. No., 9[8]; de Sá, Bol. Geral. Med. Farm. V. (8) 290-4[15]; Enderl. Wien. Ent. Zeit., XXXVIII, 47[16]; White, I. J. Med. R., VIII, 319[17]; Hacker, F. M. S. Mal. Bur. Rpt., I, 73, [Index to numerous refer.]; Kenk., Gen. Tijd. Ned. Ind., LXI, 202[14], [bionomics, not found infected]; Hacker, F. M. S. Mal. Bur. Rpt., II, 9, 17, [faunal associations].

[6]Bombay.
[11]Indore.
[10]Jeypore Hills.
[9]Andamans.
[5, 12, 17, 18]Ceylon.

---

[9]F. M. S.
[9, 13, 14, 20]Sumatra.
[9]Philippines.
[9]Siam.
[9]Japan.
[16]Madagascar.
[9]W. Africa.
[19, 20]Java.
[20]E. Malaysic.

Myzorhynchus barbirostris, Theo. Monog. Culic., III, 86, f. 25, pl. iii; *id.* Gen. Ins., Fasc., XXVI, pl. i, 3; Blanch., Moustiques, 197; James & List. Monog. Anoph. Mosq. Ind. (2nd edn.), 118-119, pl. XVI; Brun., R. I. M. I, 314[6]; *id., loc. cit.*, IV, 424; Theo. *loc. cit.*, II, 288[7]; Cruick. & Wright, I. J. Med. R., I, 773[3]; Gunasekera, Ceylon S. P., 37 of 1913, 7[18].

Myzorhyncus barbirostris, Leic. Culic. Malaya, 33; Chambers, Spol. Zeyl., II, 168[12], [Ceylon distribution]; Green, *loc. cit.*, IV, 183.

Myzorhynchus barbirostris var. pallidus, Swell., B. Ent. R., XI, 82, 86-87[20], fig. (1920), [larval variety only].

barianensis, James & Liston.

Anopheles barianensis, James & List., Monog. Anoph. Mosq. Ind. (2nd edn.), 76[1]; Brun., R. I. M., IV, 415; Patton & Cragg. T. B. Med. Ent., 230.

[1]Murree Hills.
[2]N.-W. Himalaya.

Anopheles plumbeus, Chris., (*nec.* Steph.) I. J. Med. R., III, 475, 489-496, pl. XXVIII, [larva[2]].

culicifacies, Giles.

Anopheles culiciicies, Giles, E. M. M., XXXVII, 197, (1901), [female only, male renamed *turkhudi*, List., 1901]. Theo., Monog. Culic. II, 309, [female *nec.* male, (t. Theo., *loc. cit.*, III, 48)]; Giles, Handbook, (2nd. edn.) 317, pl. IX, 12, [female *nec.* male]; James & List. Anoph. Mosq. Ind. 106, pl. IX, 1, pl. VIII, 1,

[1, 2]Burma.
[1]Assam.
[3, 5, 18]Punjab.
[6]Jeypore Hills.
[3]Central Prov.
[3, 9]Central India.
[1, 3, 4]Madras.

pl. XI; Chris., I. J. Med. R. III, 463[1]; Brun., R. I. M., XVII, 101[2]; Chalmers, Spol. Zeyl. II, 168[3], [Ceylon distribution]; Edw., B. Ent. R. III, 248; Wilson, I. J. Med. R., I, 691[4], [breeding places]; Hodgson, *loc. cit.*, II, 411[5]; Perry, *loc. cit.*, II, 469[6]; Gill, *loc. cit.*, IV, 207[7]; Sinton, *loc. cit.*, V, 192, [parasite]; *id., loc. cit.*, V, 197. 201, 208[8], [varieties]; Iyengar, *loc. cit.*, VII, Sci., Cong., No. 26[9]; Chris. *loc. cit.*, VII, 711[10]; Gough. B. Ent. R. V, 134[11]; Patton & Cragg., T. B. Med. Ent. Pl. XXXIV, 10 [egg], Pl. XXXVIII, 2, [imago], 229; Fletcher, Pusa. Sci. Rpt. (1915-16), 23[12]; Bentley, Malaria in Bombay[13]; Ross, Prevention of Malaria, 631 [carrier]; de Sá, Bol. Geral. Med. Farm, V (8), 290-294[15]; White, I. J. Med. R. VIII, 315[16].

[3]United Prov.
[12]Bihar.
[3]N. Bengal.
[8]N.-W. F. P.
[13]Bombay.
[15]Goa.
[3, 16, 17]Ceylon.

[2, 19]Algeria.
[11]Egypt.
[2, 14]Palestine.
[11]Cyprus.
[7]Muscat.
[10]Aden.

Anopheles culiciacies var. *sergenti*, Barraud, B. Ent. R. XI, 392, (1921)[14].

Myzomyia culicifacies, Theo., Proc. R.S., LXIX, 379; *id.* Monog. Culic., III, 39, figs., Pl. iii, Pl. VIII; *id., loc. cit.*, IV, 51; *id., loc. cit.*, V, 25; Blanch., Moustiques, 182; James & List., Monog. Anoph. Mosq. Ind. (2nd. edn.), 69, Pl. III, Pl. X, 1; Brun., R. I. M., I, 305[3]; *id. loc., cit.;* IV, 404, 418; Gunasekera, Ceylon S. P. 37 of 1913, 7[17], [parasites].

Myzomyia culicifacies var. *punjabensis*, James & List., Monog. Anoph. Mosq. Ind., 2nd. edn., 72, figs.[18], (1911).

Anopheles indicus, Theo., Monog. Culic., I, 183, (1901); Giles, Handbook, (2nd. edn.), 320.

Anopheles listoni, Giles [*nec.* Liston], E. M. M. XXXVII, 197, (1901); *id.*, Handbook, (2nd. edn.), 319, Pl. X, 4; Theo., Monog., Culic., II, 311, app.

Pyretophorus sergenti, Theo., Monog. Culic., IV, 68, f. 16[19], (1907).

culiciformis, Cogill.

Anopheles culiciformis, Cog., Bom. Jo., XV, 333[1], (1903); James & List., Monog. Anoph. Mosq. Ind., 122, Pl. XV; Chris. I. J. Med. R., III, 463; Brun., R. I. M., XVII, 112; Patton & Cragg. T. B. Med. Ent., 228; Black-lock & Cart, Ann. Trop. Med. Paras., XIII, 422.

[1]Karwar.

Coelodiazesis culiciformis, Chris. & Chand, I. J. Med. R., III, 638, Pl. LIX, [all stages described].

Stethomyia culiciformis, Theo. Monog. Culic., IV, 62; Brun., R. I. M., I, 312; *id., loc. cit.,* IV, 423.

Neostethopheles culiciformis, James & List., Monog. Anoph. Mosq. Ind. (2nd. edn.) 61, figs.

fuliginosus, Giles.

Anopheles fuliginosus, Giles, Handbook, 160, (1900); *id., loc. cit.,* (2nd. edn.), 298, Pl. VIII, 7; Theo. Monog. Culic., I, 132, ff. 27, 28, Pl. I. 3; James, Sci. Mem. Med. Off. Ind., II, 39, f. 18; James & List. Anoph. Mosq. Ind., 91, Pl. V, 2, Pl. X, 4, Pl. V; Chris. I. J. Med. R., III, 464; Chalmers, Spol. Zeyl., II, 169, [Ceylon distribution]; Green, *loc. cit.,* IV, 183; Brun., R. I. M., XVII, 98-99; Sinton, I. J. Med. R., V, 197, 202, 208[6], [varieties]; Theo., R. I. M. II, 288[11]; Wilson, I. J. Med. R., I, 691[2], [breeding places]; Gill, *loc. cit.,* II, 268, [minimum biting temp.]; Hodgson, *loc. cit.,* II, 411[3]; James, *loc. cit.,* II, 263[4]; Perry, *loc. cit.,* II, 469[5]; Iyengar, *loc. cit.,* VII, Sci. Cong., No. 26[7], Stanton, B. Ent. R., VI, 166, [larva]; *id., loc., cit.,* X, 333, [distribution in Far Eastern Ports]; Patton & Cragg., T. B. Med. Ent., 232, Pl. XXXIV, 8, [egg], Pl. XXXVII, [imago]; Fletcher, Pusa Sci. Rpt. (1915-16), 83[8]; Dyar, Ins. Ins. Mens., VIII, 184[9]; Fry, Malaria in Bengal, 11, [varieties], app. i, Cht. 16; Ross., Prevention of Malaria, 632, [doubtful carrier]; Iyengar, I. J. Med. R., 1920 Sci. Cong., No. 9[10]; Stanton, *loc. cit.,* III, 253[12]; Strickland, *loc., cit.,* IV, 260[13] [found in cleared hill areas]; de Sá, Bol., Geral. Med. Farm, V, (8), 290-294[14]; Hacker, F. M. S. Mal. Bur. Rpt., I, 74, [index to references]; *id., loc. cit.,* II, 10, 18, [faunal associations]; Kenk., Gen. Tijd. Ned. Ind., LXI, 202[16], [abst. R. A. E. (B) IX, 135].

[7]Central India.
[1, 18, 19]Central Prov.
[6]N.-W. F. P.
[1, 3, 20]Punjab.
[1, 8]Bihar.
[1, 10, 11]Bengal.
[1]Assam.
[1]Bombay.
[1, 2]Madras.
[1, 14]Goa.
[1]Quilon.
[4, 15]Ceylon.
[5]Jeypore Hills.

---

[13]F. M. S.
[12, 16, 17]Sumatra.
[9]Philippines.

Nyssorhynchus fuliginosus, Theo., Monog. Cylic. III, 93; *id., loc. cit.,* IV, 100, [life history]; Blanchard, Moustiques, 206, fig. 179; Brun., R. I. M., I, 317[1]; *id., loc. cit.,* IV, 428; James & List., Monog. Anoph. Mosq. Ind. (2nd. edn.), 87-89, Pl. VII; Gunasekera, Ceylon S. P. 37 of 1913, 7[15].

Anopheles fuliginosus var. *adiei,* James & List., Monog. Anoph. Mosq. Ind., (2nd. edn.) 89,[20] (1911); Chris. I. J. Med. R., III, 465; Brun., R. I. M., XVII. 99.

Nyssorhynchus jamesi, List., (*nec.* Theo.), Ind. Med. Gaz., XXXVI, 441, (1901).

Nyssorhynchus leucopus,[17] Don., Insektenborse, XVIII, 37, (1901).

Anopheles nagpori, James & List., Monog. Anoph. Mosq. Ind. 101[18], (1904).

Anopheles fuliginosus var. *nagpori, id., loc. cit.,* (2nd. edn.), 90[19] (1911).

fuliginosus var. nivipes, Theobald.

Nyssorihynchus nivipes, Theo., Entom. XXXVI, 258, (1903); *id.,* Monog. Culic., IV, 101; Brun., R. I. M., I, 319[2]; *id., loc. cit.,* IV, 429.

[1]Burma.

---

[2]F. M. S.

Nyssorhyncus nivipes, Leic., Culic., Malaya, 44.

Anopheles nivipes, Patton & Cragg, T.B. Med. Ent., 239.

Anopheles fuliginosus, James & Stant., Paludism, V, 62.

Anopheles fuliginosus var. *nivipes,* Chris. I. J. Med. R., III, 466[1]; Brun., R. I. M., XVII, 99; Swell., B. Ent. R., XI, 88; Stanton, *loc., cit.,* X, 333, [distribution in Far Eastern Ports].

funestus var. listoni, Liston.

Anopheles listoni, List., (*nec.* Giles), Ind. Med. Gaz., XXXVI, 441, (1901); Adie & Alcock, Proc. R. S., LXXVI, 319-321, [enemies in Calcutta]; Hodgson, I. J. Med. R., II, 411[2]; Perry, *loc. cit.,* II, 469[3]; Gill, *loc. cit.,* VII, 611[5]; Patton & Cragg, T. B. Med. Ent., 230; Bentley, Malaria in Bombay, 62[6]; Takaki in Ross, Prevention of Malaria, 564[7]; Ross., *loc. cit.,* 631, [carrier]; de Sá, Bol. Geral. Med. Farm, V (8), 290-294[8].

Myzomyia listoni, Theo., Monog. Culic, III, 27, ff. 12, 13, 17; James & List., Monog. Anoph. Mosq. Ind. (2nd. edn.), 73, Pl. XI, 1; Brun., R. I. M., I, 305-308; *id., loc. cit.,* IV, 420; James, Sci. Mem. Med. Off. Ind., II, 31, f. 9; Gunasekera, Ceylon S. P. 37 of 1913, 7[10].

[8]India., [5]Kashmir.
[1, 4]N.-W. F. P.
[1, 2]Punjab.
[1]United Prov.
[1]Bihar., [1]Bengal.
[1]Assam., [1]Burma.
[1]Central India.
[6]Bombay.
[3]Jeypore Hills.
[8]Goa., [1, 9, 10]Ceylon.

---

[7]Formosa., [11]Gambia.
[11]Sudan., [11]Perak.

Anopheles funestus var. *listoni,* Chris., I. J. Med. R., III, 181, 466[1]; Sinton, *loc. cit.,* VI, 192, [parasite]; 197, 201, 207[4], [varieties]; Brun., R. I. M., XVII, 107[12]; White, I. J. Med. R., VIII, 315[9].

Anopheles fluviatilis, James, Sci. Mem. Med. Off. Ind. II, 31, f. 9, (1902); James & List., Monog. Anoph. Mosq. Ind. 103, Pl. VII, I, Pl. X; *id., loc. cit.,* (2nd. edn.), Pl. IV; Theo. Proc. R. S., LXIX, 378, Pl. V, 3.

Myzomyia leptomeres, Theo., Monog. Culic., III, 38, (1903)[8]; *id., loc. cit.,* IV, 124; id., loc. cit. V, 29; Brun., R. I. M., 307; *id., loc. cit.,* IV, 419; James & List., Monog. Anoph. Mosq. Ind. (2nd. edn.), 76.

Anopheles leptomeres, Patton & Cragg, T. B. Med. Ent., 230.

Anopheles funestus var. *leptomeres,* Chris., I. J. Med. R., III, 467; Edw., B. Ent. R. III, 249; Brun., R. I. M., XVII, 107.

Myzomyia funesta var. *subumbrosa,* Theo., Monog. Culic., III, 34, (1903); *id., loc. cit.,* V, 17[11].

gigas, Giles.

Anopheles gigas, Giles, E. M. M., XXXVII, 196, (1901); Theo., Monog. Culic., II 308; James & List. Anoph. Mosq. Ind., 118; giles, Handbook, (2nd. edn.), 316, Pl. X, 2; Brun., R. I. M., I, 303[1]; *id., loc. cit.,* IV, 416; Chris., I. J. Med. R., III, 467[3]; Theo. Monog. Culic., V, 8[2]; Patton & Cragg, T. B. Med. Ent., 229; Brun., R. I. M., XVII, 100.

[3]Deesa.
[3, 4, 5]Himalayas.
[3]Khasia.
[3]Pachmari.
[1, 2, 3]Nilgiris.
[2]Ceylon.

Myzomyia gigas, Blanch., Moustiques, 184.

Patagiamyia gigas, James, R. I. M., IV, 98.

Patagiamyia simlensis, James & List., Monog. Anoph. Mosq. Ind. (2nd. edn.), 66[5], (1911).

Anopheles simlensis, Brun., R. I. M., IV, 416; Patton & Cragg, T. B. Med. Ent., 229; Acton, etc., I. J. Med. R., VIII, 952[4], (*simalensis,* laps).

Anopheles gigas var. *simlensis,* Alcock, Jo. Lop. Sch. Trop. Med., II (3), 161, (1913).

Anopheles gigas var. *rufitarsis, id., loc. cit.,* II, (3) 161, (1913).

immaculatus, James.

Anopheles immaculatus, James, Sci. Mem. Med. Off. Ind., II, 35 (1902)[1]; Theo., Monog. Culic., III, 23; James & List. Monog. Anoph. Mosq. Ind., 120; *id., loc. cit.,* (2nd. edn.), 77; Brun., R. I. M., I, 304; *id., loc. cit.,* IV, 416; Chris., I. J. Med. R., III, 468; Brun., R. I. M., XVII, 110; Theo. Monog. Culic., V, 14; Patton & Cragg. T. B. Med., Ent., 231; Edw., B. Ent. R., XII, 70 [?=*vagus,* Dön.].

[1]Ennur.

———

[2]D. E. Ind.

Myzomyia immaculata, Swell. and Graaf, B. Ent. R. XI. 77 [redescr.]

Myzomyia flava, Swell., (1917)[2]

jamesi, Theobald.

Anopheles jamesi, Theo., Monog. Culic., I, 134, Pl. I, 2, (1901); Giles, Handbook, (2nd. edn.) 299; James, Sci. Mem. Off. Ind., II, 41; James & List., Monog. Anoph. Mosq. Ind., 93, Pl. VI; Chris., I. J. Med. R., III, 468; Chatterjee, R. I. M., I, 82[2]; Hodgson, I. J. Med. R., I, 702[3]; Corn. & Patton. *loc. cit.,* II, 569, [salivary secretion]; Hodgson, *loc. cit.,* II, 411[4]; James, *loc. cit.,* II, 263[5]; Perry, *loc. cit.,* II, 469[6]; Patton & Cragg, T. B. Med., Ent., 233; Iyengar, I. J. Med. R., 1920 Sci. Cong. No. 9[7]; de Sá, Bol. Geral. Med. Farm. V, (8) 290-4[8]; White, I. J. Med. R., VIII, 315[9].

Nyssorhynchus jamesi, Blanch, Moustiques, 206; Brun., R. I. M., I, 318[1]; *id., loc. cit.,* IV, 429; James & List., Monog. Anoph. Mosq. Ind. (2nd. edn.), 91, Pl. VIII; Gunasekera, Ceylon S.P. 37 of 1913, 7[10].

[1, 4]Punjab.
[1]United Prov.
[1]Quetta., [1]Berars.
[1]Deccan.
[1]Central Ind.
[1, 2, 7]Bengal.
[1, 3]Madras.
[1, 8]Malabar Coast.
[5,9,10]Ceylon.
[6]Jeypore Hills.
[1]Bombay.

jeyporiensis, James.

Anopheles jeyporiensis, James, Sci., Mem. Med. Off. Ind., II, 32, ff. 11, 12[1]; James & List., Monog. Anoph. Mosq. Ind., 101, Pl. VII, 2, Pl. IX; Chris., I. J. Med. R., III, 468; Brun., R. I. M., XVII, 109[6]; Chalm., Spol. Zeyl. II, 169[5], [Ceylon distrib.]; Perry, I. J. Med. R., II, 469[4]; Mc. Farl., B. Ent. R., VI, 67[3]; Patton & Cragg, T. B. Med. Eng. 231.

Pyretophorus jeyporiensis, Theo., Monog. Culic., III, 66, Pl. VIII; James & List. Monog. Anoph. Mosq. Ind., (2nd. edn.), 81, Pl. VI, fig. p. 67.

Pyretophorus jeyporensis, Brun., R. I. M. I, 313[2].

[1, 2, 4]Jeypore Hills.
[2]Central Prov.
[2]Bombay., [6]Kanara.
[5]Ceylon.

---

[3]Hongkong.

karwari, James

Nyssorhynchus karwari, James in Theo., Monog. Culic., III, 102, f. 61. (1903)[1]; Brun., R. I. M., I, 318[7]; *id., loc. cit.,* IV, 429; James & List., Monog. Anoph. Mosq. Ind., (2nd. edn.), 96, Pl. XII; Leic., Culic, Malaya, 39, (*Nyssorhyncus*).

Anopheles karwari, James & List., Monog. Anoph. Mosq. Ind., 89, figs. Pl. XIV; Chris., I. J. Med. R., III, 469[6], Brun., R. I. M., XVII, 112; Perry, I. J. Med. R., II, 469[2]; Mc. Farl., B. Ent. R., VI, 67[3]; Stanton, *loc. cit.,* VI, 167, [larva descr.]; Patton & Cragg, T. B. Med. Ent., 233;

[1, 7]Karwar.
[6, 7]Malabar Coast.
[6]Central India.
[2]Jeypore Hills.
[6]Assam.,
[5]Ceylon.

[4, 8, 9, 10]F. M. S.
[3]Hongkong.

Strick., I. J. Med. R., IV, 260[4]; [in cleared hill areas]; White, *loc. cit.*, VIII, 316[5]; Lamb., B. Ent. R., XII, 96, fig. 3 f. [pupa]; Hacker, F. M. S. Mal. Bur. Rpt., I, 74[8], [index to refs.]; *id., loc. cit.*, II, 16[9], [faunal associations].

Anopheles nigrans, Stanton, Jo. Lon. Sch. Trop. Med., II, (1) 7, (1912)[10].

kochi, Dönitz.

Anopheles kochi, Dön., Insektenbörse, XVII, 36, (1901); Theo., Monog. Culic., I, 74, Pl. IV, 16; Edw., B. Ent. R., IV, 222; Gough, *loc., cit.*, V, 142; Roper, *loc. cit.*, V, 129, [life history]; Chris., I. J. Med. R., III, 469[2]; Brun., R. I. M., XVII, 102[3]; Stant., B. Ent. R., V, 129, fig. 1, 2, [early stages]; *id., loc. cit.*, VI, 168, [larva]; *id., loc. cit.*, X, 333, [distrib. in far East ports]; Swell, *loc. cit.*, XI, 85-86, [larval notes]; Takaki in Ross. Prevention of Malaria, 564[4]; Stant., I. J. Med. R., III, 253[5]; Strick., *loc. cit.*, IV, 260[3], [in cleared hill areas]; Kenk., Gen. Tijd. Ned. Ind., LXI, 202[7], [Abst. R. A. E. (B), IX, 135], [carrying]; Hacker, F. M. S. Mal. Bur. Rpt., I, 75[8], [index to refs.]; *id., loc. cit.*, II, 10[9], [faunal associations]; 14, 33, [larval distinctions].

[2, 3, 11]Assam.
[2]Burma.
———
[3, 6, 8, 9]F. M. S.
[3]Java.
[3, 5, 7]Sumatra.
[3, 2, 13]Philippines.
[4]Formosa.

Nyssorhynchus kochi, Blanch, Moustiques, 208.

Cellia kochi, Theo., Monog. Culic., III, 110; Leic., Culic. Malaya, 46; Brun., R. I. M., I, 321; *id., loc. cit.*, IV, 431.

Cellia flava, Ludl., Can. Ent., XL, 32, (1908)[12]; Brun., R. I. M., IV, 431[13].

Nyssorhynchus flavus, Ludl., Psyche, XVIII, 126.

Christophersia halli, James, Paludism, I, 33, figs., (1910)[11]; James & List., Monog. Anoph. Mosq. Ind. (2nd. edn.) 123; Brun., R. I. M., IV, 431.

Anopheles halli, Patton & Cragg, T. B. Med. Ent., 236.

Anopheles ocellatus, Theo., M. S., Monog. Culic., II, 306, [note]; Giles, Handbook, (2nd. end.), 315, Pl. IX, 5, (1902).

leucosphyrus, Dönitz.

Anopheles leucosphyrus, Dön., Insektenbörse, XVIII, 37, (1901); Theo., Monog. Culic., II, 307, app.; Roper, B. Ent. R., V, 143; Swell, *loc. cit.*, XI, 83, [larva

[1]Assam.
[1, 6]Malabar Coast.
[2, 4]Karwar.

desc.]; Stant., *loc. cit.*, VI, 168, [larva]; Brun., R. I. M., XVII, 102-103[4]; Kenk. Gen. Tijd. Ned. Ind., [abst. R. A. E. (B) IX, 135][8], [notes]; Hacker, F. M. S. Mal. Bur. Rpt., I, 75[9], [index to refs.]; *id., loc. cit.*, II, 22[10], [faunal associations], 33, [abnormal imago]; Stant, I. J. Med. R., III, 254.[5]

[3, 4]Duars. [3, 4]Andamans. Matale, (Ceylon) (White).

Anopheles leucophyrus, Chris., I. J. Med. R., III, 469.[1]

Myzomyia leucophurus, Giles, Handbook, (2nd. edn.) 312, fig. 44; James & List., Monog. Anoph. Mosq. Ind., 82, Leic., Culic. Malaya, 28; Brun., R. I. M., I, 307[2].

[4, 5, 8]Sumatra. [4, 7, 9, 10]F. M. S. [4]Borneo.

Nyssorhynchus leucophyrus, Blanch., Moustiques, 213.

Neomyzomyia leucophyrus, Brun., R. I. M., IV, 417.

Anopheles leucophyrus var. *elegans* James & List., Monog. Anoph. Mosq. Ind., 82, Pl. IX, XII, (1903).

Pyretophorus leucosphyrus var. *elegans,* Theo., Monog. Culic., IV, 77.

Neomyzomyia elegans, James, in Theo., Monog. Culic., V, 30, figs.[3] [Theo. desc.]; James & List., Monog. Anoph. Mosq. Ind. (2nd. edn.), 105, Pl. XV.

Anopheles elegans, Patton & Cragg, T. B. Med. Ent., 228; de Sá, Bol. Geral. Med. Farm, V (8), 290-4.[6]

Anopheles leucosphyrus var. *hackeri,* Edw., B. Ent. R., XII, 70, (1921).

lindesaii, Giles.

Anopheles lindesaii, Giles, Handbook, 166, (1900)[1]; *id., loc. cit.*, (2nd. edn.), 323, Pl. X, 8; Chris., I. J. Med. R., II, 470[5]; Gill, I. J. Med. R., VII. 611.[4]

[8]Ferozepore. [8]Tret., [8]Murree. [7, 8]Masuri Dist.

Anopheles lindesayi, Theo., Monog. Culic. I, 203, Pl. V, 9; James & List., Monog. Anoph. Mosq. Ind., 117, Pl. XV; Blanch., Moustiques, 169; Brun., R. I. M., I, 304[3]; *id., loc. cit.*, IV, 416; Annan., R. I. M., I, 83[2]; Patton & Cragg. T. B. Med. Ent., 229; Brun., R. I. M., XVII, 100[8]; Acton, etc., I. J. Med. R., VIII, 751.[6]

[2, 3]Naini Tal Dist. [6]Simla Dist. [4]Kashmir., [8]Kurseong. [8]Assam., [5]Nilgiris. [5]Palnis., [1]Punjab.

Patagiamyia lindesayi, James & List., Monog. Anoph. Mosq. Ind. (2nd. edn.), 62, figs., Pl. II.

Patagiamyia lindesayi, var. *maculata,* Theo., R. I. M., IV, 1; Brun., *loc. cit.*, IV, 40[7]; Theo., Monog. Culic., V, 14.

ludlowi, Theobald.

Myzomyia ludlowi, Theo., Monog, Culic., III, 42, (1903)[1]; Banks, Phil. Jo. Sci., II, 519, Pl. I-XI, [life history][2]; Brun., R. I. M., I, 309; *id., loc. cit.,* IV, 420.

Nyssomyzomyia ludlowi, James & List., Monog. Anoph. Mosq. Ind. (2nd. edn.), 101, t. fig., Strick., B. Ent. R., V, 321, Pl. XXXII, XXXIII, 2 t. fig. [morphology].

Anopheles ludlowi, Roper, B. Ent. R., V, 145; Chris., I. J. Med. R., III, 470[3]; Brun., R. I. M., XVII, 110[4]; Swell., B. Ent. R., XI, 87-91, [larva; distinctions from *rossii*]; Hodgson, I. J. Med. R., I, 702[5] [carrying]; James, *loc., cit.,* II, 263[6]; Stanton, *loc. cit.,* III, 254[7]; de Sá, Bol. Geral. Med. Farm, V (8) 290-4[8]; Balf., B. Ent. R., XII, 30 [saline water breeder]; Darling., Jo. Exp. Med., XXXII, 313-329[9], [abst. R. A. E. (B) IX, 130], [carrying]; Hacker, F. M. S. Mal. Bur. Rpt., II, 16[10], [faunal associations], 34, 43, [carrying]; Swell. & Graaf, Parasitology, XII, 180-198[11], [more dangerous than *umbrosus*].

[3]Burma., [3, 5, 4]Madras. [8]Goa., [3, 4]Andamans. [3, 6]Ceylon., ——— [7, 11]Sumatra. [9, 11]Java.,[2, 4, 10]F. M. S.

maculatus, Theobald.

Anopheles maculata, Theo., Monog. Culic. I, 71, f. 48, (1901).

Anopheles maculatus, James Sci. Mem. Med. Off. Ind., II, 47, fig. 25; Giles, Handbook, (2nd edn.), 301, Pl. IX, 2; James & List., Monog. Anoph., Mosq. Ind., 99; Chris., I. J. Med. R., III, 471[1]; Stanton, Jo. Lon. Sch. Trop. Med., II (1) V., Roper, B. Ent. R., V, 143[5]; Stanton, *loc. cit.,* VI, 169, [larva]; Patton & Cragg, T. B. Med. Ent., 232; Takaki in Ross. Prevention of Malaria, 564[6]; Stanton, I. J. Med. R., III, 254[7]; Strickland, *loc. cit.,* IV, 259[8], [not in hill jungle]; Brun., R. I. M., XVII, 104; Watson, Trop. Agric., XLII, 151, [notes in abst.]; Chalm., Spol. Zeyl., II, 169[3], [Ceylon distrib.]; de Sá, Bol. Geral. Med. Farm, V (8), 290-4[9]; White, I. J. Med. R., VIII, 315[10]; Lamb., B. Ent. R., XII, 93, fig. 1, [larva]; Hacker, F. M. S. Mal. Bur. Rpt., I, 75[12], [index to references]; *id., loc. cit.,* II, 11, 15[13], [faunal associations].

[1, 2, 4]Philippines.
[1, 2]N.-W. Himalaya.
[2, 4]Lahore.
[2, 4]Kurseong.
[1, 2]Assam.
[2, 4, 12]Duars.
[2, 4]Jeypore Hills.
[1]South India.
[9]Goa.
[1, 3, 10, 11]Ceylon.
———
[2, 7]Sumatra.
[5]Borneo.
[2, 8, 12, 13]F. M. S.
[2]Hongkong.
[6]Formosa.

Nyssorhynchus maculatus, Theo., Monog. Culic., III 96; Blanch. Moust., 207, f. 180; James & List., Monog. Anoph. Mosq. Ind. (2nd. edn.), 84-86; Leic., Culic. Malaya, 41; Brun., R. I. M., I, 318[4]; *id., loc. cit.,*

IV, 429; James & Stant., Plaudism, V, 62; Gunasekera, Ceylon S. P. 37 of 1913, 7[11].

Nyssorhyncus willmori, Leic. [*nec.* James], Culic. Malaya, 42.

Nyssorhynchus pseudo-willmori, Theo., Monog. Culic. V, 65, (1910); James & List., Monog. Anoph. Mosq. Ind. (2nd. edn.), 87, Brun., R. I. M., IV, 430.[12]

?Anopheles maculatus var. *willmori,* Alcock, Jo. Lon. Sch. Trop. Med., III, 3, (1913).

maculipalpis, Giles.

Anopheles maculipalpis, Gills, Handbook, (2nd ed.), 279, (1902); James & List., Monog. Anoph. Mosq. Ind., 95, Pl. IV X, 5. Pl. VII; Christ., I. J. Med. R., III 472[1]; Brun., R. I. M., XVII, 108[2]; Chalm., Spol. Zeyl., II, 168[4], [Ceylon distribution]; Hodgson, I. J. Med. R., II, 411[5]; Perry, *loc. cit.,* II, 469[6]; Sinton, *loc. cit.,* V, 197, 202, 206[7], [varieties]; Gill, *loc. cit.,* VII, 611[8]; Patton & Cragg. T. M. Med. Ent., 233; Ross, Prevention of Malaria, 632, [carrying].

Nyssorhynchus maculipalpis, Theo., Monog. Culic., III, 96, fig. 56, fig. p. 98, Pl. II; Brun., R. I. M., I, 319[3]; *id., loc. cit.,* IV, 429; James & List., Monog. Anoph. Mosq. Ind., (2nd. edn.), 93, Pl. IX, James, R. I. M., VI, 31[9].

Nyssorhynchus maculipalpis var. *indiensis,* Theo., Monog. Culic., III, 99, (1903); Brun., R. I. M., I, 319.

Nyssorhynchus indiensis, Theo., Monog. Culic., IV, 98; Brun., R. I. M., IV, 428[10]; James & List., Monog. Anoph. Mosq. Ind. (2nd. edn.), 95.

Anopheles indiensis, James & List., Monog. Anoph. Mosq. Ind., 95, Pl. ii. figs.

Anopheles jamesi, Steph. & Chris. [*nec.* Theo.], Rpt. R. S. Mal. Com., 1902.

[1, 2, 7]N.-W. Frontier. [1, 2]Himalaya Terai. [1, 2, 5]Indo-Gangetic Plain. [1, 2, 3, 10]Central, India. [1, 2]South India. [1, 2, 3, 10]Malabar Coast., [2]Bombay. [8]Kashmir. [6]Jeypore Hills. [9]Yunnan.,[2, 4]Ceylon.

---

[2]Mashonaland. [2]Philippines. [2]Mauritius.

minimus, Theobald.

Anopheles minimus, Theo., Monog. Culic. I, 186, fig. 55, (1901); Giles, Handbook, (2nd. edn.), 321, Pl. X, 7; Chris., I. J. Med. R., III, 473[1]; Brun., R. I. M., XVII, 105; Mc. Farl., B. Ent. R., VI, 67[2]; Stant., *loc. cit.,* X, 333, [distribution in Far East Ports].

[6]Bombay., [6]Goa. [1, 6]Assam., [1, 6]Bengal. [1, 6]Central India. [6]N. Kanara. [6]Kangra Valley.

Myzomyia minima, Swell., B. Ent. R., XI, 86, fig. 9, [larva].

Pyretophorus minimus, Giles, Jo. Trop. Med., VII, 365; Blanch., Moustiques, 188, f. 169; Brun., R. I. M., I, 313[3]; *id., loc. cit.,* IV, 423.

Anopheles aconitus var. *cohaesa,* Dön., Zeit. f. Hyg., XLIII, 233 (1903).

Anopheles christophersi, Theo., Proc. R. S., LXIX, (456) 378, (1902).

Myzomyia christophersi, James & List., Monog. Anoph. Mosq. Ind., 103, Pl. VII, 1, Pl. X; Giles, Handbook, (2nd. edn.) 512; Theo., Monog. Culic, V, 24[6]; Edw., B. Ent. R., IV, 222; Ludl., *loc. cit.,* VI, 155, [synonomy].

Anopheles febrifer, Banks, Phil. Jo. Sci., IX, (D) 405, (1914) [prelim. descr.]; Walk. & Barb., *loc. cit.,* IX, (B) 384, (1914).

Anopheles formosdensis, I. Tsutzuki, Arch. f. Schiff., IV, 9, (1902).[5]

Myzomyia funesta, Ludl. [nec. Giles], Can. Ent., XXXVII, 135, (1905).

Myzomyia mangyana, Banks, Phil. Jo. Sci., I, 901, (1906); Brun., R. I. M., I, 309.[4]

[6]Ceylon., _______
[2, 3]Hong Kong.
[3, 4]Philippines.
[5]Formosa., [6]Perak.

minimus var. aconitus, Dönitz.

Anopheles aconitus, Dön.; Beit. Kenntn. Anoph., 70[6], (1902); *id.,* Zeit. f. Hyg., XLIII, 233; Theo., Monog. Culic., III, 30, fig.; Brun., R. I. M., IV, 415, [carries]; Stant., I. J. Med. R., III, 252; Brun., R. I. M., XVII, 108[1]; Stant., I. J. Med. R., VI, 162[2], [larva]; Hacker, F. M. S. Mal. Bur. Rpt., I, 73[3], [index to references]; *id., loc., cit.,* II, 10, 18[4], [faunal associations].

Myzomyia aconita, Brun., R. I. M., I, 305.[5]

Anopheles minimus var. *aconitus,* Dön., Zeit. f. Hyg., XLI, (1902); Chris., I. J. Med. R., III, 475[7]; Brun., R. I. M., XVII, 106[8]; White, I. J. Med. R., VIII. 315.[9]

Myzomyia minima var. *aconita,* Swell. B. Ent. R., XI, 85, fig. 8, [larva].

Myzomyia albirostris, Theo., Monog. Culic, III, 24, fig. 11, (1903); Leic., Culic. Malaya, 23; Brun., R. I. M., I, 305[10]; *id., loc. cit.,* IV, 418; Strick., I. J. Med. R., I, 203.[11]

[7, 15, 16]Bengal.
[8]Assam., [8]Burma.
[8]Kangra Valley.
[13]Jeypore Hills.
[9, 12, 14]Ceylon.
________, [1, 5, 6]Java.
[1]Sumatra.
[2, 3, 4, 8, 10, 11]F. M. S.
[8]Hongkong.
[8, 17]Philippines.

Anopheles albirostris, James, I. J. Med. R., II, 263[12]; Perry, *loc. cit.*, II, 469[12]; Stant., B. Ent. R., III, 387, 6 figs., [larval changes]; Patton & Cragg, T. B. Med. Ent., 237.

Myzorhynchus albirostris, Gunasekera, Ceylon S. P. 37 of 1913, 7.[14]

Anopheles bramacharii, Chris., Paludism, V, 11, (1912); Bramachari, Ind. Med. Gaz., XLVI, 268, (1911).

Myzomyia christophersi var. *alboapicalis,* Theo., Monog. Culic., V, 25[15], (1910).

Anopheles listoni var. *alboapicalis,* Iyengar, I. J. Med. R., 1920 Sci. Cong. No. 9[16], (1921).

Myzomyia flavirostris, Ludl., Bull. 4, Surg. Gen. Off[17], (1913); *id.*, Psyche, XXI, 30.

pallidus, Theobald.

Anopheles fuliginosus var. *pallidus,* Theo., Monog. Culic, I, 134[1], fig. 28*b*, (1901); Chris., I, J. Med. R., III, 466; Brun., R. I. M., XVII, 99.

Nyssorhynchus fuliginosus var. *pallidus,* Blanch., Moustiques, 205, fig. 179*b*.

Neocellia fowleri, Chris., Paludism, II, 64, (1911); *id., loc. cit.*, III, Pl. VI.

Anopheles fuliginosus var. *fowleri,* Alcock, Jo. Lon. Sch. Trop. Med., II, (3), (1913).

Anopheles fowleri, Chris., I. J. Med. R., III, 464[2]; Brun., R. I. M., XVII, 114[5]; Kenrick, Paludism, III, 65; Hodgson, I. J. Med. R., II, 411[3]; Perry, *loc. cit.*, II, 469[4]; Patton & Cragg, T. B. Med. Ent., 236.

[1]Bihar., [2, 3]Delhi.
[4]Jeypore Hills.
[5]Amritsar.
[5]Central Prov.
[5]S. India.

pulcherrimus, Theobald.

Anopheles pulcherrimus, Theo., Proc. R. S., LXIX, 396, Pl. V, 2, (1902); James, Sci., Mem. Med., Off. Ind., II, 48, f. 26; James & List., Monog. Anoph. Mosq. Ind., 86, Pl. IV, Giles, Handbook, (2nd. edn.), 510; Chris., I. J. Med. R., III, 476[1]; Brun., R. I. M., XVII, 109[2]; Hodgson, I. J. Med. R., II, 411[3]; Sinton, *loc, cit.*, V, 197, 202, 207 [varieties]; Acton, *loc. cit.*, VI, 268[5]; Chris., *loc., cit.*, VII 713[6]; Chris. & Short, *loc. cit.*, VIII, 516.[7]

[1, 4]N.-W. F. P.
[1, 2, 3]Punjab.
[1]United Prov.
[6]Sind., [2]Purneah.
[2, 8]Goa.

[5, 6, 7]Mesopotamia.
[6]Syria., [6]Arabia.
[2]Turkestan.

Anopheles pulcherrima, Patton & Cragg, T. B. Med. Ent., 235.

Cellia pulcherrima, Theo., Monog. Culic., III, 107; Blanch. Moustiques, 215; Brun., R. I. M., I, 321; *id., loc. cit.,* IV, 432; James & List., Monog. Anoph. Mosq. Ind. (2nd. edn.), 116, Pl. XIII[8].

punctulatus var. tessellatus, Theobald.

Anopheles punctulatus, Dön., Insektenborse, XVIII, 36, (1901); James & List., Monog. Anoph. Mosq. Ind., 84, Pl. XI; Roper, B. Ent. R., V 144[1]; Brun., R. I. M., XVII, 102.[2]

Anopheles punctulata, Patton & Cragg, T. B. Med. Ent. 227.

Myzomyia punctulata, Leic., Culic. Malaya, 27; Brun., R. I. M., I, 309.[3]

Nyssorhynchus punctulatus, Blanch., Moustiques, 208.

Nyssomyzomyia punctulata, James & List., Monog. Anoph. Mosq. Ind., (2nd edn.) 104, Pl. XIV[4]; James & Stant., Trans. F. East. Ass. Trop. Med., 1912, 315; Gunasekera, Ceylon S. P. 37, of 1913, 7[5].

Cellia punctulata, Theo., Monog. Culic., IV, 109; Brun., R. I. M., IV, 132.

Anopheles tessellatum, Theo., Monog. Culic., I, 175 f. 49 Pl. XXXVII 148 (1901) [as *punctulatus,* Dön.]; Giles, Handbook, (2nd. edn.) 305, Pl. IX, 7; Edw. B. Ent. R., IV, 221; Stanton, I. J. Med. R., III, 255[6]; Chris., *loc. cit.,* III, 482[7]; Brun., R. I. M., XVII, 105[8]; James, I. J. Med. R., II, 263; Perry, *loc. cit.,* II, 469[10]; Stanton, B. Ent. R., IV, 129, figs. 1-4, [larva]; Mc. Farl., *loc. cit.,* VI 67[11]; Stanton, *loc. cit.,* VI 171, [larva]; *id., loc. cit.,* X, 333, [distribution in Far East. ports]; Lamb., *loc. cit.,* XII, 96, f. 3c, [pupa]; Darling., Jo. Exp. Med., XXXII, 313-329[13], (abs. R. A. E., (B) IX, 131]; Hacker, F. M. S. Mal. Bur. Rpt., I, 76[14], [index to references]; *id., loc. cit.,* II, 10, 19[15], [faunal associations], [larval distinctions].

Myzomyia tessellatum., Theo., Monog. Culic., III, 55.

Myzomyia tessellata Theo., Monog., Culic., IV 42; Brun., R. I. M., I, 311[16]; *id., loc. cit.,* IV, 422.

Anopheles punctulatum, Dyar, Ins. Ins. Mens. VIII, 184[12].

[4]Bombay.,
[7]Burma.
[7]Assam.
[4, 7]Central India.
[10]Jeypore Hills.
[4, 7]South India.
[7]Delhi.
[5, 7, 8, 9, 17, 19]Ceylon.

[3, 6, 8, 18]Sumatra.
[13]Java.
[14, 15, 16]F. M. S.
[11]Hongkong.
[12, 20]Philippines.
[1, 3]Borneo.
[2, 3]Papua.

Anopheles punctulatus var. *tessellatus,* Swell., B. Ent. R., XI 85, 80, 86, 89, 90; Edw., *loc. cit.,* XII, 70, 71.

Dactylomyia ceylonica, News. & Cart., Ann. Trop. Med. Paras., VI, 377, (1910).[17]

Anopheles deceptor, Dön., Zeit., F. Hyg., u. Insek, XLI, 60, (1902)[18].

Myzomyia deceptor, Theo., Monog. Culic., V, 29[19]; Brun., R. I. M., IV, 418.

Myzomyia thorntoni, Ludl., Can. Ent., XXXVI, 69, (1904)[20]; Blanch., Moustiques, 183; Brun., R. I. M., I, 311; *id., loc. cit.,* IV, 422.

rhodesiensis, Theobald.

Anopheles rhodesiensis, Theo., Monog. Culic., I, 184[1], (1901); Chris. & Chand., I. J. Med. R., III, 180[2]; Chris., *loc. cit.,* III, 476[3]; Gill, *loc. cit.,* IV, 207[4]; Sinton, *loc., cit.,* V, 197, 203, 208[5], [varieties]; Chris., *loc., cit.,* VII, 711[6]; Chris. & Short, *loc. cit.,* VIII, 516.

Myzomyia rhodesiensis, Theo., Monog. Culic., III, 35; *id., loc., cit.,* V, 26[10].

Anopheles d'thali, Patton, Bom. Jo., XVI, 623, (1905)[8].

Myzomyia d'thali, Theo., Monog., Culic., V, 20, figs. [redescr.][9].

[2, 3]Baluchistan.
[3, 5]N.-W. F. P.

---

[4]Muscat., [8, 9]Aden.
[6, 7]Mesopotamia.
[1, 10]Rhodesia.
[10]Mashonaland.
[10]Uganda.
[10]Congo Free State.
[10]Transvaal.

rossi, Giles.

Anopheles rossi. (Giles), Jo. Trop. Med., II, 63, (Oct. 1899); Theo., Monog. Culic., I, 154 ff. 37-38, Pl. A, Pl. III; 9, 10[1]; Giles, Handbook, 149; *id., loc. cit.,* 2nd. edn., 311, Pl. IX, 11; James & List., Monog. Anoph. Mosq. Ind., 109, Pl. VI, 1, Pl. X, 3 Pl. XII; Chris., I. J. Med. R., III, 476-479; Chalmers, Spol. Zeyl., II, 167[3], [Ceylon distrib.]; Green, Trop. Agric., XXVII, 84; Chatterjee, R. I. M., I, 82[4]; Hodgson, I. J. Med. R., I, 702[5]; Gill, *loc, cit.,* II, 268, [minimum biting temp.]; Corn. & Patton. *loc. cit.,* II, 569, [salivary secretion]; Hodgson, *loc. cit.,* II, 411[7]; James, *loc. cit.,* II, 263[8]; Perry, *loc. cit.,* II, 469[9]; Sinton, *loc. cit.,* V, 197, 201[10]; Soparkar, *loc. cit.,* V, 512, [parasite]; Iyengar, *loc. cit.,* VII, Sci. Cong. No. 26[11]; Patton & Cragg, T. B. Med. Ent., Pl. XXVII, 8, [salivary glands]; Pl. XXXIV, 1, 4, 6, p. 196, [larva], 303, [pupa], Pl. XXXVII, I,

[2, 10]N.-W. F. P.
[3, 7]Punjab.
[2]United Provinces.
[2]Bihar.
[1, 2, 4, 12]Bengal.
[2]Assam.
[2]Orissa.
[2, 11]Central India.
[2]Bombay.
[9]Jeypore Hills.
[2, 5]Madras.
[2, 6, 13]Malabar Coast.
[3, 8]Ceylon.
[18]Ind. or.
[16, 17]Calcutta.

[imago], p. 226; Fletcher, Pusa. Sci. Rpt. (1915-16), 83; Ross, Prevention of Malaria, 632, [not a carrier]; Iyengar, I. J. Med. R., 1920 Sci. Cong. No., 9[12]; Strickland, *loc. cit.*, IV, 260, [in cleared hill areas]; de Sá, Bol. Geral. Med. Farm., V, (8) 290-4[13]; Swell, B. Ent., R. XI, 87, 89, 91; Balf., *loc. cit.*, XII, 30, [in saline water]; Brun., R. I. M., XVII, 98; Hacker, F. M. S. Mal. Bur. Rpt., I, 75[14], [index to references]; Darling, Jo. Exp. Med., XXXII, 313-329, [abst. R. A. E. (B) IX, 130], [cattle feeder, not infected in nature].

Myzomyia rossii, Theo., Monog. Culic., III, 45, fig. 23, figs. p. 46-47, Pl. III, VI; Blanch., Moustiques, 178, ff. 162, 163; Leic. Culic., Malaya, 25; Brun., R. I. M., I, 309[2]; *id., loc. cit.,* IV, 421, (*rosisi, laps.*); Theo., R. I. M., II, 287.

Nyssomyzomyia rossii, James & List., Monog. Anoph. Mosq. Ind., (2nd. edn.), 98-101, Pl. XI; Cruick, & Wright I. J. Med. R., I, 773, 784[6]; Gunasekera, Ceylon S. P. 37 of 1913, 7; Strick, B. Ent. R., V, 321, Pl. XXXII-XXXIII, 2, t. figs., [morphology].

Myzomyia parangensis, Ludl., Psyche, XXII, 137-140[15], (1915).

? *Aldrichia, error,* Theo., Monog., Culic., III, 353[16], (1903); Brun., R. I. M., I, 322.

? *Aldrichinella error,* Brun., R. I. M., IV, 433; Theo., Monog. Culic, V, 77.

? *Anopheles error,* Brun., R. I. M., XVII, 112[17], [*A. rossii* with abdomen of some other sp.].

? *Anopheles subpictus,* Grassi, Atti, R. Acad. Lincei Rend., VIII, 1, (Feby. 1899); Brun., R. I. M., I, 322; *id., loc. cit.,* IV, 434; Kert. Cat. Dipt., I, 254[18]; Edw., B. Ent. R., X, 129, [makes *rossii* a synonym]; White, I. J. Med. R., VIII, 315.

Anopheles subpictus, Lamb. [nec. Grassi], B. Ent. R., XII, 96, fig., 3*a*, [pupa].

---

[15]Philippines.
[14]F. M. S.

rossi var. vagus, Dönitz.

Anopheles vagus, Dön., Beit, z. Kennt. Anoph., 80, (1902); *id.,* Zeit., f. Hyg., XLI, 80; Hacker, E.M.S. Mal. Bur. Rpt., II, 11, 14, [faunal associations].

[1]Burma., [2]Assam.
[2]Bengal., [3]Ceylon.

Anopheles rossii var. *vagus,* Chris., I. J. Med. R. III, 477, 479[2]; White, *loc. cit.,* VIII, 315[3].

Anopheles subpictus var. *vagus,* Lamb., B. Ent. R., XII, 96, fig. 3*b,* [pupa].

[6]Sumatra.
[5,7]Java.
[8, 9]F. M. S.
[4]Philippines.

Myzomyia rossii var. *indefinitá,* Ludl., Can. Ent., XXXVI, 299, (1904)[4]; *id.,* B. Ent. R., VI, 155, [synonomy]; Brun., R. I. M., I, 310.

Myzomyia indefinita; Theo., Monog. Culic., IV, 47, (1907); Brun., R. I. M., IV, 407; Theo., Monog. Culic, V, 24.

Anopheles indefinitus, Kenk., Gen. Tijd. Ned. Ind., LXI, 202[7], [abst. R. A. E. (B) IX, 135], [infected experimentally]; Dyar, Ins. Ins. Mens., VIII, 184; Darling, Jo. Exp. Med., XXXII, 313, 329, [abst. R. A. E. (B) (IX), 131], [cattle feeder][5].

Anopheles rossii var. *indefinita,* Stanton, I. J., Med. R. III, 254[6].

Anopheles rossii var. *indefinitus,* Stanton, B. Ent., R., VI, 159-172, figs. 13, 14, [larva]; Strick., *loc. cit.,* VI, 157, (note); Ludl., *loc. cit.,* VI, 156; Stanton, *loc. cit.,* X, 333, [distribution in Far East. ports]; Hacker F. M. S. Mal. Bur. Rpt., I, 75[8], [index to references].

Anopheles rossii, Strickland, (*nec.* Giles), B. Ent. R., V, 321[9], [comparison with *ludlowi*].

senensis, Wiedemann.

Anopheles sinensis, Wied., Auss. Zweifl. Ins. I, 547[1], (1828); Frnfld., Verh., Z.-B. Ges. Wien., XVII, 449; Theo., Monog. Culic, I, 137, fig. 30, Pl. XXXVII, 146, Pl. A; Giles, Handbook, 160; *id., loc. cit.,* (2nd. end.), 305, Pl. VIII, 9; Stanton, I. J. Med. R., III, 255[2]; Chris., *loc. cit.,* III, 480[3]; Waterst., B. Ent., R., IX, 5, Fig. 8; Brun., R. I. M., XVII, 95, 96; Hodgson, I. J. Med. R., II, 411[5]; James, *loc. cit.,* II, 263[6]; Perry, *loc. cit.,* II, 469[7]; Chris., *loc. cit.,* VII, 711[8]; Stanton, B. Ent. R., VI, 170, fig. 2, 3*c,* [larva]; *id., loc. cit.,* X, 333, [distribution in Far East, ports]; Patton & Cragg, T. B. Med. Ent., 236; Ross, Prevention of Malaria, 632, [carrier]; Iyengar, I. J. Med. R., 1920 Sci. Cong. No., 9; Strick., *loc. cit.,* IV, 260, [in cleared hill areas]; Swell., B. Ent. R., XI, 86, 91, fig., [larva] notes and distribution in D. E. Ind.]; de Sá, Bol. Geral. Med. Farm., V (8) 290-4[12]; Leger, C. R. Soc. Biol., LXXXIII,

[3]S. India.
[3]Burma.
[4, 9]Bengal.
[12]Goa.
[3, 7]Jeypore Hills.
[6, 14]Ceylon.
[17]Yunnan.
[4, 5]Punjab.

[2, 16, 19]Sumatra.
[11, 15, 20, 21]F. M. S.

1609[13]; White, I. J. Med. R., VIII, 315[14]; Hacker, F. M. S. Mal. Bur. Rpt., I, 75[15], [index to references]; Kenk., Gen. Tijd. Ned., Ind. LXI, 202[16], [abst. R. A. E., (B) IX, 135], [notes].

Myzorhynchus sinesis, Theo., Monog. Culic., III, 80, f. 53; Giles, Jo. Trop. Med., VII, 365; Blanch., Moustiques, 190, fig. 170; Leic., Culic, Malaya, 30, [*Myzorhyncus*]; Brun., R. I. M., I, 315[4]; *id., loc. cit.,* IV, 426; James, *loc. cit.,* VI, 31[17]; Gunasekera, Ceylon, S. P. 37 of 1913, 7; James & List., Monog. Anoph. Mosq. Ind. (2nd. edn.), 120. Pl. XV, ; Strick., B. Ent. R., IV, 139.

[1, 4, 23]China.
[7]Mesopotamia.
[10]Palestine., [8]Macedonia.
[13]Rhone Delta.
[18]Borneo.
[21]Caspian Sea.
[22, 24]Italy.
[24]Hungary.

Anopheles sinensis, var. *indiensis,* Theo., Monog. Culic., I, 145, (1901).

Anopheles argyropus, Swell., Gen. Tijd. Ned. Ind., LIV, 334, (1914)[19].

Anopheles jesoensis, Tsuzuki, Cent. f. Bakt., XXXI, 763, (1902); Dön., Zeit. f. Hyg., XLIII, 231.

Anopheles plumiger, Dön., Insektenborse, XVIII, 37, (1901)[23].

Anopheles pseudopictus, Grassi, Atti, R. Acad. Lincei, Rend., VIII, 109[22], (1899).

Myzorhynchus pseudopictus, Theo., Monog. Culic., V, 52[24].

Myzorhynchus separatus, Leic., Culic. Malaya, 36, (1908); Brun., R. I. M., IV, 426[20].

Anopheles separatus, Roper, B. Ent. R., V, 145[18]; Strick., *loc. cit.,* IV, 136; Brun., R. I. M., XVII, 113[11].

? *Cylex. hyrcanus,* Pallas, Reise Russ. Richs., (1771).*

? *Anopheles hyrcanus,* Edw., B. Ent. R., X, 129[21]; White, I. J. Med. R., VIII, 316.

Anopheles hyrcanus, Barraud (*nec.* Pallas)[10], B. Ent. R., XI, 391; Lamb., [*nec.* Pallas], *loc. cit.,* XII, 96, fig. 3*d*; Hacker, [*nec.* Pallas], F. M. S. Mal. Bur. Rpt., II, 10, 17[21], [faunal associations].

sinensis var. vanus, Walker.

Anopheles vanus, Wlk., P. Linn. S. Lon., IV, 91, (1860)[1]; Giles, Handbook, 158; Brun., R. I. M., XVII, 96[2].

Myzorhynchus vanus, Blanch, Moustiques, 196[4]; Brun., R. I. M., I, 316[7]; *id., loc. cit.,* IV, 427[8]; Theo., Monog. Culic, V, 52[9].

[3, 4, 7, 10]Indo-Gangetic Plain.
[4]Naini Tal.
[3, 4, 5, 9]Bengal.
[3, 4]Assam.

---

* For another possible identification of this description see *Culex mimeticus, Noë.*

Anopheles sinensis var. *vanus,* Chris., I. J. Med. R., III, 480[3]; Chris., & Short, *loc. cit.,* VIII, 516.

Anopheles sinensis, var. *annularis,* Theo., (*nec.* Wulp.), Monog. Culic., I, 142, fig. 32, Pl. V, 18, Pl. A.

Myzorhynchus sinensis annularis., Theo., Monog. Culic., III, 90.

Myzorhynchus minutus, Theo., Monog., Culic, III, 91, (1903); Brun., R. I. M., I, 314; *id., loc. cit.,* IV, 425; Theo., Monog. Culic., V. 52[10].

Anopheles minutus, Strick., B. Ent. R., IV, 136; Patton & Cragg, T. B. Med. Ent., 237.

Anopheles nigerrimus, Giles, Handbook, 161, (1900); Theo., Monog. Culic., I, 145; Chatterjee, R. I. M., I. 82[5]; Hodgson, I. J. Med. R., I, 702[6].

Myzorhynchus nigerrimus, James & List., Monog. Anoph. Mosq. Ind., 79, III, (1903); Blanch. Moustiques, 197; Brun., R. I. M., I, 314; *id., loc. cit.,* IV, 125; James & List., Monog. Anoph. Mosq. Ind. (2nd. edn.), 122, 123, [notes].

Myzorhynchus peditaeniatus, Leic., Culic, Malaya, 31, (1908)[11] (*Myzorhyncus*); Brun., R. I. M., IV, 425; Strick., B. Ent. R., IV, 135.

[3]Burma.
[3, 7]Central India.
[2, 3, 6, 7]Madras.
[2, 4, 7]Malabar Coast.
[2,9]Ceylon.
[2, 7, 9, 10, 11]F. M. S.
[2]Java., [1, 2]Celebes.
[2, 7, 9]Philippines.
[2, 9]China.
[8]Mesopotamia.
[9]Formosa.

stephensi, Liston.

Anopheles stephensi, List., Ind. Med. Gaz., XXXVI, 441, (1901); James, Sci. Mem. Med. Off. Ind., II, 45, fig. 23, 24; Giles, Handbook, (2nd. edn.), 331, (footnote); James & List., Monog. Anoph. Mosq. Ind., 113, Pl. VI, 2, Pl. X, 6; Pl. XII; Chris., I. J. Med. R., III, 481[1]; Wilson, *loc. cit.,* I, 691[2], [breeding places]; Hodgson, *loc. cit.,* II, 411[3]; Perry, *loc. cit.,* II, 469[4]; Gill, *loc. cit.,* IV, 207[5], Sinton *loc. cit.,* V, 192, [parasite], 197, 200, 207[6], [varieties]; Iyengar, *loc. cit.,* VII, Sci. Cong. No., 26[7]; Gill, *loc. cit.,* VII, 611[8]; Chris., *loc. cit.,* VII, 711[9]; Brun., R. I. M., XVII, 106[12]; Patton & Cragg, T.B. Med. Ent. 234, Pl. XXXIV, 2, [egg.], Pl. XXXVIII, 1, [imago]; Iyengar, I. J. Med. R., 1920. Sci. Cong. No. 11 *et. seq.;* de Sá, Bol. Geral. Med. Farm., V, (8), 290-4[16]; Balf., B. Ent. R., XII, 30[17] [in saline water]; Chris. & Short, I. J. Med. R., VIII, 516[18].

[1, 6]N.-W. F. P.
[11]Sind., [3, 11]Punjab.
[8]Kashmir.
[1, 11]Indo-Gangetic Plan., [1, 11, 15]Bengal.
[1, 7, 11, 13, 14]Central India., [10, 17]Bombay.
[4]Jeypore Hills.
[1, 2, 11]Madras.
[1, 16]Malabar Coast.
[11]Lushai Hills.
[11]Burma.,

Neocellia stephensi, James & List., Monog. Anoph. Mosq. Ind., (2nd. edn.) , 113, Pl. XII; Bentley, Malaria in Bombay, 24, 29, 63[10].

Nysso hynchus stephensi, Theo., Monog. Culic, III, 93, fig. 54, 55, t. figs. pp. 40 and 47; Blanch. Moustiques, 210; Brun., R. I. M., I, 319; *id., loc. cit.,* IV, 430; Ross, Prevention of Malaria, 633, [carrier].

Neocellia intermedia, Rothwell, Entom, XL, 34, (1907)[13]; Theo., Monog. Culic., IV, 115; *id., loc. cit.,* V, 73, fig. 29[14]; Brun., R. I. M. IV, 432.

Anopheles metaboles, Theo., Proc. R. S., LXIX, 374, Pl. V, 1, (1902); Giles, Handbook, (2nd. edn.), 512.

[12]Philippines.
[9, 18]Mesopotamia.
[5]Muscat.

superpictus, Grassi.

Anopheles superpictus, Grassi, Real. Acad. Lincei, 78, (1900); Chris., I. J. Med. R., VII, 712[6]; Chris. & Short, *loc. cit.,* VIII, 516[8]; Barraud, B. Ent. R., XI, 392[7].

Pyretophorus superpictus, Ross, Prevention of Malaria, 635, [doubtful carrier]; Theo., Monog. Culic., V, 39[1].

Anopheles superpictus var. *macedoniensis,* Cot. & Hor. Bull. S. Path, Ex., X, 890, (1917)[2].

Pyretophorus cardamitis, Newst. & Cart., Ann. Trop. Med. Paras., IV, 379, (1910).

Pyretophorus nursei, Theo., Monog. Culic, IV, 66, (1907)[3]; Brun., R. I. M., IV, 42[4]; James & List., Monog. Anoph. Mosq. Ind., (2nd. end.), 84.

Anopheles nursei, Chris., I. J. Med. R., III, 475[4]; Sinton, *loc. cit.,* VII, 712[5]; Patton & Cragg, T. B. Med. Ent., 232.

Pyretophorus palestinensis, Theo., Monog. Culic., III, 71, figs. 46, 17, Pl. VI, (1903)[9].

Anopheles palesinensis, Edw., Jo. R. A. S. Beng., IX, 48; Waterst., B. Ent. R., IX, 4; Brun.; R. I. M., XVII, 111[10].

[5]N.-W. F. P.
[3, 4, 6, 10]Baluchistan.
———, [6, 10]Persia.
[8]Mesopotamia.
[9]Palestine., [6]Turkestan.
[7, 10]Asia Minor.
[6]N. Africa.
[1, 2, 6, 10]S. Europe.

theobaldi, Giles.

Anopheles theobaldi, Giles, E. M. M., XXXVII, 198, (1901); *id.,* handbook, (2nd. edn.), 299; Theo., Monog. Culie., II, 311, (app.); James & List., Monog. Anoph. Mosq. Ind., 97, Pl. VIII; Perry, I. J. Med. R., II, 469[1]; Chris., *loc, cit.,* III, 482[2]; Patton & Cragg, T. B. Med. Ent., 233; Brun., R. I. M., XVII, 101[4].

[3, 5]Bombay., [3, 4, 5]Punjab.
[3, 4, 5]United, Provinces.
[2, 4, 5]Central India.
[3, 4, 5]Orissa.
[3, 4, 5]E. Bengal.
[1, 3]Jeypore Hills.
[2]Malabar Coast.

Nyssorhynchus theobaldi, Theo., Monog. Culic., III, 95; Blanch., Moustiques, 210; Brun., R. I. M., I, 320[3]; *id., loc. cit.,* IV, 430; James & List., Monog. Anoph. Mosq. Ind., (2nd. edn.), 95, Pl. X; Ross, Prevention of Malaria, 634, [carrier]; Theo., Monog Culic., V, 58[5], [carrier].

[4, 5]Aden Hinterland.
[4, 5]Philippines.

Anopheles maculatus var. *theobaldi,* Alcock, Jo. Lon. Sch. Trop. Med., II, (3), 164, (1913).

turkhudi, Liston.

Anopheles turkhudi, List., Ind. Med. Gaz., XXXVI, 441, (1901); Giles, Handbook, (2nd. edn.), 320, (footnote); James, Sci. Mem. Med. Off. Ind., II, 49, figs. 27, 28; James & List., Monog. Anoph. Mosq. Ind., 115, Pl. VIII, 2, Pl. XIV; *id., loc. cit.,* (2nd edn.), 78[1], Pl. V; Gough, B. Ent. R., V, 133, [synonomy]; Chris. & Chand, I. J. Med. R., III, 190; Chris., *loc. cit.,* III, 483[2]; Edw., B. Ent. R., III, 249; Hodgson, I. J. Med. R., II, 411[3]; Sinton, *loc. cit.,* V, 197, 202, 207[4], [varieties]; Patton & Cragg, T. B. Med. Ent., 231; Ross, Prevention of Malaria, 632, [carrier]; Barraud, B. Ent. R., XI, 392, [distinct from *A. multicolor,* Cam.]; Edw., *loc., cit.,* XII, 69; Acton, etc., I. J. Med. R., VIII, 751[5]; Brun., R. I. M., XVII, 103[6].

[1, 6, 7, 8]Kashmir.
[2, 4, 11, 12]N.-W. F. P.
[5]Dagshai.
[1, 2, 6, 7, 8]Punjab.
[1, 2, 3, 6, 7, 8]Central India.
[6, 7]Andamans.

[9, 10]Aden Hinterland.
[11, 12]Cyprus.
[13]Teneriffe.

Myzomyia turkhudi, Theo., Monog. Culic. III, 48, Pl. III; Blanch. Moustiques, 183; Brun., R. I. M., I, 311[7]; *id., loc. cit.,* IV, 422; Theo.; Monog. Culic., V, 26[8].

Myzomyia azriki, Patton, Bom. Jo., XVI, 632, (1905)[2]; Theo., Monog. Culic., V, 27, figs. 8, 9[10] [carrier].

Anopheles culicifacies, Giles, E. M. M., XXXVII, 197, (1910), [male only]; Theo., Monog. Culic., II, 309, [male only]; Giles Handbook, (2nd. edn.), 317, Pl. IX, 12, [male only]; Theo., Proc. R. S. LXIX, 379, fig. 2, [male] only; genitalia].

Pyretophorus nigrifasciatus. Theo., Monog. Culic., IV, 65, (1907); *id., loc. cit.,* V, 37[11]; Brun., R. I. M., IV, 424; Patton & Cragg, T. B. Med. Ent., 231; Brun., R. I. M., XVII, 113[12].

? *Myzomyia hispaniola,* Theo., Monog. Culic., III, 49, (1903); *id., loc. cit.,* V, 27[13].

willmori, James.

Nyssorhynchus willmori, James in Theo., Monog. Culic., III, 160, figs. 59, 60, (1903); Blanch., Moustiques, 624; Brun., R. I. M., I, 321[1]; *id., loc. cit.,* IV, 430[2]; Theo., Monog., Culic., V, 57[3].

Neocellia willmori, James & List., Monog. Anoph. Mosq. Ind., (2nd. edn.), 109[4], t. fig.

Anopheles willmori, Green, Trop. Agric., XXVI, 68[5]; Chris., I. J. Med. R., III, 484[6]; Wilson, *loc. cit.,* I, 691[7]; Prashad, *loc. cit.,* III, 498, Pl. XXIX, 1, [male genitalia]; Sinton, *loc. cit.,* V, 197, 202, 207[8], [varieties]; Prashad, *loc. cit.,* V, 610, Pl. LVII-LXI, [thorax and wing]; *id., loc. cit.,* V, 611, Pl. LXII-LXVI, [thoracic imaginal buds]; Gill, *loc. cit.,* VII, 611[9]; Patton & Cragg, T. B. Med. Ent., 234; Ross, Prevention of Malaria 631, [carrier]; Brun., R. I. M., XVII, 111-112[10].

Neocellia willmori var. *maculosa,* James & List., Monog., Anoch. Mosq. Ind., 2nd. edn.), 112, (1911)[11].

Anopheles maculatus var. *willmori,* Alcock, Jo. Lon. Sch. Trop. Med., III, (3) (1913).

Neocellia dudgeoni, Theo., Monog. Culic., IV, 112, (1907); Brun., R. I. M., IV, 432[12]; James & List., Monog. Anoph. Mosq. Ind., (2nd. edn.), 112.

Neocellia indica, Theo., Monog., Culic., IV, 97, (1907); Brun., R. I. M., IV, 432[12]; Patton & Cragg, T. B. Med., Ent., 234.

[4, 6, 11]N.-W. F. P. to Assam (foot hills).
[1, 3, 4, 9]Kashmir.
[1]Lahore.
[4, 10, 12]Kangra Valley.
[10, 13]Dehra Dun.
[3]Duars., [8]Kohat.
[7]Madras.
[2, 3, 5]Ceylon.

MEGARHINI.

Megarhininœ, Theo., Monog. Culic., V, 88.

Megalorhinina, Patton & Cragg, T.B. Med. Ent., 207.

Megarhini, Brun., R. I. M., X, 34.

4. **Toxorhynchites,** Theobald.

Toxorhynchites, Theo., Monog. Culic, I, 244, (1901); [Genotype, T. *brevipalpis,* Theo]; *id., loc. cit.,* III, 119, [notes on life history]; *id., loc. cit.,* V. 95, [notes and table of spp.]; *id.,* Gen. Ins., Fasc., XXVI, 13; Blanch., Moustiques, 229; Leic., Culic. Malaya, 59; Brun., R. I. M., I, 324; *id., loc. cit.,* IV, 435; *id., loc. cit.,* X, 72; *id., loc. cit.*, XVII, 16; Giles, Handbook, (2nd. edn.), 278.

Teromyia, Leic., Culic., Malaya, 49, (1908); [Genotype, *T. acaudata,* Le.]; Brun., R. I. M., IV, 438.

Worcesteria, Banks, Phil. Jo. Sci., I, 779, (1906); [Genotype, *W. grata,* Banks]; Theo., Monog. Culic., V, 110; Brun., R. I. M., I, 326; *id., loc. cit.*, IV, 439; *id., loc. cit.*, X, 73.

gravelyi, Edwards.

Megarhinus (Toxorhynchites) gravelyi, Edw., B. Ent. R., XII, 73, (1921)[1]. — [3, 10]F. M. S.

kempi, Edwards.

Megarhinus (Toxorhynchites) Kempi, Edw., B. Ent. R., XII, 72,[1] (1921). — [1]Darjiling. [1]Castle Rock

***leicesteri,** Theobald.

Toxorhynchites leicesteri, Theo., Entom., XXXVII, 36, (1901); *id.*, Monog. Culic,., IV, 142[2]; *id., loc. cit.*, V, 99[3]; Leic., Culic, Malaya, 59; Brun., R. I. M., I, 325; *id., loc. cit.*, IV, 437; *id., loc. cit.*, XVII, 118[4]. — (Kanara). Dungagalli. —————— [2, 3, 4]Kuala Lumpur. [4]Papua.

minimus, Theobald.

Megarhinus minimus, Theo., Bom. Jo., XVI, 237[1]; Pl. A, fig. 1, (1905); *id.*, Monog. Culic., IV, 138[2]; Brun., R. I. M., I, 324[3]. — [1, 2, 3, 4, 5]Yatiyantota. [6]Colombo. [7, 8]Matale (Ceylon).

Toxorhynchites, minimus, Theo., Monog, Culic., V, 109[4]; Brun., R. I. M., IV, 437; James, I. J. Med. R., II, 263[6]; White; Spol. Zeyl, XI, 189[7], [female described]; Brun., R. I. M., XVII, 274; White, I. J. Med. R., VIII, 319[8].

splendens, Weidemann.

Culex splendens, Wd., Zool. Mag. (2) 1, (1819); *id.*, Dipt. Exot. 7; Theo., Monog. Culic. I, 235.

Megarhinus splendens, Theo., Monog. Culic., V. 109[14]; Edw., A. M. N. H., VIII, 629.

---

* This species was included here on the strength of a specimen from Dungagalli in the Pusa collection identified a *leicesteri.* Since this Catalogue was sent to press, however, Mr. Edward's description of *albipes,* (*Ind. Journ. Med. Res.* X, 287-288; July 1922) has appeared and, on comparison with this description, the specimen in question is found to be *albipes,* Edw. The species *leicesteri* should therefore be omitted from the Indian list.—T. B. F.

Toxorhynchites argonteotarsis, Ludl., Can. Ent., XXXVIII, 367, (1908); Theo., Monog. Culic.; V, 100; Brun., R. I. M., IV, 435[7].

Megarhinus gilesii, Theo., Monog., Culic., I, 227, Pl. IX, 33, (1901).

Toxorhynchites gilesii, Theo., Monog. Culic., V, 99[8]; Brun., R. I. M., IV, 435. *Worcesteria grata,* Banks, Phil. Jo. Sci., I, 780, Pl. (1906); Brun., R. I. M., I, 326[9]; Theo., Monog. Culic., V, 111, figs. 42, 43.

Megarhinus immisericors. Wlk., P. Linn. S. Lon., IV, 90[1], (1860); Theo., Monog. Culic., I, 225, Pl. VII, 28; O-S., Berl., Ent., Zeit., XXVI, 96, [synonomy with *amboinensis,* Dol.].

Megarhina immisericors, Giles, Handbook, 119; *id., loc. cit.,* (2nd. edn.), 273.

Toxorhynchites immisericors, Theo., Monog. Culic., III, 123, fig. 67; *id., loc. cit.,* IV, 141; *id., loc. cit.,* V, 97, figs. 2; *id.,* R. I. M., IV, 4, [female redescribed]; Blanch., Moustiques, 230; Brun., R. I. M., I, 324; *id., loc. cit.,* IV, 436; *id., loc. cit.,* XVII, 117[13]; Green, Spol. Zeyl., II, 159, Pl. 3 [life history]; Paiva, R. I. M., V, 187[4], [larval habits]; Edw., B. Ent. R., IV, 422; James, I. J. Med. R., II, 263[5]; Stanton, B. Ent. R., X, 333, [distribution in Far East. ports]; White, I. J. Med. R., VIII, 319[6].

[2, 13]Bhim Tal.
[2, 13]Sikkim.
[2, 8, 13]Sylhet.
[13]Lushai Hills.
[2, 13]Burma.
[2, 4, 8, 13]Calcutta.
[2]Chittagong.
[2, 13]Mysore.
[2, 13]Travancore.
[2, 3, 5, 6, 11, 13]Ceylon.
[2, 13]Andamans.

Megarhinus lewaldii, Ludl., Can. Ent., XXXVI, 233, (1904)[20]; Brun., R. I. M., I, 324.

Culex regius, Tennent, "Ceylon", Vol. I, (2) 268, (footnote) [*nom. nud.*]; Thwaites, P. Linn., VIII, 102, (1864)[11], [description].

Toxorhynchites regius, Edw. B. Ent. R., VII, 202; Brun., R. I. M., XVII, 118, (footnote); Dyar., Ins. Ins. Mens., VIII, 183[14]; Edw., Ann. Trop. Med. Paras, XIV, 23, 40, [male genitalia].

? *Megarhinus subulifer,* Dol., Nat. Tijd. Ned. Ind.[12], XIV, 382, (1957); Edw. B. Ent. R., VII, 202; Brun., R. I. M., XVII, 118, (footnote).

[2, 8]F. M. S.
[2, 7, 9, 10, 13, 14]Philippines.
[1, 2, 13]Celebes.
[2, 13]Ceram.
[2, 13]Papua.
[12]Amboina.
[14]Java.
[14]Sumatra.
[14]Singapore.

CULICINI.

Christophers, Paludism, III, 40-54, (1911), [Aids to identification of Culicidæ other than *Anopheles,* with special reference to Indian spp.]; Edwards, B. Ent. R.,

III, 1-53, [Key to W. African genera other than *Anopheles* : synonomy]; Patton & Cragg, T. B. Med. Ent., 208, [Key to genera of *Culicina*]; R. I. M., X, 35, [Genera in *Culicini,* with key.]

5. **Mucidus,** Theobald.

Mucidus, Theo., Monog. Culic., I, 268, (1901) [Genotype, *Culex alternans,* Wstwd.]; *id.,* Gen. Ins. Fasc., XXVI, 17; *id.,* Monog. Culic., III, 132, [larva]; *id., loc. cit.,* V, 125, [table of spp.]; Giles, Handbook, (2nd. edn.) 346; Blanch., Moustiques, 243; Leic., Culic. Malaya 69; Brun., R. I. M., IV, 440; *id., loc. cit.,* X, 39, 66; *id., loc. cit.,* XVII, 119-120; Alcock, Ent. for Med. Off., 104.

Laniger, Wiedemann.

Culex laniger, Wd., Dipt. Exot., 9, (1821); *id.,* Auss. Zweifl. Ins., I, 5[1]; Mq., Dipt. Exot. I, (2) 176; Giles, Handbook, 211.

Mucidus laniger, Theo., Monog. Culic., I, 279; Blanch., Moustiques, 244; Edw., B. Ent. R., IV, 224; Theo., Monog. Culic., V, 132[2]; Brun., R. I. M., XVII, 120[3]; Stanton, I. J. Med. R., III, 256[4].

Mucidus mucidus, Leic., (nec. Karsch.), Culic. Malaya, 69[5], (1908).

[1, 2, 3]Coromandel.

[3, 4]Sumatra.
[1, 2, 3]Java., [5]F. M. S.

[Not recognized as Indian since its description.]

scatophagoides, Theobald.

Mucidus scatophagoides, Theo., Monog. Culic., I, 277, Pl. E, fig. 81, (1901); Giles, Handbook, (2nd. edn.), 348, Pl. XII, 1*a* 2*a*; Blanch., Moustiques, 245; Brun., R. I. M., I, 327[1]; *id., loc. cit.,* XVII, 120[2]; Edw., B. Ent. R., II, 246[3]; Hodgson, I. J. Med. R., II, 414[4]; Edw., B. Ent. R., III, 6[5]; Green, Spol. Zeyl., IV, 183[6].

Mucidus sudanensis, Theo., III, Rpt., Wellc. Lab., 252[7], (1908); *id.,* Monog. Culic., V, 129-131, figs. 44, 45[8].

[4]Delhi.
[1, 2]Moradabad.
[2]Bengal., [3]Trincomali.
[1]Burma.,

[3, 7, 8]Sudan.
[3]W. Africa., [5]Uganda.

6. **Uranotænia,** Arribalzaga.

Uranotænia, Arrib., Rev. Mus. la Plata, I, 405, (1891); [Genotype, *U. pulcherrima,* Arrib.]; Theo., Monog. Culic., II, 241, fig. 242, [map of distribution], 243, [table of spp.], Pl. D; *id., loc. cit.,* V, 498, [table of spp.]; *id.,* Gens. Ins. Fasc., XXVI, 36; Blanch. Moustiques, 406; Giles, Handbook, (2nd edn.), 485, 487, [table of spp.]; Brun., R. I. M., I, 363; *id., loc. cit.,* IV, 495; *id., loc. cit.,* X. 73; *id., loc. cit.,* XVII,

173; Alcock, Ent. Med. Off., 104; How. Dyar, Knab, Mosq. N. Amer., IV, 898.

Anisocheleomyia, Theo., Entom. XXXVIII, 52, (1905), [Genotype, *A. nivipes* Theo.]; *id.,* Monog, Culic, IV, 570; Brun., R. I. M., I. 363; *id., loc. cit.,* IV, 463; *id., loc. cit.,* IX, 55.

Pseudouranotania, Theo., Jo. Econ. Biol., I, 33, (1905), [Genotype, *P. rowlandii,* Theo.]; *id.,* Monog. Culic., IV, 566, fig. 262; Brun., R. I. M., IV, 494; *id., loc. cit.,* X, 70.

alboannulata, Theobald.

Anisocheleomyia alboannulata, Theo., Entom., XXXVIII, 54, (1905); *id.,* Monog. Culic., IV, 573, f. 267, Pl. VII[1]; Brun., R. I. M., IV, 493.

Uranotœnia alboannulata, Edw., B. Ent. R., IV, 238; Brun., R. I. M., XVII, 174[2].

Uranotania trilineata, Leic. Culic. Malaya, 204, (1908); Brun., R. I. M., IV, 498[3].

[1, 2]India,

[2, 3]Malaya.

atra, Theobald.

Uranotænia atra, Theo. Ann. Mus. Hung., III, 114 (1905)[1]; *id.,* Monog. Culic., IV, 563; Edw., B. Ent, R., IV, 238; Brun., R. I. M., XVII, 174[3].

Uranotania ceylonica, Theo., Monog. Culic., V, 503, f. 213[2] (1910).

[2, 3]Galle.

[1, 3]Papua.

campestris, Leicester.

Uranotœnia campestris, Leic. Culic. Malaya, 213, (1908), (Brun., R. I. M., IV, 495[1]; Stanton, B. Ent. R., X, 333[2] [distribution in far East. ports]; Brun., R. I. M., XVII, 175[3]; White, I. J. Med. R., VIII, 319[4].

[4]Matale.

[1, 2, 3]Malaysia.

lateralis, Ludlow.

Uranotœnia cœuleocephala var. *lateralis,* Ludl., Can Ent., XXXVII, 385, (1905); *id.,* Mosq. Phil. Ids., 10.

Uranotœnia lateralis, Banks, Phil. Jo. Sci., I. (9), 990, (1906); Theol.; Monog. Culic. V, 506-508[2] ; James, I. J. Med. R., II, 263[4]; Brun., R. I. M., XVII, 174[3].

[4]Colombo.

[1, 2, 3]Philippines.

testacea, Theobold.

Uranotœnia testacea, Theo., Ann. Mus., Hung., III, 113, f. 14. Pl. II, Pl. III, (1905) ; *id.,* Monog. Culic IV, 560, fig. 258[1]; Edw., B. Ent. R., IV, 239; Brun., R. I. M., XVII, 174[2].

[2]Dawnat Range.
Gauhati (White).

[2, 3]Philippines.
[2, 4]Malaysia.

Uranotœnia falcipes, Banks, Phil., Jo. Sci., I, 1004, (1906).[3] — [1, 2]Singapore.

Uranotœnia unilineata, Leic., Culic. Malaya, 220, (1908); Brun., R. I. M. IV, 498[4].

7. **Aedeomyia,** Theobald.

Aedeomyia, Theo., Monog. Culic., II, 218, f. 259, 269, [map of distribution], (1901); [Genotype, *A. squamipenna*, Arrib.]; *id.*, Jo. Trop. Med., IV, 235, [*nom. nud.*]; *id.*, Monog. Culic., V, 486; *id.*, Gen. Ins. Fasc. XXVI, 35; Blanch., Moustiques, 403, fig. 255; Leic., Culic. Malaya, 181; Brun., R. I. M., I, 366; *id.*, *loc. cit.*, IV, 488; *id.*, *loc. cit.*, X, 54; *id.*, *loc. cit.*, XVII, 177; How., Dyar & Knab, Mosq. N. Amer., IV, 893.

Aedomyia, Gile, Handbook, (2nd. edn.), 478; Edw., B. Ent. R., III. 24.

catasticta, Knab.

Aedeomyia catasticta, Knab, Ent. News, XX, 387, (1909); Brun., R. I. M., IV, 488; Edw., B. Ent. R., III, 25, [differences from *squamipenna*]; Brun., R. I. M., XVII, 177-8[1]; Edw., B. Ent. R., III, 379, figs., 4, 5, [larva]; *id.*, *loc. cit.*, VII, fig. 10*b*, [male genitalia]; Stanton, *loc.*, *cit.*, X, 333, [distribution in far East. ports]; James, I. J. Med. R., II, 263[2], [*catastricta*].

[2]Colombo.
———
[1]Philippines.
[1]Africa–general.

squamipenna, Arribalzaga.

Aedes squamipennis, Arrib., El. Nat. Argent., I, 151[1], (1878); Giles, Handbook, 347.

Aedes squamipenna, Arrib., Dept. Argent. 62.

Aedeomyia squamipenna, Theo., Monog. Culic., II, 219, fig. 261, Pl. XXXI, 124, Pl. E; *id.*, *loc. cit.*, III, 307; Leic., Culic. Malaya, 182; Paiva, R. I. M., V, 202; Green, Spol. Zeyl., IV, 483; Edw., B. Ent. R., VII, fig. 10*a*, [male genitalia]; Iyengar, I. J. Med. R., 1920 Sci., Cong., No., 9[5]; Brun., R. I. M., I, 366[2]; *id.*, *loc. cit.*, IV, 488[3]; *id.*, *loc. cit.*, XVII, 177[4].

[3, 4, 5]Calcutta.
[2, 4]Madras.
[3, 4]Dawnat Range.
[3, 4]Tuticorin.
[2, 4]Ceylon.
———, [3, 4]F. M. S.
[2, 4]Philippines.
[2]Papua., [2, 4]Sudan.
[2, 4]W. Indies.
[1, 2, 4]S. America.

Aedeomyia squamipennis, Blanch., Moustiques, 404.

Aedeomyia squammipennis, Theo., Gen. Ins. Fasc., XXVI, Pl. II, 9; Giles, Handbook, (2nd. edn.), 479.

8. **Ficalbia,** Theobald.

Ficalbia, Theo., Monog. Culic., III, 296, (1903), [Genotype

U. minima, Theo.]; *id., loc. cit.,* IV, 577, [table of spp.]; *id., loc. cit.,* V, 541, [table of spp.]; *id.,* Gen. Ins. Fasc., XXVI, 36; Blanch., Moustiques, 418; Leic., Culic. Malava, 228; Edw., B. Ent. R., IV, 237; Brun., R. I. M., I, 362; *id., loc. cit.,* IV, 494; *id., loc. cit.,* X, 60; *id., loc. cit.,* XVII, 170-1.

Dasymyia, Leic., Culic. Malaya, 113, (1908), [Genotype, *D. fusca,* Lc.]; Brun., R. I. M., IV] 453; *id., loc. cit.,* X, 59.

Etorleptiomyia, Theo., I Rpt., Wellc. Lab., 71, (1904), [Genotype, *E. mediolineata,* Theo.]; *id.,* Gen. Ins. Fasc., XXVI, 44; *id.,* Monog. Culic., IV, 505; Leic., Culic, Malaya, 178; Brun., R. I. M., I, 368; *id., loc. cit.,* IV, 484; *id., loc. cit.,* X, 60; Blanch., Moustiques, 632.

Ingramia, Edw., B., Ent. R. III, 43, (1912), (Genotype *Mimomyia malfeyti,* News.].

O'reillia, Ludl., Can. Ent., XXXVII, 101, (1905), [Genotype, O. *luzonensis,* Ludl.].

dofleini, Guenther.

Ficalbia dofleini, Guen., Zeit. f. Wiss. Insekt., IX, 204[1], figs. 1, 2, 259, figs., 3-14, (1914) [described from larva and pupa only]; Brun., R. I. M., XVII, 173.[2] — 1, 2Ceylon.

luzonensis, Ludlow.

O'reillia luzonensis, Ludl., Can. Ent., XXXVII, 101[1] (1905). — [3]Colombo.

Etorleptiomyia luzonensis, Ludl., Can. Ent., XXXVIII, 185; Theo., Monog. Culic., IV, 506. — [2, 4, 5]Singapore. [1, 2]Philippines.

Ficalbia luzonensis, Edw., B. Ent. R., IV, 238; James, I. J. Med. R., II, 263[3]; Brun., R. I. M., XVII, 172[2].

Etorleptiomyia completiva, Leic., Culic, Malaya, 178[4], (1908); Brun., R. I. M., IV, 484[5].

minima, Theobald.

Uranotœnia minima, Theo., Monog. Culic. II, 262, fig. 281[1], (1901); Giles, Handbook, (2nd. edn.), 488. — [2, 3, 4, 6]Sylhet. [2, 4, 5, 6]Calcutta. [1, 2]Quilon.

Ficalbia minima, Theo., Monog. Culic., III, 296; Blanch., Moustiques, 418; Brun., R. I. M., IV, 494; Edw., B. Ent. R. IV, 238; Brun., R. I. M., XVII, 171[2].

Mimomyia minuta, Theo., R. I. M., II, 301, (1908)[3]; *id., loc. cit.,* IV, 30[4], Pl. I, Pl. III; *id.,* Monog. Culic., V. 531[6], f. 226; Iyengar, I. J. Med. R., 1920 Sci. Cong. No., 9[5].

9. **Mimomyia,** Theobald.

Mimomyia, Theo., Monog. Cylic., III, 304, (1903) [Genotype, *M. splendens,* Theo.]; *id.,* Gen. Ins. fasc., XXVI, 36; *id.,* 1st Rpt. Wellc. Lab., 79., *id.,* Monog. Culic., IV, 581; Edw., B. Ent. R., III, 35.

Boycia, News. Ann. Trop. Med. Paras., I, 33, (1907); [Genotype, *B. mimomyiaformis,* News.].

Conopomyia, Leic., Culic., Malaya, 113, (1908) [Genotype, *C. metallica,* Leic.]; Brun., R. I. M., IV, 453; *id., loc. cit.,* X, 58.

Hispidimyia, Theo., Monog. Culic., V, 245, (1910) [Genotype, *H. hispida,* Theo.].

Ludlowia, Theo., Monog. Culic., IV, 193, (1907) [Genotype, *Mimomyia chamberlaini,* Ludl.]; *id., loc. cit.,* V, 190, [table of spp.]; Brun., R. I. M., IV, 493; *id., loc. cit.,* X, 64; Edw., B. Ent. R., II, 244.

Megaculex, Theo., Monog. Culic., IV, 282, (1907) [Genotype, *Culex albitarsis,* Theo.]; *id.,* Nov. Culic., I, 12, [*Radioculex* sunk.].

Radioculex, Theo., R. I. M., II, 295, (1908) [Genotype, *R. clavipalpus,* Theo.]; *id.,* Monog. Culic., I, 192; Brun., R. I. M., IV, 465; *id., loc. cit.,* X, 70.

chamberlaini, Ludlow.

Mimomyia chamberlaini, Ludl., Can. Ent., XXXVI, 297, (1904)[1]; James, I. J. Med. R., II, 263[2].

Ficalbia chamberlaini, Brun., R. I. M., XVII. 172[3].

Ludlowia chamberlaini, Theo., Monog. Culic., IV, 194; Edw., B. Ent. R., II, 244; Brun., R. I. M., IV, 493.

Radioculex clavipalpus, Theo., R. I. M., II, 295[4] (1908); *id., loc. cit.,* IV, 26[5], Pl. I, Pl. II; *id.,* Monog. Culic., V, 193[6], figs. 70, 71; Brun., R. I. M., IV, 465; Theo., Nov. Culic, I, 13; Cruick, Wright, I. J. Med. R., I, 773[7]; Iyengar, *loc. cit.,* 1920 Sci. Cong. No., 9[8].

Ficalbia clavipalpus, Brun., R. I. M., XVII, 172[9].

[4, 5, 6, 8, 9]Bengal. [5, 6, 9]Rangoon. [5, 6, 7, 9]Malabar Coast., [2]Ceylon.

[1, 3]Philippines.

10. **Aedes,** Meigen.

Aedes, Mg., Syst. Besch., I, 13, (1818), [Genotype, *A. cinereus,* Mg.]; R.-D., Mem. Soc. Hist. Nat., Paris, III, 403;

Mq, H.N. Dipt., I, 37; Zett., Dipt. Scand., IX, 3469; Sch., F. Aust., II, 630; Fic. B. S. Ent. Ital., 299, (1896); Theo., Monog. Culic., II, 224; *id., loc. cit.,* III, 285, [restricted]; *id., loc. cit.,* v, 482, [table of spp.]; *id.,* Gen. Ins. Fasc., XXVI, 35; Giles, Handbook, 343; *id., loc. cit.,* (2nd. edn.), 480, 346, [table of spp.]; Blanch., Moustiques, 399; Leic., Culic, Malaya, 183; Brun., R. I. M., IV, 489; *id., loc. cit.,* X, 45; *id., loc. cit.,* XVII, 141-2; Alcock, Ent., Med. Off., 102; Edw., B. Ent. R., IV, 229; *id., loc. cit.,* VII, 203; How. Dyar. & Knab, Mosq. N. Amer., IV, (2), 612; Séguy, B.S. Ent. Fr., 1920, (18) 309-311, [notes on larva of *cantans* group].

Aioretomyia, Leic., Culic. Malaya 185, (1908) [Genotype *A. varietas,* Leic.]; Brun., R. I. M., IV, *490; id., loc. cit.,* X, 55.

Catageiomyia, Theo., Monog. Culic., V, 115, (1910) [*nom. nud.*].

Heteronycha, Arrib., Rev. Mus. la Plata, I, 397, (1891) [Genotype, *Culex aestuans,* Wd. as *dolosa,* Arrib.].

Macleaya, Theo., Entom. XXXVI, 154, (1903) [Genotype. *M. tremula,* Theo.]; *id.,* Monog. Culic., IV, 203.

Neomacleaya, Theo., Monog. Culic., IV, 238, (1907) [Genotype, *N. indica,* Theo.].

Skusea, Theo., Monog. Culic., III 291, (1903) [Genotype, *S. pembaensis,* Theo.]; *id., loc. cit.,* IV, 542; *id., loc. cit.,* V, 488, [table of spp.]; *id.,* Gen. Ins. fasc., XXVI 19; Blanch., Moustiques, 416; Brun., R. I. M., I, 335; *id., loc., cit.,* IV, 450; *id., loc., cit.,* X, 71; Edw., B. Ent. R., VII, 202.

Verrallina, Theo., Monog. Culic., III, 295, (1903) [Genotype, *Aedes butleri,* Theo.]; *id., loc., cit.,* V, 494, [table of spp.]; Blanch. Moustiques, 417; Leic., Culic., Malaya, 196; Edw., B. Ent. R., IV, 229.

butleri, Theobald.

Aedes butleri, Theo., Monog. Culic., II, 230, (1901)[1]; Giles, Handbook, (2nd. edn.), 481; Edw., B. Ent. R., IV, 229; James, I. J. Med. R., II 262[2]; Brun., R. I. M., I, 367[3]; *id., loc. cit.,* IV, 489; *id., loc. cit.,* XVII, 143[4].

Verralina butleri, Theo., Monog. Culic., III, 295' Blanch., Moustiques, 417; Leic., Culic. Malaya, 196; Brun., R. I. M., IV, 491.

Gauhati (White).
[2]Colombo.,
____
[4, 6]Philippines.
[1, 3, 4, 5]Malaya.

Skusea diurna, Theo., Entom., XXXVI, 259, (1903); *id.,* Monog. Culic., IV, 547; Brun., R. I. M., I, 335[5]; *id., loc. cit.* IV, 451.

Stegomyia hatiensis, Carter, Entom., XLIII, 275[6], (1910).

ceylonicus, Edwards.

Aedes ceylonicus, Edw., B. Ent. R., VII, 221, fig. 8*a*, (1917); Brun., R. I. M., XVII, 145[2]. — [1, 2]Colombo.

indicus, Theobald.

Neomacleaya indica, Theo., Monog. Culic., IV, 238, (1907); Brun., R. I. M., IV, 458[1], [*nec.* var. *simplex,* Theo.]. — [1, 2, 3]India., [1, 2]Ceylon.

Aedes indicus, Edw., B. Ent. R., IV, 229; Brun., R. I. M., XVII, 143[2]. — [1, 2, 3, 4, 5]Philippines.

Skusea mediofasciata, Theo., Monog. Culic., IV, 544, figs. 249, 250, Pl. IX, (1907); Brun., R. I. M., IV, 451[3].

Pseudoskusea mediolineata, Ludl., Can., Ent., XL., 332, (1908)[4].

Pseudoskusea nigritarsis, Ludl., Can. Ent. XL, 52, (1908)[5].

nigrescens, Theobald.

Aedes nigrescens, Theo., Monog. Culic., IV, 540[1], figs. 246, 247, (1907); Brun., R. I. M., IV, 490[2]; *id., loc. cit.,* XVII, 144[3]. — [1, 2, 3]Castle Rock (Kanara).

pseudodiurnus, Theobald.

Skusea pseudodiurna, Theo., R. I. M., IV, 32[1], (1910); *id.,* Monog. Culic., V, 491; Brun., R. I. M., IV, 451[2]. — [1, 2, 3]Sukna (N. Bengal.).

Aedes pseudodiurna, Brun., R. I. M., XVII, 144-5[3].

pseudomediofasciatus, Theobald.

Skusea pseudomediafasciata, Theo., Monog. Culic., V, 489, figs. 211, 212[1], (1910). — [1, 3]Peradeniya, [1, 3]Hakgala., [2]Colombo. (All Ceylon.)

Aedes pseudomediofasciatus, James, I. J. Med. R., II, 262[2]; Brun., R. I. M., XVII, 145[3].

punctipes, Edwards.

Aedes (*Skusea*) *punctipes,* Edw., B. Ent. R., XII, 77[1], (1921). — [1]Maymyo.

simplex, Theobald.

Ficalbia simplex, Theo., Monog. Culic., III, 297, (1903); *id., loc., cit.,* V, 541, fig. 235[1]; Blanch., Moustiques, 418; Brun., R. I. M., XVII, 171-2[2]. — [1, 3]Trincomali. Kurunegalla. [2](Both Ceylon).

Aedes simplex, Edw., B. Ent. R., IV, 238.

Skusea simplex, Edw., B. Ent., R., VII, 224, figs. 9, *b, c.*

uncus, Theobald.

Culex uncus, Theo., Monog. Culic., II, 53, (1901); Giles, Handbook, (2nd end.), 452; Blanch., Moustiques, 350, Theo., Monog. Culic., V, 357[1]. [2, 3]Sylhet. —— [1, 2, 4]F. M. S.

Aedes uncus, Edw., B. Ent. R., IV, 229; *id., loc., cit.,* VII, 223, fig. 8, *d;* Brun., R. I. M., XVII, 143[2].

Neomacleaya, indica var. *simplex,* Theo., R. I. M., II, 291, (1908)[3].

? *Verrallina malayi,* Leic., Culic, Malaya, 198, (1908)[4].

uniformis, Theobald.

Skusea uniformis, Theo., R. I. M., IV, 33, (1910)[1]; *id.,* Monog. Culic., V, 491[2]. [1, 2, 3]Travancore.

Aedes uniformis., Brun., R. I. M., XVII, 145[3].

yerburyi, Edwards.

Aedes yerburyi, Edw., B. Ent. R., VII, 22, fig. Sc, (1917)[1]; Brun., R. I. M., XVII, 145[2]. [1, 2]Ceylon (? Kitli).

11. Armigeres, Theobald.

Armigeres, Theo., Monog. Culic., I, 322, (1901) [Genotype, *Culex obturbans,* Wlk.]; Giles, Handbook, (2nd. edn.), 384; Edw., B. Ent. R., IV, 224; *id., loc., cit.,* VII, 202, [table of spp., etc.]; Brun., R. I. M., X, 55; *id. loc. cit.,* XVII, 121.

Blanchardiomyia, Brun., R. I. M. IV, 440, (1912) [*nom. nov.* fr. *Desvoidya præocc.*]; *id., loc., cit.,* X, 156.

Desvoidya, Blanch., C. R. Soc. Biol., LIII, 1035, (1901); [*nom. nov.* for *Armigeres,* presumed præocc.]; *id.,* Moustiques, 265; Theo., Gen. Ins. Fasc., XXVI, 17; *id.,* Monog. Culic., IV, 163, [table of spp.]; *id., loc. cit.,* V, 141, [table of spp.]; Brun., R. I. M., I, 327; Leic., Culic, Malaya, 74.

Desvoidea, Theo., Monog. Culic., III, 134, [larva].

apicalis, Theobald.

Desvoidea apicalis, Theo., R. I. M., IV, 5, (1910)[1]. [1, 3, 4]Puri., [2]Siliguri.

Desvoidya apicalis, Fry, Malaria in Bengal, 36[2].

Blanchardiomyia apicalis, Brun., R. I. M., IV, 441.

Armigeres apicalis, Edw., B. Ent. R., IV, 225; Brun., R. I. M., XVII, 123[4].

Stegomyia crassipes, Theo. [*nec.* Wulp.], Monog. Culic., I, 320[5], Pl. XXXIV, 134, (1901); Giles, Jo. Trop. Med., VII, 367; *id.,* Handbook, (2nd. edn.), 381; Brun., R. I. M., I, 329[6]; *id., loc., cit.,* IV, 445[3].

Cherrapunji (White).
[5, 6]Upper Burma.
_____, [6]Philippines.

aureolineatus, Leicester.

Desvoidya aureolineata, Leic., Culic. Malaya, 79, (1908)[1].

Blanchardiomyia aureolineata, Brun., R. I. M., IV, 441[2].

Armigeres aureolineata, James, I. J., Med. R., II, 262[3]; White, *loc. cit.,* VIII, 316[4].

Armigeres aureolineatus, Edw., B. Ent. R., VII, 205, fig. 1, [genitalia]; Brun., R. I. M., XVII, 123[5].

[9, 4]Ceylon.
______, [1, 2, 5]F. M. S.

fuscus, Theobald.

Desvoidea fusca, Theo., Monog. Culic., III, 135, figs. 75-6, Pl. XVII, (1903); *id., loc., cit.,* IV, 165; *id., loc. cit.,* V, 142[1]; Brun., R. I. M., I, 327; Leic., Culic. Malaya, 78[2].

Desvoidya fusca, Banks, Phil., Jo. Sci., I, (9), 983[3].

Blanchardiomyia fusca, Brun., R. I. M., IV, 441[4].

Armigeres fusca, Brun., R. I. M., XVII, 122, 274[5].

Armigeres malayi, Edw., (*nec.* Theo.), B. Ent. R., VII, 207, fig. 3, [genitalia].

[1, 4, 5]Sylhet.
[1, 4, 5]Lushai Hills.
[1, 4, 5]Calcutta.
____
[1, 2, 5]Jugra. (F.M.S.)
[1, 3, 5]Philippines.

obturbans, Walker.

Culex obturbans, Wlk., P. Linn. S. Lon., IV, 91, (1860); Giles, Handbook, 338.

Armigeres obturbans, Theo., Monog. Culic., I, 323, figs. 104-106; Green, Trop. Agric., XXV, 97[1]; Dyar, Ins. Ins. Mens., VIII, 181[2]; Edw., B. Ent., R., VII, 205, fig. 2, [genitalia]; James, I. J. Med. R., II, 262[3]; Stanton, B. Ent. R., X, 333, [distribution in far East. ports]; Fletcher, Pusa, Sci., Rpt. (1916-17), 101, [breeding places][4]; Sharma, *loc., cit.,* (1919-20), 105, [repellents]; Brun., R. I. M., XVII, 121-2[5]; White, I. J. Med. R., VIII, 316.[6]

[12]Delhi.,
[5]Naini Tal.
[5]Kurseong.
[5]Sukna., [5]Sylhet.
[5]Lushai Hills.
[4, 5]Bihar., [5]Meerut.
[5]Madras.
[5, 7]Travancore.
[1, 3, 5, 6, 8]Ceylon.
[9]Bombay., [10]Calcutta.
____

Desvoidea obturbans, Theo., Monog. Culic., III, 138, figs. 75, 77, 78; *id., loc., cit.,* IV, figs. 47, 48; Cruick. & Wright, I. J. Med. R., I, 773[7]; Green, Trop. Agric., XXXV, 122[8], [breeding grounds]; Bentley, Malaria in Bombay, 62[9]; Iyengar, I. J. Med. R., 1920 Sci. Cong. No., 9[10].

Desvoidya obturbans, Blanch., Moustiques, 266; Leic. Culic. Malaya, 75, [notes on varieties][11]; Brun., R. I. M., I, 328; Hodgson, I. J. Med. R., II, 415[12].

Blanchardiomyia obturbans, Brun., R. I. M., IV, 442.

Culex subalbatus, Coq., Proc. U.S. Nat. Mus., XXI, 302, (1898)[13].

Culex ventralis., Wlk., P. Linn. S. Lon., V, 144, (1861), [*nec, ventralis, id., loc., cit.,* VIII, 103]; Giles, Handbook, 339.

Armigeres ventralis, Theo., Monog. Culic., Pl. XVI, 57; Giles, Handbook, (2nd. edn.), 385, Pl. XIV, II, 12).

[5]Amboina. [2, 13]Philippines. [5]Ceram., [5]Celebes. [5, 11]Fed. Malay States., [5]Java. [5]Papua., [5]China. [5]Formosa.

12. **Leicesteria,** Theobald.

Leicesteria, Theo., Entom., XXXVII, 211, (1904), [Genotype, *L. longipalpis,* Leic.]; *id.,* Gen. Ins. fasc., XXVI, 20; *id.,* Monog. Culic., IV, 201; *id., loc., cit.,* V, 212; Leic., Culic. Malaya, 94; Edw., B. Ent. R., IV, 255, 257, [table of Oriental spp.]; *id., loc., cit.,* V, 126, [probably congeneric with *Armigeres*]; Brun., R. I. M., I, 337; *id., loc., cit.,* IV, 434; *id., loc., cit.,* X, 63; *id., loc., cit.,* XVII, 145-6.

Brevirhynchus, Theo., R. I. M., II, 293, (1908) [Genotype, *B. magnus,* Theo.]; *id.,* Monog. Culic., V, 144, [table of spp.]; Brun., R. I. M., IV, 442; *id., loc., cit.,* X, 56.

Chœtomyia, Leic., Culic. Malaya, 100, (1908) [Genotype, *C. flava,* Leic.].

Leicesteriomyia, Brun., R. I. M., IV, 452, (1912) [*nom. nov.* for *Chœtomyia, præocc.*]; *id., loc., cit.,* X, 63.

annulipalpis, Theobald.

Brevirhynchus annulipalpis, Theo., R. I. M., IV, 6, (1910)[1]; *id.,* Monog. Culic., V, 148[2]; Brun., R. I. M., IV, 442.

Leicesteria annulipalpis, Edw., B. Ent. R., IV, 257, fig. 1; Brun., R. I. M., XVII, 147[3]; Edw., B. Ent. R., XII, 74[4].

[1, 2,3]Travancore. ———, [4]Sumatra.

flava, Leicester.

Chœtomyia flava, Leic., Culic, Malaya, 101, (1908)[1].

Leicesteria flava, Edw., B. Ent. R., IV, 258, fig. 1; Fry, Malaria in Bengal, 36[2]; Brun., R. I. M., XVII, 146-7[3].

Leicesteriomyia flava, Brun., R. I. M., IV, 453[3].

Leicesteria apicalis, Theo., R. I. M., II, 291, (1908)[5]; *id., loc., cit.*, IV, Pl. II, Pl. III, *id.*, Monog. Culic., V, 213[6], fig. 86; Brun., R. I. M., IV, 454[7].

Brevirhynchus apicalis, Theo., R. I. M., IV, 7[8]; *id.*, Monog. Culic., V, 149[9]; Brun., R. I. M., IV, 443[10].

[8, 9, 10]Sylhet.
[2]Siliguri.
[5, 6, 7]Lushai Hills.
———, [1, 3, 4]F. M. S.
[3]Sarawak.

magna, Theobald.

Brevirhynchus magnus, Theo., R. I. M., II, 293[1], (1908); *id., loc., cit.*, IV, Pl. ii, Pl. iii; *id.*, Monog. Culic., V, 145, figs. 51, 52[2]; Brun., R. I. M., IV, 443[3].

Armigeres magnus, Fletcher, Pusa Sci., Rpt., (1916-17), 101[4] [bionomics].

Leicesteria magna, Edw., B. Ent. R., IV, 259, fig. 2; Brun., R. I. M., XVII, 146[5].

Leicesteria annulitarsis, Leic., Culic. Malaya, 100, (1908); Brun., R. I. M., IV, 454[6].

Toxorhynchites rectirostris, Giles, in Theo., Monog. Culic., V, 214[7], (1910), [*nom. nud.*].

[1, 2, 3, 5]sylhet.
[2, 3, 5]Sukna.
[4]Pusa.
[2, 3, 5]Travancore.
———
[5, 6]F. M. S.
[5, 7]Philippines.

omissa, Edwards.

Leicesteria omissa, Edw., B. Ent. R., V, 76[1], (1914); James, I. J. Med. R., II, 262[2]; Brun., R. I. M., XVII, 147[3].

[1, 2, 3]Colombo.
[1, 3]Peradeniya.

13. Stegomyia, Theobald.

Stegomyia, Theo., in How., Mosquitos, 233, (1st Jany., 1901) [Genotype, *Culex fasciatus* F. as *calopus,* Mg.]; *id.*, Monog. Culic., I, 283, [table of spp.], 284, [map of distribution]; *id., loc., cit.*, IV, 170-1, [table of spp.]; *id., loc., cit.*, V, 151, [table of spp.]; *id.*, Gen. Ins. fasc., XXVI, 18; Giles Handbook, (2nd. edn.), 368-9, [table of spp.]; Blanch. Moustiques, 247, figs. 194, p. 248, [table of spp.]; Brun., R. I. M., I, 328; *id., loc., cit.*, IV, 443; *id., loc., cit.*, X, 39; *id., loc., cit.*, XVII, 124, 125, 274; Leic., Culic. Malaya, 81, [table of Malayan spp.]; Alcock, Ent. Med. Off., 100; Edw., B. Ent. R., VII, 203; White,

Science (n. s.), XXIII, 371; Patton & Cragg, T. B. Med. Ent., 215.

Gymnometopa, Coq., P. Ent. S. Wash., VII, 183, (1906) [Genotype, *S. mediovittata,* Coq.].

Hulecoetomyia, Theo., Entom., XXXVII, 163, (1904) [Genotype, *H. trilineata,* Leic.]; *id.,* Monog. Culic., IV, 219; Brun., R. I. M., X, 62.

Kingia, Theo., Monog. Culic., V. 135, (1910) [Genotype, *Steg. luteocephala,* News.]; Brun., R. I. M., X, 63.

Quasistegomyia, Theo., II Rpt. Wellc. Lab., 69, (1906) [Genotype, *Q. unilineata,* Theo.]; *id.,* Monog. Culic., IV, 165; *id., loc. cit.,* V, 132, [table of spp.]; Brun., R. I. M., IV, 443; *id., loc., cit.,* X, 70.

Scutomyia, Theo., Entom., XXXVII, 77, (1904) [Genotype, *Culex sugens,* Wd.]; *id.,* Monog. Culic., IV, 196; *id., loc., cit.,* V, 199), [table of spp.]; *id.,* Gen. Ins. fasc., XXVI, 19; Leic., Culic. Malaya, 105; Brun., R. I. M., I, 336; *id., loc. cit.,* IV, 452; *id., loc., cit.,* X, 71.

albolineata, Giles.

Culex albolineatus, (I), Giles, Bom. Jo., XIII, 609, (1901)[1], [provis. descr.]; *id.,* Handbook, (2nd. edn.), 430, Pl. XVII, 10*a;* Theso., Monog. Culic., V, 360[2]. — [1, 2, 3]Shahjahanpur.

Seutomyia albolineata, Brun., R. I. M., I, 336; *id., loc., cit.,* IV, 452.

Stegomyia albolineata, Brun., R. I. M., XVII, 127[3].

[Apparently not recognized since its description.]

albopicta, Skuse.

Culex albopictus, Skuse, I. M. N., III, 20, (1895)[1]; Giles, Handbook, 228. — [1, 2]India.

Stegomyia albopictus, Brun., R. I. M., I, 334[2]; Dyar, Ins. Ins. Mens., VIII, 182.

Stegomyia albopicta, Edw., B. Ent. R., VII, 209, fig. 5*b,* [genitalia]; *id.,* B. Ent. R., XI, 134; Brun., R. I. M., XVII, 126, 127; White, I. J. Med. R., VIII, 316, (footnote).

Culex scutellaris, Giles, Handbook, 224.

Stegomyia scutellaris, Theo., (nec. Wlk.), Monog. Culic., I, 298, fig. 91, Pl. XIV, 53, (1901); Giles, Handbook

[3]Philippines.
[4]Sumatra.
[5, 6]Madagascar.
[6]Amboina.

Almost universally distributed from

(2nd. edn.), 374, Pl. XIV, 4, 5; Blanch., Moustiques, 257; Leic., Culic. Malaya, 86; Brun., R. I. M., I, 333; *id., loc., cit.*, IV, 448; Green, Trop. Agric., XXV, 297; *id., loc., cit.*, XXVI, 494; *id., loc., cit.*, XXXVI, 234; *id.*, Spol. Zeyl. IV, 183; Annan, R. I. M., I, 83; Theo., *loc. cit.*, II, 289; Cruick. & Wright, I. J. Med. R., I, 773-784; Hodgson, *loc. cit.*, II, 415; James, *loc. cit.*, II, 262; Prashad, *loc. cit.*, III, 500, Pl. XXX, 4, [male genitalia]; Sen, *loc. cit.*, IV, 729, [ovulation]; Mc Farl., B. Ent. R., VI, 67; Stanton, *loc., cit.*, X, 333, [distribution in Far East. ports]; Howlett, Pusa Sci. Rpt. (1911-12), 75, [eggs]; Fletcher, *loc. cit.*, (1915-16), 83; Howlett, Proc., III Meet. Mal. Com., 1912, 31, [crosses with *fasciata*]; Bentley, Malaria in Bombay, 62; Stanton, I. J. Med. R., III, 256; Tayl. T. Ent. S. Lon., (1914), 189; *id.*, P. Linn. S., N. S. W., XXXIX, 455. — Madagascar to Japan and Honolulu.

Stegomyia scutellaris var. *samarensis*[3], Ludl., Jo. N. Y. Ent. S., XI, 138, (1903); *id.*, Can. Ent., XXXVI, 71; *id., loc., cit.*, XXXVII, 134; Banks, Phil. Jo. Sci., I, 985; Theo., Monog. Culic., IV, 179; Ludl., Mosq. Phil Ids., 10; Theo., Monog. Culic., V, 156; Brun., R. I. M., I, 333; *id., loc., cit.*, IV, 449; *id., loc., cit.*, XVII, 127.

Stegomyia samarensis, Ludl., Psyche, XVIII, 126.

Stegomyia notoscripta var. *samarensis* Theo., Gen. Ins. fasc., XXVI, 19; *id.*, Monog. Culic, V, 200; Brun., R. I. M., I, 336; *id., loc., cit.*, IV, 452.

Culex crassipes, Wulp. Dipt. Mid. Sumat. Exp., 9[4], Pl. I, 4, (1892), [*nec. Steg. crassipes,* Theo., *et auct.*].

Stegomyia lamberti, Vent., B. Mus., Paris, X, 552[5], (1904); *id.*, Arch. Paras., IX, 441; Theo., Monog. Culic., V, 184[6].

? *Culex variegatus,* Dol., Nat., Tijd. Ned. Ind., XVII, 77[7]; Giles, Handbook, 282.

annandalei, Theobald.

Stegomyia annandalei, Theo., R. I. M., IV, 10[1], (1910); Brun., *loc. cit.*, IV, 444[2]; *id., loc., cit.*, XVII, 130[5]. — [1, 2, 3, 4, 5]Sukna.

Kingia annandalei, Theo., Monog. Culic., V, 139[3]; Brun., R. I. M. IV, 443[4].

annulirostris, Theobald.

Segomyia annulirostris, Theo., Bom. Jo., XVI, 239, (1905)[1]; *id.*, Monog. Culic., IV, 173; Green, Trop. Agric., XXXVI, 234[2]; Brun., R. I. M., XVII, 129[3]. — [1, 2, 3]Peradeniya. [4]Colombo.

Aedes annulirostris, James, I. J. Med. R., II, 262[4].

argenteomaculata, Theobald.

Stegomyia argenteomaculata, Theo., Monog. Culic., IV, 184[1], (1907); Brun., R. I. M., IV, 444[2]; *id., loc., cit.,* XVII, 130[3].

[1, 2, 3]Narcondam Id.

assamensis, Theobald.

Stegomyia assamensis, Theo., R. I. M., II, 290[1], (1908); Brun., *loc. cit.,* IV, 445[2]; Theo., Monog. Culic., V, 174[3]; Brun., R. I. M., XVII, 130[4].

[1, 2, 3, 4]Sylhet.
[2, 3, 4]Travancore.

desmotes, Giles,

Stegomyia desmotes, Giles, Jo. Trop. Med., VII, 367[1], (1904); Brun., R. I. M., IV, 445; Theo., Monog. Culic., V, 185; Edw., B. Ent. R., IV, 225; Brun., R. I. M., XVII, 128.

Stegomyia albipes, Theo., R. I. M., IV, II[2], (1910), *id.,* Monog. Culic., V, 169[3].

Stegomyia gracilis, Leic., Culic. Malaya, 81[4], (1908); Brun., R. I. M., IV, 447.

[2, 3]Travancore.

---

[4]F. M. S.
[1]Philippines.

fasciata, Fabricius.

Culex fasciatus, F. Syst. Antl., 36, (1805); Wd., [*nec.* Müller], Dipt. Exot., 36, 39; *id.,* Auss. Zweifl. Ins., I, 8; R.D., [*nec.* Müller], Mem. Soc. H. Nat. Paris, III, 408; Wlk., [*nec.* Müller], List., Dipt. B.M., I. 3; V. Röder, [*nec.* Müller]; Stett. Ent. Zeit., XLVI, 338; Coq., [*nec.* Müller], U.S. Dpt. Agric. Circ., 40, (2nd. ser.), 5, 6; How. [*nec.* Müller], *loc., cit.,* Bull. 25, n. s. 1920; Giles, [*nec.* Müller], Handbook, 235; Reed, Carrol, Agrmnt. & Lazear, [*nec.* Müller], Phil Med. Jo., VI, 792; Cropper [*nec.* Müller]. Jo. Hyg., II, 51; O. S., Cat. Dipt. N. Amer., (2nd. edn.), 19; Johnson, Proc. Acad., N. Sc., Phil., 271, 319; Kert., Cat. Dipt., I, 259, 256.

Occurs throughout the Tropics and Sub-tropics.

Stegomyia fasciata, Theo., Monog. Culic., I, 289, figs. 86, 89, 292[5], [map of distribution], Pl. XIII, 49, 50, Pl. B; *id.,* Gen. Ins., fasc., XXVI, Pl. I, Pl. II; Giles, Handbook, (2nd edn.), 372, Pl. XIV, 2, 3; Leic., Culic. Malaya, 85; Brun. R. I. M., I, 330; *id., loc., cit.,* IV., 446[3]; *id., loc., cit.,* XVII, 125-6; Reed & Carrol., Med. Rec., LX, 641; Theo., Jo. Trop. Med., IV, 159, (note); How., Mosquitos, 123, 127, 135, 155, 236; Garman, Bull. 96, Ky. Ag. Expt. Sta., 211; Durham, Liv. Sch. Trop.

Med. Mem., VII, 15, 55; Nev. Lem., Arch. Paras., VI, 16; Blanch., C.R. Soc. Biol., Paris, LIV, 644; Dyé Arch. Paras., VI, 365, 367; Theo., R. I. M., II, 291[1]; *id., loc., cit.*, IV, 8[2]. Theo., Proc., R. S., LXIX, 389; Goeldi, Os. Mosq. Para., IV, 129-197; Park, Beyr. & Path., Bull, XIII, Y. F. Inst., U.S. Health Service, 21; Tayl., Revist. Med. Trop., IV, 113; Theo., Monog. Culic., III, 141; Dyar, Jo. N. Y. Ent. Soc., XI, 23; *id.*, Proc. Ent. Soc., Wash., V, 146, Pl. II, 12; Venin., "Bermuda", 335; Marlatt, Proc. Ent. Soc., Wash., V, 117; Carter, N. Y. Med. Rec., LXIV, 794; *id.*, Boston. Med. and Surg. Jo., CL., 20; March., Simond., etc., Braz.-Med., XVII, 473; *id.*, Ann. Inst. Past., XVII, 680; Guit., Rev. Med. Trop. Habana, IV, 60; Theo., Jo. Trop. Med., VI, 237; Laveran, C. R. Soc. Biol., Paris, LV, 1157; How., Pub. Hlth. Rpts., XVIII, 46, (and revised edn.); Pazos., B. S. Ent. Fr., (1904), 135; Carter, Med. Rec., LXV, 761; Balf., 1st Rpt. Wellc. Lab., 16; Theo., *loc. cit.*, 71; Bourroul, Mosq., Braz., 40; V. Dine, Bull., 6, Havana Exp. Sta., 14, *id.*, U. S. Dpt. Agric., Ann. Rpt. Exp. Stns., (1904), 377; Coffin in Shattock, "Bahamas", 280; Dyar, Jo. N. Y. Ent. Soc. XIII, 54; Theo. & Grabh., Mosq. Jamaica, 19; Felt., Bull. 97, N. Y. Mus., 489; Goeldi, C. R. VI. Cong. Int. Zool., Berne, 193; Herrick, Ent. News, XVI, 282; Banks, Phil. Jo. Sci., I, 984; Theo., Sci. Progress, I, 70; *id.*, Monog. Culic., IV, 176; Grün., Blutsaug. Dipt., 92; Becker, Mitt. Zool. Mus. Berl., IV, 81; Theo., Monog. Culic., V, 158[4]; Seidelin, Y. F. Bur. Bull., I, 365; Aust., *loc., cit.*, II, 3; Edw., B. Ent. R., III, 9; Green. Trop. Agric., XXXVI, 234[11]; *id., loc. cit.,* XXXVI, 226, (abstract); McGregor, T. Vet. Bull., III, 141, [note on rearing in Lond.]; Ann. Inst. Past., XX, 161-205, [abst. J. Trop. Vet. Sci. I, 348-350], [transmission of Y. F.]; Nuttall, Jo. Prev. Med., VIII, [abst. J. Trop. Vet. Sci., IV, 21], [*Spirochœta duttoni*]; Stanton, B. Ent. R., X, 333, [distribution in far East. ports]; James, I. J. Med. R., I, 213, 3 Pl., [Protection of India from Y. F.]; *id., loc. cit.*, 258, Pl., [reduction in Indian ports]; Hodgson, *loc. cit.*, II, 415[6]; James *loc. cit.*, II, 262[7]; Wesché, B. Ent. R., I, 25, fig. 1, [larva and pupa]; Knab, Science, I, 869, [abst. B. Ent. R., I, 156], [oviposition]; Boyce, B. Ent. R., I, 233, [prevalence, distribution, etc.]; Graham, *loc. cit.*, II, 127, [control measures]; Balf., *loc., cit.*, II, 179, [vector

Recorded Indian distribution:–
[6]Delhi.
[4]Ferozepore.
[1, 3, 4]Bhim Tal.
[1, 3, 4]Lucknow.
[1, 3, 4]Purneah.
[9]Siliguri.
[1, 3, 4]Lushai Hills.
[2, 3]Mandalay.
[2, 3]Rangoon.
[2]Chittangong.
[1, 2, 3]Calcutta.
[2, 3]Puri.
[2, 3]Coconada.
[2, 3]Madras Town.
[5]Travancore.
[10]Bombay.
[7, 8, 11]Colombo.
[11]Trincomali.
[4]Andamans.

of horse-sickness & dengue]; Doane, *loc. cit.*, IV, 265; McFie, *loc. cit.*, IV, 339, [action of salt on larvæ]; Gough, *loc. cit.*, V, 134; McFarl., *loc. cit.*, VI, 67; McFie. *loc. cit.*, VI, 205-229, [bionomics]; McGregor, *loc. cit.*, VII, 81, 3 figs., [eggs]; McFie, *loc. cit.*, VIII, 277, [larval respiration]; *id.*, *loc. cit.*, VII, 297, [larva]; McGregor, *loc. cit.*, X, 91, [occurrence in England]; McFie, *loc. cit.*, X, 61, [pupal chætotaxy]; Barraud, *loc. cit.*, X, 323; Stanton, I. J. Med. R., III, 256; Patton & Cragg, T. B. Med. Ent., 216, Pl. XXXVI, 1, [bionomics]; Howlett, Proc. 3rd Meet. Mal. Com., (1912), 31, [crosses with *albopicta*]; Fry, Malaria in Bengal, 36[9]; Bentley, Malaria in Bombay, 62[10]; Dalziel, B. Ent. R., XI, 251, 256, [larva in crab holes]; Waterst., *loc. cit.*, IX, 7; Johnson, *loc. cit.*, IX, 325-332, [notes in Nigeria]; Iyengar, I. J. Med. R., 1920 Sci., Cong. No., 9; Barraud, B. Ent. R., XI, 392; McFie., Ann. Trop. Med. Paras., XIV, 73-82, [temperature effects]; McGilchrist, I. J. Med. R., II, 322, [feeding on *Paramœcium*]; Enderl., Wien. Ent. Zeit., XXXVIII, 47; Balf., B. Ent. R., XII, 30, [in saline water]; Hill, Ann. Trop. Med. Paras., XV, 91-2, Pl., [unusual breeding places]; Cleland, etc., Jo. Hvg. Camb., XVIII, 227-254, [abst. R. A. E., (B) IX, 127], [carries dengue]; Craig., Jo. Amer. Med. Ass., LXXV, 1171-6, [abst. R. A. E. (B) IX, 128] [carries dengue].

**Inscules fasciatus*, Herrera, Anal. Inst. Med. Nat., Mexico, XI, 235.

Stegomyia fasciata var. *atritarsis*, Edw., B. Ent. R., X, 129, (1920).

Stegomyia fasciata var. *persistans*, Banks, Phil. Jo. Sci., I, 996, 984, (1906); Theo., Monog. Culic., V, 159; Brun. R. I. M., I, 331; *id.*, *loc. cit.*, XVII 126; Trop. Agric., XXXVI, 32.

**? *Culex œgypti*, Linn., Hasselquist. Plaestina Reise, 470, (1762).

Aedes (*Stegomyia*) *œgypti*, Dyar, Ins. Ins. Mens., VIII, 181, 204.

Culex albopalposus, Becker, Mitt. Zool. Mus. Berl., IV, 79, (1908).

---

* I can find no reference to the erection of this genus.

** For another possible identification of this description see *Ochlerotatus dorsalis*, Mg.

Culex anguste-alatus, Becker, Mitt. Zool. Mus. Berl., IV, 79, (1908).

Culex annulitarsis, Mq., Dipt. Exot., Supp., I, 136, (1828); Giles, Handbook, 286.

? *Culex argentenus,* Poiret, Jo. de Physique, XXX, 245 (1787); How., Dyar. & Knab, Mosq., N. Amer., IV, 824, (note).

Culex bancroftii, Skuse, P. Linn. S., N. S. W., III, 1740, (1889); Giles, Handbook, 200.

Culex calopus, Mg. Syst. Besch., I, 3, (1818); R.-D., Essai Culicides, 406; Mq., in Webb. & Berth., N. H. Canaries, II, Ins., 99; *id.,* Dipt., Exot., IV, Supp. 9, 313; V. Rod., Berl. Ent. Zeit., XXXI, 73; Fic., B. Ent. S. Ital., XXVIII, 251; Giles, Handook, 237; Becker, Mitt. Zool. Mus. Berl., II, 67; Lucas, Expl. Sci. Alg., Zool., III, 415, 3.

Stegomyia calopus, Coq., U. S. Dpt. Agric. Ent. Ser., II, 17; Dyar, U. S. Dpt., Ent., Circ., 72, 5; Goldb., Bull. 16, Y. F. Inst., U. S. Pub. Hlth. Ser., 10; Blanch., Moustiques, 249, figs. 195-201; Ludl., Mosq. Phil. Ids., 33; Dyar, Proc. U. S. N. H. Mus., XXXII, 128; Knab, Jo. N. Y. Ent. Soc., XV, 13; How. in Osler's Mod. Med., I, 386; Antran., Anal. Dpt. Nac. Hig., XIV, 17; Viereck., 1st Rpt. Comn. Hlth., Para. 469; Peryassu, Culic. Braz., 44, 165; V. Logh., Gen. Tijd. Ent., XLVIII, 577; Busck., Smiths. Mis. Col. Quart., LII, 65; Pazos, Sanid. y. Benef., II, 48, 11; News. & Thom., Ann. Trop. Med. Paras., IV, 143; Jennings, Jo. Econ. Ent., V, 132; Brinley, Science, n. s., XXXIII, 943.

Aedes calopus, How., Dyar. & Knab, Mosq. N. Amer., IV, 824; Dyar & Knab, Jo. N. Y. Ent. Soc., XIV, 190, 196; Thibault, P. Ent. Soc. Wash., XII, 19; Dyar & Knab, *loc. cit.,* XII, 81; Fullaway, An. Rpt., Hawai Exp. Sta., (1912), 9.

Cylex, elegans., Fic., B. Soc. Ent. Ital., XXVIII, 251, (1889); Lioy, Dipt. Ital., 79; Grass Stud. Mal., (2nd edn.), 132.

Culex exagitans, Wlk., Ins. Saund. Dipt., pt. V, 430, (1856); Giles, Handbook, 230.

Culex excitans, Wlk., List Dipt. B. M., I, 4, (1848); Giles, Handbook, 244.

Culex formosus, Wlk., List. Dipt. B. M., I, 4, (1848); Giles, Handbook, 283.

Culex frater, R.-D., Mem. Soc. N. H., Paris, III, 408, (1827).

Culex impatabilis, Wlk., P. Linn. S. Lon., IV, 91, (1860); Giles, Handbook, 255.

Culex inexorabilis, Wlk., List Dipt. B. M., I, 4, (1848 ; Giles, Handbook, 264.

Culex konoupi, Brullé, Exped. Morea, Zool., III, 289, Pl. XLVI, 1, (1832); Fic., B. Ent. Soc. Ital., XXVIII, 256; Giles, Handbook, 231.

Culex luciensis, Theo., Monog. Culic., I, 297, Pl. XIII, 50, (1901).

Stegomyia luciensis, Aiken, Br. Guiana Mus. Ann., (1906), 71.

Stegomyia fasciata var. *luciensis,* Theo., Monog. Culic., I, 297; James, I. J. Med. R., II, 262[8].

Culex mosquito, R.-D. Essai Culicides, 407, (1827); Arrib., Rev. la Plata, 160; Guer. & P., Gen. des Ins., II livre, No. 9, Pl. II, fig. 1; Finlay, Anal. Acad. Cien. Habana, XVIII, 153; Will., T. Ent. S. Lon., 272; Giles, Handbook, 224.

Stegomyia fasciata var. *mosquito,* Theo., Monog. Culic., I, 295, (1901).

Stegomyia nigeria, Theo., Monog. Culic., I, 303, (1901); Giles, Handbook, (2nd edn.), 375; Blanch., Moustiques, 259; Theo., Monog. Culic., IV, 171; Grün., Blutsaug. Dipt., 95.

Culex queenslandensis, Theo., Monog. Culic., I, 197, (1901).

Stegomyia fasciata var. *queenslandensis,* Theo., Monog. Culic., I, 297.

Culex rossii, Giles, Jo. Trop. Med., 64, (1899).

Culex tæniatus, Wied., Auss. Zweifl. Ins., I, 10, (1828); Wlk., List Dipt. B. M., I, 5; Giles, Handbook, 216; *id.,* Jo. Trop. Med., IV, 159; Grimsh., F. Hawai, III, 6; Ribas, Braz.-Med., XV, 412.

Culex toxorhynchus, Mq., Dipt. Exot., I, 25, (1838).

Culex variegatus, Theo., [*nec.* Dol., *nec.* Schrank, *nec.* Blanch.]. Monog. Culic., V, 159, (1910), [note].

Culex viridifrons, Wlk., List Dipt. B. M., I, 3, (1848); Giles, Handbook, 232.

Culex zonatipes, Wlk., P. Linn. Soc. Lon., Zool., V, 229, (1861); Giles, Handbook, 287.

mediopunctata, Theobald.

Stegomyia mediopunctata, Theo., Bom. Jo., XVI, 240, (1905)[1]; *id.,* Monog. Culic., IV, 187[2]; Green, Trop. Agric., XXXVI, 234[3]; Brun., R. I. M., IV, 447; *id., loc. cit.,* XVII, 130[4]. [1, 2, 3, 4]Peradeniya.

microptera, Giles.

Culex micropterus, Giles, Bom. Jo., XIII, 699, [provis. descr.]. [1, 2, 6]Allahabad. [1, 2, 6]Lucknow. [2, 6]Jhansi. [7]Colombo.

Wyeomyia micropetera, Giles in Theo., Monog. Culic., II, 281, 1, fig. 291, (1901).

Stegomyia microptera, Giles, Handbook, (2nd edn.), 380, Pl. XIV, 24-26; Theo., Monog. Culic., III, 147, [note]; Blanch, Moustiques, 263; Brun., R. I. M., I, 332[2]; *id., loc. cit.,* IV, 447, [notes]; Theo., Monog. Culic., V, 607, [type lost]; Brun., R. I. M., XVII, 127[6].

Ochlerotatus micropteron, James. I. J. Med. R., II, 262[7].

periskelata, Giles.

Stegomyia periskelata, Giles Handbook, (2nd edn.), 371, Pl. XIV, 22, (1902); Theo., Monog. Culic., III, 145; Blanch., Moustiques, 264; Brun., R. I. M., I, 332[1]; *id., loc. cit.,* IV, 448; *id., loc. cit.,* XVII, 128; Edw., B. Ent. R., XII, 77. [1]Shahjahanpur. [2, 3, 5]Jhansi. [2]Gonda.

Ochlerotatus annulifemur, Edw., B. Ent. R., V, 77, (1914)[3]; Brun., R. I. M., XVII, 141[5].

Stegomyia piperselata, Giles in Theo., Monog. Culic., II, 316, male only, (1901); *id.,* Handbook, (2nd edn.); 372, Pl. XVI, 1*a, b,* [male only; name retained for female by Edwards]; Brun., R. I. M., I, 332[2].

Ochlerotatus piperselatus, (p. p.), Edw., B. Ent. R., IV, 227, male only.

sugens, Wiedemann.

Culex sugens, Wd., Auss. Zweifl. Ins., I, 545, 4[1], (1828); Kert., Cat. Dipt., I, 268[2]. [9]Bombay., [8]Pusa. [5, 6, 10]Cylon.

Stegomyia sugens, Theo., Monog. Culic., I, 300[3], XIII, 51, 1, fig. 92; Giles, Handbook, (2nd edn.), 375; James, I. J. Med. R. II, 262[6] How. B. Ent. R., III, 214; Patton & Cragg, T. B. Med. Ent., Pl. XXXIV, 7, [egg], Pl. XXXV, 15, [larva]; Fletcher, Pusa Sci. Rpt., (1915-16), 83[8]; Bentley, Malaria in Bombay, 62[9]; Edw., B. Ent. R., III, 375, fig. 1, [larva]; Gough, *loc. cit.,* V, 134; Sen, Pusa Sci. Rpt. (1919-20) 103, [feeding on *Englena*]; White, I, J. Med. R., VIII, 316[10].

———

[4, 5]Aden.
[3, 4, 5]Africa–general.
[1, 2]Nubia.

Scutomyia sugens, Theo., Monog. Culic., IV, 199[4]; *id., loc. cit.,* V, 201[5].

Culex vittatus, Big. A. S. Ent. Fr., (4), I, (1861); Fic., B. Ent. Soc. Ital., (1886), 257.

Stegomyia vittata, Edw., B. Ent. R., VII, 210.

thomsoni, Theobald.

Stegomyia thomsoni, Theo., Gen. Ins. Fasc., XXVI, 18, (1905), [prelim. descrip.]; *id.,* Monog. Culic., IV, 174, 1907); Edw., Ent. R., VII, 210, fig. 5*d,* [male genitalia]; Brun., R. I. M., IV, 450[2]; *id., loc. cit.,* XVII, 129[3]; Howlett, Pusa. Sci. Rpt., (1911-12), 75[4], [eggs].

[1, 2, 3]N.-W. India.
[3, 4]Pusa.

trilineata, Leicester.

Hulcœtomyia trilineata, Leic., in Theo., Entom., XXXVII, 163, (1904); Theo., Monog. Culic., IV, 220[1]; Leic., Culic. Malaya, 107; Brun., R. I. M., IV, 455[6].

Stegomyia trilineata, Edw., B. Ent. R., IV, 226; Brun., R. I. M., XVII, 129[2]; White, I. J. Med. R., VI II, 317[3].

Howardina chrysolineata, Theo., Monog. Culic., IV, 218, Pl. I, (1907)[4]; Brun., R. I. M., IV, 456[5].

Coonoor (White).
[2, 6]Kurseong.
[2, 3, 4, 5]Ceylon submontane).
Cherrapunji (White).
[1, 2, 6]F. M. S.

tripunctata, Theobald.

Stegomyia tripunctata, Theo., R. I. M., II, 288[1], (1908); *id., loc. cit.,* IV, Pl. I, Pl. III; *id.,* Monog. Culic., V, 182[2], fig. 68; Brun., R. I. M., IV, 450[3]; *id., loc. cit.,* XVII, 130[4].

[1, 2, 3, 4]Lushai Hills.

w-alba, Theobald.

Stegomyia w-alba, Theo., Ann. Mus. Hung., III, 74, fig. 4 (1905)[1]; *id.,* Monog. Culic., IV, 180, fig. 151; Brun., R. I. M., I, 335[4]; Edw. Ent., R. IV, 225; James, I. J . Med. R., II, 262[5]; McFarl., B. Ent. R., VI, 67[6]; Brun., R. I. M., IV, 450; *id., loc. cit.,* XVII, 129[7].

[1, 4, 7]Matheran,
[8]Pusa., [7]Sukna.
[5]Colombo.,

———

[2, 3, 7]Kuala Lumpur.
[6]Hongkong.

Stegomyia w-album, Fletcher, Pusa Sci. Rpt., (1916-17), 101[8], [breeding place].

Stegomyia imitator, Leic., Culic. Malaya, 89[2], (1908); Brun., R. I. M., IV, 447[3].

Stegomyia minutissima, Theo., R. I. M., IV, 9, (1910); *i.,* Monog. Culic., V, 168, fig. 61; Brun., R. I. M., IV, 448.

14. Ochlerotatus, Arribalzaga.

Ochlerotatus, Arrib., Rev. Mus. la Plata, II, 143, (1891) [Genotype, *Culex albofasciatus,* Mq.]; Edw., B. Ent. R., II, 246; *id., loc. cit.,* VII, 203; Brun., R. I. M., X, 67; *id., loc., cit.,* XVII, 133-135; Carter, Pr. Zool. Soc., (1920), 623-628, fig. 1-3, [notes on Asiatic spp.]

Acartomyia, Theo., Monog. Culic., III, 251, (1903) [Genotype, *A. zammitii,* Theo.].

Aedimorphus, Theo., Monog. Culic., III, 290, (1903) [Genotype, *Uranotænia domestica,* Theo.].

? *Andersonia,* Strickland, Entom., (1911), 250, (1911) [Genotype, *A. tasmaniensis,* Strick.]; Brun., R. I. M., X, 55.

Bathosomyia, Theo., Monog. Culic., V, 267, (1910) [Genotype, *B. abnormalis.,* Theo.].

? *Cacomyia,* Coq., U. S. Bull., Tech. Ser., II, 16, (1906) [Genotype, *Hæmagogus albomaculatus,* Theo.]; Brun., R. I. M., X, 56.

Chrysoconops, Goeldi, Os. Mosq. Para, 114, (1905) [Genotype, *Culex fulvus,* Wd.]; Brun., R. I. M., X, 57; Theo., Monog. Culic., IV, 491; *id., loc. cit.,* V, 433, [table of spp.]; Brun., R. I. M., IV, 480.

Culicada, Felt, N. Y Mus. Bull., 79, Ent. 22, App. 391*b*, (1904) [Genotype, *Culex canadensis,* Theo.]; Theo., Jo. Econ. Biol., I, 26; *id.,* Monog. Culic., IV, 318 & p. 10; *id., loc. cit.,* V, 294, [table of spp.]; Brun., R. I. M., IV, 461; *id., loc. cit.,* X, 58.

Culicelsa, Felt. N. Y. Mus. Bull., 79, Ent. 22, App. 391, (1904) [Genotype, *Culex tæniorhynchus,* Wd.].

Danielsia, Theo., Entom., XXXVII, 78, (1904); [Genotype, *D. albotæniata,* Leic.]; *id.,* Gen. Ins. fasc., XXVI, 21; *id.,* Monog. Culic., IV, 240; Leic., Culic. Malaya, 117.

Duttonia, News., Ann. Trop. Med. Paras., I, 17, (1907) [Genotype, *D. tarsalis,* News.].

Ecculex, Felt., Jo. N. Y. Mus. Bull., 79, Ent. 22, App. 391*c* [Genotype, *Culex sylvestris,* Theo.].

Finlaya, Theo., Monog. Culic., III, 281, (1903) [Genotype, *F. poicilia,* Theo.]; *id., loc. cit.,* IV, 520; *id.,* Gen. Ins. fasc., XXVI, 32; Blanch., Moutiques, 415, 630; Brun., R. I. M., I, 360; *id., loc. cit.,* IV, 485; *id., loc. cit.,* X, 60.

Geitonomyia, Leic., Culic. Malaya, 134, 1908 [Genotype, *Culex cœcus,* Theo.].

? *Gilesia,* Theo., Monog. Culic., III, 233, (1903) [Genotype, *G. aculeata,* Theo.]; Brun., R. I. M., X, 60.

? *Gualteria,* Lutz in Bourroul, Mosq., Braz., 49, (?54), 1904; [Genotype, *G. oswaldii,* Ltz.]; Brun., R. I. M., X, 61; Theo., Gen. Ins. fasc., XXVI, 21; *id.,* Monog. Culic., IV, 240; Leic., Culic., Malaya, 117.

Howardina, Theo., Monog. Culic., III, 287, Pl. XV, (1903) [Genotype, *Culex walkeri,* Theo.]; Dyar, P. Ent. S. Wash., VII, 49; Theo., Monog. Culic., IV, 214; *id., loc. cit.,* V, 200, [table of spp.]; *id.,* Gen. Ins. fasc., XXVI, 21; Blanch., Moustiques, 415; Brun., R. I M., I, 338; *id., loc. cit.,* IV, 456; *id., loc. cit.,* X, 62; *id., loc. cit.,* XVII, 131.

Lepidoplatys, Coq., Science, XXIII, 314, (1906) [Genotype, *Culex squamiger,* Coq.]; Brun., R. I. M., X, 63.

Lepidotomyia, (I), Theo., Ann. Mus. Hung., III, 80, (1905) [Genotype, *L. alboscutellata,* Teho.]; Brun., R. I. M., X, 63; Theo., Monog. Culic., V, 249, (note) [sunk in *Reedomyia*].

Lepidotomyia, (II), Theo., Gen. Ins. fasc., XXVI, 22, (1905) [Genotype, *L. magna,* Theo.]; *id.,* Monog. Culic., IV, 249; *id., loc. cit.,* V, 249; Brun., R. I. M., IV, 458; *id., loc. cit.,* X, 63.

Leslieomyia, Chris., Paludism, II, 68, (1911) [Genotype, *L. tæniorhynchoides,* Chris.].

Mimetoculex, Theo., 3rd Rpt. Wellc. Lab., 258, (1908) [Genotype, *M. kingii,* Theo.]; *id.,* Monog. Culic., V, 408.

? *Molpemyia,* Theo., Monog. Culic., V, 479, (1910) [Genotype, *M. purpurea,* Theo.] Brun., R. I. M., X, 65.

Myxosquamus, Theo., Monog. Culic., V. 225, (1910) [Genotype, *M. confusus,* Theo.].

Neopecomyia, Theo., Monog. Culic., V, 261, (1910) [Genotype, *N. uniannulata,* Theo.].

Pecomyia, Theo., Jo. Econ. Biol., I, (1), 214, (1905) [Genotype, *P. maculata,* Theo.]; *id.,* Monog. Culic., IV, 265.

Phagomyia, Theo., Gen. Ins. fasc., XXVI, 21, (1905) [Genotype, *Steg. gubernatoris,* Giles]; *id.,* Monog. Culic., IV, 223; *id., loc. cit.,* V, 224 [table of spp.]; Brun., R. I. M., I, 338; *id., loc. cit.,* IV, 456; *id., loc. cit.,* X, 68.

Polyleptiomyia, Theo., Gen., Ins. fasc., XXVI, 21, (1905) [Genotype, *P. albocephala,* Theo.]; *id.,* Monog. Culic., IV, 223.

Protoculex, Felt, N. Y. Mus. Bull., 79, Ent. 22, App. 391*d,* (1904) [Genotype, *Culex serratus,* Theo.].

Protomacleaya, Theo., Monog. Culic., IV, 253, (1907) [Genotype, *Culex triseriatus,* Say.].

Pseudocarrollia, Brun., R. I. M., IV, 12, (1910) [Genotype, *P. lophoventralis,* Theo.]; *id.,* Monog. Culic., V, 186; Brun., R. I. M., IV, 454; *id., loc. cit.,* X, 69.

Pseudoculex, Dyar, P. Ent. S. Wash., VII, 45, (1905) [Genotype, *Culex aurifer,* Coq.].

Pseudograbhamia, Theo., Bom. Jo., XVI, 243, (1905) [Genotype, *P. maculata,* Theo.]; *id.,* Monog., Culic., IV, 314; Brun., R. I. M., IV, 460; *id., loc. cit.,* X, 69.

Pseudohowardina, Theo., Monog. Culic., IV, 223, (1907) [Genotype, *Culex trivittata,* Coq.].

Pseudoskusea, Theo., Monog. Culic., IV, 192, (1907) [Genotype, *Skusea multiplex,* Theo.]; Brun., R. I. M., X, 69.

Reedomyia, Ludl., Can. Ent., XXXVII, 94, (1905) [Genotype, *R. pampangensis,* Ludl.]; Theo., Monog. Culic., IV, 257 [table of spp.]; *id., loc. cit.,* V, 252; Brun., R. I. M., I, 362; *id., loc. cit.,* IV, 486; *id., loc. cit.,* X, 70.

? Stegoc nops, Lutz., Imprensa Med., (1906), [?nom. nud.]; Peryassu, Os. Culic. Braz., 34; (1908); Brun., R. I. M., X, 71.

Stenoscutus, Theo., Monog. Culic., V, 263, (1910) [Genotype, *S. africanus,* Theo.].

alboscutellatus, Theobald.

Lepidotomyia alboscutellata, Theo., Ann. Mus. Hung., III, 80, (1905)[1]; Leic., Culic. Malaya, 132[2]; Brun., R. I. M., I, 339[3].

Reedomyia alboscutellata, Theo., Monog. Culic., IV, 261, fig. 80; Brun., R. I. M., IV, 487[4]; Theo., Monog. Culic., V, 257[6], (*alboscutella laps.*).

Ochlerotatus alboscut ellatus, Brun., R. I. M., XVII, 139[6].

Matale (Celyon) (White).

[2, 4, 5]Kuala Lumpur.
[1, 3, 4, 5, 6]Papua.

albotæniatus, Leicester.

Danielsia albotœniata, Leic., in Theo., Entom., XXXVII, 111[1], (1904); Theo., Monog. Culic., IV, 241, fig. 72; Brun., R. I. M., I, 338[2]; *id., loc. cit.,* IV, 458; Liec., Culic. Malaya, 117[3].

Ochlerotatus albotœniatus, Brun., R. I. M., XVII, 139[4]; White, I. J. Med. R., VIII, 320[5].

[5]Matale, Ceylon.

[1, 2, 3, 4]F. M. S.

dorsalis, Meigen.

Culex dorsalis, Mg., Syst. Besch., IV, 242, 18, 2, 3[1], (1818); *id., loc. cit.,* VI, 242, 18, (1830); *id.,* Abbild Ent. Zweifl. Ins., I, Pl. II, 4; Ruthe, Isis, (1831), XI, 1204; Staeg., Kröj., Nat. Tidd., II, 554, 9; Wlk., List. Dipt. B. M., I, 3; Zett., Dipt., Scand., IX, 3465, 12; *id., loc. cit.,* XI, 4344, 12; Sch., F. Aust., II, 626, 3; Wulp., Dipt. Neerl., I, 325, 2; Siebke, Cat. Dipt. Norveg., 192, 7; Kow., Cat. Ins. Fam. Bohem., II, dipt., 5; Piff., E. M. M., (2) VI, (XXXI), 227; Aust., *loc. cit.,* (2) VI, (XXXI), 228; Fic., B. Ent. S. Ital., XXVIII, 264, 14; Wulp. & Meij., New. Naaml. Nederl. Dipt., 23; Thal., F. Reg. Hung., Dipt., 15, 6; Kert., Cat. Dipt., I, 259; Theo., Monog. Culic., II, 16, Pl. XXI, 84.

Grabhamia dorsalis, Theo., Monog. Culic., III, 251; *id., loc. cit.,* V, 277.

Ochlerotatus dorsalis, Edw., B. Ent. R., II, 248; *id., loc. cit.,* VII, 217, fig. 6*b,* [larva]; Waterst., *loc. cit.,* IX, 7, figs. 2, 3.

Peshawar (Pusa Coll.).

---

[1]Europe.
[2, 3, 4, 5]N. Africa.

? **Culex agypti*, Linn., Hassel quist. Palestina Reise, 470, (1762)[2].

Ochlerotatus agypti, Gough, [*nec.* L.], B. Ent. R., V, 135[3].

Grabhamia subtilis, Ed. & Et. Sergent, Bull. Mus. Paris, XI, 240, (1905)[4]; Theo., Monog. Culic., V, 278.

Grabhamia willcocksii, Theo., Monog. Culic., IV, 296, (1907); *id., loc. cit.*, V, 278.

greeni, Theobald.

Howardina greenii, Theo., Monog. Culic, III, 289[1], figs. 160, 161, (1903); *id., loc. cit.*, V, 221[2]; Brun., R. I. M., I, 338[3]; Blanch., Moustiques, 416; Brun., R. I. M., XVII, 131[4].

[1, 2, 3, 4]Peradeniya.
[5]Colombo.
[6]Matale.

Ochlerotatus greenii, James, I. J. Med. R. II, 262[5]; White, *loc. cit.*, VIII, 319[4].

gubernatoris, Giles.

Culex gubernatoris, Giles, Bom. Jo., XIII, 607, (1901), [provis. descr.]; *id.*, Entom., XXXIV, 194; Theo., Monog. Culic., I, 314, fig. 97.

Stegomyia gubernatoris, Giles, Handbook, (2nd edn.) 380; Blanch., Moustiques, 261; Bentley, Malaria in Bombay, 62[5]; Fletcher, Pusa Sci. Rpt., (1915-16), 83[4].

[1, 6]Allahabad.
[3, 4, 6, 10]Pusa.
[6, 8, 9, 10]Purnea.
[5, 6, 11, 12]Bombay.
[2]Colombo., [7]Matale.
———, [6]Philippines.

Phagomyia gubernatoris, Theo., Gen. Ins. fasc., XXVI, 22; *id.*, Monog. Culic., IV, 250; Brun., R. I. M., I, 338[1].

Ochlerotatus gubernatoris, Edw., B. Ent. R., IV, 226; James, I. J. Med. R., II, 262[2]; Fletcher, Pusa Sci. Rpt., (1916-17), 101[3], [breeding places]; Stanton, B. Ent. R., X, 333, [distribution in far East. ports]; Brun., R. I. M., XVII, 136-137[6]; White, I. J. Med. R., VIII, 317[7].

Pseudocarrolia lophoventralis, Theo., R. I. M., IV, 13[8], (1910); *id.*, Monog. Culic., V, 186; Brun., R. I. M., IV, 454[9].

Ochlerotatus lophoventralis, Edw., B. Ent. R., VII, 214, [maintained as good sp.]; Brun., R. I. M., XVII, 140-1[10].

Lepidotomyia (II) *magna*, Theo., Gen. Ins. Fasc. XXVI, 22[11], (1905); *id.*, Monog. Culic., IV, 250; *id., loc. cit.*, V, 250[12].

---

* For another possible identification of this description see *Stegomyia fasciata, F.*

imprimens, Walker.

Culex imprimens, Wlk., P. Linn. S. Lon., V, 144[1], (1861); Giles, Handbook, 262; *id.*, (2nd edn.), 411, [*imprimiens, laps.*]; Blanch., Moustiques, 306; Theo., Gen. Ins. Fasc., XXVI, 26, [*imprimiens, laps.*]; Brun., R. I. M., I, 348[2].

Ochlerotatus imprimens, Edw., B. Ent. R., IV, 228; Brun., R. I. M., XVII, 136[10].

? *Culex auratus,* Leic., Culic. Malaya, 153[3], (1908); Brun., R. I. M., IV 467.

Culicada suknaensis, Theo., R. I. M., IV, 21[7], (1910); *id.*, Monog. Culic., V, 297[8], fig. 139; Brun., R. I. M., IV, 462[9].

[7, 8, 9, 10]Sukna.

[1, 2, 10]Amboina.
[3, 10]Malaya.

jamesi, Edwards.

Ochlerotatus jamesii, Edw., B. Ent. R., V, 77[1], (1914); James, I. J. Med. R., II, 262[2]; Brun., R. I. M., XVII, 141[3].

[1, 2, 3]Colombo.
[3]Galle.

lowisi, Theobald.

Reedomyia lowisii, Theo., Monog. Culic., V, 257[1], figs. 121-124, (1910).

Ochlerotatus lowisii, Brun., R. I. M., XVII, 140[2].

[1, 2]Andamans.
[1, 2]Galle.

mediolineatus, Theobald.

Culex mediolineatus, Theo. Monog. Culic., II, 113[1], (1901); Giles, Handbook, (2nd edn.) 431; Blanch., Moustiques, 369.

Ochlerotatus mediolineatus, Edw., B. Ent. R., IV, 228; Brun., R. I. M., XVII, 137-8[2].

Culex trilineatus, Theo., Monog. Culic., II, 105[3], (1901); *id.*, *loc. cit.*, V, 359; Giles, Handbook, (2nd edn.), 464; Blanch., Moustiques, 330; Brun., R. I. M ., IV, 476[4].

[1, 2, 3, 4]Thayetmyo. (U. Burma).

nigripes, Zetterstedt.

Culex nigripes, Zett., Ins. Lapp., 807, 6, (1838); Staeg., Kröj. Nat. Tid., II, 553, 4; *id.*, *loc. cit.*, n. s., I, 349, 1; Zett., Dipt. Scand., IX, 3458, 5; *id.*, *loc. cit.*, XII, 4836, 5; Holmgr., K. Svensk. Vet. Akad. Handl., VIII, (5), 36; *id.*, Ofr. K. Vet. Akad. Förhandl, XXIX, 104; O.-S., Cat. Dipt. N. Amer., (2nd edn.) 18; Fic., B. Ent. S. Ital., XXVIII, 291, 42; Lundb., Viden.

Meddel., 296, 84; Jacob., Ins. Nov. Zemb., 31, 41, 49; Kert., Cat. Dipt., I, 263[3]; Theo., Monog. Culic., II, 93, fig., 194, 260, [map of distribution]; Brun., R. I. M., I, 350[1]; *id., loc., cit.,* XVII, 160[2].

[1, 2]Deosai Plateau (Kashmir, 13,000 ft.).

---

[3]Circumpolar.

Ochlerotatus nigripes, Lund., Biol. Dan. Culic., 79, Pl. IX, [larva].

Culex caspius, Curt., [*nec.* Pallas], 2nd Voy, J. C. Ross, (App.)., LXXVI, 26, (1831).

? *Culex impiger,* Wlk., List. Dipt. B. M., I, 6, (1848).

? *Culex implacabilis,* Wlk., List. Dipt. B. M., I, 7, (1848).

Culex pipiens, Fb., [*nec.* L.], Faun. Groenld., 209, 171, (1780).

niveoscutellatus, Theobald.

Reedomyia niveoscutellata, Theo., Jo. Econ. Biol., I, 22, Pl. III, 5[1], (1905); *id.,* Monog. Culic., IV, 259, fig. 79; Brun., R. I. M., IV, 487[2].

Ochlerotatus niveoscutellatus, Brun., R. I. M., XVII, 139[3].

[1, 2, 3]India.

[2, 3]Philippines.

niveus, Ludlow.

Stegomyia nivea, Ludl., Jo. N. Y. Ent. S., XI, 139, (1903)[1]; Leic., Culic., Malaya, 87[2].

Scutomyia nivea, Theo. Monog. Culic., V, 203; Brun., R. I. M., I 336[3]; *id., loc. cit.,* IV, 452.

Ochlerotatus niveus, Edw., B. Ent. R., IV, 227; *id., loc. cit.,* VII, 211; Brun., R. I. M., XVII, 139[4].

Aedes niveus, Edw., B. Ent. R., XII, 318[8].

Stegomyia albolateralis, Theo., R. I. M., II, 289, (1908); *id.,* Monog. Culic., IV, Pl. I, Pl. III; *id., loc. cit.,* V, 179, fig. 67; Brun., R. I. M., IV, 444; *id., loc. cit.,* XVII, 140[9].

Stegomyia pseudonivea, Theo., Ann. Mus. Hung., III, 75, (1903); *id.,* Monog. Culic., V, 176[5], fig. 64, 65; Brun., R. I. M., I, 332; *id., loc. cit.,* IV, 448[8].

[9]Lushai Hills.
[3, 5, 7]Andamans.

[9]Sylhet.

[2, 3, 4, 5, 6, 7]F. M. S.
[1, 3, 4]Philippines.
[8]Japan.

oreophilus, Edwards.

Ochlerotatus oreophilus, Edw., B. Ent. R., VI, 357[1], (1916); Pusa Bull., 89, 58; Brun., R. I. M., XVII, 141[2].

[1, 2]Murree Hills.

ostentatio, Leicester.

Aioretomyia ostentatio, Leic., Culic Malaya, 193[1], (1908); Brun., R. I. M., IV, 490.

[2, 3]Peradeniya.

[1, 2]F. M. S.

Ochlerotatus ostentatio, Edw., B. Ent. R., IV, 228; Brun., R. I. M., XVIII, 140[2].

Pseudohowardina chrysoscuta, Theo., Monog. Culic., V, 228, fig. 94[3], (1910).

pallidostriatus, Theobald.

Culex pallidostriatus, Theo., Monog. Culic., IV, 410[1], fig. 175, (1907); *id., loc. cit.,* V, 362[2].

Ochlerotatus pallidostriatus, Edw., B. Ent. R., IV, 228; Brun., R. I. M., XVII, 140[3].

Ochlerotatus ochraceus, [*p. p*], Edw. [*nec.* Theo.], B. Ent. R., II, 250, (1911).

Culex parascelos, Theo., R. I. M., IV, 18, (1910)[4]; *id.,* Monog. Culic., V, 379[5]; Brun., R. I. M., IV, 473[6].

[3, 4, 5, 6]Madras City.
[2, 3]Peradeniya.
[1, 2, 3]India.

piperselatus, Giles.

Stegomyia piperselata, Giles in Theo., Monog. Culic. II, 316, (1901); *id.,* Handbook, (2nd edn.), 372, [female only], Pl. XVI, 1*a, b;* Theo., Monog. Culic., V, 607[1]; Blanch., Moustiques, 264, [female only]; Brun., R. I. M., I, 332[2]; *id., loc. cit.,* IV, 448.

Ochlerotatus piperselatus, Edw., B. Ent. R., IV, 227; James, I. J. Med. R., II, 262[3]; Brun., R. I. M., VII, 138[4].

Pseudograbhamia maculata., Theo., Bom. Jo., XVI, 243[5], (1905); *id.,* Monog. Culic., IV, 314, fig. 109; Brun., R. I. M., IV, 460[6]; Fletcher, Pusa Sci. Rept., (1915-16), 83[7].

[1, 2, 4]Jhansi.
[1, 2, 4]Gonda., [7]Pusa.
[4, 6]Madras City.
[4, 5]Galgamuwa (Ceylon)., [3]Colombo.

pseudotaeniatus, Giles.

Stegomyia pseudotaeniatus, Giles, Bom. Jo., XIII, 606, (1901), [provis. descr.]; *id.,* Entom., XXXIV, 192.

Stegomyia pseudotaeniata, Giles, Handbook, (2nd edn.), 379, Pl. XIV, 8, 10; Theo., Monog. Culic. I, 312 fig. 96, [larva]; *id., loc. cit.,* III, fig. 16, [larva]; Blanch., Moustiques, 261.

Hulecoetomyia pseudotaeniata, Theo., R. I. M., II, 291; Brun., R. I. M., I, 337; *id., loc. cit.,* IV, 455[2]; Theo., Monog. Culic., V, 222[3], figs. 89-91.

Ochlerotatus pseudotaeniatus, Prashad, I. J. Med. R., III, 500, Pl. XXX, fig. 3, [male genitalia]; *id., loc. cit.,* III, 503, Pl. XXXI, [halteres].

Aedes pseudotaeniatus, Brun., R. I. M., XVII, 142[4].

[1, 4]Bakloh.
[2, 3, 4]Dehra Dun.
[1, 4]Naini Tal.
[2, 3, 4]Lushai Hills.

[1, 3, 4]Philippines.

pulchriventer, Giles.

Culex pulchriventer, Giles, Bom. Jo., XIII, 608[1], (1901) [provis. descr.]; *id.,* Handbook, (2nd edn.), 449, Pl. XVII; Theo., Monog. Culic., II, 48, Pl. XXIII, figs. 92, 170, 173, [larva, etc.]; Blanch., Moustiques, 338; Brun., R. I. M., 351[2].

[1, 2, 3, 4, 5]Naini Tal.

Ochlerotatus pulchriventer, Edw., B. Ent. R., IV, 228; Brun., R. I, M., XVII, 137[3].

Howardina himalayana, Giles, Jo. Trop. Med., VII, 384, (1904); Brun., R. I. M., IV, 456[4]; Theo., Monog. Culic., V, 221[5].

taeniorhynchoides, Christophers.

Leslieomyia tæniorhynchoides, Chris., Paludism, II, 68, (1911).

[1, 2]India.

Ochlerotatus taeniorhynchoides Edw., B. Ent. R., IV, 227; Brun., R. I. M., XVII, 141[1].

Pecomyia maculata, Theo.,]*nec.* Mg.], Jo. Econ. Biol., I, 24[1], Pl. IV, 7, (1905); *id.,* Monog. Culic., IV, 266; figs. 82-84.

vexans, Meigen.

Culex vexans, Mg., Syst. Besch., VI, 241[1], (1830).

Ochlerotatus vexans, Edw., B. Ent. R., VII, 218; Brun., R. I. M., XVII, 135-136[2] [notes on varieties].

Aedes (*Ecculex*) *vexans,* Dyar, Ins. Ins. Mens., VIII, 183[3].

Culex articulatus, Rond., B. Ent. S. Ital., IV, 70[4], (1832).

Aedes enochrus, How., Dyar & Knab. Mosq. N. Amer., IV, 716[5], (1907).

[2, 9]India.
[2, 6, 13, 14, 15]Ceylon.
[17]Madras City.

[1, 4]Europe.
[5, 11, 16]N. America.
[12]Fiji., [8]S. Africa.
[3, 7]Philippines.
[10]Japan., [17]China.

Culicada eruthrosops Theo., Monog. Culic. V 299, fig. 140[6], (1910).

Culex hirsutum, Ludl., Psyche, XVIII, 126, (1911)[7]; Theo., Monog. Culic. V, 344[8].

Culicada minuta, Theo., Monog. Culic., IV, 338, (1907)[9].

Culex montcalmi, Blanch., Moustiques, 407, (1905)[11]; Theo., Monog. Culic., V, 347, [sinks in *sylvestris*].

Culicada nipponii, Theo., Monog. Culic., IV, 337, (1907)[10].

? *Culex nocturnus.,* Theo., Monog. Culic., III, 159, (1903)[12].

Culex pseudostenœtrus, Theo., Monog. Culic., V, 343, fig. 154, (1910)[15].

Culex stenœtrus, Theo., Monog. Culic., IV, 395, (1907)[13].

Ochlerotatus stenœtrus, Edw., B. Ent. R., IV, 228; James, I. J. Med. R., II, 262[14].

Culex sylvestris, Theo., Monog. Culic., I, 406, (1901)[16].

Culex vagans, Theo., (*nec* Wied.), Monog. Culic., I, 411, (1901)[17].

15. Theobaldia, Neveu-Lemaire.

Theobaldia, Nev.-Lem., C. R. Soc. Biol., Paris, LIV, 1331, (1902) [Genotype, *Culex annulatus,* Schrank.]; Theo., Monog. Culic., III, 148, Pl. X; *id., loc. cit.,* V, 270, [table of spp.]; *id.,* Gen. Ins. fasc., XXVI, 23; Edw., B. Ent. R., II, 251; Brun., R. I. M., I, 339; *id., loc. cit.,* IV, 459; *id., loc. cit.,* X, 72; *id., loc. cit.,* XVII, 169.

Quliceta, Felt, Jo. N. Y. Mus., Bull. 79; App. 391*e*, (1904) [Genotype, *Culex absobnnus,* Ft.].

Pseudotheobaldia, Theo., Monog. Culic., IV, 271, (1907) [Genotype, *P. niveitœniata,* Theo.].

Theobaldinella, Blanch., Moustiques, 390, (1905), [nom. nov. for *Theobaldia,* supposed præocc.]; Brun., R I. M., X, 72.

annulata, Schrank.

Culex annulatus, Schrank, Beit, Naturg., 97[1], (1776); *id.,* Ins. Aust., 984; Fb., Ent. Syst., IV, 400; *id.,* Syst. Antl., 35; Mg. Syst. Besch., I, 4; *id.,* Klass., 2; Mq. H. N. Dipt., I, 35; Sch., F. Aust., II, 626; O. S., Cat. Dipt. N. Am., 18[2]; Wulp. Dipt. Neerl, 325; Theo., Monog. Culic., I, 331, fig. 108, Pl. XV, 58; Meinert, Euceph. Myggelarv., 376, Pl. I, 16, [larva]; Giles, Handbook, (2nd edn.), 391, Pl. XV; Blanch., Moustiques, 280, fig. 206, Coq., Circ. No. 40; *id.,* Bull., 25, 20, (note); How., Mosq., 80; Ludl., Jo. N. Y. Ent. S., X, 31; O.-S., Biol. Cent. Amer., Dipt., I, 5[5].

Theobaldia annulata, Theo., Monog. Culic., III, 148; *id.,* Gen. Ins. fasc., XXVI, Pl. I, 12; Brun., R. I. M., I, 339[4]; *id., loc. cit.,* IV, 459; *id., loc. cit.,* XVII, 169[6]; Ald., Cat. N. Amer. Dipt., 126[7]; Waterst., B. Ent. R., IX 9[8]; Barraud, *loc. cit.,* X, 323[9]; *id., loc. cit.,* XI, 395[10]; Edw., *loc. cit.,* X, 137.

[4, 6]Bakloh (Punjab).

[1, 6, 8, 13, 14]Europe.
[2, 7]N. America.
[3, 6]Mexico.
[10]Palestine.
[9]Mesopotamia.

Culex affinis, Steph., Zool. Jo., IV, (1825)[13]; Staeg., Dipt. Dan., 554; Zett., Ins. Lapp., 806; *id.,* Dipt. Scand., IX, 3460.

Culex variegatus, Schrank, Enum. Ins. Aust., 482[14], (1781).

glaphyroptera, Schiner.

Culex glaphyropterus, Sch., F. Aust., II, 628[1], (1864); Fic., B. Ent. S. Ital., XXVIII, 247, 4; *id., loc. cit.,* XXXI, 192, 9, figs. 70, 71; Kert., Cat. Dipt., I, 260; Giles, Handbook, 216; Theo., Monog. Culic., I, 347, 1.

Theobaldia, glaphyroptera, Theo., Monog. Culic., V, 276[2]; Edw., B. Ent. R., IV, 237; Brun., R. I. M., XVII, 169-170[3].

Pseudotheobaldia niveitæniata, Theo., Monog. Culic., IV, 272[2], figs. 87-88; Brun., R. I. M., IV, 478.

[2, 3]Dehra Dun.
[2, 3]Theog.

[1, 2, 3]C.&W. Europe.

indica, Edwards.

Theobaldia indica, Edw., B. Ent. R., X, 137, (1920)[1].

[1]Bakloh., [1]Ambala.
[1]Dalhousie.

spathipalpis, Rondani.

Culex spathipalpis, Rond., B. Ent. S. Ital., IV, 31, (1872); Theo., Monog. Culic., I, 339; Giles, Handbook, (2nd edn.), 392[3], Pl. XV, 23-25; Fic., B. Ent. S. Ital., XXI, 86; *id., loc., cit.,* XXVIII, 242, Pl. II, 8, 9, Pl. III, 7, 12, Pl. IV, 18, 21, 22, Pl. V, 31, 34; *id., loc., cit.,* XXXI, 94, figs. 72-75, 146; Blanch., Moustiques, 283, f. 209.

Theobaldia spathipalpis, Theo., Monog. Culic., III, 154, Pl. X; *id., loc., cit.,* IV, 276, [larva described]; *id., loc., cit.,* V, 273[1]; Brun., R. I. M., I, 340[2]; *id., loc., cit.,* IV, 459; *id., loc., cit.,* XVII, 170[4]; Prashad, I. J. Med. R., III, 500, Pl. XXX, fig. 5, [male genitalia]; Pringault, B. S. Path. Exot., XIV, 163, [abst. R. A. E. (B) IX, 107][5] [carrying *Micrococcus melitensis*].

Culex serratipes, Becker, Mitt. Zool. Mus. Berl., IV, 78[6], (1908).

[2, 3]Naini Tal.

[1, 2, 4, 5]S. Europe.
[1, 4 6]Teneriffe.
[1, 4]Maderia.
[1, 2, 4]N. & S. Africa.
[4]Palestine.

16. Orthopodomyia, Theobald.

Orthopodomyia, Theo., Entom., XXXVII, 236, (1904) [Genotype, *O. albipes,* Theo.]; *id.,* Monog. Culic., IV, 527; *id., loc., cit.,* V, 470, [table of spp.]; Leic., Culic. Malaya, 175; Brun., R. I. M., I, 362; *id., loc., cit.,*

IV, 486; *id., loc., cit.*, X, 67; *id., loc., cit.*, XVII, 176; Edw., B. Ent. R., IV, 239; How. Dyar & Knab, Mosq. N. Amer., IV, 778.

Bancroftia, Lutz in Borroul, Mosq. do Braz., 59, (1904) [Genotype, *B. albicosta,* Ltz.]; Theo., Monog. Culic., V, 469.

anopheloides, Giles.

Mansonia anopheloides, Giles, Jo. Trop. Med. VI, 315, (1903).

Finlaya anopheloides, Brun., R. I. M., I, 361.

Orthopodomyia anopheloides, Edw., B. Ent., R., IV, 239[2]; James, I. J. Med. R., II, 263[3]; Brun., R. I. M., XVII, 176, 177[3]; White, I. J. Med. R., VIII, 317[4].

Orthopodomyia albipes, Leic. in Theo., Entom., XXXVII, 237[5], (1904); Theo., Monog. Culic., IV, 527; Leic., Culic., Malaya, 176; Brun., R. I. M., I, 362[6]; *id., loc., cit.*, IV, 486; *id., loc., cit.*, XVII, 177[7].

Orthopodomyia albipes var. *nigritarsis,* Leic., Culic. Malaya, 177[8], (1908); Brun., R. I. M., IV, 486[9].

Orthopodomyia maculata, Theo., R. I. M., IV, 29[11], (1910); *id.*, Monog. Culic., V, 473[12]; Brun., R. I. M., IV, 486[13].

Orthopodomyia maculipes, Theo., Monog. Culic., V, 470[10], figs. 208-210, (1910); Brun., R. I. M., IV, 486.

[3, 10, 11, 12, 13]Travancore.
[2, 4, 10]Ceylon.
[10]Andamans.

[3, 5, 6, 7, 8, 9]F. M. S.

17. Tæniorhynchus, Arribalzaga.

Tæniorhynchus, Arrib., Rev. Mus. la Plata, II, 147, (1891) [Genotype, *Culex titillans,* Wlk., as *tæniorhynchus,* Wd.]; How., Dyar & Knab. Mosq. N. Amer., III (1), 502, [regarding genotype]; Theo., Monog. Culic., II, 190, [restricted]; *id., loc., cit.*, IV, 483; *id., loc., cit.*, V, 419, [table of spp.]; *id.*, Gen. Ins. fasc., XXVI, 30, [modified]; *id.*, Nov. Culic., I, 18, [sinks *Mansonia*]; Giles, Handbook, (2nd edn.), 358; Blanch., Moustiques, 381, fig. 244; Leic., Culic. Malaya, 163; Edw., B. Ent. R., II, 251; Brun., R. I. M., I, 356; *id., loc., cit.*, IV, 478; *id., loc., cit.*, X, 72; *id., loc., cit.*, XVII, 147-8; Patton & Cragg, T. B. Med. Ent., 214, [general account].

Coquilletidia, Dyar, P. Ent. S. Wash., VII, 47, (1905) [Genotype, *Culex perturbans.,* Wlk.].

Mansonia, Blanch., C. R. Soc. Biol., Paris, LIII, 1046, [*nom. nov.* for *Panoplites,* prœcc.]; *id.,* Moustiques, 375, 356, [table of spp.]; Theo., Gen. Ins. fasc., XXVI, 31; *id.,* Monog. Culic., IV, 494; *id., loc. cit.,* V, 446, [table of spp.]; Leic., Culic. Malaya, 171; Brun., R. I. M., I, 358; *id., loc. cit.,* IV, 481; *id., loc. cit.,* X, 64; Martin, Le Boeuf & Roub., B. S. Path. Exot., [abst. Jo. T. Vet. Sci., III, 484-6], [trans. of 'nagana' by this genus].

Panoplites, Theo., Rpt. Mosq. Coll. B. M., 5, (1900) [Genotype, *Culex titillans,* Wlk. as *T. tœniorhynchus,* Wd.]; Giles, Handbook, (2nd edn.), 350.

Rhynchotaenia, Brethes, Ann. Mus. Buen. Ayers, XX, 470, (1910) [Genytope, *T. fasciolatus,* Arrib.].

brevicellulus, Theobald.

Taeniorhynchus brevicellulus, Theo., Monog. Culic., II, 212, figs. 255, 256, Pl. XXIII, 89, (1901); *id., loc. cit.,* III, 268; Leic., Culic. Malaya, 163[6]; Brun., R. I. M., I, 357[1]; Edw., B. Ent. R., IV, 230; Brun., R. I. M., XVII[2], 148-9; James, I. J. Med. R., II, 262[3]; Stanton, *loc. cit.,* III, 257[4].

Tœniorhynchus brevicellula, Giles, Handbook, (2nd edn.), 363; Blanch., Moustiques, 389.

Chrysoconops brevicellulus, Theo., Monog. Culic., V, 438; Brun., R. I. M., IV, 481[5].

Tœniorhynchus acer, Theo., (*nec.* Wlk.), Monog. Culic., II, 21, (1901)[7].

Chrysoconops fuscopteron, Theo., Tijd., v. Ent., LIV, 239, (1911)[8].

Chrysoconops pygmœus, Theo., R. I. M., IV, 300, (1908)[9]; *id.,* Monog. Culic., IV, 25, Pl. I, Pl. III; *id., loc. cit.,* V, 435[10], figs. 192-195; Brun., R. I. M., IV, 481[11].

[2, 5, 9, 10, 11]Sylhet.
[2, 5]Manipur.
[1, 2]U. Burma.
[2, 5, 10, 11]Bengal.
[2, 3, 5]Ceylon.

[1, 2, 6]F. M. S.
[2, 5]Philippines.
[2, 4, 8]Sumatra., [7]Papua.

ochraceus, Theobald.

Tœniorhynchus ochraceus, Theo., Monog. Culic., III, 263, fig. 140[1], (1903); Leic., Culic. Malaya, 164[2]; Brun., R. I. M., 358[3].

Chrysoconops ochraceus, Theo., Monog. Culic., V, 435[4]; Brun., R. I . M., IV, 481.

Ochlerotatus ochraceus, (*p. p.*), Edw., B. Ent., R., II, 250; Brun., R. I. M., XVII, 138[5], (*p. p.*).

Cherrapunji (Mrs. Fletcher.

[1, 2, 3, 4, 5]Kuala Lumpur.

18. **Mansonioides,** Theobald.

Mansonioides, Theo., Monog. Culic, IV, 498, (1907) [Genotype, *M. 7-guttata,* Theo.]; *id., loc. cit.,* V, 452, [table of spp.]; Brun., R. I. M., IV, 483; *id., loc. cit.,* X, 64; *id., loc. cit.,* XVII, 149; Edw., B. Ent. R., II, 253; Patton & Cragg, T. B. Med. Ent., 215.

annulifera, Theobald.

Panoplites annulifera, Theo., Monog. Culic., II, 183, fig. 224, Pl. XXX, 120, (1901); Giles, Handbook, (2nd edn.), 356, Pl. XIII, 8.

Mansonia annulifera, Theo., Monog. Culic., III, 274; Blanch., Moustiques, 380; Leic., Culic. Malaya, 174; Brun., R. I. M., I, 359[1]; *id., loc. cit.,* IV, 482; Green, Spol. Zeyl., IV, 183[2].

Mansonioides annuliferus, Edw., B. Ent. R., IV, 230; James, I. J. Med. R., II, 262[3]; Stanton, *loc. cit.,* III, 257; Iyengar, *loc. cit.,* 1920 Sci., Cong. No., 9[4]; Stanton, B. Ent. R., X, 333, [distribution in far East. ports].

Mansonioides annulifera, Brun., R. I. M., XVII, 150[5].

Mansonioides 7-guttata, Theo., Monog. Culic., IV, 499[6], figs. 226, 227, (1907); Cruick. & Wright, I. J. Med. R., I, 773[7]; Brun., R. I. M., IV, 483[8].

Mansonia 7-punctata, Theo., Ann. Mus. Hung., III, 187[9], (1905); *id.,* Monog. Culic., IV, 494[10]; Brun., R. I. M., I, 359[11].

[5]Sylhet.
[5]Manipur.
[1, 4, 5]Bengal.
[1, 5]Madras.
[1, 5, 7]Malabar Coast.
[1, 2, 3, 5]Ceylon.
[1, 5]F. M. S.
[5, 6, 8]Borneo.
[1, 5]Philippines.
[5, 9, 10, 11]Papua.

annulipes, Walker.

Culex annulipes, Wlk., P. Linn., S. Lon., I, 6, (1857).

Panoplites annulipes, Theo., Monog. Culic., II, 185, Pl. XXX, 119.

Mansonia annulipes, Blanch., Moustiques, 380; Leic., Culic. Malaya, 172; Brun., R. I. M., I, 359[1]; *id., loc. cit.,* IV, 482; Theo., Monog. Culic., V, 447[2].

Mansonioides annulipes, Stanton, I. J. Med. R., III, 257[3]; *id.,* B. Ent. R., X, 333, [distribution in far East. ports].

Tœniorhynchus annulipes, Brun., R. I. M., XVII, 148[4].

Culex dives, Sch., Novara Reise, 31[6], (1868); Giles, Handbook, 261.

[4]Calcutta.
Assam (White).

[1, 2, 4] F. M. S.
[1, 2, 4, 5, 6]Java
[3]Sumatra.
[1, 2, 4]Philippines.

Culex nero, Dol., Nat. Tijd. Ned. Ind., XIV, 383[5], Pl. V, 3, (1857); Giles, Handbook, 284.

uniformis, Theobald.

Panoplites uniformis, Theo., Monog. Culic., II, 180, Pl. XXX, 118, (1901); Giles, Handbook, (2nd edn.), 353, Pl. XIII, *a-d.*

Mansonia uniformis, Theo., Monog. Culic., III, 270, fig. 144; Blanch., Moustiques, 379; Brun., R. I. M., I, 359[2]; *id., loc. cit.,* IV, 483[3]; Leic., Culic. Malaya, 171; Green, Spol. Zeyl., IV, 183[4]; Bentley, Malaria in Bombay, 62[5]; Theo., Monog. Culic., V, 448[6].

Mansonioides uniformis, James, I. J. Med. R., II, 262[7]; Simpson, B. Ent. R., II, 216; Edw., *loc. cit.,* II, 254; *id., loc. cit.,* IV, 51, fig. 1, [male genitalia]; McFarl., *loc. cit.,* VI, 67[9]; Stanton, *loc. cit.,* X, 333, [distribution in far East. ports]; Edw., *loc. cit.,* XI, 135; Heck. & Blanch., T. Vet. B., I, 287, [trypanosomes]; Johnson, B. Ent. R., IX, 325-332, [notes in Nigeria]; Stanton, I. J. Med. R., III, 257; Ingram, B. Ent. R., III, 75, fig. 1, [larva and pupa].

[3, 6]Manipur.
[3, 6]Sylhet.
[3, 6]Bengal., [3]Orissa.
[1, 2]Shahjahanpur.
[2, 3, 6]Malabar Coast.
[4, 6, 7]Ceylon.
[3]Rangoon., [5]Bombay.

[2, 3, 6]Philippines.
[3]Sumatra.
[2]Papua.
[9]Hongkong.
[6]Madagascar.
[3, 6, 11]Africa generally.
[3, 6]F. M. S.
[3, 6, 12]Australia.
[13, 14]Tonkin.

Panoplites africanus var. *reversus,* Theo., Monog. Culic., II, 189, (1901)[11].

Panoplites australiensis, Giles, Handbook, (2nd edn.), 355, (1902)[12].

? *Panoplites seguini,* Lav., C. R. Soc. Biol., Paris, LIII, 992 (1901)[13].

? *Mansonia seguini,* Blanch., Moustiques, 380; Theo., Monog. Culic., IV, 496[14].

19. Banksinella, Theobald.

Banksinella, Theo., Monog. Culic. IV, 468, (1907) [Genotype, *Culex luteolateralis,* Theo.]; Brun., R. I. M., IV, 477, [*Banksiella, laps.*]; *id., loc. cit.,* X, 56; Theo., Monog. Culic., V, 402, [table of spp.].

luteolateralis, Theobald.

Culex luteolateralis, Theo., Monog. Culic, II, 71[2], Pl. XXVII, 108, (1901); Giles, Handbook, (2nd edn.), 488; Blanch., Moustiques, 278; Leic., Culic. Malaya, 160[4]; Brun., R. I. M., I, 349[1].

[3, 6]Sylhet., [5]Colombo.

[1, 3, 4, 6]F. M. S.
[1, 6, 7]Philippines.

Banksinella luteolateralis, Theo., Monog. Culic. IV, 469; Brun., R. I. M., IV, 477[3], [*Banksiella, laps.*]; Edw., B. Ent. R., II, 245; *id., loc. cit.*, III, 6; *id., loc. cit.*, V, 274, [? *lineatopennis,* a good sp.]; James, I. J. Med. R., II, 263[5]; Brun., R. I. M., XVII, 124[6]. — [1, 2, 6]Africa generally.

Tœniorhynchus lineatopennis, Ludl., Can. Ent., XXXVII, 133[7]; (1905); Theo., Monog. Culic., IV, 489; Brun., R. I. M., IV, 479.

Banksinella lineatopennis, Stanton, B. Ent. R., X, 333, [distribution in far East. ports].

20. **Culiciomyia,** Theobald.

Culiciomyia, Theo., Monog. Culic., IV 227, (1907) [Genotype, *Culex inornata,* Theo.]; *id., loc. cit.*, V, 229, [table of spp.]; Brun., R. I. M., IV, 456; *id., loc. cit.*, X, 58; Edw., B. Ent. R., IV, 234; Brun., R. I. M., XVII, 151.

Neomelanoconion, Theo., Monog. Culic., IV, 514, (1907) [male only] [Genotype, *Culex rima,* Theo.]; Edw., B. Ent. R., II, 254.

Pectinopalpus, Theo., A. M. N. H. (8), V, 375, (1910) [Genotype, *P. fuscus,* Theo.].

Trichorhynchus, Theo., Bom., Jo., XVI, 241, (1905) [Genotype, *T. fuscus,* Theo.]; *id.*, Monog. Culic., IV, 270; Brun., R. I. M., I, 356; *id., loc. cit.*, X, 73.

Trichorhynchomyia, Brun., R. I. M., IV, 477, (1912) [*nom. nov.* for *Trichorhynchus,* præomcc.]; *id., loc. cit.*, X, 73.

bahri, Edwards.

Culiciomyia bahri, Edw., B. Ent. R., V, 79[1], (1914); Brun., R. I. M., XVII, 152[2]. — [1,2]Badulla. [1, 2]Hakgalla. [1, 2]Peradeniya.

fragilis, Ludlow.

Culex tragilis, Ludl., Jo. N. Y. Ent. S., XI, 142, (1903); Theo., Monog. Culic., IV, 424; Brun. R. I. M., I, 345[1]; *id., loc. cit.*, IV, 469[2]; *id., loc. cit.*, XVII, 164[3]. — [4, 5, 6, 7, 8]Ceylon (generally)., —— [9]Sarawak. [1, 2, 3]Philippines.

Culiciomyia fragilis, Edw., B. Ent. R., V, 79, [note].

Culiciomyia ceylonica, Theo., Monog. Culic., IV, 236, fig. 70, (1907); Brun., R. I. M., IV, 457[4]; James, I. J. Med. R., II, 262[5]; Brun., R. I. M., XVII, 152[6].

Trichorhynchus fuscus, Theo., Bom. Jo., XVI, 242, Pl. A, 2, (1905); *id.*, Monog. Culic., IV, 270; Edw., B. Ent. R., III, 33; *id., loc. cit.*, IV, 235; Brun., R. I. M., I, 356[7].

Trichorhynchomyia fuscus, Brun., R. I. M., IV, 477.

Culiciomyia fusca, Edw., B. Ent. R., IV, 235; Brun., R. I. M., XVII, 152; White, I. J. Med. R., VIII, 318[8].

Culiciomyia inornata, Theo., Monog. Culic., IV, 227, figs. 61-63, (1907); Brun., R. I. M., IV, 457[9].

pallidithorax, Theobald.

Culex pallidithorax, Theo., Jo. Econ. Biol., I, 32, (1905); Brun., R. I. M., I, 350[1].

Culex pallidothorax, Theo., Monog Culic., IV, 446 [*emend.*]; Brun., R. I. M., IV, 473.

Culiciomyia pallidothorax, Edw., B. Ent. R., IV, 235; James, I. J. Med. R., II, 262[3]; McFarl., B. R. VI, 67[4]; Brun., R. I. M., XVII, 151[5]; White, I. J. Med. R., VIII, 318[6].

Culex albopleura, Theo., Monog. Culic., IV, 456, (1907); Brun., R. I. M., IV, 466[7].

Culiciomyia annuloabdominalis, Theo., Monog. Culic., V, 236[8], figs. 102, 103, (1910).

[1, 5, 7]India.
[3, 5, 6, 8]Ceylon (generally).,———
[4]Hongkong.

pulla, Theobald.

Culex pullus, Theo., Ann. Mus. Hung., III, 87[1], fig. 6, (1905); Brun., R. I. M., I, 351[2].

Culiciomyia pulla, Theo., Monog. Culic., IV, 232[3], fig. 66; Brun., R. I. M., IV, 457; James, I. J., Med. R., II, 262[4]; Brun., R. I. M., XVII, 152[5].

[4]Colombo.

[1, 2, 3, 5]Papua.

viridiventer, Giles.

Culex viridiventer, Giles, Bom., Jo., XIII, 609, (1901)[1] [provis. desc.]; Theo., Monog. Culic., II, 128, figs. 219, 220, Pl. XXXIX, 116; Giles, Handbook, (2nd edn.,) 345, Pl. XVII, 12; Blanch., Moustiques, 346; Brun., R. I. M., I, 354[2].

Culiciomyia viridiventer, Edw., B. Ent. R., IV, 235; Brun., R. I. M., XVII, 151[3]; White, I. J. Med. R., VIII, 319[4].

Culex angulata, Theo., Monog. Culic., II, 324, (1901); Brun., R. I. M., I, 342[5].

[1, 2, 3, 5]Naini Tal.
[3, 6, 7]Nepal.
[4]Matale, Ceylon.

Culex angulatus, Blanch., Moustiques, 362; Brun., R. I. M., IV, 466.

Culex longifurccatus, Theo., [*nec* Becker], R. I. M., IV, 19[6], (1910).

Culex pseudolongifurcatus, Theo., Monog. Culic., V, 366, (1910)[7], [*nom. nov.*]; Brun., R. I. M., IV, 474.

21. **Lophoceratomyia,** Theobald.

Lophoceratomyia, Theo., Ann. Mus. Hung., III, 93, (1905) [Genotype, *L. fraudatrix,* Theo; *id.,* Bom. Jo., XVI, 245; *id.,* Monog. Culic., IV, 471; Leic., Culic., Malaya, 119; Brun., R. I. M., X, 64; Edw., B. Ent. R., VII, 226, [key to genus]; Brun., R. I. M., XVIII, 152; Edw., B. Ent. R., XII, 79.

Cyathomyia, Meij., Ann. Bot. Jard. Buitenzg, 2), Supp. III, 921, (1910) [Genotype, *Culex jenseni,* Meij.]; Edw., B. Ent. R., IV, 59; Brun., R. I. M., XVII, 168.

Oculeomyia (p. p.), Theo., Monog. Culic, IV, 515, (1907) [Genotype, *O. sarawaki,* Theo.].

Philodendromyia, Theo., Monog. Culic., VI, 515, (1907) [Genotype, *P. barkerii,* Theo.].

mammilifer, Leicester.

Lophoceratomyia mammilifer, Leic., Culic., Malaya, 128[1], (1908); Brun., R. I. M., IV, 464; Edw., B. Ent. R., IV, 236; Brun., R. I. M., XVII, 154[2]; White, I. J. Med. R., VIII, 317[3].

Lophoceratomyia bicornuta, Theo., R. I. M., IV, 25[4], (1910) *id.,* Monog. Culic., V, 412[5], figs. 178-182; Brun., R. I. M., IV, 463[6].

[2, 4, 5, 6]Kawkareik (L. Burma).
[3]Matale (Ceylon).

[1, 2]F. M. S.

minutissima, Theobald.

Culiciomyia minutissima, Theo., Monog. Culic., IV, 235, (1907); Brun., R. I. M., IV, 457[2].

Lophoceratomyia minutissima, Edw., B. Ent. R., IV, 235; James, I. J. Med. R., II, 262[3]; Edw., B. Ent. R., V, 80, (note); McFarl., B. Ent. R., VI, 67[4]; Brun., R. I. M., XVII, 153[5]; White, I. J. Med. R., VIII, 317[6].

Melanoconion juxtapallidiceps, Theo., Monog. Culic., V, 456, (1910); Brun., R. I. M., IV, 484[8].

[1, 2, 3, 5, 6, 7, 8, 9, 10]Ceylon (general).

---

[4]Hongkong.

Culiciomyia nigerima, Theo., Monog. Culic., V, 233, fig. 100, (1910); Brun. R. I. M., IV, 457[10].

tæniata, Leicester.

Lophoceratomyia tæniata, Leic., Culic., Malaya, 127, (1908)[1]; Brun., R. I. M., IV, 465[2]; James, I. J. Med. R., II, 262[3]; Brun., R. I. M., XVII, 154[4].

[3]Colombo.

[1, 2, 4]F. M. S.

uniformis, Theobald.

Lophoceratomyia, uniformits, Theo., Ann. Mus. Hung., III, 93[1], Pl. A, Pl. B, (1905); *id.,* Bom. Jo., XVI, 245; *id.,* Monog. Culic., IV, 473; Edw., B. Ent. R., V, 80; *id., loc. cit.,* VII, 227; James, I. J. Med. R., II, 262[3]; Brun., R. I. M., XVII, 153[4]; White, I. J. Med. R., VIII, 317[5].

[1, 2, 3, 4, 5]Ceylon (general).

22. Protomelanoconion, Theobald.

Protomelanoconion, Theo., Monog. Culic., V, 462, (1910) [Genotype, *P. fusca,* Theo.]; Brun., R. I. M., X, 69, Edw., B. Ent. R., XII, 79.

brevipalpis, Giles.

Stegomyia brevipalpis, Giles, Handbook, (2nd edn.), 384[1], Pl. XIV, 17-20, (1902); Blanch., Moustiques, 264; Brun., R. I. M., I, 329[2]; *id., loc. cit.,* IV, 415.

Culex brevipalpis, Theo., Monog. Culic., III, 146; *id., loc. cit.,* V, 607.

Cyathomyia brevipalpis, Edw., B. Ent. R., IV, 237; James, I. J. Med. R., II, 262[3]; Stanton, B. Ent. R., X, 333[7], [distribution in far East. ports]; Fletcher, Pusa Sci. Rpt., (1916-17), 101[4], [bionomics]; Brun., R. I. M., XVII, 168-9[5]; White, I. J. Med. R., VIII, 317[6].

Protomelanoconion brevipalpis, Edw., B. Ent. R., XII, 79.

Melanoconion uniformis, Leic., Culic, Malaya, 136[8]; Brun., R. I. M., IV, 484.

[1, 2, 5]Shahjahanpur.
[4]Pusa.
[3]Colombo.
[6]Matale (Ceylon).

---

[7]Siam.
[7]Cochin China.
[8]F. M. S.

23. Culex, Linnaeus.

Culex L., Syst. Nat. Ed., X, 602, (1758) [Genotype, *Culex pipiens,* L.]; Mg., Illig. Mag., II, 260; *id.,* Syst. Besch., I, 1; R.-D., Mem. S. H. Nat., Paris, III, 403; Zett., Dipt. Scand., IX, 3453; Sch., F. Aust., II, 625; Meinert, Eucephale Mygelarver, in Vidensk, Selsk., 6 R., Nat. og Math., Afd., III, (4), 1886, 375, [literature of transformations, etc.]; Giles, Handbook, 186; *id., loc. cit.,* (2nd edn.), 386; Theo., Monog. Culic., I, 326; *id., loc. cit.,* IV, 487; *id., loc. cit.,* V, 323, [table of spp.]; *id.,* Gen. Ins. fasc., XXVI, 24, 25, [list of spp. of World]; Blanch., Moustiques, 267, 269, [table of spp.], 627, [supply, table], 372, [list of uncertain spp.]; Leic., Culic., Malaya, 138, [table of groups]; Brun., R. I. M., I, 342; *id., loc. cit.,* IV, 465; *id., loc. cit.,* X, 39, 42, 43; *id., loc. cit.,* XVII, 155-157; Alcock, Ent. Med. Off. 98; How., Dyar & Knab, Mosq. N. Amer., III, (1), 215, 217; Edw., B. Ent. R., V, 63, [male hypopygium descr.]; Patton & Cragg, T. B. Med. Ent., Pl. XX, 1, [respiratory system], Pl. XXII, 7, [œsophagus], Pl. XXIX, 2, [ovarian tube] Pl. XXIII, [alimentary tract], Pl. XXXVIII, 3, [male genitalia] p. 214, [general account].

Aedinus, Lutz in Peryassu, Os. Culic. Braz., 36, (1908) [Genotype, *A. amazonensis,* Ltz.].

Aporoculex, Theo., Monog. Culic., IV, 316, (1907) [Genotype, *A. punctipes,* Theo.].

Climacura, How., Dyar & Knab, Mosq. N. Amer., III, 452, (1905) [Genotype, *Culex melanurus,* Coq.].

Culicella, Felt., N. Y. Mus. Bull., 79, Ent., 22, App. 391*c,* (1904) [Genotype, *Culex dyari,* Coq.]; *id.,* How., Dyar & Knab, Mosq. N. Amer., III, 457.

Diceromyia, Theo., IV Rpt. Wellc. Lab., B, 151, (1911) [Genotype, *D. africana,* Theo.].

Eumelanomyia, Theo., Monog. Culic., V, 240, (1910) [Genotype, *E. inconspicua,* Theo.].

Gnophodeomyia, Theo., Jo. Econ. Biol., I, (1), 21, (1905) [Genotype, *G. inornata,* Theo.].

Gnophodromyia, Theo., Monog. Culic.,V, 114, (1910) [*nom. nud.,* ? misprint for *Gnophodeomyia*].

Heptaphlebomyia, Theo., Monog. Culic., III, 336, (1903) [Genotype, *H. simplex,* Theo.]; *id., loc. cit.,* IV, 531; id., loc. cit., V, 481; *id.,* Gen. Ins. fasc., XXVI, 41; *id.,* Entom., XXX, 156.

Jamesia, Chris., Sci. Mem. Med. Off. Ind., n. s., XXV, 12, (1906) [Genotype, *Culex concolor,* R.-D., erected on larval characters only]; Brun., R. I. M., X, 62.

Lasiconops, Theo., Monog. Culic., III, 235, (1903) [Genotype, *L. poicilipes,* Theo.].

Leucomyia, Theo., Monog. Culic., IV, 372, Pl. IX, (1907) [Genotype, *Culex gelidus,* Theo.]; *id., loc. cit.,* V, 311, [table of spp.].

Lutzia, Theo., Monog. Culic., III, 155, (1903) [Genotype, *Culex bigotii,* Bell.].

Maillotia, Theo., Monog. Culic., IV, 274, (1907) [Genotype, *M. pilifera,* Theo.].

Melanoconion, Theo., Monog. Culic., III, 238, (1903) [Genotype, *Culex atratus,* Theo.]; *id., loc. cit.,* IV, 507; *id., loc. cit.,* V, 455, [table of spp.]; *id.,* Gen. Ins. fasc., XXVI, 32; Herrick., Ent. News, 382; Leic., Culic. Malaya, 136; Brun., R. I. M., I, 360; *id., loc. cit.,* IV, 484; *id., loc. cit.,* X, 65; Dyar, P. Ent. S., Wash., VII, 49 [genotype].

Melanoconium, Blanch, Moustiques, 395 [emended name].

Melanoconops, Theo., Monog. Culic., III, 178, (1903) [*nom. nud.*].

Microculex, Theo., Monog. Culic., IV, 461, (1907) [Genotype, *M. argentioumbrosus,* Theo.].

Mimeteomyia, Theo., Monog. Culic., V, 210, [1910) [Genotype, *M. apicotriangulata,* Theo.].

Mochlostyrax, Dyar & Knab., Jo. N. Y. Ent. S., XIV, 223, (1906) [Genotype, *M. caudelli,* D. & K.].

Neoculex, Dyar, P. Ent. Soc. Wash., VII, 47, (1905) [Genotype, *Culex territans,* Wlk.].

Neomelanoconion, Theo., Monog. Culic., IV, 514, (1907) [female only] [Genotype, *N. rima,* Theo.]; News, Dutt. & Todd, Ann. Trop. Med. Paras., I, 31; Brun., R. I. M., X, 66.

Pneumaculex, Theo., Monog. Culic., IV, 523, (1907) [Genotype, *Culex signifer,* Coq.]; Dyar, P. Ent. S. Wash., VII, (1), [*nom. nud.*]; Brun., R. I. M., X, 68.

Popea, Ludl., Can. Ent. XXXVII, 95, (1905) [Genotype, *P. lutea,* Ludl.]; Theo., Monog. Culic., IV, 211; Brun., R. I. M., IV, 485; *id., loc. cit.,* X, 69.

Pseudoheptaphlebomyia, Vent., B. Mus., Paris, XI, 427, [*nom. nud.*; Genotype, *P. madagascarensis,* Vent.].

Rachisoura, Theo., Monog. Culic., V, 207, (1910) [Genotype, *R. sylvestris,* Theo.].

Theobaldiomyia, Brun., R. I. M., IV, 462, [1912] [*nom. nov.* for *Leucomyia,* prœocc.].

Thomascina, News: & Cart., Ann. Trop. Med. Paras., IV, 553, fig. 1, (1910-11) [Genotype, *Mansonia longipalpis,* N. & C.].

Tinolestes, Coq., P. Ent. S. Wash., VII, 185, [1906] [Genotype, *T. lutisquama,* Coq.].

Trichopronomyia, Theo., Ann. Mus. Hung., III, 98, (1905) [Genotype, *T. annulata,* Theo.]; *id.,* Monog. Culic., IV, 479; Brun., R. I. M., I, 356; *id., loc. cit.,* IV, 477; *id., loc. cit.,* X, 73.

bitæniorhynchus, Giles.

Culex bitæniorhynchus, Giles, Bom., Jo., XIII, 607[1], (1901) [prov. descr.]; *id.,* Entom., XXXIV, 196; Edw., B. Ent. R., IV, 231; James, I. J. Med. R., II, 262[2]; Stanton, *loc. cit.,* III, 257[3]; McFarl., B. Ent. R., VI, 67[4]; Stanton, *loc. cit.,* X, 333, [distribution in far East. ports]; White, I. J. Med. R., VIII, 317[5].

Grabhamia bitaeniorhynchus, Brun., R. I. M., XVII, 132[6].

? *Culicelsa abdominalis,* Tayl., Rpt. Aust. Inst. Trop. Med., 1911, 53, (1913).

Tæniorhynchus ager, Giles in Theo., Monog. Culic., II, 199, fig. 248, (1901); *id.,* Handbook, (2nd edn.), 365; Blanch., Moustiques, 385; Leic., Culic. Malaya, 168; Brun., R. I. M., I, 357[7]; *id., loc. cit.,* IV, 478[8]; Green, Spol. Zeyl., IV, 182[9].

Grabhamia ambiguus, Theo., Monog. Culic., III, 248, (1903); Brun., R. I. M., I, 341[10]; *id., loc. cit.,* IV, 461, [*ambigua*].

? *Tæniorhynchus domesticus,* Leic., Culic. Malaya, 169[11], (1908); Brun., R. I. M., IV, 479[12]; Edw., B. Ent. R., X, 134.

Culex infula, Theo., Monog. Culic., I, 370, (1901); Blanch., Moustiques, 298; Leic., Culic. Malaya, 146; Brun., R. I. M., I, 348[13]; Giles, Handbook, (2nd edn.), 407 [*infulus*].

[1, 6, 7]Shahjahanpur.
[6]Pusa., [6, 8]Bengal.
[6, 8, 18]Sylhet.
[6, 8, 18]Puri.
[6]Mandalay.
[6, 18]Manipur.,
[8, 7]Madras.
[6, 7, 8, 10]Travancore.
[2, 5, 6, 7, 9, 16, 19]Ceylon.

[6, 11, 12, 17, 18]F. M. S.
[3, 6, 13]Sumatra.
[6, 14, 15, 20]Borneo.
[6, 8]Philippines.
[4]Hongkong.,
[17, 20]China.
[17]S. & W. Africa.

Oculeomyia sarawakii, Theo., Monog. Culic., IV, 5151[4], fig. 236, 237, Pl. VI, (1907); Brun., R. I. M., IV, 485[15].

Grabhamia taeniarostris, Theo., Monog. Culic., IV, 299, (1907); Brun., R. I. M., IV, 461[16].

Taeniorhynchus tenax, Theo., Monog. Culic., II, 198, Pl. XVII, 65, (1901); *id., loc. cit.,* III, 259, fig. 236; Giles, Handbook, (2nd edn.), 365; Blanch., Moustiques, 386; Leic., Culic, Malaya, 167; Brun., R. I. M., 358[17]; *id., loc. cit.,* IV, 480[18]; Fletcher, Pusa Sci. Rpt.; (1915-16), 83.

? *Taeniorhynchus tenax,* Green, Spol. Zeyl., IV, 182[19].

Taeniorhynchus tenax var. *ocellata,* Theo., Monog. Culic., IV, 488, (1907); Brun., R. I. M., IV, 480[20]; Theo., Monog. Culic., III, 259.

concolor, Robinean-Desvoidy.

Culex concolor, R.-D., Mem. Soc. H. Nat., Paris, IV, 405, (1825); Theo., Monog. Culic., II, 107, fig. 203, Pl. XXXVIII, 109, 110; Giles, Jo. Trop. Med., VII, 365; *id.,* Handbook, (2nd edn.), 454, Pl. XVII, 8*a, b;* Blanch., Moustiques 365; Leic., Culic. Malaya, 154; Brun., R. I. M., I, 343[1]; *id., loc. cit.,* IV, 468[2]; *id., loc. cit.,* XVII, 157; James, I. J. Med. R., II, 262[7]; Stanton, B. Ent. R., X, 333, [distributin in far East. ports]; Patton & Cragg, T. B. Med. Ent., Pl. XVIII, 6, [clasper], Pl. XXXIV, 3, [larva]; Fletcher, Pusa Sci. Rpt., (1916-17), 101[4], [bionomics]; Bentley, Malaria in Bombay, 62[5]; Iyengar, I. J. Med. R., 1920 Sci. Cong. No., 9[6].

[2]Manipur., [1, 2]U.P. [1, 4]Bihar., [1, 2, 6]Bengal. [1, 2, 3]Sylhet. [1]Orissa., [5]Bombay. [1, 2]Malabar Coast. [1]Madras., [1, 2, 3]Burma. [2]Andamans. [7]Ceylon., ——— [1, 2, 3]Siam. [1, 2, 3]Malaysia (general). [1, 3]Philippines.

epidesmus, Theobald.

Tœniorhynchus epidemus, Theo., R. I. M., IV, 22[1]; (1910); *id.,* Monog. Culic., V, 429[2]; Brun., R. I. M., IV, 479[3].

Culex epidesmus, Edw., B. Ent. R., IV, 231.

Grabhamia epidesmus, Brun., R. I. M., XVII, 133.

Tœniorhynchus luteoabdominalis, Theo., R. I. M., IV, 23[4]; *id.,* Monog. Culic., V, 423[5]; Brun., R. I. M., IV, 479[6].

Grabhamia ochracea, Theo., Jo. Econ. Biol., I, 25, (1905); *id.,* Monog. Culic., IV, 300; Brun., R. I. M., I, 341[7].

[1, 2, 3]China. [1, 2, 3]Bhogaon. [4, 5, 6]Katihar., [7]India.

fatigans, Weidemann.

Culex fatigans, Wd., Auss. Zweifl. Ins., I, 10, 17[1], (1828); Big., Cat. Orient. Dipt., 251; Wulp., Cat. Dipt. S. Asia, 33; Kert., Cat. Dipt., I, 259; Giles, Handbook, 298; *id., loc., cit.,* (2nd edn.), 438, fig. 45, 440, [list of sub-spp.]; Theo., Monog. Culic., II, 151, fig.; 234-236, Pl. XXIX, 114, 115, Pl. D, 155, [map of distribution], fig. 238; *id.,* Gen. Ins. fasc., XXVI, Pl. II, 2; *id.,* Monog. Culic., IV, Pl. XVI [larva]; Blanch., Moustiques, 353, fig. 230, [as pungens], 231-234; Leic., Culic. Malaya, 157; Brun., R. I. M. I., 344; *id., loc. cit.,* VI, 468; *id., loc. cit.,* XVII, 158-9; Trop. Agric., XXXVI, 32, [in abstract]; Green, Spol. Zeyl., IV, 183; Chris., Rpt. King Inst., 1907, [abst. J. T. Vet. Sci., IV, 228, [flagellates]; Gill, I. J. Med. R., II, 268, [minimum biting temp.]; Cruick. & Wright, *loc., cit.,* I, 773, 784; James, *loc., cit.,* II, 262; Prashad, *loc. cit.,* II, 500, Pl. XXIX, 2, [male genitalia]; Edw., B. Ent. R., IV, 55, fig. 4, [genitalia]; Doane, *loc. cit.,* IV, 265; McFie, *loc. cit.,* VII, 277, [larval respiration]; McFie & Ingram, *loc. cit.,* VII, 73, [pupa]; Barraud, *loc. cit.,* X, 323; Stanton, B. Ent. R., X, 334; Patton & Cragg, T. B. Med. Ent., Pl. II, 4 [male antenna], Pl. XIV, 1, [exoskeleton]; Pl. XXVII, 7, [salivary gland], Pl. XXXIV, 59, [egg], Pl. XXXVI, 2, [imago], Pl. XXXV, 16, [larva]; Fletcher, Pusa Sci. Rpt., (1915-16), 83; McDonald, B. Ent. R. VII, 259, [in latrines]; Edw., B. Ent. R., XI, 137; Dalziel, *loc. cit.,* XI, 253, [in crab holes]; Johns., *loc. cit.,* IX, 325-332, [Nigerian notes]; Iyengar, I. J. Med. R., 1920 Sci. Cong. No., 9; Barraud, B. Ent., R., XI, 394, [does not occur in Egypt]; Gill, R. S. Trop. Med. Hyg., XIV, 77, [transmission expts.]; Stanton, I. J. Med. R., III, 257; Gough, B. Ent. R., V, 135; Lough, Jo. R. A. M. C., XXXVI, 137; White, I. J. Med. R., VIII, 320; Gill. *loc. cit.,* VIII, 666, [transmission expts.]; Chris., Sci. Mem. Med. Off. Ind., XLVI; 118-125, [*Proteosoma* transmission]; Cleland, etc., Jo. Hyg. Camb. XVIII, 217-254, [abst. R. A. E. (B), IX, 127]. [not a dengue vector] Craig, Jo. Amer. Med. Ass., LXXV, 1171-6 [abst. R. A. E. (B) IX, 128] [not proved guiltless in dengue].

Circumterrestrial in Warmer Regions. Universally distributed in Oriental region.

Culex aestuans, Wd., Auss. Zweifl. Ins., I, 11, (1828).

Culex anxifer, Coquerel in Big., A. S. Ent, Fr., (1958), VII, 117, (1859).

Type localities.
[1]Ind. Or.
[2]Madagascar.
[3]India, [4]Argentina.

Culex cartroni, Vent., B. Mus., Paris, XI, 429, (1905)[2].

Culex christophersi, Theo., Monog. Culic., IV, 453[3], (1907); Brun., R. I. M., IV, 467.

**Heteronycha delosa,* Arrib., Dipt. Argent., 56[4], (1891).

Culex foochowensis, Theo., Monog. Culic., II, 137, fig. 224, 225, (1901)[5]; Brun., R. I. M., I, 345; Blanch., Moustiques, 344. [*fouchowensis*].

Culex goughii, Theo., S. Af. Ag. Dpt., 1st Vety. Rpt., 26, 8[6], (1911) [male only].

Culex osakensis, Theo., Monog. Culic., IV, 453[8], (1907).

Culex pallipes, Mg., Syst. Besch. VII, 1, (1938).

Culex panalectoros, Giles, Bom. Jo., XIII, 608[7], (1901).

Armigeres panalectoros, Giles, Handbook, (2nd edn.), 386.

Desvoidya panalectoros, Blanch., Moustiques, 266.

Desvoidea panalectoros, Theo., R. I. M., IV, 5, [sinks].

Blanchardiomyia panalectoros, Brun., R. I. M., IV, 442.

Culex quasipipiens, Theo., Monog. Culic., II, 136[9], fig. 223, (1901); Giles, Handbook, (2nd edn.), 438; Blanch., Moustiques, 344; Brun., R. I. M., I, 351.

Culex quinquefasciatus, Dyar & Knab, [? *nec.* Say], P. Ent. Soc., Wash., XI, 34[10], (1909); How., Dyar & Knab, Mosq. N. Amer., III, 345; Dyar, Ins. Mens., IX, 32 [males swarming round man].

Culex reesii, Theo., Monog. Culic., II, 145, fig. 232[11], (1901); Giles, Handbook, (2nd edn.), 449; Blanch., Moustiques, 361; Brun., R. I. M., I, 351.

Culex sericeus, Theo., Monog. Culic., II, 147, fig. 223, (1901)[12]; Giles, Handbook, (2nd edn.), 452; Blanch., Moustiques, 362; Brun., R. I. M., I, 352; *id., loc. cit.,* IV, 475.

Culex skusii, Giles, Handbook, 292, (1900).

fatigans var. luteoannulatus, Theobald.

Culex fatigans var. *luteoannulatus,* Theo., Monog. Culic., II, 159, (1901); Brun., R. I. M., XVII, 159.

[5]China., [6]S. Africa.
[7]Calcutta., [8]Japan.
[9]Sambalpur, C.P.
[10]N. America.
[11, 12]Hongkong.

---

* I can find no reference to the erection of this genus.

? *Culex pungens.*, Wd., Auss. Zweifl. Ins., I, 9, (1828).

? *Culex macleayi,* Skuse, P. Linn. S. N. S. W., (1896), 1745, (1896).

fuscanus, Wiedemann.

Culex fuscanus, Wd., Dipt., Exot., 1, 9, 8, (1821)[1]; *id.,* Auss. Zweifl. Ins., I, 10; R.-D., Mem. S. H. Nat. Paris, III, 404, 6; Wlk., List Dipt. B. M., I, 9, *id.*[2], P. Linn. S. Lon., I, 5, 3 & 105, 6[3]; Big., Cat. Orient. Dipt., 251; Wulp., Cat. Dipt. S. Asia, 33[4]; Kert., Cat. Dipt., I, 260; Giles, Handbook, 301; *id., loc. cit.,* (2nd edn.), 455; Theo., Monog. Culic., II, 167; *id.,* Gen. Ins., Fasc. XXVI, 30; Blanch., Moustiques, 275; Brun., R. I. M., I, 345; *id., loc. cit.,* XVII, 158.

[1, 4]Ind. Or.
[2, 4]Hindustan.

[3, 4]Malacca.
[3, 4]Borneo.

[Apparently not recognized as Indian since Walker, (1848).].

fuscocephalus, Theobald.

Culex fuscocephalus, Theo., Monog. Culic., IV, 420, (1907); Brun., R. I. M., IV, 469[1]; Edw., B. Ent. R., IV, 234; *id., loc cit.,* VII, 225; Brun., R. I. M., XVII, 166[2]; Stanton, B. Ent. R., X, 333, [distribution in far East. ports].

Culex luteola, Theo., Monog. Culic., V, 378[3], fig. 172, (1910); Brun., R. I. M., IV, 471.

Culex minor, Theo., R. I M., II, 298[4], (1908); *id., loc. cit.,* IV, Pl. I; *id.,* Monog. Culic., V, 363, fig. 150; Brun., R. I. M., IV, 473.

Culex taytayensis, Banks, Phil. Jo. Sci., IV, 545[5], (1910); Brun., R. I. M., IV, 475.

Culex uniformis, Leic., Culic. Malaya, 159[6], (1908); Brun., R. I. M., IV, 476.

[2, 4]Lushai Hills.
[2, 4]Sylhet.
[2, 4]Calcutta.
[1, 2]Travancore.
[1, 2, 3]Ceylon.

[2, 6]Malaysia.
[2, 5]Philippines.

gelidus, Theobald.

Culex gelidus, Theo., Monog. Culic., II, 20, Pl. XXXIV, 93, fig. 158, (1901); Giles, Handbook, (2nd edn.), 421; Blanch., Moustiques, 316; Brun., R. I. M., I, 346[1]; *id., loc. cit.,* XVII, 163[2]; James, I. J. Med. R., II, 262[3]; Stanton, B. Ent. R., X, 333[3], [distribution in far East. ports]; Howlett, Pusa Sci. Rpt. (1914-15) 76[4], [breeding places]; Fletcher, *loc. cit.,* (1915-16), 83[5]; Fry, Malaria in Bengal, 36[6]; White, I. J. Med. R., VI I, 318[7].

[2]Rangoon.
[1, 2, 6, 8]Bengal.
[4, 5]Pusa.
[2]Madras City.
[2]Malabar Coast.
[1, 2, 3, 7]Ceylon.
_____, [1, 2, 9]F.M.S.,
[2]Borneo.
[1, 2]Philippines.

Leucomyia gelida, Theo., Monog. Culic., IV, 372; Leic., Culic. Malaya; 147[9]; Iyengar, I. J. Med. R., 1920 Sci. Cong. No., 9[8].

Theobaldiomyia gelida, Brun., R. I. M., IV, 462.

gelidus var. bipunctatus, Theobald.

Leucomyia gelida var. *bipunctata,* Theo., Monog. Culic., IV, 374[1], (1907); Cruick. & Wright, I. J. Med. R., I, 773[2].

Culex gelidus var. bipunctatus, *Brun.,* R. I. M., XVII, 163[3].

[1, 3]India., [2]Cochin.
____
[1, 3]Sarawak.

gelidus var. cuneatus, Theobald.

Culex gelidus var. *cuneatus,* Theo., Monog. Culic., II, 22, fig. 159, (1901); Giles, Jo. Trop. Med., VII, 368; Brun., R. I. M., I, 346[1]; *id., loc. cit.,* XVII, 163[2].

[2]Sylhet., [1, 2]Bengal.
[2]Puri., [2]Madras.
[1, 2]Travancore.
[1, 2]Ceylon.
______, [1, 2]F. M. S.
[1, 2]Philippines.

halifaxi, Theobald.

Culex halifaxi, Theo., Monog. Culic., III, 231, (1903); *id., loc. cit.,* V, 390[1]; Brun., R. I. M., IV, 470[2]; *id., loc. cit.,* XVII, 164[3]; Edw., B. Ent. R., IV, 234; Stanton, I. J. Med. R., III 258[4]; *id.,* B. Ent. R., X, 333, [distribution in far East. ports]; White, I. J. Med. R., VIII, 318[5].

Culex multimaculosus, Leic., Culic. Malaya, 155[6], (1908); Brun., R. I. M., IV, 473.

[5]Matale, Ceylon.
____
[1, 2, 3, 6]F. M. S.
[4]Sumatra.

japonicus var. ceylonicus, Theobald.

Culex japonicus var. *ceylonica,* Theo., Monog. Culic., V, 391[1], (1910).

Culex japonicus var. *ceylonicus,* Brun., R. I. M., IV, 471[3]; *id., loc. cit.,* XVII, 167[4].

Culex japonicus, Brun., R. I. M., I 348 [*partim.*][2].

[1, 2, 3, 4]Peradeniya.

mimeticus, Noë.

Culex mimeticus, Noë, B. S. Ent. Ital., XXXI, 240[1], (1899); Giles, Handbook, 202; *id., loc. cit.,* (2nd edn.), 389, Pl. XV, 16-18; Theo., Monog. Culic., I, 329, Pl. XVI, 63; Blanch., Moustiques, 271; Leic., Culic., Malaya, 139[2]; Brun., R. I. M., I, 349[3]; *id., loc. cit.,* IV, 472[4]; *id., loc. cit.,* XVII, 161[5]; Fletcher, Pusa Bill. 89, 58; James, I. J. Med. R., II, 262[6]; Edw., B. Ent. R., VI, 359[7]; Waterst., *loc. cit.,* IX, 9[8]; Barrraud., *loc, cit.,* XI, 395[9]; McFarl., *loc., cit.,* VI, 67[10].

[3, 4, 5]Theog.
[3, 5]Punjab, 6,000′.
[4, 5]Nepal.
[4, 5]Lushai Hills.
[3, 5, 7]Nilgiris.
[4, 5, 6]Ceylon.
[3, 5]Kanara District.
____, [1, 5, 8]S. Europe.
[5]Cyprus.
[11]Caspian Sea.
[11]Caspian Sea.
[5, 9]Palestine.
[2, 3, 5]F. M. S.

*? *Culx hyrcanus,* Pallas, Reise Russ. Reich.[11], (1771).

[Many of these references probably refer to *C. mimulus,* Edw. True *mimeticus* in the Orient seems confined to the hills.]

mimulus, Edwards.

Culex mimulus, Edw., B. Ent. R., V, 284[1], (1915); Brun., R. I. M., XVII, 167[2]; White, I. J. Med. R., VIII, 318[3].

[2]India., [2, 3]Ceylon.

---

[2]Malaysia., [1]Sarawak.

nilgiricus, Edwards.

Culex nilgiricus, Edw., B. Ent. R., VI, 358, fig. 1[1], (1916); Brun., R. I. M., XVII, 167[2]; Fletcher, Pusa Bull, 89, 58.

[1, 2]Utakamund (Nilgiris).

sinensis, Theobald.

Culex gelidus var. *sinensis,* Theo., Monog. Culic., III, 180[1], (1903).

Leucomyia sinensis, id., R. I. M., IV, 20[2], (1910); *id.,* Monog. Culic., V, 313[3].

Theobaldiomyia sinensis, Brun., R. I. M., IV, 463[4].

Culex sinensis, Edw., B. Eng. R., IV, 231; Brun., R. I. M., XVIII, 164[5].

Culex sepositus, Leic., Culic. Malaya, 152, (1908); Brun., R. I. M., IV, 475[6].

Tœniorhynchus tenax, Leic., (*nec.* Theo.), Culic. Malaya, 167, (1908).

[2, 3, 4, 5]Puri.

[5, 6]F. M. S.
[1, 3, 5]China.

sitiens, Wiedemann.

Culex sitiens, Wd., Auss. Zweifl. Ins., I, 542[1], (1828); Wulp; Cat. Dipt. S. Asia, 33[2]; Cruick. & Wright, I. J. Med. R., I, 773[3]; James, *loc. cit.,* II, 262[4]; Edw., B. Ent. R., IV, 232; McFarl., *loc. cit.,* VI, 67[5]; Stanton, *loc. cit.,* X, 333, [distribution in far East. ports]; Edw., *loc. cit.,* XI, 137; Balf., *loc. cit.,* XII, 129, [in saline water]; Brun., R. I. M., XVII, 159-160[6]; Fletcher, Pusa Sci. Rpt., (1915-16), 83[9].

Culex annulirostris, Skuse, Proc. Linn. S. N. S. W., (2) III, 1737, (1889).

Culex gnophodes, Theo., Monog. Culic., III, 163, (1903); Leic., Culic., Malaya, 145; Brun., R. I. M., I, 347[7].

[8]Punjab.
[8]United Provinces.
[9]Bihar.
[8, 10, 14]Bengal.
[6, 10]Sylhet.
[8]Madras.
[3, 8]Malabar Coast.
[4, 6, 8]Ceylon.
[10]Orissa.

[6, 7, 8, 10]F. M. S.
[1, 2]Sumatra.
[12, 13]Celebes.

---

* For another possible identification of this description see *Anopheles sinensis,* Wd.

Culex impellens, Wlk., (*nec.* Theo.), P. Linn S. Lon., IV, 91, (1860)[12]; Wulp., Cat. Dipt. S. Asia, 34[13]; Iyengar, I. J. Med. R., 1920 Sci. Cong. No., 9[14].

Culex microannulatus, Theo., Monog. Culic., I, 353, Pl. XVIII, 69, fig. 118, *b, d,* (1901); Blanch., Moustiques, 292; Leic., Culic. Malaya, 140; Brun., R. I. M., I, 349[8]; *id., loc. cit.,* IV, 472[10].

Tœniorhynchus microannulatus, Neave, B. Ent. R., III, 313[11].

Culex ronaldi, Charm., Ann. Trop. Med., II, 259, (1908)[15].

Culex salus, Theo., III Rpt. Wellc. Lab., 256, (1909)[16].

Culex somaliensis, Nev.-Lem., Arch. Paras., X, 254, (1906)[17].

[6, 8]Philippines.
[5]Hongkong.
[6, 10, 15]Mauritius.
[11, 16, 17]E. Africa.

tigripes, Grandpré and Charmoy.

Culex tigripes, Grand. & Charm., Planters, Gaz. Press, (1900); Giles, Handbook, (2nd edn.), 407, Pl. XVI, 11; *id.,* Jo. Trop. Med., VII, 368; Theo., Monog. Culic., III, figs. 120-122, [early stages]; Edw., B. Ent. R., II, 261; Brun., R. I. M., I, 352[3]; *id., loc. cit.,* IV, 475[4]; Green, Spol. Zeyl., IV, 183; Bentley, Malaria in Bombay, 62[6]; Johns., B. Ent. R., IX, 325-332[7], [Nigerian notes]; Enderl., Wien. Ent. Zeit., XXXVIII, 47[8]; Theo., Nov. Culic., I, 17, fig. 6, [male genitalia]; Brun., R. I. M., XVII, 161[9]; Theo., Monog. Culic., V, 392[10]; Charm., Ann. Trop. Med. Paras., II, (3), 262; Banks, Phil. Jo. Sci., I, (9), 998.

Culex maculicrura, Theo., Monog. Culic., II, 34, Pl. XXII, 85, (1901).

[6]Bombay.
[4, 9, 10]Bengal.
[4, 9]Sylhet., [4]Manipur.
[9, 10]U. Burma.
[4, 5, 9]Ceylon.

———

[3, 9, 10]Philippines.
[3, 9]F. M. S.
[3, 9, 10]Queensland.
[3, 4, 7, 9, 10]Africa (general).
[3, 8]Madagascar.
[1]Mauritius., [10]Arabia.

tipuliformis, Theobald.

Culex tipuliformis, Theo., Monog. Culic., II, 325, fig. 206[1], (1901); Giles, Handbook, (2nd edn.), 443; Blanch., Moustiques, 363; Brun., R. I. M., I, 353; Edw., B. Ent. R., III, 31; Brun., R. I. M., XVII, 164[2]; Barraud, B. Ent. R., X, 323[8]; *id., loc. cit.,* XI, 394[9].

Culex creticus, Theo., Monog. Culic., III, 189[3], (1903).

Culex onderstepoortensis, Theo., S. Af. Dpt. Ag., 1st Vety. Rpt., 265[4], (1911).

Culex pettigrewi, Theo., R. I. M., IV, 5 (1910)[5]; *id.,* Monog. Culic., V, 351; Brun., R. I. M., 475; *id., loc. cit.,* XVII, 167[6].

[1, 2]Bakloh (Punjab).
[5, 6]Manipur.
——— [4, 7, 11]Africa.
[11]Madeira.
[8]Mesopotamia.
[9, 12]Egypt., [3, 9]Levan.

Culex theileri, Theo., Monog. Culic., III, 187, (1903); *id., loc. cit.*, V, 356[7]; Edw., B. Ent. R., II, 262[11]; Gough, *loc. cit.*, V, 135[12].

trimaculatus, Theobald.

Culex trimaculatus, Theo., Ann. Mus. Hung., III, 86, fig. 5[1], (1905); *id.,* Monog. Culic., IV, 427[2], fig. 187; Brun., R. I. M., XVII, 163[3].

[1, 2, 3]Bombay.

[Apparently not recognized since its description.]

tritæniorhynchus, Giles.

Culex tritœniorhynchus, Giles, Bom. Jo., XIII, 606, (1901)[1], [provis. descr.]; *id.,* Entom., XXXIV, 192; *id.,* Handbook, (2nd edn.), 401; Theo., Monog. Culic., I, 364, fig. 124; Blanch., Moustiques, 294; Brun., R. I. M., I, 355[2]; Edw., B. Ent. R., IV, 233; *id., loc. cit.*, VII, 224, [distinction of *sitiens, tritœniorhynchus & vishnui*]; Brun., R. I. M., XVII, 161[3]; James, I. J. Med. R., II, 262[4]; Stanton, *loc. cit.*, III, 257[5]; Ingram & McFie, B. Ent. R., VII, 73[6], [pupa]; *id., loc. cit.*, VII, 149, [early stages]; Barraud, *loc. cit.*, X, 323; Stanton, *loc. cit.*, X, 333, [distribution in far East. ports]; White, I. J. Med. R., VIII, 318[8]; Aust., B. Ent. R., XII, 116[9].

Culex annulus, Theo., Monog. Culic., I, 358, (1901); Giles, Handbook, (2nd edn.), 405; Blanch., Moustiques, 293; Leic., Culic, Malaya, 144; Brun., R. I. M., I, 343[11].

Culex biroi, Theo., Ann. Mus. Hung., III, 82, Pl. I, (1905); *id.,* Monog. Culic., IV, 390, figs. 164, 165; Brun., R. I. M., I, 343[10].

Culex sitiens, Theo., (*nec.* Wd.), Monog. Culic., I, 360, fig. 121, (1901); Giles, Handbook, (2nd edn.), 400; Blanch., Moustiques, 293; Leic, Culic., Malaya, 143; Brun., R. I. M., I, 352; *id., loc. cit.*, IV, 475.

Culex vishnui, Theo., Monog. Culic., I, 355, figs. 119, 120*a*, (1901) [male only] ; Giles, Handbook, (2nd edn.), 399, Pl. XVI, 5*a, b,* [male only]; Blanch., Moustiques, 293, [male only]; Theo., Monog. Culic., IV, 387, fig. 163, [male only], Brun., R. I. M., I, 354; *id., loc. cit.*, IV, 476.

[3]Bengal.
[3]Central India.
[3, 10]Bombay., [3]Madras.
[1, 2, 3][Travancore.]
[3, 4, 8]Ceylon.
———, [3, 11]F.M.S.
[3, 5]Sumatra.
[3]Malaysia., [3]Celebes.
[3]Philippines., [3, 11]China.
[3, 7]Mesopotamia.
[6]Zanzibar.
[6]Gold Coast.
[9]Palestine.

univittatus, Theobald.

Culex univittatus, Theo., Monog. Culic., II, 29, fig. 161, Pl. XXII, 86, [1901] ; Giles, Handbook, (2nd edn.),

Shillong (White).

———

[1]Singapore.

428, [*univitatus, laps.*]; Blanch., Moustiques, 321; Brun., R. I. M., I, 353[1]; Theo., Monog. Culic., V, 356; Brun., R. I. M., XVII, 163; Edw., B. Ent. R., IV, 58, [genitalia]; *id., loc. cit.*, V, 67, fig. 5, [genitalia]; *id., loc. cit.*, XI, 137[3]; Barraud., *loc. cit.*, XI, 394[2].

[3, 4]Madagascar.
[1]Africa, S. and E.
[2]Palestine.
[2]Egypt.

Heptaphlebomyia montforti, Vent., Arch. Paras., IX, 448[4], (1905).

vagans, Wiedemann.

Culex vagans, Wd., Auss. Zweifl. Ins., I, 545[1], (1828); Wlk., List. Dipt. B. M., I, 9; Wulp., Cat. Dipt. S. Asia, 33[2]; Theo., Monog. Culic., I, 411, fig. 146; *id., loc. cit.*, 14; *id., loc. cit.*, V, 347[3], [male described]; Giles, Handbook, 279; *id., loc. cit.*, (2nd edn.), 414; Pl. XVI, 14; *id.*, Jo. Trop. Med., VII, 368; Blanch.; Moustiques, 304; Brun., R. I. M., I, 353[4]; *id., loc. cit.*, IV, 476[5]; *id., loc cit.*, XVIII, 160[6].

[3, 5, 6]Madras City.

[3, 4, 6]Philippines.
[1, 2, 3, 4, 6]China.

vishnui, Theobald.

Culex vishnui, Theo., Monog. Culic., I, 355, figs. 119, 120, Pl. XVIII, 66, (1901) [female only]; *id., loc. cit.*, IV, fig. 162, Pl. V; Blanch., Moustiques, 293, [female only]; Brun., R. I. M., I, 354[2]; *id., loc. cit.*, IV, 476[3]; Edw., B. Ent. R., IV, 233; *id., loc. cit.*, VII, 225; Green, Spol., Zeyl., IV, 180[4], [sucking a moth]; James, I. J. Med. R., II, 262[5]; Stanton, *loc. cit.*, III, 257[6]; McFarl., B. Ent. R., VI, 67[7]; Stanton, *loc. cit.*, X, 333, [distribution in far East. ports]; Edw., *loc. cit.*, XII, 77; Brun., R. I. M., XVII, 162[10].

Culex impellens, Theo., (*nec.* Wlk.), Monog. Culic., I, 362, figs. 122, 123, (1901); *id., loc. cit.*, IV, 15; Giles, Handbook, (2nd edn.), 399, Pl. XVI, 3*a*, [female only]; Blanch., Moustiques, 294; Leic., Culic., Malaya, 142[8]; Brun., R. I. M., I, 347[9]; *id., loc. cit.*, IV, 470[11].

? *Culex mic otaeniatus,* Theo., Tijd. v. Ent., LIV, 236[12], (1911).

Culex perplexus, Leic., Culic. Malaya, 150, (1908), Brun., R. I. M., IV, 473[13].

Culex pseudoinfula, Theo., Tijd. v. Ent., LIV, 237[14], (1911).

[9, 11]United Provinces.
[3]Punjab., [10, 11]Nepal.
[3, 10]Sylhet., [2, 11]Bengal.
[3, 11]Orissa.
[1, 2]Central India.
[2]Madras.
[2, 11]Malabar Coast.
[2, 4, 5, 10, 9] Ceylon.
[10, 11]Burma.

[3, 10, 8, 9, 13]F. M. S.
[6]Sumatra.
[7]Hongkong.
[10, 9]Celebes.
[10, 11, 12, 14]Java.
[10, 9]Philippines.

[The Indian localities, (Nilgiris, Madras, North India), given by Brun., R. I. M., I, 343, for *Culex* (*Aedes*) *cantans,* Mg. are almost certainly the result of misidentification.]

SABETHINI.

Brun, R. I. M., X, 48; Theo., Monog. Culic., V, 554, 574, [*Metanota richœ*].

24. **Rachionotomyia,** Theobald.

Rachionotomyia, Theo., Bom. Jo., XVI, 248, (1905) [Genotype, *Wyeomyia aranoides,* Theo., as *R. ceylonensis,* Theo.]; *id.,* Monog. Culic., IV, 518; Edw., B. Ent. R., IV, 240; Brun., R. I. M., I, 368; *id., loc. cit.,* IV, 485; *id., loc. cit.,* X, 70; *id., loc. cit.,* XVII, 180.

Binotia, Blanch., Arch. Paras., VIII, 478, (1904) [*nom. nov.* for *Runchomyia,* supposed præocc.].

Colonemyia, Leic., Culic. Malaya, 233, (1908) [Genotype, *C. caeruleocephala,* Leic.]; Brun., R. I. M., IV, 498; *id., loc. cit.,* X, 58.

Polylepidomyia, Theo., Ann. Mus. Hung., III, 118, (1905) [Genotype, *P. argen teiventris,* Theo.]; *id.,* Monog. Culic., IV, 625; Brun., R. I. M., I, 366; *id., loc. cit.,* IV, 502; *id., loc. cit.,* X, 68.

Runchomyia, Theo., Monog. Culic., III, 319, (1903) [Genotype, *R. frontosa,* Theo.]; *id., loc. cit.,* IV, 585; *id., loc. cit.,* V, 555; *id.,* Gen. Ins. fasc., XXVI, 38; Brun., R. I. M., I, 365; *id., loc. cit.,* IV, 493; *id., loc. cit.,* X, 71.

Skeiromyia, Leic., Culic. Malaya, 248, (1908) [Genotype, *S. fusca,* Leic.]; Brun., R. I. M., IV, 500; *id., loc. cit.,* X, 71.

Squamomyia, Theo., R. I. M., IV, 28, (1910) [Genotype, *S. inornata* Theo.]; *id.,* Monog. Culic., V, 529; Brun., R. I. M., IV, 489; *id., loc. cit.,* X, 71.

affinis, Edwards.

Phoniomyia caeruleocephala, Theo., Monog. Culic., V, 577[1], fig. 252, (1910); Brun., R. I. M., IV, 502. — [1, 2]Hakgala (Ceylon).

Rachionotomyia affinis, Edw., B. Ent. R., IV, 241, (1913), [*nom., nov.* præocc. by *Colonemyia cœruleocephala,* Leic.]; Brun., R. I. M., XVII, 182[2].

aranoides, Theobald.

Wyeomyia aranoides, Theo., Monog. Culic., II, 274, (1901); Giles, Handbook, (2nd edn.), 499; Blanch., Moustiques, 425; Brun., R. I. M., I, 365[1]. — [2, 3, 4, 5]Ceylon. [3, 7, 8, 9]Dawnat Hills (Lower Burma).

Rachionotomyia aranoides, Edw., B. Ent. R., IV, 241; James, I. J. Med. R., II, 263[2]; Stanton, B. Ent. R., X, 333, [distribution in far East. ports]; Brun., R. I. M., XVII, 181[3]; White, I. J. Med. R., VIII, 320[4], [systematic position discussed].

Rachionotomyia ceylonensis, Theo., Bom. Jo., XVI, 248, Pl. B, 6, (1905); Brun., R. I. M., I, 369[5].

Skeiromyia fusca, Leic., Culic. Malaya, 248, (1908)[6]; Brun., R. I. M., IV, 500.

Squamomyia inornata, Theo., R. I. M., IV, 28[7], (1910); *id.,* Monog. Culic., V, 529[8]; Brun., R. I. M., IV, 489[9].

[1, 6]Malaysia.

25. Wyeomyia, Theobald.

Wyeomyia, Theo., Monog. Culic., II, 267, (1901) [Genotype, *W. grayii,* Theo.]; *id.,* Jo. Trop. Med., II, 235, [*nom. nud.*]; *id.,* Monog Culic., III, 310, [restricted]; *id., loc. cit.,* V, 579; *id.,* Gen. Ins. fasc., XXVI, 38; Blanch., Moustiques, 423; Giles, Handbook, (2nd edn.), 495; Leic., Culic. Malaya, 250; Brun., R. I. M., I, 365; *id., loc. cit.,* IV, 501; *id., loc. cit.,* X, 73; *id., loc. cit.,* XVII, 182-3; How., Dyar & Kanb. Mosq. N. Amer., IV, (1), 49.

Bolbodeomyia, Theo., R. I. M., IV, 31, (1910) [Genotype, *B. complex,* Theo.]; *id.,* Monog. Culic., V, 581; Brun., R. I. M., IV, 492; *id., loc. cit.,* X, 56.

Dendromyia, Theo., Monog. Culic., III, 213, [1903] [Genotype, *D. ulocoma,* Theo.]; *id., loc. cit.,* IV, 603; *id., loc. cit.,* V, 587, [table of spp.]; *id.,* Gen. Ins. fasc., XXVI, 39; Blanch., Moustiques, 426; Leic., Culic. Malaya, 25; Brun., R. I. M., IV, 502; *id., loc. cit.,* X, 59.

Heinzmannia, Ludl., Can. Ent. XXXVII, 130, (1905), (*Heizmannia, laps.*) [Genotype, *H. scientillans,* Ludl.]; Banks, Phil. I, Jo. Sci., I, 99, Brun., R. I. M., I, 366; *id., loc. cit.,* X, 61.

Phoniomyia, Theo., Monog. Culic., III, 311, (1903) [Genotype, *Wyeomyia longirostris,* Theo.]; *id., loc. cit.,* IV, 598; *id.,* Gen. Ins., fasc., XXVI, 38; Blanch., Moustiques, 425; Leic., Culic., Malaya, 253; Brun., R. I. M., I, 365]; *id., loc. cit.,* IV, 502; *id., loc. cit.,* X, 68.

complex, Theobald.

Bolbodeomyia complex, Theo., R. I. M., IV, 31[1], (1910); *id.,* Monog. Culic., V, 581[2] figs. 253-255; Brun., R. I. M., IV, 492[3].

Wyeomyia complex, Brun., R. I. M., XVII, 184[4].

[1, 2, 3, 4]Wawnat Hills. (L. Burma).

greeni,

Wyeomyia greenii, Theo., Bom., Jo., XVI, 247[1], Pl. B. 5, (1905); *id.,* Monog. Culic., IV, 596; Brun., R. I. M., I, 366; *id., loc. cit.,* IV, 501; *id., loc. cit.,* XVII, 183[2].

[1, 2]Peradeniya.

26. Topomyia Leicester.

Topomyia, Leic., Culic. Malaya, 238, (1908) [Genotype, *T. minor,* Leic.]; Brun., R. I. M., IV, 499; *id., loc. cit.,* X, 72; *id., loc. cit.,* XVII, 179.

Pseudograhamia, Theo., R. I. M., IV, 26, (1910) [Genotype, *P. auroventer,* Theo.]; *id.,* Monog. Culic., V, 551; Brun., R. I. M., IV, 460; *id., loc. cit.,* X, 69.

argenteoventralis, Leicester.

Topomyia argenteoventralis, Leic., Culic. Malaya, 240[1], (1908).

Topomyia argyroventralis, Brun., R. I. M., IV, 499[2], [*laps.*].

Topomyia argenteoventralis, Edw., B. Ent. R., IV, 240; Brun., R. I. M., XVII, 179[3].

Pseudograhamia aureoventer, Theo., R. I. M., IV, 27[4], (1910); *id.,* Monog., Culic., V. 551; Brun., R. I. M., IV, 460, 489[5].

[3, 4, 5]Travancore.

[1, 2, 3]F. M. S.

27. Hodgesia, Theobald.

Hodgesia, Theo., Jo.Trop. Med., VII, 17, (1904) [Genotype, *H. sanguinæ,* Theo.]; *id.,* Monog. Culic., IV, 579' *id., loc. cit.,* V, 544; *id.,* Gen. Ins. fasc., XXVI, 40; Giles, Jo. Trop. Med., VII, 368; Leic., Culic. Malaya 229; Blanch, Moustiques, 631; Brun., R. I. M., I, 367; *id., loc. cit.,* IV, 500; *id., loc. cit.,* X, 62; *id., loc. cit.,* XVII, 179.

malayi, Leicester.

Hodgesia malayi, Leic., Culic. Malaya, 231[1], (1908); Brun., R. I. M., IV, 500; James, I. J. Med., R., 263[3]; Brun., R. I. M., XVII, 179[2].

[3]Colombo.

[1, 2]Kuala Lumpur.

28. **Harpagomyia,** de Mejiere.

Harpagomyia, Meij., Tijd. v. Ent., LII, 165, (1909) [Genotype, *H. genurostris,* Leic. as *H. splendens,* Meij.]; *id., loc. cit.,* LIV, 126, [metamorphoses]; Theo., Monog. Culic., V, 547; Brun., R. I. M., IV, 504; *id., loc. cit.,* X, 61; *id., loc. cit.,* XVII, 178; Edw., B. Entt. R., IV, 240.

Grahamia, Theo., Rpt. Dr. Graham's coll., (1909) [Genotype, *G. trichorostris,* Theo.]; *id.,* Monog. Culic., V, 497, (note); Brun., R. I. M., X, 61.

Malaya, Leic., Culic, Malaya, 258; (1908) [Genotype, *M. genurostris,* Leic.] (Præocc. *Malaia,* Heller]; Brun., R. I. M., IV, 503; *id., loc. cit.,* X, 64.

genurostris, Leicester.

Malaya genurostris, Leic., Culic, Malaya, 258, (1908); Brun., R. I. M., IV, 503[1].

Harpagomyia genurostris, Edw., B. Ent. R., IV, 240; James, I. J. Med. R., II, 263[2]; Stanton, *loc. cit.,* III, 258[3]; Brun., R. I. M., XVII, 178[4].

Harpagomyia cœruleovittata, Ludl., Psyche, XVIII, 131[5], (1911).

Harpagomyia splendens, Meij., Tijd. v. Ent., LII, 167[6], Pl. X, 1-9, (1909); Theo., Monog., Culic., V, 550[7]; Brun., R. I. M., IV, 504.[8]

[2]Colombo.

[1, 4]F. M. S.
[6, 7, 8]Java.
[4, 5]Philippines.
[3]Sumatra.

PART III
BOMBYLIIDAE

General account: Howlett, Ind. Ins. L., 597-600, figs. ; Will., Dipt. N. Amer., 210-213, figs.; Brun., F. I. Dipt. Brachy. I, 173-174 [characterized]; Verrall, Br. Flies, V, 474-478 [general account and systematic order of genera]: Brun., R. I. M., II, 437-492 [revised cat. of Oriental spp.]; Villen., Ann. S. Ent. Fr. 1905, 304-306 [Meigen's types in Paris Mus.].

BOMBYLIINAE

Verrall, Br. Flies, V, 483; Brun., F. I. Dipt. Brachy. J, 259.

1. Empidideicus, Becker.

Empidideicus, Beck., Zeit. f. Hym. u. Dipt., VII, 97, fig. (1907), [Genotype, *E. carthaginensis,* Beck.]; Verr., Br. Flies, V, 484, fig. 272; Brun., F. I. Dipt. Brachy. I, 260.

indicus, Brunetti.

Empidideicus indicus, Brun., R. I. M., XIII, 77, fig. 1 (1917)[1]; *id.,* F. I. Dipt. Brachy. I, 260[2].

[1,2] Simla.

2. Geron, Meigen.

Geron, Mg., Syst. Besch., II, 223, (1829) [Genotype, *Bombylius hybridus,* Mg.]; Sch., F. Aust., I 68; Coq., Tr. Amer. Ent. Soc., XXI; Mik, Wien. Ent. Zeit., XIV, 106 [parasitic on Pyralid larva]; Ald., Cat. N. Amer. Dipt., 245; Wulp, Cat. Dipt. S. Asia, 75; Kert., Cat. Dipt., V, 98; Brun., R. I. M., II, 459; *id.,* F. I. Dipt. Brachy. I, 279.

albescens, Brunetti.

Geron albescens Brun., R. I. M., III, 229 (1909)[1]; *id.* F. I. Dipt. Brachy. I, 280, figs. 17, 18[2].

[1, 2]Orissa.
[2]Satara.
Abbottabad.

argentifrons, Brunetti.

Geron argentifrons, Brun., R. I. M., II, 459 (1909)[1]; *id.* F. I. Dipt. Brachy. I, 279[2]; Howl., Ind. Ins. L., 598 [parasite of *Laspeyresia*].

[1, 2]Lahore.
[1, 2]Bombay., [1, 2]Pusa.
[2]Almora., [2]Ceylon.

3. Usia, Laterille.

Usia, Latr., H. N. Crust. Ins., III, 430 (1802) [Genotype, *Volucella florea,* Fb.]; *id., loc.* cit., XIV, 300; Kert., Cat. Dipt., V, 100; Brun., F. I. Dipt. Brachy. I, 281.

Volucella, Fb., (*nec* Geoff.), Ent. Syst., IV, 412, (1794); *id.,* Syst. Antl., 114; Mg., Klass, I, 194.

marginata, Brunetti.

Usia marginata, Brun., R. I. M., III, 228 (1909)[1]; *id.,* F. I. Dipt. Brachy. I, 283[2]. [1, 2]Simla.

sedophila, Brunetti.

Usia sedophila, Brun., R. I. M., III, 228 (1909)[1]; *id.,* F. I. Dipt. Brachy. I, 282, fig. 19.[2] [1, 2]Simla. [2]Phagu.

4. Phthiria, Meigen.

Phthiria, Mg., Illig. Mag. II, 268 (1803) [Genotype, *Bombylius pulicarius* Mikan]; Sch. F. Aust. I, 66; Coq. Trans. Amer. Ent. Soc., XXI, 66; Will., Biologia, Dipt. I, 288; Ald., Cat. N. Amer. Dipt., 239; Wulp, Cat. Dipt. S. Asia, 75; Kert., Cat. Dipt., V, 112; Verr., Br. Flies, V, 487; Brun., R. I. M., II, 459; *id.,* F. I. Dipt. Brachy. I, 284.

Cyclorhynchus, Mq., H. N. Dipt., I, 392 (1834); *id.,* Dipt. Exot., II, (1) 114.

Dasypalpus, Mq., Dipt. Exot., II, (1) 112 (1840).

Megapalpus, Mq., H. N. Dipt., I, 394 (1834).

Phtyria, Rond., Arch. Zool. Modena, III, 56 (1863).

Poocilognathus, Jaenn., Abh. Senck. Nat. Ges., VI, 350 (1863); *id.,* N. Exot. Dipt., 43.

gracilis, Walker.

Phthiria gracilis, Wlk., Dipt. Saund., 184 (1852)[1]; Wulp, Cat. Dipt. S. Asia, 75[2]; Kert., Cat. Dipt., V, 113[3]; Brun., R. I. M., II, 459[4]; *id.,* F. I. Dipt. Brachy. I, 285, fig. 20[5]. [2, 3, 4]E. India. [5]Simla.

5. Systoechus, Loew.

Systoechus, Lw., Neu Beitr., III, 34 (1855) [Genotype, *Bombylius sulphureus,* Mik.); Kert., Cat. Dipt., V, 120; Wulp, Cat. Dipt. S. Asia, 74; O. S., W. Dipt., 250; Ald., Cat. N. Amer. Dipt., 237; Brun., R. I. M., II, 458; *id.,* F. I. Dipt. Brachy. I, 275, fig. 16.

eupogonatus, Bigot.

Systoechus eupogonatus, Big., Ann. S. Ent. Fr., (7) II, 365 (1892); Wulp, Cat. Dipt. S. Asia, 74[2]; Kert., Cat. Dipt., V, 121[3]; Brun., R. I. M., II, 458[4]; *id.,* F. I. Dipt. Brachy. I, 278[5]. — [2, 3, 4, 5]Hindostan. [3]Ceylon.

flavospinosus, Brunetti.

Systoechus flavospinosus, Brun., F. I. Dipt. Brachy. I, 277 (1920)[1]. — [1]S. Shan States.

nivalis, Brunetti.

Systoechus nivalis, Brun., R. I. M., VII, 472 (1912)[1]; *id.,* F. I. Dipt. Brachy. I, 277[2]. — [1, 2]Kumaon.

socius, Walker.

Bombylius socius, Wlk., Dipt. Saund., I, 201 (1852)[1]; Wulp, Cat. Dipt. S. Asia, 74[2]; Kert., Cat. Dipt., V, 143.[3]

Systoechus socius, Brun., R. I. M., II, 458[4]; *id., loc. cit.,* III, 226[5]; *id.,* F. I. Dipt. Brachy. I, 276[6].

[2, 3]E. India., [4]Masuri. [5, 6]Simla District. [6]Kashmir. Abbottabad. [6]Dehra Dun. [6]Kumaon. [6]Sikkim. [6]Kangra Valley. Mohanpur (Bengal). Nilgiris. [6]Coimbatore. [6]Bellary District.

6. Anastoechus, Osten-Sacken.

Anastoechus, O. S., W. Dipt., 251 (1877) [Genotype, *Bombylius nitidulus* F., as *barbatus,* O-S.]; Ald., Cat. N. Amer. Dipt., 237 ; Kert., Cat. Dipt., V, 124; Wulp, Cat. Dipt. S. Asia, 74; Brun., R. I. M., II, 458; *id.,* F. I. Dipt. Brachy. I, 278.

longirostris, van der Wulp.

Anastoechus longirostris, Wulp, Leyd. Mus. Notes, VII, 85 (1885)[1]; *id.,* Cat. Dipt. S. Asia, 74[2]; Kert., Cat. Dipt., V, 124[3]; Brun., R. I. M., II, 458[4]; *id.,* F. I. Dipt. Brachy. I, 278[5]. — [2, 3, 4, 5]Himalaya.

7. Bombylius, Linnaeus.

Bombylius, L., Syst. Nat., edn. X, 606 (1758) [Genotype, *B. major,* L.]; *id.,* F. Suec., 1918; Latr., H. N. Crust. Ins. III, 429; *id., loc. cit.,* XIV, 298; Fb., Syst. Antl., 128; Mg., Syst. Besch., II, 141; Mq. H. N. Dipt. I 377; Sch., F. Aust., I, 58; O-S., W. Dipt., 247; Ald., Cat. N. Amer.

Dipt., 235; Wulp, Cat. Dipt. S. Asia, 73; Kert., Cat. Dipt., V, 125; Verr., Br. Flies, V, 491; Brun., R. I. M., II, 456; *id.*, F. I. Dipt. Brachy. I, 261.

Choristus, Wlk., Dipt. Saund., I, 196, (1852).

Parisus, Wlk., Dipt. Saund., I, 197, (1852).

albosparsus, Bigot.

Bombylius albosparsus, Big., Ann. S. Ent. Fr., (7) II, 362 (1892)[1]; Wulp, Cat. Dipt. S. Asia, 74[2]; Kert., Cat. Dipt., V, 125[3]; Brun., R. I. M., II, 457[4]; *id.*, F. I. Dipt. Brachy. I, 270[5].

[2, 3, 4, 5]Pondicherry.

brunettii, Senior-White.

Bombylius brunettii, S.-W., Spol. Zeyl., XII, 203 (1922)[1].

[1]Matale., [1]Opalgalla. [1]Habarana., [1]Kanthalai. [1]Trincomali. (All Ceylon).

comastes, Brunetti.

Bombylius comastes, Brun., R. I. M., III, 225 (1909)[1]; *id.*, F. I. Dipt. Brachy. I., 268[2].

[1]Trivandrum. [2]Bombay Presidency., Tanjore., Samalkota.

erectus, Brunetti.

Bombylius erectus, Brun., R. I. M., II, 457, 480, t. 12, f. 27 (1909)[1]; *id.*, F. I. Dipt. Brachy. I, 267[2].

[1, 2]Bangalore.

fulvipes, Bigot.

Bombylius fulvipes, Big., Ann. S. Ent. Fr., (7) II, 362 (1892); Wulp, Cat. Dipt. S. Asia, 74[2]; Kert., Cat. Dipt., V, 132[3]; Brun., R. I. M., II, 457, 479[4] [redescr.]; *id.*, F. I. Dipt. Brachy. I, 271[5].

[2, 3, 4]Pondicherry. [4, 5]Allahabad. [5]Belgaum., [5]Lahore. [5]N. Kanara.

maculatus, Fabricius.

Bombylius maculatus, Fb., Syst, Ent., 803, 6 (1775); Wd., Auss. Zweifl. Ins., I, 342, 19; Wulp, Cat. Dipt. S. Asia, 73[2]; Kert, Cat. Dipt. V, 134[3]; Brun., R. I. M., II, 456; *id.*, F. I. Dipt. Brachy. I, 267[5].

[2,4]Transquebar. [5]Orissa., [5]Bellary. [5]Coimbatore. Trichinopoly District. [3]S. Asia.

major, Linnaeus.

Bombylius major, L., Syst. Nat. Edn. X, 606 (1758)[1]; *id.*, F. Suec., 1918; *id.*, Syst. Nat. edn. XII, 11, 1009; Fb., Spec. Ins., II, 472; *id.*, Ent. Syst., IV, 407 ; *id.*, Syst. Antl., 128; Gmel., Syst. Nat., V, 2902; Mg., Syst. Besch., II, 144 [bibliog]; Mq., H. N. Dipt., I, 377;

[2, 3, 4, 5]Simla District. Murree Hills.

---

[2]N. America. [2]Europe.

Zett., Dipt. Scand., I, 188; Sch., F. Austr., I, 60; Kirby, F. Bor.-Amer., Ins., 312; *id.*, Can. Ent., XIII, 166; Lw., Neue Beitr., III, 14 [notes]; O.-S., W. Dipt., 248; Coq., Tr. Amer. Ent. Soc., XXI, 110; Ald., Cat. N. Amer. Dipt., 236; Verr., Br. Flies, V, 497, fig. 290; Kert., Cat. Dipt., V, 134[2]; Brun., R. I. M., I, 168[3]; *id.*, *loc. cit.*, II, 457[4]; *id.*, F. I. Dipt. Brachy. I, 265[5].

[2]N. Africa.
[2]Japan.

Bombylius albipectus, Mq., Dipt. Exot., supp. V, 82, t. 4, f. 10 (?1846).

Bombylius fratellus, Wd., Auss. Zweifl. Ins., I, 583 (1828).

Bombylius sinuatus, Mg., Klass, 179 (?1800); Mik, Monog. Bombyl. Behem. 35, t. 2, f. 4.

Bombylius variegatus, de Geer, Mem. H. N. Ins., VI, 107, t. 14, f. 10 (1782).

Bombylius vicinus, Mq., Dipt. Exot., II, i, 98 (1840).

orientalis, Macquart.

Bombylius orientalis, Mq., Dipt. Exot., II, i, 90, t. 6, f. 2 (1840)[1]; Wulp, Cat. Dipt. S. Asia 74[2]; Kert. Cat. Dipt., V. 139[3]; Brun., R. I. M., II, 456[4]; *id.*, F. I. Dipt. Brachy. I, 263[5], fig. 14; Howl., Ind. Ins. L., t. 63, f. 1.

Bombylius tricolor, Guer., Icon. Regne Anim., III, 538, t. 95 (1828-1838); Wulp, Cat. Dipt. S. Asia, 74[6]; Kert., Cat. Dipt., V, 144[7].

[3, 7, 4, 5]Bengal.
[4, 5]Bengalore.
[5]Coorg., [5]Cawnpore.
[5]Karikkupam.
[5]Dehra Dun.
[5]Bellary District.
Palamau.
Chapra., [2, 3]E. India.
_________, [6]Java.

propinquus, Brunetti.

Bombylius propinquus, Brun., R. I. M., III, 226 (1909)[1]; *id.*, Spol. Zeyl., VI, 171[2]; *id.*, F. I. Dipt. Brachy., I, 272[3].

[1, 2, 3]Cylon.

pseudoterminalis, *nom. nov.*

Bombylius vicinus, Brun. [*nec.* Mq.], R. I. M., II, 457, 480 (1909)[1]; *id.*, *loc. cit.*, III, 226[2]; *id.*, F. I. Dipt. Brachy. I, 270[3].

[1,3]Allahabad.
[2]Nepal Terai.

terminalis, Brunetti.

Bombylius terminalis, Brun., R. I. M., II, 457, 479 (1909)[1]; *id.*, F. I. Dipt. Brachy. I, 270[2].

[1, 2]Allahabad.
[2]Lahore.

wulpii, Brunetti.

Bombylius pulchellus, Wulp [*nec.* Lw.], Tijd. v. Ent. XXIII, 164, t. 10, f. 8 (1880)[1]; *id.*, Leyd. Mus. Notes, VII, 15.

[3, 4, 5, 6, 7]Cylon.
[5]Tenmalai.
[7]Dehra Dun.

Comastes pulchellus, Wulp, Cat. Dipt. S. Asia, 74[2].

Eucharimyia pulchellus, Kert., Cat. Dipt. V, 147[3].

Eucharimyia dives, Big., Ann. S. Ent. Fr., (6) VIII, Bull. p. exl. (1888); Wulp, Cat. Dipt. S. Asia, 73[4].

Bombylius scintillans, Brun., R. I. M., III, 224 (1909)[8].

Bombylius wulpii, Brun., R. I. I., II, 457 (1909)[5]; *id., loc. cit.,* III, 226[6]; *id.,* F. I. Dipt. Brachy. I, 264[7].

[7]Bellary District. [7]Orissa. [7]N. Kanara. [7, 8]Travancore. [7]Cochin State. [7]Coimbatore. District. [7]Upper Burma. ———, [2, 3, 5]Java.

8. Dischistus, Loew.

Dischistus, Lw., Neue Beitr., III. 45, (1855) [Genotype, *Bo. bylius minimus,* Schrk.]; Kert., Cat. Dipt., V, 116; Burn., F. I. Dipt. Brachy. I, 273.

Bombylisoma, Rend., Dipt. Ital. Prodr., i, 164 (1856) [*nec. id.,* Arch. Zool. Modena, III, 67 (1863)].

Bombyliosoma, Marschall, Nomencl. Zool., 323 (1873).

Bombyliosoma, Verr. *apud* Scudder, Nomencl. Zool., 47 (1882).

ardens, Walker.

Bombylius ardens, Wlk., List Dipt. B.M., II, 284 (1849)[1]; Wulp, Cat. Dipt. S. Asia, 74[2]; Kert., Cat. Dipt., V, 126[3].

[2, 3, 4]E. India.

Systoechus ardens, Brun., R. I. M., II, 458.

Dischistus ardens, Brun., F. I. Dipt. Brachy. I, 275[4].

resplendens, Brunetti.

Dischistus resplendens, Brun., R. I. M., II, 458, 481 (1909)[1]; *id., loc. cit.,* III, 226; *id.,* F. I. Dipt. Brachy. I, 273, fig. 15[3].

[1, 2, 3]Simla Hills. [1, 3]Nepal. [1,3]Naini Tal., [1, 3]Assam. [3]Dehra Dun. [3]Cylon. E. Prov.

LOMATINAE.

Verrall, Br. Flies, V, 506; Brun., F. I. Dipt. Brachy. I, 255.

9. Aphoebantus, Loew.

Aphoebantus, Lw., Cent., X, 39, (1872) [Genotype, *A. cervinus,* Lw.]; O.-S., Biologia, I, 143-146; Coq., Can. Ent., XVIII, 84; *id.,* Trans. Amer. Ent. Soc., XXI, 104; *id.,* W. Amer. Sci., 1891, 256; Will., Biologia, I, 285, [notes]; Ald., Cat. N. Amer. Dipt., 242; Brun., F. I. Dipt. Brachy. I, 256.

Triodites, O.-S., Bull. U.S. Geol. Surv., III, 245 (1877).

ceylonicus, Brunetti.

Argyramoeba ceylonica, Brun., R. I. M., II, 450, 471, t. 12. f. 24 (1909)[1]; *id.,* Spol. Zeyl., VI, 171[2].

Aphoebantus ceylonicus, Brun. F. I. Dipt. Brachy. I, 257[3].

[1, 2, 3]Ceylon. [1]Trivandrum., [1,3]Pusa. [3]Darjiling District. [3]Allahabad., [3]Chapra. Abbottabad.

TOXOPHORINÆ.

Verr., Br. Flies V, 509; Brun., F. I. Dipt., Brachy. I, 286.

10. Toxophora Meigen.

Toxophora, Mg., Illig. Marg., II, 270, (1803) [Genotype, *Asilus maculatus,* Rossi]; *id.,* Klass., 1804; *id.,* Syst. Besch., II, 178; Sch., F. Austr., I, 69; Coq., Entomol. Amer., I, 1886, 221; *id.,* W. Amer. Sci. 1891, 199; Glover, Entomol. Rpt. 1866, 45 [reared from *Eumenes* nest]; *id., loc. cit.,* 1870, 78; *id., loc. cit.,* 1877, 105, [preying on *Bembex*]; Wulp, Cat. Dipt. S. Asia, 74 ; Kert., Cat. Dipt., V, 91; Brun., R. I. M., II, 458; *id.,* F. I. Dipt. Brachy. I, 287.

javana, Wiedemann.

Toxophora javana, Wd., Dipt. Exot., 179 (1821); *id.,* Auss. Zweifl. Ins., I, 363; Mq., Suit. A Buff., 387, 2; Wulp, Cat. Dipt. S. Asia, 74[1]; Meij., Tijd. v. Ent., LIII, 66; Brun., R. I. M., II, 458, t. 12 f. 28[3]; *id.,* F. I. Dipt. Brachy. I, 287, t. 3 f. 16[4].

[3, 4]Calcutta. [4]Allahabad. [4]Madras. [2, 4]Java.

SYSTROPINÆ.

Verr., Br. Flies, V, 513; Brun., F. I. Dipt. Brachy. I, 289.

11. Systropus, Wiedemann.

Systropus, Wd., Nov. Gen. Dipt., 18, (1829) [Genotype, *S. macilentus,* Wd.] ; *id.,* Auss. Zweifl. Ins., I, 359; Mq., H. N. Dipt., I, 391; O.-S., Biologia, I, 157 [notes]; Towns., Trans. Amer. Ent. Soc., XXVII, 160; Will., Biologia, I. 292; Ald., Cat. N. Amer. Dipt., 244; Wulp, Cat. Dipt. S. Asia, 74; Kert., Cat. Dipt., V, 93; Brun., R. I. M., II, 459; *id.,* F. I. Dipt. Brachy. I, 289.

Cephenus, Latr., Fam. Nat., 496 (1825); Karsch, Zeit. Berl. Ent. Ges., LIII, 657; Wandoll., Ent. Nacht., XXIII, 198.

Xystropus, Verr. *apud* Scudder, Nomencl., 355 (1882); Verr., Br. Flies, V. 514. [note].

edwardsi, Brunetti.

Systropus edwardsi, Brun., F. I. Dipt. Brachy. 1, 294 (1920)[1]. — [1]Burma.

eumenoides, Westwood.

Systropus eumenoides, Wstwd., Guérin Mag. Zool., 1842, 4, t. 90 (1842)[1]; *id.,* Trans. Ent. Soc., Lon., 1876, 575, t. 10, ff. 13-15; Wulp, Cat. Dipt. S. Asia, 74[2]; Kert., Cat. Dipt., V, 94[3]; Brun., R. I. M., II, 459[4]; *id.,* F. I. Dipt. Brachy. I, 294[5]. — [2, 3, 4]Ind. Or. [5]N. India.

flavipleura, Brunetti.

Inset 5 year Systropus flavipleura, Brun., F. I. Dipt. Brachy., I, 295 (1920)[1]. — Lebong., [1]Sikkim. [1]Khasis., Shillong.

ophioneus, Westwood.

Systropus ophionius, Wstwd., Trans. Ent. Soc., Lon., V, 233, t. 23, f. 6 (1849)[1]; *id., loc. cit.,* 1876, 574; Wulp, Cat. Dipt. S. Asia, 74[2]; Kert., Cat. Dipt., V, 95[3]; Brun., R. I. M., II, 459[4]; *id.,* F. I. Dipt. Brachy. I, 291, fig. 21[5]. — [2, 3, 4]Ind. Or. [5, 6]Nepal., [5, 6]Masuri. [5]Darjiling District. [5]Shillong.

Systropus nigricaudus, Brun., R. I. M., II, 460, 483, (1909)[6].

ANTHRACINÆ

Verr., Br. Flies, V, 515; Brun., F. I. Dipt. Brachy. I, 176-178 [tab. genera].

12. Anthrax, Scopoli.

Anthrax, Scop., Ent. Carn., 358, (1763) [Genotype, A. Morio, Fb.]; Fb., Syst. Antl., 118, (1805); Mg., Syst. Besch., II, 107; Sch. F. Austr., I, 48; Coq., Trans. Amer. Ent. Soc., XIV 159; *id., loc. cit.,* XIX, 170; *id., loc. cit.,* XXI, 97; Ald., Cat. Dipt. N. Amer., 228; O.-S., Biologia, I, 111-140 [subgenera]; Wulp, Cat. Dipt. S. Asia, 71; Kert., Cat. Dipt., V, 25; Verr., Br. Flies, V, 518; Brun., R. I. M., II, 451-456; *id.,* F. I. Dipt. Brachy. I, 234-236.

Aspiloptera, Kunckel, Bull. Soc. Fr. Belg. XXXIX, 145 [note] (1905).

Chrysanthrax, O.-S., Biologia, 121 (1886).

Dipalta, O.-S., Bull. U.S. Geol. Surv., III, 236 (1877); Coq., Can. Ent., 1886, 157; *id.*, Trans. Amer. Ent. Soc., XIV, 159; *id.*, *loc. cit.*, XIX, 169; Will., Biologia, I, 275; Ald., Cat. N. Amer., Dipt., 227.

Hemipenthes, Lw., Berl. Ent. Zeit., XIII, 28 [note] (1869).

Hyalanthrax, O.-S., Biologia, 134 (1877).

Isopenthes, O.-S., Biologia, 80, 96 (1886).

?*Paranthrax*, Big., Bull. Soc. Ent. Fr., (5) VI, lxvi (1876).

Poecilanthrax, O.-S., Bologia, 119 (1886).

Stonyx, O.-S., Biologia, 80, 94 (1886).

Thyridanthrax, O.-S., Biologia, 123 (1886).

Villa, Lioy, Atti. Istit. Veneto, (3) IX, 732 (1864).

afra, Fabricius.

Anthrax afra, Fb., Ent. Syst., IV, 258 (1794); *id.*, Syst. Antl., 122, 15; Sch., F. Austr., I. 50; Wulp. Cat. Dipt. S. Asia, 71[2]; Kert., Cat. Dipt., V, 27[3]; Brun., R. I. M., II, 451[4]; *id.*, *loc. cit.*, III, 222[5]; *id.*, F. I. Dipt. Brachy. I, 240[6].

Anthrax fimbriata, Mg., Klass., I, 205, (1804); *id.*, Syst. Besch., II, 154, t. 17, f. 13; Mq., Suit. à Buff., 405, 16; *id.*, Dipt. Exot., II, 162, 18.

[2, 3]E. India.
[4]Jubbulpore.
[4]Poona., [4, 6]Meerut.
[4]Jhansi.,[4, 6]L. Burma.
[4, 6]Orissa
[4, 5, 6]Gonda, U. P.
[4, 6]Allahabad.
[4, 6]N. Bihar.
[6]Dehra Dun.
[6]Anasmalais.
[6]Satara District.
[6]Madhupur.
[6]Ferozepore.
[6]Madras.
[6]Trincomali.
[6]Andamans.

[2, 3] Europe
[2, 3] N. & W. Africa
[3] Oceania.

albofulva, Walker.

Anthrax albofulva, Wlk., Dipt. Saund., (3) 182 (1852); Wulp, Cat. Dipt. S. Asia, 72[2]; Kert., Cat. Dipt., V, 28[3]; Brun., R. I. M., II 454[4]; *id.*, F. I. Dipt. Brachy. I, 251.

[2,3]E. India.
[4]Muttra.
[4]Meerut.
[4]Ferozepore.
[4]Allahabad.

aperta, Walker.

Anthraxaperta, Wlk., Dipt. Saund., I, 180 (1852); Wulp, Cat. Dipt. S. Asia, 72[2]; Kert., Cat. Dipt., V, 29[3]; Brun., R. I. M., II, 453, 475[4] [redescr.]; *id., loc. cit.,* III, 223[5]; *id.,* F. I. Dipt. Brachy. I, 245[6].

[2, 3]E. India. [4, 6]Masuri., [5, 6]Simla. [6]Dehra Dun., [6]Khasis. [6]Dawnat Range.

Argyramoeba aperta, Howl., Ind. Ins. L., t. 63, f. 2.

approximata, Brunetti.

Anthrax approximata, Brun., R. I. M., XIII, 74 (1917)[1]; *id.,* F. I. Dipt. Brachy. I, 244[2].

Anthrax (?) clara var., Brun., R. I. M., II, 474[3].

[1, 2]Simla., Abbottabad. [1, 2]Siliguri., Pusa. [1, 2, 3]Tenasserim. [1, 2]Khasis. [1,2]Ceylon, E. P.

aureohirta, Brunetti.

Anthrax aureohirta, Brun. R. I. M., III, 223 (1909)[1]; *id.,* F. I. Dipt. Brachy. I, 252[2].

[1, 2]Calcutta.

baluchianus, Brunetti.

Anthrax baluchianus, Brun., F. I. Dipt. Brachy. I, 250[1] (1920).

[1]Baluchistan [1] Kashmir, [1]Bauhist Khal. [1]Bhittani.

clara, Walkar.

Anthrax clara, Wlk., Dipt. Saund., (3) 179 (1852); wulp, Cat. Dipt. S. Asia, 72[2]; Kert., Cat. Dipt. V, 32[3]; Brun., R. I. M., II, 452, 473[4] [redescr]; *id.,* F. I. Dipt. Brachy. I, 243[5].

[2, 3]E. India. [4, 5]Dehra Dun. [4, 5]Gahrwal. [5]W. Himalaya. [5]Khasis.

clausa, Brunetti.

Anthrax clausa, Brun., R. I. M., II, 454, 477[1] (1909); *id.,* F. I. Dipt. Brachy. I, 253.

[1, 2]L. Burma.

dia, Wiedemann.

Anthrax dia, Wd., Anal. Ent., 23 (1824)[1]; *id.,* Auss. Zweifl. Ins., I, 293; Wulp, Cat. Dipt. S. Asia, 71[2]; Kert., Cat. Dipt., V, 34[3]; Brun, R. I. M. II, 453[4]; *id.,* F. I. Dipt. Brachy. I, 250[5].

[2, 3, 4, 5]Tranquebar.

fletcheri, Brunetti.

Anthrax fletcheri, Brun., F. I. Dipt. Brachy. I, 236 (1920)[1].

[1]Mysore., Habarane. Trincomali (White)

fuscolimbata, Brunetti.

Anthrax fuscolimbata, Brun., R. I. M., XIII, 75 (1917)[1]; *id.,* F. I. Dipt. Brachy. I, 246[2].

Hangu (N.-W.F.P.). [1, 2]Simla., [1, 2]Masuri. [1, 2]Madras.

gutta tipennis, Brunetti.

Anthrax guttatipennis, Bruna., F. I. Dipt. Brachy. I, 237 (1920)[1]. — [1]Trincomali.

himalayensis, Brunetti.

Anthrax himalayensis, Brun., R. I. M., III, 222, fig. (1909)[1].

Anthrax himalayanus, Brun.., F. I. Dipt. Brachy., I, 238[2] [*lapsus*].

Anthrax maura, Brun. [*nec.* L.], R. I. M., II. 451[3].

[1, 2, 3]Naini Tal. [2]Simla District. [2]Khasis.

hottentota, Linnæus var. **claripennis,** Brunetti.

Anthrax hottentota var. *claripennis,* Brun., R. I. M., II, 452 (1909)[1]; *id.,* F. I. Dipt. Brachy. I, 247[2]. — [1, 2]Naini Tal (plains).

leucopygus, Macquart.

Anthrax leucopyga, Mq., Dipt. Exot. II, 74, t. 21, f. 1 (1842)[1]; Wulp, Cat. Dipt. S. Asia, 71[2]; Kert., Cat. Dipt., V, 43[3]; Brun., R. I. M., II, 453[4]. [Omitted from the "Fauna of India" volume] — [4]Jubbulpore. [4]Itarsi., [4]Poona. [4]Allahabad. ———, [2, 3]Timor.

limpida, Walker.

Authrax limpida, Wlk., Dipt. Saund., (3) 179 (1853); Wulp, Cat. Dipt. S. Asia, 72[2]; Kert., Cat. Dipt., V, 44[3]; Brun., R. I. M., II, 453[4]; *id.,* F. I. Dipt. Brachy. I, 255[5]. — [2, 3, 4, 5]E. India.

lucida, Walker.

Anthrax lucida, Wlk., Dipt. Saund., (3), 179 (1852); Wulp, Cat. Dipt. S. Asia, 72[2]; Kert., Cat Dipt., V, 44[3]; Brun., R. I. M., II, 453[4]; *id.,* F. I. Dipt. Brachy. I, 254[5]. — [2, 3, 4, 5]E. India.

manifesta, Walker.

Anthrax manifesta, Wlk., Dipt. Saund., I, 178 (1852); Wulp, Cat. Dipt. S. Asia, 72[2]; Kert., Cat. Dipt., V, 45[3]; Brun., R. I. M., II, 453, 476[4]; *id.,* F. I. Dipt. Brachy. I, 243[5]. — [2, 3, 5]E. India. [4]Masuri., [5]Himalaya. [5]Dehra Dun. [5]Garhwal, [5]Khasis.

paniscus, Rossi.

Bibio paniscus, Ross., F. Etrus., II, 256 (1790)[1].

Anthrax paniscus, Verr., Br. Flies, V, 526; Kert., Cat. Dipt., V, 50[2]; Brun., R. I. M., II, 452[3]; *id.,* F. I. Dipt. Branchy. I, 248[4].

Anthrax bimaculatus, Mq., Suit., à Buff., I, 403 (1834).

[3, 4]Simla, [3, 4]Naini Tal. [3]Masuri.

Anthrax cingulatus, Mg., Syst. Besch., II, 145 (1820). [2]Europe.

Anthrax hottentotus, Wlk. [*nec.* L.], List Dipt. B. M., II, 259 (1849).

referrens, Walker.

Anthrax referrens, Wlk., Dipt. Saund., (3) 189 (1852); Wulp, Cat. Dipt. S. Asia, 72[2]; Kert., Cat. Dipt., V, 52[3]; Brun., R. I. M., II. 452[4]; *id.,* F. I. Dipt. Brachy. I, 255[5]. [2, 3, 4, 5]E. India.

Semifuscata, Brunetti.

Anthrax semifuscata, Brun., F. I. Dipt. Brachy., I, 241 (1920)[1]. [1]Trincomali.

troglodyta, Febricius.

Bibio troglodyta, Fb., Syst. Ent., 759 (1775)[1].

Anthrax troglodyta, Wd., Auss. Zweifl. Ins., I, 306; Wulp, Notes Leyd. Mus., VII, 83 ; *id.,* Cat. Dipt. S. Asia, 71[2]; Meij., Tijd. v. Ent. L, 240; Brun., R. I. M., 452[3]; *id.,* F. I. Dipt. Brachy. I, 254[4]; Kert., Cat. Dipt., V, 56[5]. [3, 5]E. Indies. [3, 4]E. India.

Anthrax hyalina, Wd., Dipt. Exot., I, 141, 34 (1821); *id., Auss.* Zweifl. Ins., I, 297; Wulp, Tijd. v. Ent., XXIII, 165. [2, 3]Java., [2]Sumatra. [3, 5]Papua., [5]Ind. Archip.

Anthrax lucens, Wlk., dipt. Saund. 180, (1852).

13. Lepidanthrax, Osten-Sacken.

Lepidanthrax, O.-S., Biologia, I, 107 (1886) [Genotype, *Anthrax disjuncta,* Wd.]; Coq., Trans. Amer. Ent. Soc., XIX, 169; Ald., Cat. N. Amer. Dipt., 227; Brun., F. I. Dipt. Brachy. I, 232.

compacta, Brunetti.

Lepidanthrax compacta, Brun., F. I. Dipt. Brachy. I, 233 (1920)[1]. [1]Allahabad.

transversa, Brunetti.

Lepidanthrax transversa, Brun., F. I. Dipt. Brachy. I, 233 (1920)[1]. [1]Dehra Dun.

14. Argyromoeba, Schiner.

Argyromoeba, Sch., Wien. Ent. Monat., IV, 5 (1860) [Genotype, *Anthrax tripunctata,* Wd.]; *id.,* F. Austr., I, 52; Lw., Besch. Europ. Dipt., I, 228.

Argyramoeba, Lw., Centur., II, 290 [emend.]; Wulp, Cat. Dipt. S. Asia, 723; Kert., Cat. Dipt., V, 59 [*Agyramoeba, laps.*]; Brun., R. I. M., II, 448, 450; *id.*, F. I. Dipt. Brachy. I, 214.

Spogostylum, Will. [*nec.* Mq.], Man. N. Amer. Dipt., (2nd edn.), 65, (1896); *id.*, *loc. cit.*, (3rd edn.) ; 214, Ald., Cat. N. Amer. Dipt., 221.

argentiapicalis, Brunetti.

Argyramoeba argentiapicalis, Brun., R. I. M. VII, 470, t. 37, f. 6 (1912)[1]; *id.*, F. I. Dipt. Brachy. I, 227, Pl. III, 14[2].

[1, 2]Mahableshwar.

bigoti, Brunetti.

Argyramoeba appendiculata, Big. [*nec.* Mq.], Ann. S. Ent. Fr. (7) II, 347 (1892)[1]; Wulp, Cat. Dipt. S. Asia, 73[2]; Kert., Cat. Dipt., V. 61[3]; Brun., R. I. M., II, 449[4].

Argyramoeba bigo i, Brun., R. I. Dipt. Brachy. I, 222 (1920)[5].

[1, 2, 3, 4, 5]Pondicherry.

bipunctata, Fabricius.

Anthrax bipunctata, Fb., Syst. Antil, 118 (1805)[1]; Wd., Auss. Zweifl. Ins., I, 286; Wulp, Cat. Dipt. S. Asia, 71[2].

Argyramoeba bipunctata, Kert., Cat. Dipt. V, 62[3]; Brun., R. I. M., II, 448[4]; t. 12, f. 23; *id.*, F. I. Dipt. Brachy. I, 217[6].

Anthrax carbonaria, Wlk. Dipt. Saund., 173 (1852); Wulp, Cat. Dipt. S. Asia, 72[5]; Kert., Cat. Dipt., V, 31[7].

[2, 3, 4]Tranquebar.
[4, 6]Karachi.
[4, 6]Puraneah.
[5, 7]E. India.,
[6]Calcutta.
[6]Kumaon., [6]Pusa.
[6]Chapra., [6]Bombay.
[6]Assam.
[6]Konan Kuppam.

claripennis, Brunetti.

Argyramoeba claripennis, Brun., R. I. M., III, 218 (1909)[1]; *id.*, F. I. Dipt. Brachy. I, 228[2].

[1, 2]Lahore.

distigma, Wiedemann.

Anthrax distingma, Wd. Auss. Zweifl. Ins., I, 309 (1828)[1]; Mq., Dipt. Exot, II, I, 58, 9.

Spogostylum distigma, Coq., Proc. U. S. Nat. Mus., XXI, 318.

Argyramoeba distigma, Sch., Novara Reise, 122, 14; Wulp, Tijd. v. Ent., XXIII, 166, 19; *id.*, Leyd. Mus. Notes, VII, 83, 38; *id.*, Cat. Dipt., S. Asia, 73[2]; O.-S., Berl. Ent. Zeit., XXVI, 122; Kert., Cat. Dipt., V, 63[3]; Brun., R. I. M., II. 449[4]; *id.*, *loc. cit.*, III, 221[5]; *id.*, Spol. Zeyl., VI, 171[6]; *id.*, F. I. Dipt. Brachy. I,

[8]Matheran.
[4]Jullundur.
[4, 7]Bengal, [4]Allahabad.
[4]Bombay.
[5, 7]Paresnath.
[5, 7]Naini Tal Dist.
[6,7]Ceylon., [7]Hamirpur.
[7]Madras., [7]Utakamund.
[7]Mettupalaiyam.
[7]Khasis., [7]L. Burma.

218, t. 3, f. 10[7]; Meij., Tijd. v. Ent. L, 241 (1907); Howl., Ind. Ins. L., 598; Longst., Butterfly Hunting, 393[8]; Dutt, Pusa Ent. Mem., IV, 212 [notes].

[4, 7]Darjiling.
[4]Dehra Dun.
[4, 7]Lucknow.
[4, 7]Bangalore.
[3, 4]Nicobars.

[2, 4]Java., [2, 4] Sumatra
[2, 4]Celebes.
[2, 3, 4]Philippines.
[2, 3, 4]New Guinea.

duvaucelii, Macquart.

Anthrax duvaucelii, Mq., Dipt. Exot., II, 63, t. 20, f. 7 (1840)[1]; Wulp, Cat. Dipt. S. Asia, 71[2]; Kert., Cat. Dipt., V, 35[3]; Brun., R. I. M., II, 452[4].

Argyramoeba duvaucellii, Brun., R. I. M., III, 219[5] [redescr.]; *id.,* F. I. Dipt. Brachy. I, 225[6].

[2, 3, 4]Bengal.
[4, 5, 6]Purneah.
[4]Cawnpore.
[4, 5]Chapra.
[5, 6]Baraich, U. P.
[6]Kangra Valley.
[6]Pusa., [6]Lahore.
[6]Murree Hills.
Peshawar.
[6]Allahabad.

gentilis, Brunetti.

Argyramoeba gentilis Brun. R. I. M., II, 449, 470, t. 12, f. 25 (1909)[1]; *id.,* F. I. Dipt. Brachy. I, 230[2].

[1, 2]Laksam.
[1, 2]Margherita.

gestroi, Brunetti.

Argyramoeba gestroi, Brun., R. I. M., VII, 470, t. 37, f. 7 (1912)[1]; *id.,* F. I. Dipt. Brachy. I, 222[2].

[1]Bombay Presidency.
[2]Satara District.
[2]Travancore.
[2]Bangalore., [2]Chapra.
Pusa.,
Peshwar.

instituta, Walker.

Anthrax instituta, Wlk., Dipt. Saund., 183 (1852)[1]; Wulp, Cat. Dipt. S. Asia, 72[2]; Kert., Cat. Dipt., V; 42[3].

Argyramoeba instituta, Brun., R. I. M., II. 450[4]; *id.,* F. I. Dipt. Brachy., I, 227[5].

[2, 3, 4, 5]E. India

intermedia, Brunetti.

Argyramoeba interemedia, Brun., F. I. Dipt. Brachy., I, 221 (1920)[1].

[1]Orissa.,
[1]Allahabad.
[1]Trincomali.

limitarsis, Brunetti.

Argyramoeba limitarsis, Brun., R. I. M., III, 217 (1909).

[1, 2]Garhwal.

Argyramoeba limitaris, Brun., F. I. Dipt. Brachy. I, 220[2], [emend.].

nigrofemorata, Brunetti.

Argyramoeba nigrofemorata, Brun., R. I. M., III, 219 (1909)[1]; *id.*, F. I. Dipt. Brachy. I, 230[2]. — [1,2]Garhwal District. [1,2]Satara Distric.

niveicauda, Brunetti.

Argyramoeba niveicauda, Brun., F. I. Dipt. Brachy. I, 224 (1920)[1]. — [1]Travancore. [1]Calcutta., [1]Sylhet. [1]Chapra.

niveisquamis, Brunetti.

Argyramoeba neveisquamis, Brun., R. I. M., II. 472 (1909)[1]; *id.*, F. I. Dipt. Brachy. I, 229[2]; — [1,2] Baluchistan.

obscurifrons, Brunetti.

Argyramoeba obscurifrons, Brun., R. I. M., III, 216 (1909)[1]; *id.*, F. I. Dipt. Brachy. I, 220[2], t. 3. f. 11. — [1,2]Simla District.

varia, Fabricius.

Anthrax varia, Fb., Ent. Syst., IV, 259 (1794)[1]; *id.*, Syst. Antl., 122; Mg., Syst. Besch., II, 124. — [3]Bhim Tal.

Argyramoeba varia, Sch., F. Austr., I, 54; Kert., Cat. Dipt., V, 68[2]; Brun., R. I. M., II, 450[3]; Coq., Trans. Amer. Ent. Soc., XXI, 96. — [2]Europe. [2]N. America.

Spogostylum varium, Ald., Cat. N. Amer. Dipt., 223.

[Omitted from the "Fauna of India" volume.]

15. Spogostylum, Macquart.

Spogostylum, Mq., Dipt. Exot., II, 1, 53 (1841) [Genotype, *Sp. mystaceum*, Macq.]; Kert., Cat. Dipt., V, 58 [*Spongostylum*]; Brun, F. I. Dipt. Brachy. I, 213.

austeni, Brunetti.

Spogostylum austeni, Brun., F. I. Dipt. Brachy. I, 213 (1920)[1]. — [1]Lahore.

16. Exoprosopa, Macquart.*

Exoprosopa, Mq., Dipt. Exot., II, 35, (1840) [*Genotype, Bibio capucina*, Fb. as *pandora*, Fb.]; Sch., F. Austr.,

* Whilst this catalogue has been passing through Press a paper by Nurse has appeared (*Jo. Bomb. N. H. S.* xxviii, 630, tab., June 1922), in which ten new species are described and three Palæarctic species added to the Indian Fauna. Revised key for determination of all described species, pp. 631-632.

I, 54; Coq., Can. Ent., XXIV, 123; O.-S., Biologia, I, 77; Will., *loc. cit.*, I, 269; Ald., Cat. N. Amer. Dipt., 225; Wulp, Cat. Dipt. S. Asia, 69; Kert., Cat. Dipt., V, 5; Brun, R. I. M., II, 443, 447, *id.*, F. I. Dipt., Brachy. I, 189.

Argyrospyla, Rond., Dipt. Ital. Prodr., I, 162 (1856).

Argyrospila, Verr., *apud* Scudd., Nomencl., 32 (1882).

Defilippia, Lioy, Atti. Istit. Veneto, (3) IX, 733 (1864).

Exoptata, (p. p.), Coq., Can. Ent., XIX, 13 (1887).

Hoteralonia, Rond., Arch. Zool. Modena, III, (sep.) 57 (1863).

Litorhynchus, Mq., Dipt. Exot., II, 78 (1840).

Trinaria, Muls., Mem. Acad. Lyon. (2) II, 20 (1852).

albida, Walker.

Anthrax albida, Wlk., Dipt. Saund., I,. 171 (1852)[1]; Wulp, Cat. Dipt. S. Asia, 72[2]. [1, 2, 3, 4, 5]E. Indies.

Exoprosopa albida, Kert., Cat. Dipt., V, 6[3]; Brun., R. I. M., II, 447; *id.,* F. I. Dipt. Brachy. I. 212.

alexon, Walker.

Anthrax alexon, Wlk., List Dipt. B. M., II, 246 (1849)[1]; Wulp, Cat. Dipt. S. Asia, 71[2]. [1, 2, 3, 4, 5]E. Indies.

Exoprosopa alexon, Kert., Cat. Dipt., V, 6[3]; Brun. R. I. M., II, 445[4]; *id.,* F. I. Dipt. Brachy., I, 211[5].

annandalei, Brunetti.

Exoprosopa annandalei Brun., R. I. M., II. 445, 469, t. 12, f. 20 (1909)[1]; *id.,* F. I. Dipt. Brachy. I, 199, t. 3, f. 8[2]. [1, 2]Moulmein.

auripleura, Walker.

Anthrax auripleura, Wlk., Dipt. Saund., I, 171 (1852)[1]; Wulp, Cat. Dipt. S. Asia, 72[2]. [1, 2, 3, 4, 5]E. Indies.

Exoprosopa auripleura, Kert., Cat. Dipt., V, 7[3]; Brun., R. I. M., II, 445[4]; *id.,* F. I. Dipt. Brachy. I, 212[5].

basifascia, Walker. [2, 4]Bengal.

Anthrax basifascia, Wlk., List Dipt. B. M., II, 248 (1849)[1]; Wulp, Cat. Dipt. S. Asia, 71[2]. [3, 5]N. Bengal.

Exoprosopa basifascia, Kert., Cat. Dipt., V, 7[3]; Brun., R. I. M., II, 447[4]; *id.,* F. I. Dipt. Brachy. I, 211[5].

bengalensis, Macquart.

Exoprosopa begalensis, Mq., Dipt Exot., II, 49, t. 18, f. 4 (1840)[1]; Wulp, Cat. Dipt. S. Asia, 70[2]; Kert., Cat. Dipt., V, 7[3]; Brun., R. I. M., II, 445, t. 12, f. 18[4]; *id.,* F. I. Dipt. Brachy. I, 206, t. 3. f. 5[5].

[2, 3, 4]Bengal.
[6]Bellary.
[5]Bellary District.
[5]Ceylon, E. P.

brahma, Schiner.

Exoprosopa brahma, Sch., Novara Reise, 118 (1868)[1]; Wulp; Cat. Dipt. S. Asia, 71[2]; Kert., Cat. Dipt., V, 8[3]; Brun., R. I. M., II, 446, t. 12, f. 22[4]; *id.,* F. I. Dipt. Brachy. I, 206, t. 3, f. 6[5].

[2, 3, 4, 5]Ceylon.

flammea, Brunetti.

Exoprosopa flammea, Brun. R. I. M., II, 444, 466 (1909)[1] ; *id.,* Spol. Zeyl., VI, 171[2]; *id.,* F. I. Dipt. Brachy. I, 193[3]; Howl., Ind. Ins. L., t. 63, f. 3 (1909).

[1, 2]Pusa., [2]Ceylon.
[1]Trivandrum,
[3]Trincomali.

flavipennis, Brunetti.

Exoprosopa flavipennis, Brun., R. I. M. II, 444, 465 (1909)[1]; *id.,* F. I. Dipt. Brachy. I, 192[2].

[1, 2]Pusa.

insulata, Walker.

Anthrax insulata, Wlk., Dipt. Saund., 172 (1852)[1]; Wulp, Cat. Dipt. S. Asia, 72[2]; Brun., R. I. M., III, 222[3].

Exoprosopa insulata, Kert., Cat. Dipt., V, 14[4]; Brun., R. I. M., II, 446, t. 12, f. 21[5]; *id.,* F. I. Dipt. Brachy. I, 203, t. 3, f. 4.

Exoprosopa octonotata, Big., *nom. nud.*

[2, 4, 5]E. India.
[3, 5, 6]Nepal Terai.
[3, 6]Gonda, U. P.
Simla., [5]Masuri.
[6]Dehra Dun.
[6]Naini Tal.
[6]Allahabad.
Belgachia.
[6]Bhandhara, C. P.
[6]Madhupur.
Chitorgarh.
Surat., [6]N. Kanara.
[6]Utakamand.
[6]Ceylon., [6]N. Khasis
[6]L. Burma.

lar, Fabricius.

Bibio lar, Fb., Spec. Ins., II, 414(1781)[1].

Anthrax lar, Fb., Syst. Antl., 110; *id.,* Ent. Syst, IV, 257, 4; Wd., Auss. Zweifl. Ins., I, 268.

Exoprosopa lar, Wulp, Cat. Dipt. S. Asia, 70[2]; Kert., Cat. Dipt., V, 15[3]; Brun., R. I. M., II, 443, 444[4]; *id.,* F. I.

[2, 3, 4]Bengal., [4]Bombay.
[5]United provinces.
[5]Central. provinces.
[7]Central India.
[6, 7]Madras.
[5, 9]Kangar Valley.

Dipt. Brachy. I, 194-7, t. 3, f. 1[5].

Exoprosopa binotata, Mq., Dipt. Exot., Supp. 5, 69, 66 (1855); Wulp, Cat. Dipt. S. Asia, 70[8]; Kert., Cat. Dipt., V, 8[12].

[5, 9]Sikkim.
[5, 10]Ceylon, E. P.
[5, 7]Travancore.
[8]E. India.

**Anthrax collaris,* Wd., Auss. Zweifl. Ins., I, 271 (1828); Wlk., List Dipt. B. M., II, 247; Wulp, Cat. Dipt. S. Asia, 71[6].

Litorhynchus collaris, Mq., Dipt. Exot., I, 1, 80.

Exoprosopa collaris, Kert., Cat. Dipt., V, 9[11]; Brun., R. I. M., II, 444, t. 12, f. 16[7]; *id., loc. cit.,* III, 216[9]; *id.,* Spol. Zeyl., VI, 171[10].

[11]Australia.

Anthrax ruficollis, Saund., T. Ent. Soc. Lon., III, 59, Pl. V, 5 (1841).

lateralis, Brunetti

Exoprosopa lateralis, Brun., R. I. M., II, 445, 467-8, t. 12, f. 7[1] (1909); *id.,* F. I. Dipt. Brachy. I, 198, t. 3, f. 7[2].

[1, 2]Calcutta.
[2]N. Kanara.

latipennis, Brunetti.

Exoprosopa latipennis, Brun., R. I. M., II, 444, 464-5 (1909)[1]; *id.,* F. I. Dipt. Brachy. I, 198[2].

[1, 2]Shillong.
[2]Khasis.

maculiventris, Brunetti.

Exoprosopa maculiventris, Brun., F. I. Dipt. Brachy. I, 200 (1920)[1].

[1]Garhwal.

niveiventris, Brunetti.

Exoprosopa niveiventris, Brun., R. I. M., III. 214-216 (1909)[1]; *id.,* F. I. Dipt. Brachy. I, 207[2].

[1, 2]Paresnath.
[2]S. Coorg.
[2]Ceylon. E. P.

pennipes, Wiedemann.

Anthrax pennipes, Wd., Dipt. Exot. I, 129 (1821)[1]; *id.,* Auss. Zweifl. Ins., I, 272.

Hyperalonia pennipes, Rend., Ann Mus. Genoa, VII, 452.

Exoprosopa pennipes, Mq., Dipt. Exot., II, 1, 49; Wulp, Cat. Dipt. S. Asia, 70[2]; Kert., Cat. Dipt., V, 19[3]; Brun., R. I. M., II, 443[3]; *id.,* F. I. Dipt. Brachy. I. 197[5].

[4, 5]Karachi.
[4, 5]Assam., [4, 5]Pusa.
[3]Madras.

[2, 3, 4]Borneo.
[3]Java., [3]Oceania.
[2] ?Cape of Good Hope.

puerula, Brunetti.

Exoprosopa puerula Brun. F. I. Dipt. Brachy. I, 205, t. 3, f. 9 (1920)[1].

[1]Madras.

---

* Nurse (*Bmb. Jo.,* xxviii, 630) now regards this species as distinct, placing it and per in *Litorhynehus* as a valid genus.

semilucida, Walker.

Anthrax semilucida Wlk. Dipt. Saund. III, 170 (1852)[1]; Wulp, Cat. Dipt., S. Asia, 72[2].

Exoprosopa semilucida Kert., Cat. Dipt., V, 21[3]; Brun., R. I. M., II, 447[4]; *id.,* F. I. Dipt. Brachy. I, 212[5].

[1, 2, 3, 4, 5]E. Indies.

stylata, Brunetti.

Exoprosopa stylata Brun. F. I. Dipt. Brachy. I, 201 (1920)[1].

[1]Bellary District.

vitrea, Bigot.

Exoprosopa vitrea Big., Ann. Soc., Ent. Fr., LXI, 344 (1892)[1]; Wulp, Cat. Dipt. S. Asia, 71[2]; Kert., Cat. Dipt. V, 24[3]; Brun., R. I. M., II, 447[4]; *id., loc. cit.,* VII, 467-8, t. 37, f. 5[5]; *id.,* F. I. Dipt. Brachy. I, 210.

[2, 3, 4]Pondicherry.
[5, 6]Satara District.
[6]W. Ghatas.
Mohapur (Bengal).

vitripennis, Brunetti.

Exoprosopa vitripennis, Brun. R. I. M. VII, 467-9 (1912)[1]; *id.,* F. I. Dipt. Brachy. I, 210[2].

[1, 2]Mahableshwar.
[2]Ceylon E. P.

17. Hyperalonia, Rondani.

Hyperalonia, Rond., Arch. Zool. Modena, III, 1, (1864) [Genotype, *Stomoxys morio,* Fb., (*Anthrax erythrocephalus,* Fb.]; O.-S., Biologia, I, 89, [revised chars.] ; Ald., Cat. N. Amer. Dipt., 224; Brun., R. I. M., II, 438, 442; *id.,* F. I. Dipt. Brachy. I, 178; Kert., Cat. Dipt., V, I, Will. Biologia, I, 273 [notes].

Velocia, Coq., Can. Ent., XVIII., 158. (1886).

aurantiaca, Guerin.

Anthrax aurantiaca, Guer., Icon. Regne Anim., III, (Ins.) 539. t. 95, f. 6, Chapra. (?1843)[1].

Exoprosopa aurantiaca, Wulp, Cat. Dipt. S. Asia, 70[2]; Kert., Cat. Dipt., V. 7[3].

Hyperalonia aurantiaca, Brun., R. I. M., II, 440, 460 [redescr.][4]; *id.,* F. I. Dipt. Brachy. I, 186[5].

[2, 3, 4, 5]Bengal.
Chapra.

chrysolampis, Jaennicke.

Exoprosopa chrysolampis, Jaen., Abh. Senck. Nat. Ges., VI, 344, t. 43, f. 8 (1867)[1]; O.-S., Ann. Mus. Gen., XVI, 433; Wulp, Cat. Dipt. S. Asia, 70[2].

Hyperalonia chrysolampis, Kert., Cat. Dipt., V, 2[3]; Brun., R. I. M., II, 439[4]; *id.,* F. I. Dipt. Brachy. I, 182[5].

[4, 5]Sikkim.
[4, 5]Bhim Tal.

[2, 3, 4, 5]Java.
[2, 3, 4, 5]Moluccas.
[4, 5]Borneo.

dives, Walker.

Anthrax dives. Wlk., List Dipt. B. M., II, 40 (1849)[1]; Wulp; Cat. Dipt. S. Asia, 71[2].

[2, 3, 4, 5]Sylhet.

Exoprosopa dives, Kert. Cat. Dipt., V, 10[3].

Hyperalonia dives, Brun. R. I. M., II, 439[4]; *id.,* F. I. Dipt. Brachy. I, 180.

flaviventris, Doleschall.

Anthrax flaviventris, Dol., Nat. Tijd. Ned. Ind., XIV, 400, (1857)[1];

Exoprosopa flaviventris, Wulp, Cat. Dipt. S. Asia, 70[2]; Kert., Cat. Dipt., V, 12[3].

Hyperalonia flaviventris, O.-S., Berl. Ent. Zeit., XXVI, 112; Brun., R. I. M., II, 441[4]; *id.,* F. I. Dipt. Brachy. I, 183.

[4, 5]Tenassarim.
[4, 5]Trivandrum.
Coog.

[2, 3, 4, 5]Amboina.
[3]Philippines.

oenomaus, Rondani.

Hyperalonia oenemaus, Rond., Ann. Mus. Gen., VII, 453 (1875)[1]; Brun., R. I. M., II, 442[2]; *id.,* F. I. Dipt. Brachy. I, 185[3].

Exoprosopa oenomaus, O.-S., Berl. Ent. Zeit., XXVI, 112; Wulp, Cat. Dipt. S. Asia, 71[4].

[3]Darjiling District.
[3]Sikkim., [3]Mergui.
[3]Sadiya., Shillong.
———, [2, 3, 4]Borneo.
[2, 3, 4]Phillippines.

semifuscata, Brunetti.

Hyperalonia semifuscata, Brun., R. I. M., VII, 466, t. 37, f. 4 (1912)[1]; *id.,* F. I. Dipt. Bracy., I, 188[2].

[1, 2]Mahableshwar.

sphinx, Fabricius.

Bibio sphinx, Fb. Mantissa, II, 329, 20 (1787)[1].

Anthrax sphinx, Fb. Syst. Antl., 126; Wd., Auss. Zweifl. Ins., I 258; Mq., Dipt. Exot., II, L, 37.

Exoprosopa sphinx, Wulp, Cat. Dipt. S. Asia, 70[2]; Kert., Cat Dipt., V, 22[3].

Hyperalonia sphinx. Brun., R. I. M., II, 440[4]; *id.,* F. I. Dipt. Brachy. I. 187[5]. Howl., Ind. Ins. L., 600 [parasite of *Sceliphron*]; Dutt, Pusa Ent. Mem., IV, 210, fig. 8 [pupa], fig. 9 [bionomics].

[2, 3]E. India.
[4, 5]Trivandrum.
[4, 5]Bombay.
[5]Orissa., [5]Chapra.
[5]Coimbatore.
[4, 5]Ceylon.

suffusipennis, Brunetti.

Hyperalonia suffusipennis, Brun., R. I. M., II, 462, t. 12, f. 15 (1909)[1]; *id.,* F. I. Dipt. Brachy I, 188, t. 3, f. 1[2].

[1, 2]Purneah., [2]Naini Tal.
[2]Dehra Dun., [2]Coorg.
Tranquebar.

tantalus, Fabricius.

Anthrax tantalus, Fb., Ent. Syst., IV, 260 (1794)[1]; Wd. Auss. Zweifl. Ins., I, 255, 4; Mq., H. N. Dipt., I, 401; Kirby., A. M. N. H., 1884, 458.

Exoprosopa tantalus, Mq., Dipt. Exot., I, L. 37; Kert., Cat. Pal. Dipt., II, 169[4]; Meij., Tijd. v. Ent., L, 239; Wulp, Cat. Dipt. S. Asia, 70[2].

Hyperalonia tantalus, Rond., Ann. Mus. Gen. VII, 453; Brun., R. I. M., II, 438[3]; *id.,* F. I. Dipt. Brachy., I, 181[5]; Kert., Cat. Dipt., V, 4[8]; Longstaff, Butterfly Hunting, 372[9].

Anthrax caeruleopennis, Dol., Nat. Tijd. Ned. Ind., XIV, 400, t. 9, f. 2[6] (1857).

Exoprosopa caeruleopennis, O.-S., Ann. Mus. Gen., XVI, 433; Wulp, Cat. Dipt. S. Asia, 70[7].

Hyperalonia hyx Big., *nom. nud.*

[2, 3 8]Hindostan.
[3, 5]Tranquebar.
[3, 5]Sikkim., [3, 5]Nepal.
[3, 5]Assam.
[3, 5]Tennaserim.
[5]Naini Tal District.
[9]Ceylon.

———

[2, 3, 5]Java., [2, 3, 5]Borneo.
[2, 3, 5]Celebes.
[2, 3, 5, 8]China.
[3, 5, 8]Pepua.
[3, 5]Sumatra.
[3, 4, 5, 8]Japan.
[6, 7]Amboina.
[8]Malay Archipelago.

tristis, van der Wulp.

Exoprosopa tristis, Wulp, Tijd. v. Entr., XI, 107, t. 3, f. 11 (1868)[1]; *id.,* Cat. Dipt. S. Asia, 71[2]; Kert., Cat. Dipt., V, 23[3].

Hyperalonia tristis, Brun., R. I. M., II, 439[4]; *id.,* F. I. Dipt. Brachy. I, 180[5].

[4, 5]Tenmalai (Travancore).

———

[2, 3, 4]Timor.

PART IV
TRYPETIDÆ (TRYPANEIDAE)

Howl., Ind. Ins. L., 631-634, ff. [general acct.] ; Frogg., "Fruit-Fly," 73-115, tab. i-viii [economic]; Zerny, Verh. z. b. ges. Wien, lix, 252 [nomenclature]; Bezzi, Boll. Lab. Zool. Portici, V, 2-4 [classification]; *id.*, Mem. Ind. Mus., iii, 52-175, tab. viii-x [full acct. Indian spp.]; Zerny, Wien. Ent. Zeit., xxxiii, 74 [synopsis of genera of World].

DACINAE

Bezzi, Mem. Ind. Mus., iii, 84-87, 92 [classificatory]; *id.*, B. Ent. R., vii, 99-121 [revision, list of lost plants]; *id.*, Phil. Jo. Sci., xv, 411-442, tab. i, ii [list of Indian spp., key to genera]; anon., Trop. Agric., xxxv, 222 [baits]; Howl., B. Ent. R., vi, 297 [chemical reactions].

1. **Dacus,** Fabricius.

Dacus, F., Syst. Antl., 272, (1805) [genotype, *Dacus podagricus,* Fb.]; Wulp, Cat. Dipt. S. Asia, 186; Frogg., "Fruit-Fly," 78; Bezzi, B. Ent. R., vi, 85, ff. 1, 2, 3 [chaetotaxy, wing]; *id., loc. cit., vii,* 101[revision]; Hend., Suppl. Ent. (1), Aug. 1912, 13-24, Tab. 1.

Leproxyda, Macq., H. N. Dipt., ii, 452, 2, (1835) [genotype, *D. longistylus,* Wd., as *L. testacea,* Mq.]; Bezzi, Mem. Ind. Mus., iii, 92.

brevistylus, Bezzi.

Dacus brevistylus, Bezzi, Boll. Lab. Zool. Portici, 111, (1908)[1]; *id.,* B. Ent. R., v, 154[2]; *id., loc. cit.,* vi, 100[3]; *id., loc. cit.,* vii, 101[4].

[4]Hagari., [4]Cuddapah.
[4]Coimbatore.

[1, 2, 3]Africa (General)
[2]Europe.

longistylus, Wiedemann.

Dacus longistylus, Wd., Auss. Zweifl. Ins., ii, 522-523, (1830)[1]; Bezzi, B. Ent. R., v, 154[2]; *id., loc. cit.,* vi, 99[3]; *id., loc. cit.,* vii., 101[4]; Surc. Insecta, (1911), 269, f.[5]; Adair, Ag. Jo. Egypt, x, 18[7].

[6]Karachi.
[2, 4]Coimbatore.
[4]Puri.,
Nagpur.
Matale (Ceylon).

Leptoxyda longistyla, Bezzi, Mem. Ind. Mus, iii, 92, t. viii. f. 1.[1] — [3,7]Egypt.

Leptoxyda testacea, Mq., H. N. Dipt., ii, 452, (1835).[8] — [3, 5, 8] Senegal

Oleae, Rossi.

Musca oleae, Rossi, F. Etrus., ii, 317, (1790); F., Ent. Syst. iv, 349, (1794[1]), Latr; Gen. Crust. Ins., iv, 351; Coqueb., Illuster., 110, tab. xxiv, f. 16; Gmel. in Linn., Syst. Nat., ed. xiii, 2844, (1790); Petagna, Inst. Ent. 685, (1792). — [1]S. Europe. [2]N. Africa.

Dacus oleae, F., Syst. Antl., 215; Latr., Dict. Hist. Nat., 24, 196, 577; Mg., Syst. Besch., vi, 22, tab. lvi, fig. 11-13; Mq., Suit. a Buff., ii, 451; Guer.-Men., Bull. S. Ent. Fr., (2) iii, p. lxiii; *id., Gard.* Chron., 203, col. 2, (1847); Lucas, Expl. Sci. Alg., iii, 496[2]; Goureau, Bull. S. Sci. Yonne, xv, pt. 2, 226-228; Lw., Trypet., 124, tab. xxvi. fig. 2; Schin., F. Austr., ii, 175; Rond., Dipt. Ital. Prodr., vii, 48; Kaltenb., Pflanzenf., 437, (1872); Tasehenb., Prak. Inseck. Kund., iv, 141, (1880); Peragallo, Ins. Nuisibles, 88-103, 157-160; Brauer, Zweifl. K. K. Mus., (1883); 88. Tryon, queensland, Rpt. Ins. Fungus Pests, I, 73-74; *id.,* Queensland Dept. Agric., bull. x, (1896); Neville-Rolfe, Queens. Ag. Jo., ii, 377; Leonardi, Inset. Noc., iii, 272-283; Berlose, Redia, iii, 386-388 [control methods]; Fischer, Tropenpflz., ix, 668; *id,* Redia, iv, 2-44; Silv., Atti Ist. Nap., lix, 657-677 [parasites]; de Stef., Nat. Sicil., xix, 174-176; Silv., Coltivatore, nos. 23 & 24, 1907 [parasites]; Berlese[1] Redia, iv, 90-180; *id., loc. cit.,* iv[1], 193-197; Martelli, *loc. cit.,* iv, 261-275; Guercio, *loc. cit.,* iv, 276-284; Paoli, *loc. cit.,* iv, 285-288 [parasites determined]; Berlese, *loc. cit.,* iv, 289-328 [parasites]; Martelli, Boll. Lab. Portici, ii, 3-12; Silvestri, *loc. cit.,* ii, 13-17; Silv., Masi & Mart., *loc. cit.,* ii, 18-82 [parasites]; Masi, *loc., cit.,* ii, 185-194 [parasites]; Paoli, Redia, v, 27-30; Guercio, Atti Acc. Firenzo, (5) v. disp. 1, 18-63; Berlese, Atti. Ist. Nap., lx, 193-224; Comes, *loc. cit.,* lx, 3-12; Berlese, *loc. cit.,* (5) v, no. 3, 9 pp.; *id.,* Ann. Fac. Sci. Marseille, xviii, 149-153; *id.,* Quinz. coloniale, xiii, 525; *id.,* Atti. Acc. Firenze, (5) vii, disp. 3, 203-216; Marchal Bull. S. Ent. Fr., 1910, 243 [parasite]; Martelli, Boll. Lab. Zool. Portici, iv, 73-104, (1910); Berlese, Redia, vii, 111-155; *id., loc. cil.,* vii, 470 [parasites]; Marchal, C.R. Acad. Sci. Paris clii, 215-218 [paras; ites in Tunis];

Berlese, Atti. Acc. Firenze, (5) viii, disp. 2, 53-100, 2 ta. ; Guiccard, *loc. cit.*, (5) viii, disp. 2, 101-113; *id.*, Quinz. Colon., xv, 202; Campbell, II Coltivat. Casale Monferrato, lvii, 48-52; Berlese, Agric. toscana, Firenze, ii, 49-53; *id.*, Atti. Acad. Firenze, (5) ix, disp. 3, 180-195; *id.*, Quinz. Colon., xvi, 333; Truflaut, Enem. Plant. Cult., 387; Macdonald, Jo. Ag. Vict., x, 405; *id.*, *loc. cit.*, x, 470: Gabriele, Giorn. Agric. Ital., xxi, no. 13, 19 pp.; Chapelle, Vie Agric., iii, 80, 82; Silvestri, Boll. Lab. Zool. Portici, viii, 70-88; Berlese, Inset. Noc., 134, 143-162, (1915); Mantero, Boll. Lab. Zool. Portici, ix, 389; Guen., ent. Par. Agr., 280-283; Silvestri, Boll. Lab. Zool. Portici, xi, 170-182, 9, ff. [hyperparasite]; *id.*, Boll. Soc. Naz. olivicoltor., xii, 16-18 [parasites]; Cyprus. Agrie. Jo., xiv, 76 (legislation].

oleae var. **asiatica,** Silvestri.

Dacus oleac var. asiatica, Silv., Att. Acad. Lincei, xxv, (2) 425, (1916), [description and new parasite].[1]

[1]Peshawar District.

2. Chaetodacus, Bezzi.

Chaetodacus, Bezzi, Mem. Ind. Mus., iii, 89, (1913) [genotype *Ch. ferrugineus,* F., p. 93]; *id.*, B. Ent. R., vii, 101 [key to Indian spp.], 113 [partial key].

Dasyneura, Saund, nec Rond., Tr. Ent. Soc. Lon., iii, 60, 1849 [genotype, *D. zonata,* Saund.]; wulp, Cat. Dipt. S. Asia, 194.

biguttatus, Bezzi.

Chaetodacus biguttatus, Bezzi, B. Ent. R., vii, 111, (1916).[1]

[1]Darjiling.

bipustulatus, Bezzi.

Bactrocera bipustulata, Bezzi, B. Ent. R., v, 153, (1914).[1]

Chaetodacus bipustulatus, id., loc. cit., vii, 112[2].

[1, 2]Mysore Hills.
[2]Coonoor.

caudatus, Fabricius

Dacus caudatus, F., Syst. Antil., 276, 16, (1805)[1]; Wd., Auss. Zweifl. Ins., 11, 518, 8 (pp.)[2]; Wulp, Cat. Dipt. S. Asia, 186[3]; Meij., Tijd. v. Ent., li, 179, 5[4]; Bezzi, Boll. Lab. Zool. Portici, 111, 294[5]; Enderl. Zool. Jahrb., xxxi, 409[6]; Hend., Supp. Ent., i, 16,2, tab. 1, 2; Green, Trop. Agric, xxxvii, 240.[14]

[9]Kumaon.
[9]Nepal.
[8, 9]Bengal.
[9]Assam.
[9]Dehra Dun.
[13]Pusa.

Dasyneura caudata, Walk., List. Dipt. B. M., iv, 1073.[8]

[10]Madras Hills.

Bactrocera caudata, Bezzi, Mem. Ind. Mus., iii, 97, tab. viii, f. 8.[9]

[9]Burma.

Chaetodacus caudatus, id., B. Ent. R., vii, 110[10]; *id.,* Phil. Jo. Sci., (1913), 321[11]; *id., loc. cit.,* xv, 411, 428[12]; Fletcther, iv, Ent. Mtg., 15[13]; White, Ind. Agric. Ent. Mem., vii, 97.[14]

[9, 11]Ceylon.
[1, 2, 3]Java.
[1, 2, 3]Philippines.
[6, 7]Formosa.

Correctus, Bezzi.

Bactrocora zonata, Bezzi [nec. Saund.], Mem. Ind. Mus., iii, 94, (1913), tab. viii, fig. 4.

[2]Pusa.
[2]Hagari.

Chaetodacus correctus, id., B. Ent. R., v., 107, (1916).[2]

[2]Guindy.

eucurbitae, Coquillet.

Dacus cucurbitae, Coq., Ent. News, 1899, 129, (1899)[1]; Grim., F. Hawaii Dipt., 45, 4[2]; v. Dine, Haw, Forest, and Agric., 111, 127[3]; Lefroy, Mem. Dept. Ag. Ind., 1, 130, 228[4]; Bezzi Boll. Lab. Zool. Portici, iii, 291[5]; Frogg., "Fruit-Fly," 84, tab. ii, f. 6, 7[6]; Howl., Ind. Ins. L., 633, fig. 418[7]; Green, Trop. Agric., xxxvii, 240 [host plants].[8]

Throuhgout India, Burma and Ceylon.
[11, 12]Philippines.
[1, 2, 3, 14, 5] Hawaii.

Bactrocera cucurbitae, Bezzi, Mem. Ind. Mus., iii, 96, tab. viii, fig. 7.[9]

Chaetodacus cucurbitae, Bezzi, B. Ent. R., vii, 109 [variations][10]; *id.,* Phil. Jo. Sci., 1913, 321[11]; *id., loc. cit.,* xv, 411, 428[12]; Feltcher, iv, Ent. Mtg., 15[13]; Fullaway, iii, Ent. Mtg., 625-629 [control].[14]

diversus, Coquillet.

Dacus diversus, Coq., Proc. Ent. S. Wash, vi, 139, (1904)[1]; Frogg., "Fruit-Fly," 83[2]; Bezzi, Boll. Lab. Zool. Portici, iii, 294[3]; Howl., Ind. Ins. Life, 632, tab. lxvi, fig. 2.[4]

[5]Bengali., [2, 6, 7]Pusa.
[6]Dehra Dun.
[7]Ahmedabad.
[7]Nagpur.

Bactrocera diversa. Bezzi, Mem. Ind. Mus., iii 94, tab. viii, figs. 2, 3.[5]

Chaetodacus diversus, id., B. Ent. R., vii, 108[6]; Fletcher, iv, Ent. Mtg., 15 [at bug exerement].[7]

[7]Shillong.
[7]Godavari District.
[1, 2, 7]Bangalore.
[7]Coimbatore.
[1]Colombo.

duplicatus, Bezzi.

Chactodacus duplicatus, Bezzi, B. Ent. R., vii, 107, (1916).[1]

[1]Pachmari.
Banhar (Inglis).

ferrugineus, Fabricius.

Musca ferruginea, F., Ent. Syst., iv, 342, (1794).[2]

Dacus ferruguineus, id., Syst. Antl., 274, 5[2]; Wd., Auss. Zweifl. Ins., ii, 515 5[3]; Mq., Dipt. Exot., supp. 3, 64[4]; *id., loc. cit.,* supp. 4, 284[5]; Wulp, C. R. S. Ent. Belg., 1884, 296, 14[6]; *id.,* Cat. Dipt. S. Asia, 186[7]; v. Rod., Ent. Nacht., xix, 235, 22[8]; Cotes Ind. Mus. Notes, iii, 17[9]; Meij., Tijd. v. Ent., li, 126[10]; Frogg., "Fruit-fly," 86, tab. ii, fig. 8[11]; Bezzi, Boll. Lab. Zool. Portici, iii, 293[12]; Green, Trop. Agric., xxv, 297[13]; *id., loc. cit.,* xxxvii, 240 [host plants][14]; Jarvis, Queens. Ag. Jo., xvii, 309 (abs. R. A. E. [A] x, 477) [control].[15]

Bactrocera ferruginea, Bezzi, Mem. Ind. Mus., iii, 95, tab. viii, fig. 5.[16]

Chaetodacus ferrugineus, id., B Ent. R., vii, 104, [redefined *sens. strict.*], (1916)[17]; Fletcher, iv, Ent. Mtg., 14.[18]

Bactrocera conformis, Dol., Nat. Tijd. Ned. Ind., xvii, 122. 78, (1859).[19]

Dacus conformis, Konigs., Med. uit. s'Lands Plant., xx, 24, (1897).[20]

Tephritis tryoni, Frogg., Ag. Gaz. N.S. W., 1917, 410, tab. viii, fig. 1, (1907).[21]

Dacus tryoni, id., "Fruit-Fly," 79, tab. i, fig. 1, tab. vi.[22]

[18]Nagpur.
[16]Sikkim.
[16]Assam.
[16]Tenasserim.
[17]U. Burma.
[16, 17]N. Bihar.
[16]Bangalore.
[9, 16]Calcutta.
[7, 11, 16, 14]Ceylon.
[7]Java.
[7, 19, 22]Amboina.
[21, 22]Queensland.

ferrugineus var. **dorsalis,** Hendel.

Dacus dorsalis, Hend., Supp. Ent., i, 1912, 18, tab. i. fig. 3,. (1912).[1]

Chaetodacus ferrugineus dorsalis, Bezzi, B. Ent. R., vii, 104[2]; *id.,* Phil. Jo. Sci. xv, 412, 423.[3]

[2]Peshawar.
[2]Pusa., [2]Coimbatore.
[2, 3]Burma
[2, 3]Ceylon., [3]Java.
[3]Philippines,

ferrugineus var **incisus,** Walker.

Dacus incisus, Wlk., T. Ent. S. Lon., n.s., V, 323, (1860)[1]; Wulp, Cat. Dipt. S. Asia, 187.[2]

Chaetodacus ferrugineus incisus, Bezzi, B. Ent. R., vii, 105[3]; Fletcher, iv, Ent. Mtg., 14.[4]

[1, 3]Formosa.
[4]Nagpur., [3]Kumaon.
[3]S. India Hills.
[1, 2, 3, 4]Burma.
Ceylon, (White).

ferrugineus var. **versicolor,** Bezzi.

Chaetodacus ferrugineus versicolor, Bezzi, B. Ent. R., vii, 105, (1916).[2]

[1]Pusa., [1]Coimbatore.
[1]Peradeniya.

garciniae, Bezzi.

Bactrocera garciniae, Bezzi, Mem. Ind. Mus., iii, 97, tab. viii, fig. 9, (1913).[1]

Dacus garciniae, Green, Trop. Agric., xxxvii, 240 [host plants].[2]

Chaetodacus garciniae, Bezzi, B. Ent. R., vii, 110.[3]

[1, 2, 3]Peradeniya.

hageni, de Meijere.

Dacus hageni, Meij., Tijd. v. Ent., liv, 375, (1911).[1]

Chaetodacus hageni, Bezzi, B. Ent. R., vii, 109.[2]

[2]Meiktila.
[1]Sumatra.

maculipennis, doleschall.

Bactrocera maculipennis, Dol., Nat. Tijd. Ned. Ind., x, 412, 36, tab. ii, fig. 1, (1856)[1]; Bezzi, Mem. Ind. Mus., iii, 99, tab. viii, fig. 11.[2]

Dacus maculipennis, Meij., Tijd. v. Ent., li, 127, 2[3]; Bezzi, Boll. Lab. Zool. Portici, iii, 293.[4]

Chaetodacus maculipennis, id., B. Ent. R., vii, 109[6]; Fletcher, iv, Ent. Mtg., 15.[5]

[2]Kurseong.
[2]Calcutta., [2]Pachmari.
[2]Coimbatore.
[2]Minbu.
[1, 2, 3]Java.

scutellaris, Bezzi.

Bactorcera scutellaris, Bezzi, Mem. Ind. Mus., iii, 98, tab. viii, fig. 10, (1913).[1]

Chaetodacus scutellaris, id., B. Ent. R., vii, 113.[2]

[1]Sikkim.
[1]Khasia
[1, 2]Kumaon.
[2]Shan States.

scutellarius, Bezzi.

Chaetodacus scutellarius, Bezzi, B. Ent. R., vii, 110, (1916)[1]; White, Ent. Mem. Dept. Ag. Indi., vii, 97, (1922).[2]

[1]S. Mysore.
[2]Shillong.

tuberculatus, Bezzi.

Chaetodacus tuberculatus, Bezzi, B. Ent. R., vii, 106, (1916).[1]

[1]Shan States.

zonatus, Saunders.

Dasyneura zonata, Saund., T. Ent. S. Lon., iii, 61, tab. v, fig. 3. (1841)[1]; Wlk., List. Dipt. B. M., iv, 1075[2]; Wulp, Cat. Dipt. S. Asia, 194.[3]

Dacus zonatus, Bezzi, Boll. Lab. Zool. Partici, iii, 293[4]; Ender., Zool. Jahrb. xxxi, 408.[5]

Chaetodacus zonatus, Bezzi, B. Ent. R., vii, 105[6]; Fletcher, iv, Ent. Mtg., 15[7]; Adair, Ag. Jo. Egypt, x, 18.[8]

[6, 7]Peshawar.
[6]Moradabad.
[7]Nagpur.
[6]Pachmari.
[17]Chota Nagpur.
[6, 21, 23, 7]Pusa.
[1, 2, 3, 13, 14]Bengal.
[6]Coorg.

Bactrocera maculigera, Doll., Nat. Tijd. Ned. Ind., xvii, 122, 79, (1859).[9] — [21]Bangalore. [24, 25]India.

Dacus maculiger, Frogg., Rpt. Par. Ins. In.s, 1917, 8. 94[10]; Wulp, Cat. Dipt. S. Asia, 187[11]; Bezzi, Boll. Lab. Zool Portici, iii, 294.[12] — [9, 11]Amboina. [8]Egypt.

Dacus mangiferae, Cotes, Ind. Mus. Notes, iii, 17 fig., (1893).[13]

Bactrocera ferruginea mangiferae, Bezzi, Mem. Ind. Mus., iii, 96, tab. viii, fig. 6.[14]

Dacus ferrugincus, Lefroy [nec. F.], Ind. Ins. Pests, 170, figs. 190-192[15]; *id.,* Mem. Dept. Ag. Ind., i, 227, 129, fig. 71.[16]

Rivellia persicae, Big., Ind. Mus. Notes, i, 192, (1889)[17]; Wulp, Cat. Dipt. S. Asia, 179[18]; Cotes, Ind. Mus. Notes, i, 195[19]; Lefroy, Ind. Ins. Pests, 170, fig. 193.[20]

Dacus persicae, Frogg., "Fruit-fly," 82, tab. ii, fig. 5[21]; Bezzi, Boll. Lab. Zool. Portici, iii, 293[22]; Howl., Ind. Ins. Life, 634.[23]

Dacus squalidus, Wlk., T. Ent. S. Lon., n. s., v, 323, (1860)[24]; Wulp, Cat. Dipt. S. Asia, 187.[25]

Bactroceera squalidus, Bezzi, Mem. Ind. Mus., iii, 79.[26]

3. **Mellesis,** Bezzi.

Mellesis, Bezzi, B. Ent. R., vii, 114, (1916) [genotype, *Monacrostichus crabroniformis,* Bezzi] [key to Indian spp.].

Axania, Ender., Zool. Jahrb., Xliii, 336, (1920) [genotype, A *ichneumonea,* Enderl.].

Polistomimetes, Enderl., Zool. Jahrb., xliii, 336, (1920) [genotype, *P. minax,* End.].

brachyeera, Bezzi.

Mellesis brachyeera, Bezzi. B. Ent. R., vii, 116 (1916).[1] — [1]Dehra Dun.

crabroniformis, Bezzi.

Monacrostichus crabroniformis, Bezzi, B. Ent. R., v, 153, (1914).[1] — [1, 2]Yereaud.

Mellesis crabroniformis, id., loc. cit., vii, 117.[2]

destillatoria, Bezzi.

Mellesis destillatoria, Bezzi. B. Ent. R., vii, 118, (1916).[1] — [1]Bhamo.

eumenoides, Bezzi — [1] U. Burma.

Mellesis eumenoides, Bezzi, B. Ent. R., vii, 119, (1916).[1]

icariiformis, Enderlein.

Callantra icariiformis, Enderl., Zool. Jahrb., xliii, 336 (1920).[1] — [1]. Burma, [1]Sikkim.

ichneumonea, Enderlein.

Axania ichneumonea, Enderl., Zool. Jahrb, xliii, 336, (1920).[1] — [1]N. India.

minax, Enderlein.

Polistomimetes minax, Enderl., Zool. Jahrb., xliii, 336, (1920).[1] — [1]Sikkim.

polistiformis, Senior-White.

Mellesis polistiformis, S-W., Ent. Mem. Dpt. Ag. Ind., vii, 156-157, t. 15, fig., (1922).[1] — [1]Sukna

sphaeroidalis, Bezzi,

Mellesis sphaeroidalis, Bezzi, B. Ent. R., vii, 115, (1916).[1] — [1]Dehra Dun.

4. Adrama, Walker.

Adrama, Wlk., P. Linn. Soc., iii, 118, (1858) [genotype, *Dacus determinate,* Wlk.]; Wulp., Cat. Dipt. S. Asia, 183; Bezzi, Mem. Ind. Mus., iii, 87; *id.,* B. Ent. R., vii, 120.

Acanthipeza, Rond., Ann. Mus. Gen., vii, 438, (1875). [genotype, *A maculifrons,* Rond.].

austeni, Hendel.

Adrama austeni, Hend., Wien. Ent. Zeit., xxxi 12, (1912)[1]; Bezzi, Mem. Ind. Mus., iii, 83[2]; *id.,* B. Ent. R., vii, 120.[3] — [1, 2, 3]Peradeniya.

5. Meracanthomyia, Hendel.

Meracantha, Mq., Dipt. Exot. supp. 4, 285, (1851) [genotype, *M. maculipennis,* Mq.]; Lw, Mon. N. Amer. Dipt., iii, 23, 70; Wulp, Cat. Dipt. S. Asia, 193.

Meracanthomyia, Hend., Wien. Ent. Zeit, xxix, 107, (1910) [praeocc. emend.]; Bezzi, B. Ent. R., vii, 120.

gamma, Hendel.

Meracanthomyia gamma, Hend., Wien. Ent. Zeit., xxix, 107, tab. i, fig. 13, (1910)[1]; Bezzi, Mem. Ind. Mus., vii, 120[2]; *id.,* B. Ent. R., vii, 120[3]. — [1, 2, 3]Ceylon.

maculipennis, Macquart.

Meracantha maculipennis, Mq., Dipt. Exot. supp. 4, 285, (1851)[1]; Wulp, Cat. Dipt. S. Asia, 193.[2] — [1, 2, 3]E. India.

Meracanthomyia maculipennis, Bezzi, B. Ent. R., vii, 120.[3]

TRYPETINAE, (TRYPANEINAE)

Bezzi, Mem. Ind. Mus., iii, 86 [systematic arrangement, tribes].

6. **Anoplomus,** Bezzi.

Anoplomus, Bezzi, Mem. Ind. Mus., iii, 100, (1913) [genotype *A. flexuosus,* Bezzi].

flexuous, Bezzi.

Anoplomus flexuosus, Bezzi, Mem. Ind. Mus., iii, 100, tab. viii, fig. 12, (1913).[1] — [1]Bengal. [1]Sikkim. [1]Paresnath.

Tephrilis fasciventris, Mq., Dipt. Exot. supp. 3, 225, (65), tab. vii, fig. 7, (1847), [nec. Mq., 1843].[2]

Trypeta fasciventris, Wulp, Cat. Dipt. S. Asia, 192.[3] — [2, 3]Java.

7. **Chaetellipsis,** Bezzi.

Chaetellipsis, Bezzi, Mem. Ind. Mus., iii, 126, (1913) [genotype, *C.* paradoxa, Bezzi].

paradoxa, Bezzi.

paradoxa, Bezzi, Mem. Ind. Mus., iii, 127, tab. ix, fig. 37 (1913).[1] — [1]Paresnath.

7. **Poecillis,** Bezzi.

Poecillis, Bezzi, Mem. Ind. Mus., iii, 128, (1913) [genotype, *P, judicauda,* Bezzi].

judicauda, Bezzi.

Poecillis judicauda, Bezzi, Mem. Ind. Mus. iii, 128, tab. ix, fig. 38, (1913).[1] — [1]Paresnath.

9. **Acanthoneura,** Macquart.

Acanthoneura, Mq., Dipt. Exot., iii, 377, (220), 7, (1843) [genotype *A. fuscipennis,* Mq.]; Wulp, Cat. Dipt. S. Asia, 190; Enderl., Zool. Jahrb., xxxi, 415; Bezzi, Mem. Ind. Mus., iii, 116; Edw., Jo. F. M. S. Mus., viii, 50.

fuscipennis, Macquart.

Acanthoneura fuscipennis, Mq., Dipt. Exot., iii, 378, (221), tab. xxx. fig. 2, (1843)[1]; Wulp, Cat. Dipt. S. Asia. 190[2]; Enderl., Zool. Jahrb. xxxi, 417[3]; Bezzi, Mem. Ind. Mus., iii, 117, tab. ix, fig. 29.[4] — [1, 2]Bengal. [4]Sadiya.

10. Xanthorrhachis, Bezzi.

Xanthorrhachis, Bezzi, Mem. Ind. Mus., iii, 137, (1913) [genotype, *X. annandalei,* Bezzi].

annandalei, Bezzi, Mem. Ind. Mus., iii, 138, tab. ix, fig. 43, (1913).[1] — [1]Dawnat Range.

11. Rioxa, Walker.

Rioxa, Wlk., P. Linn. S., i, 35, (1857) [genotype R. *lanceolata* Wlk.]; Wulp, Cat. Dipt. S. Asia, 189; Bezzi, Mem. Ind. Mus., 111.

contraria, Walker.

Trypeta contraria, Wlk., Dipt. Saund., 385, (1856)[1]; Wulp, Cat. Dipt. S. Asia, 193.[2] — [1, 2, 3, 6]E. India.

? Rioxa contraria, Bezzi, Mem. Ind. Mus., III, 69.[3]

dunlopi, van der Wulp.

Ptilona dunlopi, Wulp, Tijd. v. Ent., xxiii, 186, 45, tab. xi, figs. 8, 9, (1880); *id.,* Cat. Dipt. S. Asia, 191. — [5]Sylhet. [4]Kharagpur.

Rioxa dunlopi, Meij., Bijdr. Dierk. xvii, 110[3]; Enderl., Zool. Jahrb., xxxi, 448[4]; Bezzi, Mem. Ind. Mus., iii, 113, tab. ix, fig. 25.[5] — [5]Java. [1, 2, 6]Sumatra.

Acanthoneura dunlopi, Edw., F. M. S. Mus. Jo., viii, 50.[6]

magnifica, Senior-White.

Rioxa mangifica, S-W., Spol. Zeyl., xi, 390, tab. i, fig. 2, (1921).[1] — [1]Matale (Ceylon).

mutyca, Walker.

Trypeta mutyca, Wlk., List, Dipt. B.M., iv, 1036, (1849)[1]; Wulp, Cat. Dipt. S. Asia, 192.[2] — [1, 2,] E. India. [3]Sadiya.

Rioxa mutyca, Bezzi, Mem. Ind. Mus., iii, 112, tab. viii, fig. 23[3]; S-W., Ent. Mem. Dpt. Ag. Ind., vii, 97, (1922), [note].

punctatipleura, Senior-White.

Rioxa punctatipleura, S-W., Ent. Mem. Dpt. Ag. Ind., vii, 160-161, t. 13, f. 4, (1922).[1] — [1]Colombo.

quinquemaculata, Bezzi.

Rioxa quinquemaculata, Bezzi, Mem. Ind. Mus., iii, 115, tab. ix, fig. 27, (1913).[1] — [1]Tenasserim. Ceylon.

soluta, Bezzi,

Rioxa, soluta, Bezzi, Mem. Ind. Mus., iii, 114, tab. ix, fig. 26, (1913).[1] — [1]Tenasserim.

stellata, Macquart.

Acinia stellata, Mq., Dipt. Exot., supp. 4, 266, (137), 9, Tab. xxvii, fig. 7, (1851)[1]; Wulp, Cat. Dipt. S. Asia, 194[2]; Enderl., Zool. Jahrb., xxxi, 433, fig. 4.[3] — [5]Tenasserim. Ceylon.

Trypeta stellata, O-S., Berl. Ent. Zeit., xxvi, 227.[4] — [1, 2, 3, 4]Philippines.

Rioxa stellata, Bezzi, Mem. Ind. Mus., iii, 115, tab. ix, fig. 28.[5]

vaga, Wiedemann.

Trypeta vaga, Wd., Auss. Zweifl. Ins., ii, 490, (1830)[1]; Wulp, Cat. Dipt. S. Asia, 192.[2] — [1, 2]Bengal. [3]Tenasserim.

Rioxa vaga, Bezzi, Mem. Ind. Mus., iii, 112 tab. viii, fig. 22[3]; S-W., Mem. Dpt. Ag. Ind., vii, 97.[4] — [3]Gauhati.

vidua, Bezzi.

Rioxa vidua, Bezzi, Mem. Ind. Mus., iii, 113, tab. viii, fig. 24, (1913).[1] — [1]Sikkim.

12. Diarrhegma, Bezzi.

Diarrhegma Bezzi, Mem. Ind. Mus., iii, 108, (1913) [genotype, *Dacus modestum.* F.].

modestum, Fabricius.

Dacus modestus, F., Syst. Antl. 278, 29, (1805).[1] — [1, 2, 3, 4, 9, 10, 11]Bengal.

Trypeta modesta, Wd., Auss. Zweifl. Ins., ii, 493, 26[2]; O-S., Berl. Ent. Zeit., xxvi, 227[3]; Wulp, Cat. Dipt. S. Asia, 192.[4] — [8]Calcutta. [8]Kharagpur. [12, 13]Amboina.

Ptilona modesta, id., Tijd. v. Ent. xli, 219, 6, tab. x, fig. 17.[5]

Rioxa modesta, Meij., Bijdr. Dierk., xvii, 111[6]; Enderl., Zool. Jahrb., xxxi, 449.[7]

Diarrhegma modestum, Bezzi, Mem. Ind. Mus., iii, 109, tab. viii, fig. 19.[8]

Trypeta incisa, Wd., Anal. Ent. 53, 117, (1824)[9]; *id.,* Auss. Zweifl. Ins., ii, 500, 37[10]; Wulp, Cat. Dipt. S. Asia, 192.[11]

Tephritis paritii, Dol., Tijd. Ned. Ind., x, 412, 38, tab. 1, fig. 2, (1856).[12]

Trypeta paritii, Wulp, Cat. Dipt. S. Asia, 192.[13]

Rioxa paritii, Enderl., Zool. Jahrb., xxxi, 449.[14]

13. **Ptilona,** van der Wulp.

Ptilona, Wulp, Tijd. v. Ent., xxiii, 183, (1880) [genotype, *P. brevicornis,* Wulp]; *id.,* Cat. Dipt. S. Asia 191; Bezzi, Mem. Ind. Mus., iii, 109.

brevicornis, van der Wulp.

Ptilona brevicornis, Wulp, Tijd. v., Ent. xxiii, 185, 44, tab. xi, fig. 7, (1880)[1]; O-S., Berl. Ent. Zeit., xxvi 226[2]; Wulp Cat. Dipt. S. Asia, 191[3]; Enderl. Zool. Jahrb., xxxi, 415[4]; Bezzi, Mem. Ind. Mus., iii, 111, tab. viii, fig. 21.[5]

[5]Lushai Hills.
[1, 3]Java.
[2]Philippines.

nigriventris, Bezzi.

Ptilona nigriventris, Bezzi, Mem. Ind. Mus., iii, 110, viii, 20, (1913).[1]

[1]Assam.
[1]Sylhet.

14. **Callistomyia,** Bezzi.

Callistomyia, Bezzi, Mem. Ind. Mus., iii, 124, (1913) [genotype, *C. pavonina,* Bezzi].

pavonina, Bezzi.

Callistomyia pavonina, Bezzi, Mem. Ind. Mus., iii, 125, tab. ix, fig. 36 (1913)[1]; Dever. R. I. M. xxii, 395.[2]

[1]Calcutta.
[1, 2]Chilka Lake.
[1]Rajmahal.
[1]Naini Tal (Plains).

15. **Gastrozona,** Bezzi.

Gastrozona Bezzi, Mem. Ind. Mus., iii, 105, (1913) [genotype, *Tephritis fasciventris, Macq.*]

fasciventris,

Tephritis fasciventris, Mq., Dipt. Exot; iii, 382, (225). 4. tab. xxxi fig. 2, (1813), [*fusciventris* in index, p. 459, (302)[1]].

[1, 2]E. India.
[3]Sylhet.

Trypeta fasciventris, Wulp, Cat. Dipt. S. Asia, 192.[2] — [4]Sumatra.

Gastrozona fasciventris Bezzi, Mem. Ind. Mus., iii, 105, tab. viii, fig. 16[3]; Edw., F. M. S. Mus. Jo., viii, 49.[1]

melanista, Bezzi.

Gastrozona melanista, Bezzi, Mem. Ind. Mus., iii, 107, tab. viii. fig. 18, (1913).[1] — [1]Calcutta.

montana, Bezzi.

Gastrozona montana, Bezzi, Mem. Ind. Mus., iii, 106, tab. viii, fig. 17, (1913).[1] — [1]Kurseong.

16. Taeniostola, Bezzi.

Taeniostola, Bezzi, Mem. Ind. Mus., iii, 119, (1913) [genotype, *T. vittigera,* Bezzi].

gracilis, Bezzi.

Taeniostola gracilis, Bezzi, Mem. Ind. Mus. iii, 120, tab. ix, fig. 32, (1913).[1] — [1]Dawnat Hills.

vittigera, Bezzi.

Taeniostola vittigera, Bezzi, Mem. Ind. Mus., iii, 119, tab. ix, fig. 31, (1913)[1]; S-W; Ent. Mem. Dpt. Ag. Ind., vii, 97, (1922).[2] — [1,2]Assam., [1]Sylhet. [1]Lungleh. [2]Gauhati.

17. Chelyophora, Rondani.

Chelyophora, Rond., Ann. Mus. Gen., vii, 4333, (1875) [genotype, *C. borneana,* Rond.]; Wulp, Cat. Dipt. S. Asia, 185; Bezzi, B. Ent. R., viii, 229.

Stictaspis, id., Mem. Ind. Mus., iii, 102, (1913) [genotype *S. ceratitina,* Bezzi].

ceratitina, Bezzi.

Stictaspis, ceratitina, Bezzi, Mem. Ind. Mus., iii, 103, tab. viii, fig. 13, (1913).[1] — [1]Paresnath.

separata, Bezzi.

Stictaspis separata, Bezzi, Mem. Ind. Mus., iii, 104, tab. viii, fig. 15, (1913).[1] — [1]Kohima.

striata, Froggatt.

Ceratitis striata, Frogg., "Fruit-Fly," iii, tab. v, fig. 17, (1909)[1]; Green, Trop. Agric, xxxiii, [breeding place].[2] — [1,2,3]Feradeniya.

Stictaspis striata, Bezzi, Mem. Ind. Mus., iii, 104, tab. viii, fig. 14.[3]

Chelyophora striata striata, End., Zool. Jahrb., xliii, 348 [notes].

18. Phaeospila, Bezzi.

Phaeospila, Bezzi, Mem. Ind. Mus., iii, 117, (1913) [genotype, *P. varipes,* Bezzi].

varipes, Bezzi.

Phaeospila varipes, Bezzi, Mem. Ind. Mus., iii, 118, tab. ix, fig. 30, (1913).[1] — [1]Darjiling.

19. Acidia, Robineau-Desvoidy.

Acidea, R-D., Myod., 720, xiv, (1830) [genotype, *Tephritis cognata* Wd.] Wulp, Cat. Dipt. S. Asia, 189; Bezzi, Mem. Ind. Mus., iii, 139. *Ocneros,* O. G. Costa.

Hemiliea.

alboscutellata, van der Wulp.

Anomoea alboscutellata, Wulp, Tijd. v. Ent. xli, 217, 4, tab. x. fig. 15, (1898)[1]; *id., loc. cit.,* xliii, 51.[2] — [3]Tenasserim.

Acidia alboscutellata, Bezzi, Mem. Ind. Mus., iii, 145, tab. x, fig. 49.[3] — [1,2]Sumatra.

apicalis, Bezzi.

Acidia apicalis, Bezzi, Mem. Ind. Mus., iii, 144, tab. ix, fig. 47, (1913).[1] — [1]Darjiling.

discalis, Brunetti.

Acidia discalis, Brun., R. I. M., xiii, 97. fig. 4, a, b, (1917).[1] — [1]Phagu.

erythraspis, Bezzi.

Acidia erythraspis, Bezzi, Mem. Ind., Mus., iii, 145, tab. x, fig. 50, (1913).[1] — [1]Kurseong.

fossata, Fabricius.

Tephritis fossata, F., Syst. Antl, 320, 20, (1805).[1] — Sukna.

Trypeta fossata, Wd., Auss. Zweifl. Ins., ii, 503, 41[2]; Wulp, Cat. Dipt. S. Asia, 192[3]; Big., Jo. A. S. B., lxi, 224.[4] — Tenasserim. Assam.

Anomea fossata, Wulp, Tijd., v. Ent., xli, 216, 3, tab. x, fig. 14[5]; *id., loc. cit.,* xliii, 54, 57.[6] — Travancore. [1,2,3]Tranquebar.

Acidia fossata, Bezzi, Mem. Ind. Mus., iii, 144, tab. ix, fig. 48.[7] — Peradeniya (Petch).

Anomoia elimia, Wlk., List Dipt. B.M. iv, 1033, (1849).[8] — [5, 6, 8, 10]Java.

Trypeta elimia, O-S., Berl. Ent. Zeit., xxvi, 227[9]; Wulp Cat. Dipt. S. Asia, 192.[10] — [10, 11]Amboina. [9, 10]Philippines.

Ortalis regularis, Dol., Nat. Tijd. Ned. Ind., xvii, 119, 75, (1859).[11]

himalayensis, Bezzi.

Acidia himalayensis, Bezzi, Mem. Ind. Mus., iii, 142, tab. ix, fig. 45, (1913).[1] — [1]Kurseong.

praestans, Bezzi.

Acidia (Ocneros) praestans, Bezzi, Mem. Ind. Mus., iii, 141, tab. x, fig. 51, (1913).[1] — [1]Bhowali.

rioxaeformis, Bezzi.

Acidia, rioxaeformis, Bezzi, Mem. Ind. Mus., iii, 143, tab. ix, fig. 46, (1913).[1] — [1]Simla.

20. Vidalia, Robineau-Desvoidy.

Vidalia, R-D., Myod., 719, xii, (1830) [genotype, *V. impressifrons,* R-D.]; Bezzi, Mem. Ind. Mus., iii, 135.

ceratophera, Bezzi.

Vidalia ceratophera, Bezzi, Mem. Ind. Mus., iii, 136, tab. ix, fig. 42, (1913).[1] — [1]Siliguri.

cervicornis, Brunetti.

Vidalia, cervicornis, Brun., R. I. M., xiii, 95, figs. 2, 3, (1917).[1] — [1]Phagu., Murree.

melanonotum, Brunetti.

Vidalia melanonotum, Brun., R. I. M., xiii, 96, (1917).[1] — [1]Phagu.

triceratops, Bezzi.

Vidalia, triceratops, Brun., Mem. Ind. Mus., iii, 137, tab. ix, fig. 43, (1913).[1] — [1]Darjiling.

21. Zonosema, Loew.

Zonosema, Lw., Monog. Trypet., 43, viii, (1862) [genotype, *Tephritis alternata,* Fall.]; Bezzi, Mem. Ind. Mus., iii, 134.

Dubium, Bezzi

Zonosema dubium, Bezzi, Mem. Ind. Mus., iii, 135, tab. ix, fig. 41, (1913).[1] — [1]Nainital

22. Carpomyia, Costa.

Carpomyia, Costa, Annal. Sci., i, 87, (1854) [genotype, *C. vesuviana,* Costa] ; Rond., Bull. S. Ent. Ital., i, 164; *id.,* Dipt. Ital. Prodr., i, 111; Wulp, Cat. Dipt. S. Asia, 195; bezzi, Mem. Ind. Mus., iii, 133.

vesuviana, Costa.

Carpomyia vesuviana, Costa, Annal Sci., i, 87, 10, (1854)[1]; Rond., Bull. S. Ent. Ital., i, 164[2]; *id., loc. cit.,* ii, 23, 1[3]; Bezzi, Boll. Lab. Zool. Portici, v, 10, 4, figs. 5, 6[4]; *id.,* Mem. Ind. Mus., iii, 134, tab. ix, fig. 40.[5] — [5]Calcutta. [8, 9, 10, 7, 1, 2, 3, 4,] S. Europe.

Orellia vesuviana, V. Rod., Ent. Nacht., xvi, 210[6]; Beck., Cat. Pal. Dipt., iv, 116.[7]

Orellia bucchichi, Frauenfld., Verh. z. b. Ges. Wien, xvii, 500, tab. xii, fig. 23, (1867)[8]; *id., loc. cit.,* xviii, 154, 5[9]; Kaltenbach, Pflanzenf., 776, 1.[10]

23. Myiopardalis, Bezzi.

Myiopardalis, Bezzi, Boll. Lab. Zool. Portici, v, 8, (1910) [genotype, *Carpomyia pardalina,* Big.] ; *id.,* Mem. Ind. Muse., iii, 132.

pardalina, Bigot.

Carpomyia pardalina, Big., Ind. Mus. Notes, ii, 77, tab. v, fig. I, (1891)[1]; Wulp, Cat. Dipt. S. Asia, 195[2]; Cleghorn. Ind. Mus. Notes, ii, 24[3]; Lefroy, Ind. Ins. Pests, 171, fig. 194[4]; *id.,* Mem. Dpt. Ag. Ind., i, 129, fig. 72[5]; Frogg., "Fruit-Fly," 112. — [1, 2]India. [3, 6]Pishin (5000') [8]Baluchistan. Pusa.

Myiopardalis paradalina, Bezzi, Boll. Lab. Zool. Portici, v, 9, 2, figs. 1, 3, 4,[7]; *id.,* Mem. Ind. Mus., iii, 132, tab. ix, fig. 39[8]; Rao, iv, Ent. Mtg., 170[9]; Adair. Ag. Jo. Egypt, x, 20.[10] — [9]Mesopotamia. [10]Palestine. [10]Egypt.

24. Ceratitis, Mac Leay.

Ceratitis, McL., Zool. Jo., iv, 475, (1829) [genotype, *Trypeta capitata* Wd.] Wulp, Cat. Dipt. S. Asia, 189; Frogg., "Fruit-Fly," 99; Bezzi, Mem. Ind. Mus., iii, 129.

Halterophora, Rond., Dipt. Ital. Prodr., iv, 10, (1861) [genotype, *Trypota capitata,* Wd.).

Petalophora, Macq. H. N. Dipt., ii, 454, 5, (1835) [genotyoe, *Trypeta, capitata,* Wd.].

asiatica, Becker.

Ceratitis asiatica, Beck., Ann. Mus. Zool. Petersb., xii, 291, 64, tab. i, fig. 10 (1907)[1]; Bezzi, Mem. Ind. Mus., iii, 67.[2] — [1, 2]N. E. Tibet.

[In spite of the record by van der Wulp of *C. capitata* from "E. India," this species does not seem to occur within the Oriental Region.]

25. **Phagocarpus,** Rondani.

Phagocarpus, Rond., Bull. S. Ent. Ital., iii, 171, xix, (1871) [genotype, *Musca permundus,* Harris]; Bezzi, Mem. Ind. Mus., iii, 130.

Anomoia, Wlk., E. M. M., iii, 80, (1836) [genotype, *M. permundus,* Har.] [praeocc.].

immsi, Bezzi.

Phagocarpus immsi, Bezzi, Mem. Ind. Mus., iii, 131, tab. x, fig. 72, (1913).[1] — [1]Bhowali.

26. **Spheniscomyia,** Bezzi.

Sphaeniscus, Beck, Mitt. Zool. Jus. Berl., iv, 138, (1908) [genotype, *Trypeta quadrincisa,* Wd., as *S. brevicauda,* Back.].

Spheniscomyia, Bezzi, Mem. Ind. Mus., iii, 146, (1913) [emend]; *id.,* B. Ent. R., ix, 121.

quadrincisa, Wiedemann.

Trypeta quadrincisa, Wd., Anal. Ent., 55, 122. (1824)[1]; *id.,* Auss. Zweifl. Ins., ii, 508, 50.[2] — [1, 3, 5, 6]E. India.

Acidia quadrincisa, Sch., Novara Reise, 264, 99[3]; O-S., Ann. Mus. Gen., xvi, 459[4]; Wulp, Cat. Dipt. S. Asia, 189.[5]

Spheniscomyia quadrincisa, Bezzi, Mem. Ind. Mus. iii, 147, tab. x, fig. 52.[6]

Sphaeniscus brevicauda, Beck., Mitt. Zool. Mus. Berl., iv, 138, 401, tab. iii, fig. 42, (1908).[7]

Aciura filiola, Lw., Zeit. f. ges. Nat., xxxiv, 12, 1, (1869).[8]

[6]Calcutta Dist.
[6]Puri Dist.
[12, 13]Bengal.
[3]Nicobar.
[6]Ceylon., [6, 9, 10, 11]Java
[6]Teneriffe., [6]Spain.

Euxesta parvula, Wulp, Termes. Fuzet., xx, 141, 20, tab. iii, fig. 2, (1897)[9]; Meij., Tijd. v. Ent., xli, 126.[10]

Acidia parvula, Hend., Ann. Mus. Nat. Hung., vii, 172 12.[11]

Trupeta tucia, Wlk., List Dipt. B. M., iv, 1021, (1849)[12]; Wulp, Cat. Dipt. S. Asia, 192 [*tacia, laps*].[13]

sexmaculata, Macquart.

Urophora sexmaculata, Mq., Mem. Soc. Lille, 1842, 379, 2, tab. xxx, fig. 5, (1843).[1]

Spheniscomyia sexmaculata, Bezzi, Mem. Ind. Mus., iii, 148, tab. x, fig., 53.[2]

Trypeta atilia, Wlk., List Dipt. B. M., iv, 1021, (1849).[3]

Acidia atilia, Wulp, Cat Dipt. S. Asia, 189.[4]

Trypeta formosana, Enderl., Zool. Jahrb., xxxi, 427, fig. H, (1911).[5]

Trypeta melaleuca, Wlk., P. Linn. Soc., vii, 238, 40, (1864).[6]

Acidia melaleuca, O-S., Ann. Mus. Gen., xvi, 459[7]; *id.,* Berl. Ent. Zeit. xxvi, 227.[8]

Acidia melania, Bezzi, Denk. med-nat. Ges. Jena, xiii, 194, 34, fig. 3, (1908).[9]

Ortalis sanctae-mariae, Big., Bull. S. Ent. Fr., (3) vii, 458, tab. xiii, fig. 3, (1859).[10]

Acidia, sanctae-mariae, Hend., Ann. Mus. Nat. Hung., vii, 172, 12.[11]

Trypeta sexincisa, Thoms, Eugen, Reise, 579, 252, (1868).[12]

[2]Calcutta.
[2]Naini Tal.
N. Bihar (Inglis).
[2]Java.
[2]Broneo.
[2,4]Celebes.
[2,7,8]Philippines.
[2,3,4]Ceram.
[2,5]Formosa.
[2,4]China., [2]Reunion.
[2]Madagascar.
[2]S. Africa.

27. **Aciura,** Robineau-Desvoidy.

Aciura, R-D., Myod., 773, xiii, (1839) [genotype, *Musca coryli,* Rossi]; Lw., Besch. Europ. Behrfl., 29; Sch., F. Asutr., ii, 112; Will., Cat. Dipt. N. Amer., 607; Bezzi, Mem. Ind. Mus, iii, 149; *id.,* B. Ent. R., ix, 18.

monochaeta.

Aciura monochaeta, Bezzi, Mem. Ind. Mus., iii, 150, tab. x, fig. 54, (1913).[1]

[1]Calcutta.

xanthotricha, Bezzi.

Aciura xanthotricha, Bezzi, Mem. Ind. Mus., iii, 151, tab. x, fig. 55, (1913).[1]

[1]Garhwal District.
[1]Tenasserim.

28. **Staurella,** Bezzi.

Staurella, Bezzi, Mem. Ind. Mus., iii, 121, (1913) [genotype, *Musca crux,* F.].

crux, Fabricius.

Musca crux, F., Ent. Syst., iv, 358, 190, (1794).[1] — [1, 2, 3, 4]E. India.

Dacus crux, id., Syst. Antl., 277, 23.[2] — [5]Calcutta.

Trypeta crux, Wd., Auss. Zweifl. Ins., ii, 488, 19[3]; Wulp, Cat. Dipt. S. Asia, 192.[4]

Stavurella crux Bezzi, Mem. Ind. Mus., iii, 122, tab. ix, fig. 33.[5]

dissoluta, Bezzi.

Staurella dissoluta, Bezzi, Mem. Ind. Mus., iii, 123, tab. ix, fig. 34, (1913).[1] — [1]Naini Tal District.

Luteifasciata, Senior-White.

Staurella luteifasciata, S-W., Ent. Mem. Dpt. Ag. Ind., vii, 158-159, t. 14, fig. 2, (1922).[1] — [1]Matale (Ceylon).

nigripeda, Bezzi.

Staurella nigripeda, Bezzi, Mem. Ind. Mus., iii, 123, tab. ix, fig. 35, (1913).[1] — [1]Kurseong.

zeylanica, Senior-White.

Staurella zeylanica, S-W., Spol. Zeyl., xi, 392, tab. i, fig. 3. (1921)[1]; *id.,* Ent. Mem. Dpt. Ag. Ind., vii, 159, (1922) [note].[2] — [1]Matale (Ceylon).

29. **Platensina,** Enderlein.

Platensina, Enderl., Zool. Jahrb., xxxi, 453, (1911) [genotype]; Bezzi B. Ent. R., ix, 24.

Tephrostola, Bezzi, Mem. Ind. Mus., iii, 153, (1913) [genotype, *Trypeta acrostacta,* Wd.].

acrostacta, Wiedemann.

Trypeta acrostacta, Wd., Anal. Ent. 54, 119, (1824)[1]; *id.,* Auss. Zweifl. Ins., ii, 501, 39[2]; Wulp, Cat. Dipt. S. Asia, 192.[3] — [1, 2, 3,] E. India. [1]Sikkim. [4]Bengal. [4]Ranchi. [4]Puri. [4]Moulmein.

Tephrostola acrostacta, Bezzi, Mem. Ind. Mus., iii, 153, tab. x, fig. 57.[4]

Ensina guttat, Mq., Mem. Soc. Lille, 1842, 387, 1, tab. xxxi, fig. 10. (1843)[5]; Wulp, Cat. Dipt. S. Asia, 194.[6] — [5, 6]Coromandel. Ceylon (White).

reinhardi, Wiedemann.

Trypeta reinhardi, Wd., Anal. Ent., 54, 121, (1824)[1]; *id.,* Auss. Zweifl. Ins., ii, 507, 48[2]; Wulp, Cat. Dipt. S. Asia, 192.[3] — [1, 2, 3]E. India. [4]Calcutta. Ceylon (White)

Tephrostola reinhardi, Bezzi, Mem. Ind. Mus., iii, 154.[4]

30. Tephrella, Bezzi.

Tephrella, Bezzi, Mem. Ind., Mus., iii, 151, (1913) [genotype *T. decipiens,* Bezzi]; *id.,* B. Ent. R., ix, 22.

Decipiens, Bezzi, Mem. Ind. Mus., iii, 152, tab. x, fig. 56, (1913).[1] — [1]Darjiling District.

31. Elaphromyia, Bigot.

Elaphromyia, Big., Rev. Mag. Zool., vii, 10, (1859) [genotype. *E. melas,* Big.]; Bezzi, B. Ent. R., ix, 27.

Paralleloptera, id., Mem. Ind. Mus., iii, 154, (1913). [genotype, *P. pterocalaeformis,* Bezzi].

pterocallaeformis, Bezzi.

Paralleloptera pterocallaeformis, Bezzi, Mem. Ind. Mus., iii, 155, tab. x, fig. 58, (1913).[1] — [1]Simla Hills. [1]Bhowali. [2]Cherrapunji.

Elaphromyia pterocallaeformis, S-W., Mem. Dpt. Ag. Ind., vii, 97.[2]

32. Craspedoxantha, Bezzi.

Craspedoxantha, Bezzi, Mem. Ind. Mus., iii, 156, (1913) [genotype, *C. octopunctata,* Bezzi].

octopunctata, Bezzi,

Craspedoxantha octopunctata, Bezzi, Mem. Ind. Mus., iii, 156, tab. x, fig. 59, (1913)[1]; S-W., Mem Dpt. Ag. Ind., vii, 98.[2] — [1]Dawnat Range. [2]Nagpur.

33. Sphenella, Robineau-Desvoidy.

Sphenella, R-D., Myod., 773, xiv, (1830) [genotype, *Tephritis, marginata,* Fallen]; Bezzi, Mem. Ind. Mus., iii, 157.

Indica, Schiner.

Sphenella indica, Sch., Novara Reise, 267, 110, (1868)[1]; wulp, Cat. Dipt. S. Asia, 194[2]; Bezzi, Mem. Ind. Mus., iii, 158, tab. x, fig. 60.[3] — [1, 2, 3]Madras, [3]Puri. Ceylone (White).

34. Ensina, Robineau-Desvoidy.

Ensina, R-D., Myod., 751, (1830) [genotype, *Ensina sonchii,* Linn.]; Lw., Europ, Bohrfl., 64; *id.,* Mongo. N. Amer. Dipt., iii, 291; Sch., F. Austr., ii, 143; Wulp, Biologia, Dipt., ii, 416; Ald., Cat. N. Amer. Dipt., 610; Bezzi, B. Ent. R., ix, 31.

Antiqua, Walker.

Trypeta antiqua, Wlk., Dipt. Saund., 378, (1856)[1]; Wulp, Cat. Dipt. S. Asia, 192.[2] — [1, 2, 3]E. India.

? *Ensina antiqua,* Bezzi, Mem. Ind. Mus., iii, 67.[3]

contingens, Becker.

Oxyna contingens, Beck., Mem. Acad. Sci. Petersb., xii, 288, 58, tab. i, fig. 5, (1907)[1]; Bezzi, Mem. Ind. Mus., iii, 69.[2] — [1, 2]E. Tibet.

evanescens, Becker.

Oxyna evanescens, Beck., Mem. Acad. Sci. Petersb., xii, 189, 59, tab. i, fig. 6, (1907)[1]; Bezzi, Mem. Ind. Mus., iii, 71.[2] — [1]N. E. Tibet. [2]Turkestan.

sororcula, *Wiedemann.*

Trypela sororcula, Wd., Auss. Zweifl. Ins., ii, 509, 52 (1830).[1] — [3]Bengal.

Oxyna sororcula, Beck., Mitt. Zool. Mus. Berl., iv, 144, 420[2]; Bezzi, Mem. Ind. Mus., iii, 159, tab. x, fig. 61.[3] — [3]Simla Hills. Ceylon.

Ensina sororcula, id., B. Ent. R., ix, 32.[4]

Ensina bisetosa, Enderl., Zool. Jahrb., xxxi, (1911) [1 ♂ only).[5] — [5]Formosa. [4]E. Africa.

Ensina vacillans, Woll., A. M. N. H., (3) i, 115, (1858)[6]; Bezzi, Boll. S. Ent. Ital., xxxix, 159, 127.[7] — [4]S. Africa.

Leptomyza variipennis, Wulp, Termes. Fuzet., xx, 143, 28, tab. iii, figs. 3, 4. (1897).[8]

Oxyna variipennis, Zerny, Wien. Ent. Zeit., xxi, 256[9]; *id., loc. cit.,* xxv, 251[9], figs. 1, 2.[10]

35. **Oxyna,** Robineau-Desvoidy.

Oxyna, R-D., Myod., 755, iii, (1830) [genotype, *Tephritis absinthii* F.]; Bezzi, Mem. Ind. Mus., iii, 158.

diluta, Becker.

Oxyna diluta, Beck., Mem. Acad. Sci. Petersb., xii, 289, 60, tab. i, fig. 7, (1907)[1]; Bezzi, Mem. Ind. Mus., iii, 70. [1]Turkestan.

36. **Spathulina,** Rendani.

Spathulina, Rond., Dipt., Ital. Prodr., i, 113, (1856) [genotype, *Tephritis tristis,* Lw., as *Litricha sicula,* Rond.]; Bezzi, B. Ent. R., ix, 28.

parca, Bezzi.

Oxyna parca, Bezzi, Mem. Ind. Mus., iii, 159, tab. x, fig. 62, (1913).[1] [1]Calcutta.

Spathulina parca, id., B. Ent. R., ix, 28.[2]

37. **Campiglossa,** Rondani.

Compiglossa, Rond., Boll. S. Ent. Ital., ii, 121, (1870) [genotype, *Tephritis irrorata,* Fallen]; Bezzi, Mem. Ind. Mus., iii 160; *id.,* B. Ent. R., ix, 38.

cribellata, Bezzi.

Campiglossa cribellata, Bezzi, Mem. Ind. Mus., iii, 161, tab. x, fig. 63, (1913).[1] [1]Kurseong.

38. **Tephritis,** Latreille.

Tephritis, Latr., H. N. Crust. Ins., xiv, 389, (1804) [genotype, *Musca leontodontis,* de Geer]; Lw., Europ. Behrfl., 96; *id.,* Monog. N. Amer. Dipt., iii, 295; Sch., F. Asut., ii, 151; Wulp, Biolegia Dipt., ii, 419; Ald., Cat. N. Amer. Dipt., 611; Bezzi, Mem. Ind. Mus., iii, 162; *id.,* B. Ent. R., ix, 15.

Urophora, R-D, Myod., 769, (1830) [genotype, *Musca solstitialis,* L.]; Sch., F. Austr., ii, 134; Ald., Cat. N. Amer. Dipt., 614.

brahma, Schiner.

Tephritis brahma, Sch., Novara Reise Dipt., 272, (1868)[1]; Wulp, Cat. Dipt. S. Asia, 195[2]; Bezzi, Mem. Ind. Mus., iii, 68.[3] [1, 2]Madras.

euryptera, Bezzi.

Tephritis euryptera, Bezzi, Mem. Ind. Mus., iii, 162, tab. x, fig. 64, (1913).[1] — [1]Tenasserim.

lyncea, Bezzi.

Tephritis lyncea, Bezzi, Mem. Ind. Mus., iii, 165, tab. x, fig. 67, (1913).[1] — [1]Darjiling.

mixta, Walker.

Trypeta mixta, Wlk. Dipt. Saund., 385, (1856)[1]; Wulp, Cat. Dipt. S. Asia, 193.[2] — [1, 2]E. india.

Tephritis mixta, Bezzi, Mem. Ind. Mus., iii, 75.[3]

spiloptera, Bezzi

Tephritis spilotera, Bezzi, Mem. Ind. Mus., iii, 165, tab. x, fig. 68, (1913).[1] — [1]Calcutta. Ceylon.

stella, Walker.

trypeta stella, Wlk., List Dipt. B. M., iv, 1030, (1849)[1]; Wulp, Cat. Dipt. S. Asia, 192.[2] — [1, 2]Bengal

Tephritis stella, Bezzi, Mem. Ind. Mus. iii, 79.[3]

tribulicola, Senior-White.

Tephritis tribulicola, S-W., Ent. Mem. Dpt. Ag. Indi., vii, 161-164, fig., (1922).[1] — [1]Khasis.

vittata, Macquart.

Tephritis vittata, Mq., Dipt. Exot., ii, supp. iv, 290, (1851)[1]; Bezzi, Mem. Ind. Mus., iii, 81.[2] — [1, 3]E. Indian

Trypeta vittata, Wulp, Cat. Dipt. S. Asia, 192.[3]

zodiacalis, Bezzi.

Tephritis zodiacalis, Bezzi, Mem. Ind. Mus., iii, 163, tab. x, fig. 65, (1913).[1] — [1]Calcutta.

zonogastra, Bezzi.

Tephritis zonogastra, Bezzi, Mem. Ind. Mus., iii, 164 tab. x, fig. 66, (1913).[1] — [1]Puri. N. Bihar

39. **Trypanea,** Schrank.

trypanea, Schrk., Briefe Donaumoor., 147, (1795) [genotype, *Musca stellata,* Fuessly]; Bezzi, Mem. Ind. Mus., iii, 166; *id.,* B. Ent. R., ix, 41.

Urellia, R-D., Myod., 774, xv, (1830) [genotype, *cluta,* Meig.]; Lw., Europ. Behrfl, 117; Wulp, Biologia Dipt., ii, 426; Ald., Cat. N. Amer. Dipt., 613.

amoena, Frauenfeld.

Trypeta amoena, Frnfld., Sitz. Akad. Wiss. Wien., xxii, 542 fig. 2, (1856).[1]

Trypanea amoena, Bezzi, Mem. Ind. Mus., iii, 167, tab. x, fig. 70.[2]

[2]Calcutta., [2]Paresnath. [2]Lahore., Ceylon. [1, 2]Europe. [2]N. Africa.

asteria, Schiner.

Tephritis asteria, Sch., Novara Reise Dipt. 270, 118, (1868)[1]; Wulp, Cat. Dipt. S. Asia, 195[2]; Meij., Tijd. v. Ent., li, 132, 2, tab. iv, fig. 6.[3]

[1, 2, 4]Madras. [4]Paresnath. [3, 4]Java.

Trypanea asteria, Bezzi, Mem. Ind. Mus., iii, 167, tab. x, fig. 71.[4]

aucta, Bezzi.

Trypanea aucta, Bezzi, Mem. Ind Mus., iii, 167, tab. x, fig. 69, (1913).[1] — [1]Puri.

Trypanea aucta var. repleta, Bezzi, B. Ent. R., ix, 45, fig. 3, (1918).[2] — [2]Africa.

cribrata, Becker.

Urellia cribrata, Beck., Mem. Acad. Sci. Petersb., xii, 287, 53, tab. i, fig. 3, (1907).[1] — [1]Thibet.

Trypanea cribrata, Bezzi, Mem. Ind. Mus., iii, 69.[2]

ferruginea, Walker.

Trypeta ferruginea, Wlk., Dipt. Saunds., 387, (1856)[1]; Wulp, Cat. Dipt. S. Asia, 193[2]; Bezzi, Mem. Ind. Mus., iii, 72.[3] — [1, 2]E. India.

nebulosa, Becker.

Urellia nebulosa, Beck., Mem. Acad. Sci. Petersb. xiii, 286, tab. i, fig. 2, (1907).[1] — [1]E. Thibet.

Trypanea nebulosa, Bezzi, Mem. Ind. Mus., iii, 77.[2]

pentadactyla, Senior-White.

Trypanea pentadactyla, S-W., Ent. Mem. Dpt. Ag. Ind., vii, 164, t. 15, fig. 10, (1922).[1] — [1]N. Bihar.

PART V
NITIDULIDAE

Nitidulariae propriae Latreille (1807), Gen.

Crust. Ins. II, 10-19 (pars).

Nitidulariae Murray (1864), M. N. 211-414.
Nitidulidae Ganglb. (1899), K. M. 411 & 442

565 [key to tribes]—Grouv. (1908),— C. R. I. 329-397 [Key to tribes] Reitter

(1911), F. G. K. 10-41 [Key to tribes].

Biology: Erichson (1845), Naturg. Ins. Deutschl. III, 124—Rupertsberger (1880), Biol. Käf. Eur. 127-130—Thomson (1862), S. C. IV. 137—Rey [1886 (87)], Essai d'ètudes sur certaines larves de Col. Ann. Soc. Linn. Lyon, XXXIII, 173-174; Sep. ed. André, Beaune 43 & 44—Ganglb. (1899), K. M. 445—Reitter (1911), F. G. K. 10—Peyerimhoff (1921), B.S.E.F., 284—Verhœff (1923), A. N. 2-109, pl. I-III & VII.

Catalogues; Schönherr (1808), Synon. Ins. I, 2, 135-150— Gemminger & Harold (1868), C. C. III, 802-838— Grouv. (1913), Col. Cat. Junk. Pars. 56.

I. CATERETINAE

Cateretes Erichson (1843), G. Z. IV, 226-227.

Brachyptèrides Lacord. [1854 (1855)], G. C. 290 & 291 [key to tribes and genera].

Brachypterid Murr. (1864). M. N. 230-231 [key to tribes and genera].

Cateretini Ganglb. (1899), K. M. 447 & 448 [key to genera]—Grouv. (1908), C.R.I. 325—Reitter (1911), F. G. K. 11-12 [key to genera].

Biology: Verhœff (1923), A. N. I. 8-11, 18-22, 66-72.

Heterhelus Jacq. du Val.

Jacq. du Val. (1858), (Subg. *Cercus*) Gen. Col. d'Eur. II 137—Murr. (1864), (Subg. *Cercus*) M. N. 23 & 239— Ganglb. (1899), K. M. 448 & 451.

Amartus Reitt. (1875), Eur. Nitid. 2 (pars).

Cateretes Heer (1841), F. H. I, 410 (pars).

kraatzi Grouv. (1903), A. S. E. F. LXXII, 108[1]. Darjiling[1]

II. MELIGETHINAE.

Meligithina Thoms. (1859), S. C. I, 67—(1862), IV, 151—(1867), IX, 368.

Meligethini Seidl. (1872), Fn. Balt. ed. I, 49— Reitt (1911), F. G. K. 11 & 15 [key to genera].

2. **Pria** Steph.

Steph. (1830), I. B. E. III. 30-49—Lacord.

[1854 (1855)], G. C. 301 & 312— Ganglb. (1899), K. M. 464 & 535 —Grouv. (1908), C. R. I. 340 & 367 [key to species].

Cormyphora Cast. (1840), Hist. Nat. Col. II, 12.

Laria Scop. (1763), Entom. Carn. 21 (pars).

apicalis Grouv. (1894), A. S. E. B. XXXVIII, 582[1]—(1908), C. R. I. 367. Kurseong.[1]

ceylonica Grouv. (1902), A. S. E. F. LXXI, 481[1]—(1908), C. R. I. 367. Ceylon[1]

diluticollis Grouv. (1913), R. I. M. VII, 2, 102[1]. North-East Assam[1]

indica Grouv. (1894), A. S. E. B. XXXVIII, 581[1]—(1908), C. R. I. 366 & 367[2]. Kurseong[1] Pegu.[2], Sumatra[2]

latipes Grouv. (1908), C. R. I. 366 & 367[1], t. 6. f. 7. Nilgiris[1]

mirmidon Grouv. (1913), R. I. M. VIII, 2, 101.[1] Dibrugarh[1]

pallida F. (*Nitidula*) (1801), S. E. I, 349—Ol. (*Nitidula*) (1811), Encycl. méth. VIII, 216—Motsch. (1858), E. E. VII, 131—Reitt. (1873), S. E. N. 59 [note]—Grouv. (1908), C. R. I. 365[1] & 367 [footnote.] Ceylon[2]

plagiata Grouv. A. S. E. B., XXXIII, 582[1] —(1908), C. R. I. 367. Kurseong[1]

3. **Meligethes** Steph.

Steph. (1830), I. B. E. III, 30 & 45 (pars)—Lacord. [1854 (1855)], G. C. 301 & 313—Ganglb. (1899), K. M. 464, 492 & 535 [key to species]—Grouv. (1908), C. R. I. 340 & 367-377 [key to species].

Subgen. *Acanthogethes* Reitt. (1911), F. G. K. 16— Ganglb. (1899), K. M. 494.

Subgen. *Odontogethes* Reitt. (1911), F. G. K. 16—Ganglb. (1899), K. M. 494.

Biology.— Omerod (1874), E. M. M. 46-52— Perris [1875 (1876)]. A. S. L. L. 298, sep. ed. Deyrolle (1867), 40— Krassilstchik (1915), Report of Bio-Entom. Station Bessarabia, 1914, Kishinev, 49 [control]— Lind, Rostrup & Kølpin Ravn (1915), Oversogt over Landbrugsplanternes Sygdomme i 1914, Beretning fra Statens Forsøgsvirksomhed i Plantekultur 31, København [control]— Kemner (1918), Rapsbaggen Central—anstalten för Jordbruksförsök, 1917, Flygblad, 64, Entom. Avdelningen, 17, 4, fig. 3 [control]—Hukkinen (1919). Om Rapsbaggen Och dess Avvarjände, Meddelanden till Landtmän No. 58 fran Agrikulture ekonomiska Försöksanstalten Finland, Entom. avdeln. 8 figs. 1-6 (bionomics and control] Voss (1919), Rapsglanzkäfer und Rapsverborgenrüssler, Schaffnits Flugblattsammlung über Pflanzenschutz, Bonn-Poppelsdorf, Flugblatt 14, April 1919 [control]— Heikertinger (1920), Unterssuchungen über die standpflanzen der Blütenkäfergattungen, E. B. XVI, 4-9, 126-143 [bionomics]- Friederichs (1920), Untersuchungen über Rapsglanzkäfer in Mecklenburg, Z. A. E. VII, 1, 1-36, figs. 1-13, plates 1-2 [bionomics & control]— Burkhardt & Lengerken (1920), Z. A. E. VI, 270-295, figs. 1-32 [egg, larva, pupa]. Verhoeff (1923), A. N. 10, 63-64, 68-70, pl. III, figs. 31, 32, 36, 41 & 42.

andrewesi Grouv. (1908), C. R. I., 370[1] & 377 — Nilgiris[1]

arrowi Grouv. *l.c.* 373[1] & 375 (*Acanthogethes*). — Nilgiris[1]

binotatus Grouv. (1894) A. S. E. B. XXXVIII, 460[1]—(1908) C. R. I .375[2]. — Chachuba[1] Pagu[2]

bræti Grouv. *l.c.* 583[1]—*l. c.* 375. — Kurseong[1]

cardoni Grouv. *i. c.* 584[1]— *l.c.* 376. — Barway[1]

castanescens Grouv. (1903), A. S. E. F. LXII, 114[1]—(1908), C. R. I. 376 (*Odontogethes*). — Darjiling[1]

crassus Grouv. (1908), C. R. I. 369[1] & 377. — Nilgiris[1]

cribrosus Grouv. *l. c.* 368[1] & 375. — Bombay[1] Nilgiris[1]

curvipes Grouv. *l. c.* 373[1] & 374 (*Acanthogethes*). Madura[1] Nilgiris[1]

ferrugineus Reitt. [1872 (1873)], V. N. V. B. 51, sept. 3—Grouv. (1908) C. R. I. 376[1]. Sikkim[1]

foedus Grouv. (1908), C. R. I. 370[1] & 377. Nilgiris[1]

luminosus Reitt. (1873), S. E. N. 72—Grouv. *l. c.* 376.[1] Ceylon[1]

luteolus Reitt. *l. c.* 75— Grouv. *l. c.* 377.[1] Ceylon[1]

melleus Grouv. *l. c.* 371[1] & 376. Pegu[1]

mixtus Grouv. *l. c.* 368[1] & 375. Nilgiris[1]

?politus Motsch. (1863), B. M. XXXVI, II 442—Reitt. (1873), S. E. N. 71.[1] Ceylon[1]

rubricollis Grouv. (1903), A. S. E. F. LXXII, 125[1]—(1908), C. R. I. 375. Nilgiris[1]

rufithorax Grouv. (1894), A. S. E. B. XXXVIII, 460[1]—(1908), C. R. I. 376. Belgaum[1]

waterhousei Grouv. (1908), C. R. I. 372[1] & 375. Nilgiris[1]

III. CARPOPHILINAE.

Carpophilinœ Er. (1843), G. Z. IV, 226 & 233.

Carpophilides Lacord. [1854 (1855)], 290 & 293.

Carpophilini Ganglb. (1899), 447 & 457— Grouv. (1908), C. R. I. 326—Reitt. (1911), F. G. K. 11 & 14.

Carpophilidæ Murr. (1864), M. N. 230 & 255— Reitt. (1873), S. E. N. 7.

4. **Brachypeplus** Er.

Erichs. (1843), G. Z. IV 245—Lacord. [1854 (1855)], G. C. 293 & 297—Murr. (1864), M. N. 257 & 286-309 [key to species]—Grouv. (1908), C. R. I. 326-328 [key to species].

Nitidulopsis Wlk. (1858), A. M. N. H. (3) II, 206—Olliff (1885), P. L. W. X, 70.

Subgen. *Leiopeplus* Murr. (1864). M. N. 287 & 304.

Subgen. *Liparopeplus* Murr. *l. c.* 287 & 306.

Subgen. *Onicotis* Murr. *l. c.* 287 & 288.

Subgen. *Selis* Murr. *l. c.* 287 & 302.

Subgen. *Tasmus* Murr. *l. c.* 287 & 290.

aequalis Walk. (*Nitidulopsis*), 1858, A. M. N. H. (3) II, 206[1]—Olliff (1885), P. L. W. X, 70— Grouv. (1908), C. R. I. 328.	Ceylon[1]
birmanicus Grouv. (1892), A. M. G., XXXII, 834[1].	Burma[1]
dubreuili Grouv. (1908), C. R. I, 327 & 328[1].	Madura[1]
feai (*feœ*) Grouv. (1892), A. M. G. XXXII, 835.[1]	Burma[1]
latus Murr. (1864) M. N. 301[1] & 309.	Singapore[1] Sarawak[1]
lowei Murr. *l. c.*, 301[1] & 309.	Borneo[1]
notatus Murr. *l. c.* 300[1] & 309— Grouv. (1908), C. R. I. 328.	Ceylon[1]
obesus Grouv. (1895) B. S. E. F. CCLVIII.[1]	East Indies[1]
patruelis Murr. (1864), *l. c.*, 299[1] & 309.	East Indies[1]

5. **Cillæus** Laporte.

Laporte, (1835) E. E. 138—Eriches. (1843), G. Z. IV 247— Lacord. [1854 (1855)], G. C. 293 & 297— Murr. (1864), M. N. 311—Grouv. (1906), A. S. E. F. LXXV, 68— (1908), C. R. I. 327 & 329.

latus Grouv. (1908), C. R. I. 329.[1]	Nilgiris[1]
simoni Grouv. (1893), A. S. E. F. LXII, 383[1].	Ceylon[1]

6. **Platynema** Rits.

Ritsema (1885), N. L. M. VII, 29—Grouv. (1908), C. R. I. 327.

Orthogramma Murr. (1864), M. N. 318.

breviceps Murr., *l. c.*, 323.[1]	Nicobars[1]
olliffi Rits. (1885), N. L. M. VII, 29[1].	Andamans[1]

7. **Carpophilus** Steph

Steph. (1830), I. B. E. III, 50—Lacord. [1854 (1855)], G. C. 293 & 295—Murr. (1864), M. N. 341-398 [key to species]—Ganglb. (1899), K. M. 458— Grouv. (1908), C. R. I. 327 & 330-338 [key to species]—Blair (1922), E. M. M. LVIII, 3, 65 (Key to species].

Biology: Perris, (1853), A. S. E. F. (3) I, 593, sep. I (1863), 71, t. 18, figs. 69-76— Murtfeldt (1903), Ent. News,

XIV, 293—Steb. (1914), I. F. I. 109— Dash. (1916), Rept. Dept. Agric. Barbados, 1914-15, 38-44—Rutgers (1918), Verslag Van den Directeur Van het Algemeen Proefstation der A. V. R. O. S., 1-VII-17—30-VI-18, 44—Zwaluwenburg (1918), Rept. Porto Rico Agric. Expt. Sta. 1917, 31-34— Roepke (1918), Entomologische onderzoekingen 1917-18, 13-14 [control]—Ghosh (1921), R. P. E. M. 108.

Subgen. *Tribrachys,* Lec. (1861), C. N. A. 83.

Subgen. *Ecnomorphus* Motsch. (1858), E. E. VII, 42—Murr. (1864), M. N. 385.

Subgen. *Eidocolastus* Murr. *l. c.* 399—Grouv. (1908), C. R. I. 327 (note).

Idocolastus Gemm. & Harold (1868), C. C. III, 814.

Subgen. *Endomerus* Murr., *l. c.,* 384.

Nitops Murr. l. c. 382—Grouv. (1896), A. S. E. F. LXV, 178.

Subgen. *Heterodontus* Murr. *l. c.* 342.

Subgen. *Microxanthus* Murr. *l. c.* 389.

Subgen. *Myothorax* Murr. *l. c.* 372.

Subgen. *Stauroglossicus* Murr. *l. c.* 398— Grouv. (1908), C. R. I. 327 [note].

Subgen. *Urophorus* Murr. *l. c.* 342.

bicolor Grouv. (1892), A. M. G. XXXII, 836[1] & 838—Grouv. (1897), *l. c.* XXXVIII, 351 (*Urophorus*). — Burma[1]

bifenestratus Murr. (1864), M. N. 364[1] & 396—Grouv. (1908), C. R. I. 330[2] & 337. — Ceylon[1], Kurseong[2] Mahe[2]

biguttatus Motsch. (1858), E. E. VII, 43—Murr. (1864), M. N. 387— Grouv. (1908), C. R. I. 333[1] & 338 (*Ecnomorphus*). — India[1]

cribratus Murr. *l. c* 369[1] & 397. — East Indies[1]

cylindricus Murr., *l. c.* 382[1] & 397—Grouv. (1908), C. R. I. 337 (*Myothorax*). — Ceylon[1]

dimidiatus F. (*Nitidula*) (1792), E. S. I, 1, 261—Murr. *l. c.* 379 & 397— Ganglb. (1899), K. M. 459 & 460—Grouv. (1908), C. R. I. 332 & 337. — Cosmoplitan.

Biology: Morrill (1913), J. E. E. VI. 2, 185-195—Wimhurst (1920), Ad. Rept. Agric. Directorate [Mesopotamia], 1919, 39-41—Buxton (1920), B. E. R. 294—Essig (1921), M. B. C. S. X, 4, 140-143, 3 figs.

auropilosus Woll. (1854), I. M. 117.

leuisi Reitt. (1884), N. J., 258 & 300—Grouv. [1912 (1913)], A. S. E. F. LXXXI, 388.

pusillus Steph. (1830), I. B. E., 51. var. *contigens* Wlk. (*Nitidula*), (1858), A. M. N. H. (3) II, 206— Olliff (1885), P. L. W. X, 69— Grouv. (1908), C. R. I. 332.

biguttatus Gemm. & Harold (1868), C. C. 810.

dilutus Murr. (1864), M. N. 373.

limbalis Murr. *l. c.* 373.

nigritus Murr. *l. c.* 373.

robustus Murr. *l. c.* 373.

testaceus Murr. *l. c.* 373.

vittiger Murr. *l. c.* 373—Grouv. (1908), C. R. I. 332.

var. *fumatus* Boh. (1851), Ins. Caffr. I, 2, 564—Murr. *l. c.* 375 & 397.

var *luridus* Murr. *l. c.* 377 & 397.

var. *mutilatus* Er. (1843), G. Z. IV, 258—Murr. *l. c.* 378 & 397—Grouv. (1908), C. R. I. 322-Steb. (1914), I. F. I. 108.

hemipterus F. (*Nitidula*) (1792), E. S. I. 261.

bimaculatus Montr. (*Ips*) (1860), A. S. E. F. (3) VIII. 262.

puberulus Montr. (Ips) *l. c.* 263.

var. *ochropterus* Klug (1862), Peter's Reise, 209, t. 12, f. 9—Murr. (1864), M. N. 377 & 397—Fairm. (1869), A. S. E. F. (4) IX, 200—Grouv. (1908), C. R. I. 332.

tempestivus Jacq. du Val. (non Er.) (1856), Hist. Cuba, Ins. 42, ed. 8°, 1857, 96—Chevr. (1863), A. S. E. F. III, 602.

dubitabilis Grouv. (1897), A. M. G. XXXVIII, 355[1]—Grouv. (1908), C. R. I. 336[2] & 338 (*Stauroglossicus*). Ceylon[1] Burma[2]

erythrogonus Grouv. (1898), A. S. E. B. XLII, 110[1]—(1908), C. R. I. 338. Bengal[1]

familiaris Grouv. (1897), A. M. G. XXXVIII, 353[1].
Tenasserim[1]

feai (*feae*) Grouv. (1890), *l. c.* XXIX, 120—(1892), XXXII, 338—(1897), XXXVIII, 351 (*Urophorus*).
Burma.

flavipes Murr. (1864), M. N. 359[1] & 395—Grouv. (1908), C. R. I. 337—Steb. (1914), I. F. I., 109[2], f. 72 [biol.]
Celebes[1], Singapore[1]
Siwaliks[2]

fulvipes Motsch, (1858), E. E. VII 42—Murr. *l. c.* 352[1] & 395—Grouv. (1908), C. R. I. 338.
Ceylon[1]

var. *fusculus* Motsch. *l. c.* 42[1].

funereus Murr. *l. c.* 358[1] & 396—Grouv. *l. c.* 338.
Ceylon[1]

hebetatus Grouv. *l. c.* 334 & 338, t. 6, f. 1. *fulvipes* Grouv. (1894), (non Motsch.), A. S. E. B. XXXVIII, 578—(1908), C. R. I. 335[1].
Bengal[1]
Belgaum[1]
Nilgiris

hemipterus L. (*Dermestes*) (1758), S. N. X, 358— (1767), XII, 1, 2, 565—Murr. (1864), M. N. 362 & 395—Ganglb. (1899), K. M. 439 & 460—Grouv. (1908), C. R. I. 337—Blatchley (1910), Cat. Col. Indiana, 631 & 632, fig. (*Myothorax*).
Cosmopolitan.

Biology: Perris [1875 (1876)], A. S. L. L. XXII, 303, f. 33 & 34, sep. ed. Deyrolle (1877), 45, f. 33 & 34—French (1911), Jl. Dept. Agr. Victoria, IX, 640—Steb. (1914), I. F. I. 108, f. 71—Essig (1915), J. E. E. VII, 396-400, 4 figs. [control]-Morrill (1917), 9th Ann. Rept. Arizona Commiss. Agric. & Hortic. 30-VI-17, 15-61—De Ong (1918), M. B. C. S. VII, 6, 429 (control]—Gurney (1918), Agric, Gaz., N. S. W., XXIX, 9, 641-650, 11 figs.—Maskew (1919), M. B. C. S. VII, 6, 429-432–Morris (1920), Ann. Trop. Med. Parasit. XII, 4, 339-342—Essig (1920), M. B. C. S. IX, 3, 119-125, 5 figs. [control]—Maskew & Strong *l. c.* 12, 721-735—Poutiers (1921), Bull. Soc. Path. Vèg. France, VIII, 1, 45-46—De Ong (1921), J. E. E. XIV, 5, 444-447 [control].

bimaculatus L. (1767). S. N. ed. XII, 1, 2, 569—Ol. (*Nitidula*) (1790). Ent. II, 12, p. 4, t. 2, f. 11—Melsh. (*Cercus*) (1844), II, 105.

brevipennis Germain (1835), Anales Unio. Chile, 397.

cadaverinus F. (*Nitidula*) (1801), S. E. I. 354.

dimidiatus Heer (*Cateretes*) (1841), F. H. I, 413.

ficus F. (*Stenus*) (1801), S. E. 603.	
flexuosus Herbst. (*Nitidula*) (1790), Käf, V, 246, t. 54, f. 5.	
pictus Heer (*Cateretes*) (1841), F. H. 413.	
quadriguttatus Thunb. (*Nitidula*) (1794), Ins. Suec. V. 70.	
var. *quadratus* F. (*Nitidula*) (1798), Suppl. E. S. 74.	
var *quadrisignatus* Er. (1843), G. Z. 257—Murr. (1864), M. N. 363 & 396.	
heros Grouv. (1892), A M. G. XXXII, 836[1] & 838—(1897), *l. c.* XXXVIII, 351.	Burma[1]
humeralis F. (*Nitidula*) (1798), E. S. Suppl. 74— Murr. (1864), 343 & 394—Grouv. (1892), A. M. G. XXXII, 838—(1897), XXXVIII, 351—(1908), C. R. I. 330 & 336 (*Urophorus*)— Muir (1918), T. E. S. L. 224, pl. X, f. 7 [genitalia].	India.[1] ——————— Trop. Africa Madagascar Seychelles
foveicollis Murr. *l. c.* 344 & 394—Grouv. (1892) A. M. G. XXXII, 838—(1897), XXXVIII, 351—(1908), C. R. I. 330— Lefroy (1909), Ind. Ins. Life-297 [biol.]—Steb. (1914), I. F. I. 109 [biol].	
picinus Boh. (*Brachypterus*) (1851), Ins. Caffr. I. 2, 560—Murr. (1864), M. N. 343.	
var. *punctatus* Fleut. (1887), A. S. E. F. (6), VII, 61—Grouv. [1912 (1913)]. *l. c.* LXXXXI, 388.	
lepidus Murr. (1864), M. N. 399[1] (*Stauroglossicus*).	East Indies.[1]
ligatus Murr. *l. c.* 363[1] & 396—Grouv. (1908), C. R. I. 337[2].	Ceylon[1,] Burma[2] Sumatra[2],
marginellus Motsch. (1858), E. E. VII, 40—Murr. *l. c.* 347[1] & 394, t. 33, f. 1—Grouv. (1908), C. R. I. 330[2] & 336.	India[1] Ceylon[1,] China[1] Madagaskar[2] Southern Asia[2]
murrayi Grouv. (1892), A. M. G. XXXII, 837—(1908), C. R. I. 333[2] & 338.	Burma[1] Nilgiris[2]
obesus Murr. (1864), M. N. 361[1] & 396.	East Indies[1]
oblongopunctatus Grouv. (1903), A. S. E. F. LXXII, 108[1]—(1908), C. R. I. 337.	Darjiling[1]
obsoletus Er. (1843), G. Z. 259—Murr. (1864), M. N. 368[1] & 396—Grouv. (1908), C. R. I. 336—Steb. (1914), I. F. I. 109.	Ceylon[1], East Indies[1] ——————— Siam[1]
cribellatus Motsch. (1858), E. E. VII, 41.	

funerus Reitt. (non Murr.) (1884), N. J. 259—Grouv. [1912 (1913)], LXXXI, 388.

var. *strigipennis* Motsch. *l. c.* 41—Murr. *l. c.* 368.

pilosellus Motsch. *l. c.* 41—Murr. *l. c.* 380[1] & 397—Grouv. (1908), C. R. I. 333 & 337 (*Myothorax*).
Ceylon[1], East Indies[1] Japan[1], Tongking[1] Celebes[1],

plagiatipennis Motsch. (*Colastus*) (1858), E. E. VII, 39—Murr. *l. c.* 400[1], t. 33, f. 6—Grouv. *l. c.* 336[2] & 338 (*Eidocolastus*).
Ceylon[1] Belgaum[2],

dilutus Motsch. (*Colastus*) *l. c.* 39—Murr. *l. c.* 400—Grouv. [1912 (1913)], A. S. E. F. LXXXI, 388.

bosschai (*bosschae*) (1892), N. L. M. XIV, 43.

pubescens Murr. *l. c.* 383[1] & 398 (*Endomerus*).
Ceylon[1]

sericeus Motsch. (1858), E. E. VII, 41—Murr. *l. c.* 368[1] & 396—Grouv. (1908), C. R. I. 331[2] & 336.
East Indies[1] Kurseong[2]

signatus Grouv. (1908), C. R. I. 333[1] & 337. var. *ornatus* Grouv. *l. c.* 334[2].
Kurseong[1] Sikkim[2]

tectus Murr. (1864), 359[1] & 396—Grouv. *l. c.* 332[2] & 336.
Ceylon[2], India[2] ________, Borneo[1] Singapore[1], Sarawak[1]

undulatus Grouv. *l. c.* 331[1] & 336.
Siwaliks[1], ________ Java[1]

8. **Tetrisus** Murr.

Murr. (1864), M. N. 404—Grouv. (1908), C. R. I 327 & 338.

curvipes Grouv. *l. c.* 338[1].
Nilgiris[1]

epuraeoides Grouv. (1892), A. M. G. XXXII, 839[1].
Burma[1]

hydroporoides Murr. (1864), M. N. 405[1], t. 35, f. 10.
East Indies[1]

IV. NITIDULINAE.

Nitidulinæ Er. (1843), G. Z. IV, 226 & 266.

Nitidulides Lecord, [1854 (1855)], G. C., 290 & 300.

Nitidulites Jac. du Val. (1858), Gen. Col. d'Eur II, 140 & 157.

Nitidulina Thoms. (1859) S. C. I, 67— (1862), IV, 162—(1867), IX, 370.

Nitidulini Lec. (1861), C. N. A. 82 & 83— Ganglb. (1899), K. M. 447 & 462 [key to genera]—Grouv. (1908), C. R. I. 326 & 339 [key to genera]

Nitidulidae Murr. (1864), M. N. 230.

Nitiduli Mars. (1885), N. R. 19 & 20.

Nitiduliens Lameere (1900), Fn. Belg. 343 & 345.

9. **Echomaeus** Er.

Er. (1843), G. Z., IV, 264—Lacord. [1854 (1855)], G. C. 293 & 299—Murr. (1864), M. N. 408—Grouv. (1908), C. R. I. 340 & 342.

haroldi Reitt. (1873), S. E. N. 182[1]—Steb. (1914) I. F. I. 110[2]. — Himalaya[1] Siwaliks[2]

waterhousei Grouv. (1908), C. R. I. 342[1], t. 6, f. 5—Steb. (1914), I. F. I. 110[2]. — North India[1] Siwaliks[2]

10. **Trimenus** Murr.

Murr. (1864), M. N. 405—Grouv. (1908), C. R. I. 340 & 343.

adpressus Murr. *l. c.* 406[1], t. 36, f. 4. *angustatus* Murr. *l. c.* 407[2]—Grouv. (1908), *l. c.* 343[3]. — East Indies[2], Ceylon[3] Nilgiris[3], ——— Malay Archipelago[1]

parallelopipedus Motsch.

(*Epuræa*) (1863), B. M. I, 438[1]—Reitt. (1873), S. E. N. 29—Grouv. (1908), C. R. I. 343[1]. — Ceylon[1]

11. **Haptoncus** Murr.

Murr. (1864), M. N. 401—Reitt. (1884), N.J. 258—Grouv. (1908), C. R. I. 340 & 343-346 [key to species].

Haptoncura Reitt. (1875), V. N. V. B. XIII, 61 & 64, Sep. 11 & 14—(1884), N. J. 259.

concolor Murr. (1864), M. N. 402[1]—Grouv. (1908), C. R. I. 346[2]. — Nilgiris[2] ——— New Guinea[1]

epuraeoides Grouv. *l. c.* 344[1] & 346. — Ceylon[1]

luteolus (*Epuræa*) (1843), G. Z. 272—Chevr. (*Epuræa*), (1863), A. S. E. F. 4 (III), 602—Reitt. (*Haptoncura*) (1875), V. N. V. B. XIII, 62, Sep. 12—Grouv. (1906), A. S. E. F. LXXV, 75[2]— (1908), C. R. I. 346 [note]. — Ceylon[3, 4] ——— Cuba.[1] Madagascar[2]

floreolus Sharp (1890), B. C. A. II, 1, 305.

intendens Walk. (*Nitidula*) (1858), A. M. N. H. (3) II, 206[3]—Olliff (1858), P. L. W. X. 70.

pauperculus Reitt. (1873), S. E. N. 179.

pubescens Murr. (1864) M. N. 403[4].

subquadralus Reitt. (1877), Mittheil. Münch. Ent. Ver. I, 22—Blackb. (*Haptoncura*) (1891). Trans. R. Soc. S. Austr. XIV. 105—Grouv. (1906), A. S. E. F. LXXV, 75[2].

texanus Crotch (1874), Trans. Amer. Ent. Soc. III 76.

motschulskyi Reitt. (1873). S. E. N. 29—Grouv. (1908), C. R. I. 344.

augustulus Motsch. (*Epuræa*) (1863), B. M. I. 439[1]. — Ceylon[1],

ocularis Fairm. (*Epuræa*) (1849), Rev. Zool. (2) I, 363, sep. Col. Polynesie, 28—Reitt. (1873), S.E. N. 27—(*Haptoncura*) (1875), V. N. V. B. 62, sep. 12—Grouv, (1908), C. R. I. 343[1]-345. — Ceylon[1] Calcutta[1], Sikkim[1] ——— Japan[1] Tahiti[1]

bifaciatus Kraatz. (*Epuræa*) (1895), D. E. Z. 148—Grouv. [1912 (1913)], A. S. E. F. LXXXI, 395[1]. — East Africa[1]

decoratus Reitt. (*Epuræa*) (1873), S. E. N. 28 & 41—Grouv. (1906), A. S. E. F. LXXV, 75[1]. — Madagascar[1]

significans Wlk. (*Nitidula*) (1858), A. M. N. H. 3 (II), 206[1]—Olliff (1885), P. L. W. X. 70. — Ceylon[1]

tetragonus Murr. (1864), M. N. 401[1], t. 33, f. 7—Reitt. (1884), N. J. 259. — Ceylon[1]

thiemei Reitt. (*Epuræa*) (1873), S. E. N. 28 & 41—(1884), N. J. 260[1]. — Japan[1]

pulsillus Grouv. (*Macroura*). (1894), A. S. E. B. XXXVIII, 585[1]—(1907), A. M. G. XLIII, 575—(1908), C. R. I. 344 & 345. — Kurseong[1]

reflexicollis Motsch. (*Epuræa*) (1863), B. M. II, 431[1]—Reitt. (1873), S. E. N. 27[2]— (1877), D. E. Z. XXI, 189—Grouv. (1908), C. R. I. 344[3] & 346. — Ceylon[1], Nilgiris[3] ——— Japan[2] Ceylon[2]

testaceus Murr. (1864), M. N. 403[1]—Grouv. (1908), C. R. I. 346[2]. — ——— Macassar[1]

vulpecula Redt. (1867). R. N. 34—Reitt. (1873), S. E. N. 27.

12. **Propetus** Reitt.

Reitt. (1875), V. N. V. B. XIII, 62 & 64, sep. 12 & 14—Grouv. (1908), C. R. I. 340 & 345.

nigripennis Redtb. (*Epuræa*) (1867), R. N. 34[1]—Reitt. (1873), V. N. V. B. XIII 63, sep. 13. Ceylon[1]

13. **Amystrops** Grouv.

Grouv. (1906), A. M. G. XLII, 312.

montana Grouv. (1913), R. I. M. VIII, 2, 103[1]. Assam[1]

14. **Omosita** Er.

Er. (1843) G. Z. IV, 298—Lacord. [1854 (1855)], G. C. 301 & 309—Ganglb. (1899), K. M. 464 & 488—Grouv. (1908), C. R. I. 341 & 361— Verhoeff. (1923), A. N. 13, 78-80, pl. I, figs. 1-3 & 15.

ornata Grouv. (1903), A. S. E. F. LXXII, 112[1]. Darjiling[1]

15. **Epuraea** Er.

Er. (1843), G. Z. IV, 267—Lacord. [1854 (1855)], G. C. 301 & 302— Ganglb. (1899), K. M. 464 & 470—Grouv. (1908), C. R. I. 340 & 346-352 [key to species].

Biology: Perris (1862), A. S. E. F. 2 (IV), 184-189. pl. 5—Bagnall (1906), E. M. M. 42 (1906), Trans. Nat. Hist. Soc. Northumberland. 1, 416-420—Scott (1920), T. E. S. L. 99-127, figs. 1-8—Verhoeff (1923), A. N. 13, 72-78, pl. II, figs. 16-18 & 20-21; pl. VII, figs. 87-91.

Subgen. *Dadopora* Thoms. (1859), S. C. I, 68—(1862), IV, 168—(1867), IX, 378—Reitt. (1911), F. G. K. 29 & 30— Ganglb. (1899), K. M. 473.

Subgen. *Epuræanella* Crotch (1874) Trans. Amer. Ent. Soc. V, 76.

Omosiphora Reitt. (1875), V. N. V. B. XIII 56 & 63, sep. 6 & 13.

Subgen. *Micrurula* Reitt. (1884), W. E. Z. III, 209—Ganglb. (1899), K. M. 472.

Micruria Reitt. (1874), V. N. V. B. XIII, 56 & 64, Sep. 8 & 14.

andrewesi Grouv. (1908), C. R. I. 350[1] & 352. Nilgiris[1]

apicalis Grouv. (1903). A. S. E. F. LXXII, 110[1]—(1908), C. R. I. 352. Sikkim[1]

arcuata Grouv. (1908) *l. c.* 350 & 352, t. 6 f. 6. Nilgiris[1]

birmanica Grouv. (1892), A. M. G. XXXII, 840[1].	Burma[1]
braeti Grouv. (1894), A. S. E. B. XXXVIII, 578[1]—(1908), C. R. I. 355 (*Micrurula*).	Kurseong[1]
brunnea Wiedem (1823). Z. M. II, 72[1]— Grouv. (1908), C. R. I. 352 [note].	India[1]
cerina Grouv. (1894), A. S. E. B. XXXVIII, 579[1]—(1908), *l. c.* 355 (*Micrurula*).	Kurseong[1]
consobrina Grouv. (1892), A. M. G. XXXII 840[1].	Burma[1]
convexa Grouv. (1908), C. R. I. 354[1] & 355 (*Micrurula*).	Darjiling[1]
cribrata Grouv. (1903), A. S. E. F. LXXII, 111[1]—(1908), *l. c.* 351.	Darjiling[1]
dubitabilis Grouv. (*Haptoncura*) (1890), A. M. G. XXIX 121[1].	Tenasserim[1]
emarginata Grouv. (1906), *l. c.* XLII, 316[1].	Burma[1]
exculta Grouv. [1912 (1913)], A. S. E. F. LXXXI, 394.	Darjiling[1]
ornata Grouv. (non Reitt.) (1903), *l. c.* LXXII, 109[1]— (1908), C. R. I. 352.	
fossicollis Grouv. *l. c.* 347[1] & 351.	Assam[1]
indica Grouv. (1894), A. S. E. B. XXXVIII, 459[1]— (1908), *l. c.* 346 & 351, t. 6, f. 9.	Belgaum[1]
insolita Grouv. *l. c.* 353[1] & 355 (*Micrurula*).	Khasis[1], Sumatra[1]
latissima Reitt. (1880), N. L. M. II, 42[1]— Grouv. *l. c.* 351[2].	———, Ceylon[2]
omissa Grouv. *l. c.* 349 & 352[1].	Assam[1]
persimilis Grouv. (1913), B. S. E. F. IV, 253.	Assam[1]
affinis Grouv. (non Steph.) (1908) C. R. I. 352[1] & 355 (*Micrurula*).	
propinqua Grouv. (1906), A. M. G. XLII, 365[1].	Burma[1]
pumila Grouv. (1897), *l. c.* XXXVIII, 359[1]—(1908), C. R. I. 351.	India[1]
reticulata Grouv. (1892), A. M. G. XXXII, 841[1].	Burma[1]
submaculata Wlk. (*Nitidula*) (1859), A. M. N. H. (3) III, 52[1]—Olliff (1885), P. L. W. 70—Grouv. (1908), C. R. I. 352.	Ceylon[1]
subochracea Grouv. (1903), A. S. E. F. LXXII, 111[1]—(1908), C. R. I. 352.	Darjiling[1]

Subtilis Grouv. (1894), A. S. E. B. XXXVIII, 579[1]— (1908) C. R. I. 355 (*Micrurula*). — Kurseong[1]

waterhousei Grouv. *l. c.* 318[1] & 352. — N. W. Himalayas[1]

16. Stelidota Er.

Er. (1813), G. Z. 300—Lacord [1854 (1855)], G. C. 301 & 310— Ganglb. (1899), K. M. 463 & 466— Grouv. (1908), C. R. I. 339 & 361.

multiguttata Reitt. (1877), D. E. Z. XXI, 110[1]— (1884), N. J. 262[1]—Grouv. (1908), C. R. I. 361[2]. — Japan[1]
———————
Ceylon[2]
Nilgiris[2]

nigrovaria Fairm.

(*Omosita*) (1849), Rev. Mag. Zool. (2) I, 364, sep. Col. Polynésie, 28. — Tahiti[1]
———————

multiguttata Grouv. (non Reitt.) (1892), A. S. E. B. XXXVI, 60[1]—(1894), C. R. I. 361. — Bengal[1]

17. Ipidia Er.

Er. (1843), G. Z. IV, 289—Lacord. [1854 (1855)], G. C. 301 & 304—Ganglb. (1899), K. M. 463 & 465.

sibirica Reitt. (*Stelidota*) (1879), D. E. Z. XXIII, 216—(1884), N. J. 262—Mars. (*Stelidota*) (1885), N. R. 77. — Japan, Siberia
————, Burma

18. **Prometopia** Er.

Er. (1843), G. Z. IV, 279—Lacord. [1854 (1855)], G. C. 301 & 306—Murr. (1867), A. M. N. H. (3) XIX, 168, sep. Col. Old. Cal. 85—Grouv. (1908), C. R. I. 340 & 358-359 [key to species].

catilina Olliff (1883), T. E. S. L. 177[1]—Grouv. (1897), A. M. G. XXXVIII, 366—(1908), C. R. I. 358[2] & 359 [note]. — Borneo[1]
———————
Kandhesh[2], Nilgiris[1], East India[1], Ceylon[1]

gracilis Grouv. (1908), C. R. I. 358[1] & 359, t. 6, f. 11.

quadrimaculata Motsch. (1863), B. M. XXXVI, I, 440[1]— Grouv. (1897), A. M. G. XXXVIII, 366—(1908), C. R. I. 358[2] & 359—Arrow (1900), Monogr. Christm. Isl. 92[3]. — Sikkim[2]
———————
Cochin-China[3]

setosula Grouv. (1894), A. S. E. B. XXXVIII, 580[1]— (1908), C. R. I. 359. — Barway, Bengal[1]

19. **Parametopia** Reitt.

Reitt. (1884), N. J. 259 & 262—(1885), 17—Grouv. (1908), C. R. I. 340 & 359-361 [key to species].

concolor Grouv. (1891), C. R. Soc. Ent. Belg. XXXV, CCXXXVII[1]—(1908), C.R. I. 361. — Kunbir, Bengal[1] Kurseong[1]

indica Grouv. (1894), A. S. E. B. XXXVIII, 581[1]—(1908), C. R. I. 361.

tripunctata Grouv. (1908), C. R. I. 360[1]. var. *trinotata* Grouv. *l. c.* 360[1] & 361, t. 6, f. 12. — Nilgiris[1] Mahe[1]

20. Axyra Er.

Er. (1843), G. Z. IV, 285—Lacord. [1854 (1855)], G. C. 301 & 305—Murr. (1867), A. M. N. H. (3) XIX, 169, sep. Col. Old. Cal. 87—Grouv. (1908), C. R. I. 341 & 356.

Galaor Thoms. (1858), Arch. Ent. II, 43.

Subgen. *Axyrodes* Murr. (1867), A. M. N. H. (3) XIX, 170, sep. Col. Old. Cal. 88.

feai (*feæ*) Grouv. (1890), A. M. G. XXIX, 122[1]— (1908), C. R. I. 356 [note]. — Burma[1]

perplexa Reitt. (1873), S. E. N. 53[1]—Grouv. (1908), C. R. I. 356-357 [add. descr.]. — East India[1]

21. Megauchenia M'Leay.

M'Leay (1825), A. J. 45, ed. Lequien (1833), 155—Arrow (1909), A. M. N. H. (8) IV, 191—Heller (1923), E. B. 19, 1-43 [key to species].

Ischæna Er. (1843), G. Z. IV, 287—Lacord. [1854 (1855)], G. C. 301 & 306—Reitt. (1873), S. E. N. 10 & 53—Grouv. (1908), C. R. I. 340 & 357-358 [key to species].

angustata Er. (*Ischæna*) (1843), G. Z. 288[1] Grouv. (1890), A. M. G. XXIX, 124—(1908), C. R. I. 358[2] (*Ischæna*). — East Indies[1], Pegu[2] Burma[3], Tenasserim[4]

var *peninsularis* Grouv. (1890), A. M. G. XXIX, 123 & 124[3]—Heller (1923), E. B. 19, 1, 41[4] (*angustata*). — Malay Archipelago[2] Malacca[1], Tongking[2]

feai Grouv. (*feæ*) emend. (*Ischæna*) (1892), A.M.G. XXXII. 842[1]—Heller (1923), E. B. 19, 1, 42. — Burma[1]

indica Grouv. (*Ischæna*) (1908), C. R. I. 357[1] t. 6, f. 5—Heller (1923), E. B. 19, 1. 42. — Nilgiris[1]

interstitialis Reitt. (*Ischæna*) (1880), A. M. G. XV, 456—Grouv. (*Ischæna*) (1890), A. M. G. XXIX, 123[1]—Heller (1923), E. B. 19, 1, 41[2]. — India[1] New Guinea[2]

22. **Soronia** Er.

Er. (1843), G. Z. IV, 277—Lacord. [1854 (1855)], G. C. 301 & 304—Ganglb. (1899), K. M. 464 & 468—Grouv. (1908), C. R. I. 340 & 356 [key to species]—Verhoeff (1923), A. N. 13. 88-91, pl. II, figs. 25-26.

bræti Grouv. (1894), A. S. E. B. XXXVIII, 580[1]—(1908), C. R. I. 356. — Kurseong[1]

imperialis Grouv. (1903), A. S. E. F. LXXII, 113[1]— (1908), C. R. I. 356. — Darjiling[1]

minima Grouv. (1903), A. S. E. F. LXXII, 113[1]—(1908), C. R. I. 356. — Darjiling[1]

23. **Atarphia** Reitt.

Reitt. (1884), N. J. 260 & 263—(1885), *l. c.* 39.

Pseudoplatychora Grouv. (1890), N. L. M. XII, 15 & 31.

convexiuscula Grouv. (*Pseudoplatychora*) *l. c.* 15[1] & 31. — East Indies[1]

24. **Physoronia** Reitt.

Reitt. (1884), N. J. 260 & 263—(1885), 18.

reitteri Grouv. (1892), A. M. G. XXXII, 843[1] (*Physorhina*) — Burma[1]

25. **Amphicrossus** Er.

Er. (1843), G. Z. IV, 346—Lacord. [1854 (1855)], G. C. 319 & 323—Grouv. (1908), C. R. I. 341 & 381-384 [key to species].

Lobostoma Fairm. (1892), Rev. d'Ent. XI 90—Grouv. (1908), B. S. E. F. 109.

Rhacostoma Berg. (1898), Com. Mus. Buenos Aires, I, 18.

densatus Grouv. (1908), C. R. I. 383[1] & 384. — Assam[1], Bengal[1], Burma[1], ________

discolor Er. (1843), G. Z. IV, 348—Reitt. (1873), S. E. N. 99—Grouv, (1908), C. R. I. 381[1] & 384. — Annam[1], East Indies[1]

muticus Grouv. (1908), C. R. I. 381[1] & 384. — Nilgiris[1], Sumatra[1] ________

oblongus Grouv. (1897), A. M. G. XXXVIII, 368—(1908), C. R. I. 381[1] & 384. — Pegu[1] Travancore[1]

acopus Grouv. (1892), A. S. E. B. XXXVI, 60—(1908), C. R. I. 381[1] & 384. — Bengal[1], Calcutta.[1] Pegu[1]

pilosus Grouv. (1908), C. R. I. 382[1] & 384. Ceylon[1]

plagiatus Grouv. (1890), A. M. G. XXIX, 125[1]. Burma[1]

26. Aethina Er.

Er. (1843), G. Z. IV, 306—Lacord. [1854 (1855)], 301 & 302—Murr. (1857), A. M. N. H. (3), XIX, 176—(1868) sep. Ins. Old. Cal. 94—Grouv. (1908) C. R. I. 341 & 380 [key to species]

Subgen. *Aethinopa* Reitt. (1875), V. N. V. B. XIII, 109, sep. 13.

argus Grouv. (1890), A. M. G. XXIX, 125[1]. Burma[1]

elongata Reitt. (1875), V. N. V. B. XIII, 109[1], sep. 13. East Indies[1]

nigrocastanea Grouv. (1903), A. S. E. F. LXXII, 115[1]—(1908), C. R. I. 380. Darjiling[1]

obscura Reitt. (1873), S. E. N. 86[1]—Grouv. (1908), C. R. I. 380. Himalayas[1]

puncticollis Grouv. (1897), A. M. G. XXXVIII, 366[1]. Tenasserim[1]

vicina Grouv. (1894), A. S. E. B. XXXVIII, 585[1]—(1908), C. R. I. 380. Kurseong[1]

27. Aethinopsis Grouv.

Grouv. (1908), C. R. I. 341 & 379.

antennata Grouv. (1908), C. R. I., 379[1]. Sikkim[1]

28. Macroura Reitt.

Reitt. (1873), S. E. N. 56 & 80—Grouv. (1907), A. M. G. XLIII, 552—(1908), C. R. I. 341 & 377-378 [key to species].

Idæthina Olliff (1884), N. L. M. VI. 73—Grouv. (1908), A. M. G. XLIV, 78.

cyaneipennis Grouv. (*Aethina*) (1903), Darjiling[1] A. S. E. F. 116— (1907), A. M. G. XLIII, 572[1] & 574—(1908), C. R. I. 378[1]. Darjiling[1] Sikkim[1]

densita Reitt. (1875), V. N. V. B. 110[1], sep. 14—Grouv. (1908), C. R. I. 378[1]. Ceylon[1]

longipennis Motsch. (*Carpophilus*) (1858), E. E. VII, 40[1]—Grouv. (1907), A. M. G. XLIII. 564 & 574. Ceylon[1]

meligethoides Reitt. (1873), S. E. N. 82[1]— Grouv. (1907), A. M. G. XLIII, 574— (1908), C. R. I. 378[1]. — Himalayas[1]
________, Australia[1]

nigritula Reitt. (1873), S. E. N. 82—Grouv. (1907), A. M. G. XLIII, 575—(1908), C. R. I. 378[1]. — ________, India[1]
________, Malacca[1]
Sumatra[1]

dubia Grouv. (1894), A. S. E. B. XXXVIII, 461—(1907), A. M. G. XLIII, 575. — Java[1]

orientalis Nietner (*Meligethes*), (1856), Ent. Papers (*Colombo*), II, 11—(1857), A. M. N. H. (2), XIX, 384—Grouv. (1907), A. M. G. XLIII, 574—(1908), C. R. I. 378.

nigritula Olliff (*Idæthina*) (non-Reitt.) (1885), P. L. W. X. 70.

respondens Wlk. (*Meligethes*) (1859), A. M. N. H. (3) III, 53[1]. Grouv. (1907), A. M. G. XLIII, 562[1]. — Ceylon[1], Bombay[1]

subrugosa Grouv. (1894), A. S. E. B. XXXVIII, 461[1]—(1907), A. M. G. XLIII, 575—(1908), C. R. I. 378. — Belgaum[1]

29. **Pocadius** Er.

Er. (1843), G. Z. IV, 318—Lacord. [1854 (1855)], 301 & 317—Ganglb, (1899), K. M. 464 & 540—Grouv. (1908), C. R. I. 341 & 380.

harmandi Grouv. (1903), A. S. E. F. LXXII, 114[1]. — Darjiling[1]

testaceus Grouv. (1892), A. S. E. B. XXXVI, 60[1]. — Bengal[1]

yunnanensis Grouv. (1910), N. L. M. XXXII, 251[1]. — Yunnan[1]

30. **Pocadites** Reitt.

Reitt. (1884), N. J. 264 & 267—(1885), 44—Grouv. (1908), C. R. I. 341 & 380.

dubitabilis Grouv. (1892), A. M. G. XXXII, 843[1]. — Burma[1]

insularis Grouv. (1908), C. R. I. 380[1]. — Andamans.[1]

31. **Cyllodes** Er.

Er. (1843), G. Z. IV, 342—Lacord. [1854 (1855)], G. C. 319 & 320—Ganglb. (1899), K. M. 464 & 543—Grouv. (1908), C. R. I. 341 & 385-391 [key to species].

Pseudocamptodes Grouv. (1896), A. S. E. F. LXV, 76.

Strongylus Herbst. (1792), Käf. IV, 179 (pars)—Steph. (1830), I. B. E. III, 30 & 42 (pars)— Reitt. (1878), S. E. N. 98 & 138—Sharp. (1891), B. C. A. 355.

Volvoxis Kugel (1794), Schneid. Mag. I, 5, 355 (pars).

ænescens Grouv. (1892), A. M. G. XXXII, 847[1] & 848. — Burma[1]

andamanensis Grouv. (1908), C. R. I., 387[1] & 390. — Andamans[1]

angustipes Grouv. *l. c.* 385-386[1] & 390. — Sikkim[1]

bifascies Wlk. *(Tritoma)* (1859), A. M. N. H. (3) III, 259[1]—Olliff (*Camptodes*) (1885), P. L. W. X, 71—Grouv. (1908), C. R. I. 385[1].

opponens Wlk. (*Chilocorus*) (1859), A. M. N. H. (3) IV, 219—Harold. (*Camptodes*) (1875), Col. Hefte XIV, 211.

ornatus Motsch. (*Comptodes*) (1863), B. M. II, 442—Reitt. (1884), N. J. 268—Grouv. (1908), C. R. I. 391. — Ceylon[1], ________ East Indies Cochin China Japan

notatus Reitt. (*Strongylus*) (1873), 129—(1885), W. E. Z. IV, 174.

dubitabilis Grouv. (1913), Col. Cat. Junk, pars, 56, 163. — Burma[1]

dubius Grouv. (non. Reitt.), (1892), A. M. G. XXXII, 846[1] & 849.

humeralis Grouv. *l. c.* 847[1] & 849. — Burma[1], Bengal[1]

indicus Grouv. (1894), A. S. E. B. XXXVIII, 462—(1908), C. R. I. 385[1] & 391. — Kanara[1] Nilgiris[1]

montanus Grouv. (*Pseudocamptodes*) (1903), A. S. E. F. LXXII, 126—(1908), C. R. I. 385[1]. — Kashmir[1]

insularis Grouv. (1908), *l. c.* 388[1] & 391, t. 7, f. 9. — Andamans[1]

limbatus Grouv. (1892), A. M. G. XXXII, 847[1] & 849. — Burma[1]

marginatus Grouv. (1908), C. R. I. 386[1] & 391. — Cachar[1] Assam.

nigropunctatus Grouv. (1892), A. M. G. XXXII, 846[1] & 849. — Burma[1]

quadrimaculatus Grouv. (1908), C. R. I. 388[1] & 391. — Andamans[1]

quadrinotatus Grouv. (1892), A. M. G. XXXII, 845[1] & 849. — Burma[1]

sexpunctatus Grouv. (1908), C. R. I. 389[1] & 391. — Nilgiris[1]

undulatus Grouv. (1892), A. M. G. XXXII, 845[1] & 849. — Burma[1]

variegatus Grouv. (1892), A. M. G. XXXVIII, 370[1]. — East Indies[1]

32. **Oxycnemus** Er.

Er. (1843), G. Z. IV, 351—Lacord. [1854 (1855)], G. C.

319 & 324—Reitt. (1873), S. E. N. 98 & 113—Sharp (1891), B. C. A. 362.

Eugoniopus Reitt. (1884), N. J. 264 & 267—(1885), *l. e.* 76—Sharp (1891), B. C. A. 364.

Psilopyga Lec. (1853), Proc. Acad. Philad. VI, 286—Lacord. [1854 (1855)]. G. C. 301 & 315.

reitteri Grouv. (*Eugoniopus*) (1892), A. M. G. XXXII, 852.[1] Burma[1]

33. **Tricanus** Er.

Er. (1843), G. Z. IV, 352—Lacord. [1854 (1855)], G. C. 319 & 325—Reitt. (1873), S. E. N. 98 & 138—Grouv. (1908), C. R. I. 341 & 392.

nigripennis Reitt. (1873), S. E. N. 139[1]—Grouv. (1892), A. M. G. XXXII, 854—(1908), C. R. I. 392[1]. Ceylon[1]

punctatissimus Grouv, (1892), A. M. G. XXXII, 853[1]—(1908), C. R. I. 392[2]. Pegu[1], Burma[1] Andamans[2]

striatopunctatus Grouv. (1892), A. M. G. XXXII, 853[1] & 854. Pegu[1] Burma[1]

34. **Pallodes** Er.

Er. (1843), IV 348—Lacord. [1854 (1855)], G. C. 319 & 323—Reitt. (1873), S. E. N. 98 & 132—Sharp (1891), B. C. A. 364.

Neopallodes Reitt. (1884), N. J. 264 & 269—(1885), 78—Grouv. (1908), C. R. I. 341.

birmanicus Grouv. (1892), A. M. G. XXXII, 851[1]. Burma[1]

cardoni Grouv. (1894), A. S. E. B. XXXVIII, 585—(*Neopallodes*) (1908), C. R. I. 392[1]. Bengal[1]

dentatus Grouv. (*Neopallodes*) (1892), A. M. G. XXXII, 849[1]. Burma[1]

falsus Grouv. [1912 (1913)], A. S. E. F. LXXX [1912 (1913)], 398[1]. *harmandi* Grouv. (non. Grouv. 1902), (1903), A. S. E. F. LXXII, 117.—(1908), C. R. I. 392. Darjiling[1]

feai (*feæ* emend.) Grouv. (1892), A. M. G. XXXII, 851[1]. Burma[1]

vicinus Grouv. (*Neopallodes*), (1892), A. M.G. XXXII, 850[1]. Burma[1]

35. **Lasiodactylus** Perty.

Perty (1830-1834), D. A. A. 34—Lacord. [1854 (1855)],

301 & 317—Grouv. (1908), C. R. I. 341 & 361-365 [key to species].

Biology: Ghosh (1921), R. P. E. M. 108 (*Lordites*).

Lordites Er. (1843), G. Z. IV, 316—Lacord. [1854 (1855)], 301 & 316—Murr. (1867), A. M. N. H. (3) XIX, 175, Sep. Col. Old. Cal. (1878), 93.

æthinodes Reitt. (1873), S. E. N. 91[1]—Grouv. (1908), C. R. I. 364.	India[1]
chevrolati Reitt. (1873), S. E. N. 90[1]—Grouv. (1908), C. R. I. 364[2]—De Peyerimhoff (1916), R. I. M. XII. 3, 109-113[3], 3 figs. [descr. larva & biol.].	South Africa[1] ———, Bengal[2] Coromandel[3]
monticola Grouv. (1910), N. L. M. XXXII, 250[1].	Yunnan[1]
pictus M'Leay (*Nitidula*) (1825), A. J. 40—Lequien (1833), A. J. 147—Arrow (1909), A. M. N. H. 8 (IV), 191.	Malay Archipelago. ———, East Indies
costulatus Fairm. (Lordites) (1868), A. S. E. F. 4. (VIII), 775—Reitt. (1873), 88—Grouv. (1908), C. R. I. 365.	Bourbon Isd. Madagascar ———
glabricola Candèze (*Lordites*) (1861), Mém. Liége XVI, 340—Reitt. (1873), S. E. N. 88—(1884), N. J. 266—Grouv. (1908), C. R. I. 361 & 365.	Ceylon, Bengal Nilgiris, Andamans Pagu, ———
testudinarius Reitt. (1873), S. E. N. 87—Grouv. (1895), B. S. E. F. CCLVII—(1908), C. R. I. 365.	Madagascar
pubescens Grouv. (1908), C. R. I. 362[1] & 364 (*Lordites*).	Murree Hills[1]
quadrinotatus Grouv. (*Lordites*) (1908), A. S. E. F. LXVIII, 143[1].	East Indies[1]
reitteri Grouv. (1908), C. R. I. 363 & 364[1].	Bengal[1], Ceylon[1]

36. **Circopes** Reitt.

Reitt. (1873), S. E. N. 56 & 79—Grouv. (1908), C. R. I. 340 & 365 [key to species].

indicus Motsch. (*Pocadius*) (1858), E. E. VII, 43—Reitt. (1873), S. E. N. 79—Grouv. (1908), C. R. I. 365[1].	Kurseong[1] Nilgiris[1]
marginatus Grouv. (1894), A. S. E. B. XXXVIII, 584[1]—Grouv. (1908), C. R. I. 365.	Kurseong[1]
subquadratus Motsch. (*Pocadius*) (1858), E. E. VII, 43—Reitt. (1873), S. E. N. 79—Grouv. (1908), C. R. I. 365[1].	Ceylon[1] Nilgiris[1]

37. **Xenostrongylus** Woll.

Woll. (1854), I. M. 127—Ganglb. (1899), K. M. 464 & 536—Grouv. (1908), C. R. I. 341 & 378.

Subg. *Strongylolasius* Reitt. (1911), F. G. K. 26 [note].

Subg. *Oxystrongylus* Reitt. *l. c.* [note].

dorsalis Grouv. (1908), C. R. I. 378[1]. North Kanara[1]

38. **Cametis** Motsch.

Motsch. (1863), B. M. XXXVI, I, 2, 440—Grouv. (1908), C. R. I. 384.

picea Motsch. *l. c.* 441[1]—Grouv. *l. c.* 384[1]. Ceylon[1]

V. CRYPTARCHINAE

Ipinæ Er. (1843), G. Z. IV, 355 (—*Rhizophagus).*

Ipides Lacord. [1854 (1855)], G. C. 291 & 326.

Ipites Jacq. du Val (1858), Gen. Col. d'Eur, 153 & 157.

Ipina Thomas. (1859), S.C.I, 66&69 (—*Cyllodes*, +*Cryptarchina*).

Ipideas Philippi. (1864), An. Univ. Chile, 454.

Ipini Lec. (1861), C. N. A. 82 & 84.

Ipidæ Reitt. (1873), S. E. N. 7 & 139 (—*Ipsomorpha*).

Ipsi Mars. (1885) N. R. 20.

Ipsinæ Sharp (1891), B. C. A. 873.

Pityophagini Fauc. (1894), Gen. Col. 27.

Ipiens Lameere (1900), Fn. Belg. II. 342 & 354.

Cryptarchini Reitt. (1884), N. J. 257 & 269—Ganglb. (1899), K. M. 447 & 548 [key to genera]—Grouv. (1908), C. R. I. 326 & 395—Reitt. (1911), F. G. K. 11 & 36 [key to genera].

39. **Cryptarcha** Shuck.

Shuck (1839), E. B. E. 165—Lacrod. [1854 (1855)], G. C. 326—Ganglb. (1899), K. M. 549—Grouv. (1908), C. R. I. 395-397 [key to species].—Reitt. (1911) F. G. K. 37.

Cryptarchus Heer (1841), F. H. I. 409.

Subg. *Arhina* Murr. (1867), A. M. N. H. (3) XIX, 178—Col. Old. Cal. (1868), 96.

Subg. *Cryptarchula* Ganglb. (1899), K. M. 550.

andrewesi Grouv. (1908), C. R. I. 396[1] & 397. — Nilgiris[1]

bicolor Grouv. (1892), A. M. G. XXXII, 854 & 856[1]. — Burma[1]

dubia Grouv. (1890), A. M. XXIX, 126—(1892), *l. c.* XXXII, 856[1]. — Burma[1]

feai (*feæ* emend.) Grouv. (1892), A. M. G. XXXII, 855 & 856[1]—Grouv. (1908), C. R. I. 397[1]. — Bengal[1], Pegu[1]

flavipennis Reitt. (1873), S. E. N. 120[1], sep. 24—Grouv. (1908), C. R. I. 397[2]. — East Indies[1] Sikkim[2]

Flavoguttata Reitt. (1875), V. N. V. B. XIII, 120[1], sep. 24. — East Indies[1]

fraterna Grouv. (1908), C. R. I. 395[1] & 397. — Mahé[1]

lesnei Grouv. (1903), A. S. E. F. LXXII, 118[1]—(1908), C. R. I. 397. — Darjiling[1]

nigropunctata Grouv. (1892), A. M. G. XXXII, 854[1] & 856. — Burma[1]

octopunctata Grouv. (1897), A. M. G. XXXVIII, 371[1]. — Burma[1]

ocularis Reitt. (1875), V. N. V. B. XIII, 121, sep. 25—Grouv. (1892), A. M. G. XXXII, 855[1] & 856. — East Indies[1]

rajah Motsch. (1863), B. M. XXXVI, II, 443[1]—Grouv. (1908), C. R. I. 397 [note]. — Ceylon[1]

ritsemai Olliff (1884), N. L. M. VI, 246[1]. — East Indies[1]

rufomaculata Grouv. (1897), A. M. G. XXXVIII, 373[1]. — Tenasserim[1]

40. **Glischrochilus** Reitt.

Reitt. (subg. *Ips.*) (1873), S. E. N. 162—Ganglb. (1899), K. M. 549 & 551.

Biology : Munro, 1917, Proc. R. Physical Soc., Edinburgh, XX. 3, 123-158, 5 pl. 28 fis.—Ritchie (1917), Trans. R. Soc. Edinburgh, LII. part 1, No. 10, 213-234, plates—Blackman (1919), Psyche, XXVI, 4, 85-96, 1 plate, I fig. —Verhoeff (1923), A. N. 12, 80-88, pl. I, figs. 4-6, 13 & 14, pl. II, figs. 22-24; pl. VII, fig. 92.

Ips F. (1776), Gen. Ins. 23—Lacord, [1854 (1855)], G. C. 226 & 327 (pars).

Librodor Reitt. (1884), N. J. 269 & 270.

Subg. *Cryptarchips* Reitt. (1911), F. G. K. 37.

Subg. *Librodor* Ganglb. (1899), K.M. 553—Reitt. (1911), F. G. K. 37 & 38.

egregius Grouv. (*Librodor*) (1892), A. M. G. XXXII, 856[1]. Burma[1]

japonius Motsch. (*Ips*) (1857), E. E. VI, 28—Grouv. (1892), A. M. G. XXXII, 856[1] (*Librodor*). Burma[1]

chinensis Reitt. (*Ips*) (1873), S. E. N. 160—(1879), Verh. Zool-bot. Ges. Wien. XXIX, 508.

davidis (*Ips*) Fairm. (1878), A. S. E. F. (5) VIII, 93—Reitt. *l. c.* 508. Japan[1] China[1]

nankineus Fairm. *l. c.* 93—Reitt. *l. c.* 508.

VI. CYBOCEPHALINAE

Cybocephalites Jac. du Val. (1858), Gen. Col. d'Eur. II, 151 & 157.

Cybocephalini Horn (1879), Nitid. U. State, 268 & 319—Ganglb. (1899), K.M. 447 & 545—Grouv. (1908), C. R. I. 326 & 392.

Cybocephalinæ Sharp (1891), B. C. A. 372—Reitter (1911), F. G. K. 10.

41. **Cybocephalus** Er.

Er. (1844), G. Z. V. 441—Lacord. [1854 (1855)], G. C. 319 & 322—Gangb. (1899), K. M. 546—Grouv. (1908), C. R. I. 392-395 [key to species]

Biology : Rutherford (1914), Trop. Agric. XLIII, 4 319-323—Malenotti (1917), Redia, XIII, 1, 17-53, 2 pl.—Herbert (1920), U. S. Dept. Agric. Bull. 838, 22, 6 pl. 5 figs.

Acribis Waterh. (1877), Proc. Zool. Soc. Lond. 78—Champ. (1913), T. E. S. L. 71.

Phantazomerus Jac. du Val. (1854), B. S. E. F. 307.

Stagnomorpha Woll. (1854), I. M. 482.

æneonitens Champ. (1923), E. M. M. LIX, 77-78[1]. West Almora[1]

binotatus Grouv. (1908), C. R. I. 394[1] & 395—Champ (1923), E. M. M. LIX, 78[2] [addl. descr.] Assam[1] West Almora[2]

chlorocephalus Er. (1844), G. Z. V. 442— Motsch. (1863), B. M. XXXVI, II, 479—Reitt. (1874), V. N. V. B. XII, 6—Grouv. (1908), C. R. I. 395[1]—Champ. (1923), E. M. M. LIX, 77[2]—[addl. descr.]	India[1], Ceylon[1] Nilgiris[1], Haldwani[2] West Almora[2] ———— Singapore[2]
flavipes Motsch. (1863), B. M. XXXVI, II, 478[1]—Reitt. (1874), V. N. V. B. XII, 7—Grouv. (1908), C. R. I. 392[1] & 395.	India[1] Ceylon[1]
major Motsch. (1863), B. M. XXXVI, II, 478[1]—Reitt. (1874), V. N. V. B. XII, 12—Grouv. (1908), C. R. I. 395[1].	Ceylon[1] Bombay[1]
nigripennis Motsch. (1863), B. M. XXXVI, II, 479[1]—Reitt. (1874), V. N. V. B. 12.	Ceylon[1]
puncticeps Grouv. (1908), C. R. I. 393[1] & 395.	Kanara[1], Nilgiris[1]
smaragdicollis Motsch. (1866), B. M. XXXIX, I, 398—Reitt. (1874), V. N. V. B. XII, 6—Grouv. (1908), A. S. E. F. LXXVII, 395.	Ceylon.
splendens Grouv. (1908), A. S. E. F. LXXVII, 392[1] & 394.	Nilgiris[1]
subquadratus Motsch. (1863), B. M. XXXVI, II, 479[1]—Reitt. (1874) V. N. V. B. XII, 7[1].	Ceylon[1] India[1]
subrotundatus Motsch. *l. c.*—479[1].	India[1]

Total 228 species.

ADDENDA.

Pseudomystrops Grouv.

Grouv. [1912 (1913)], A S. E. F. LXXXI, 389

Concolor Grouv. [1912 (1913)], A. S. E. F. 390[1].	Tenasserim[1]
humeralis Grouv. (1890), A. M. G. XXIX, 124[1] [1912 (1913)], A. S. E. F. 389	Burma[1]

230 species.

PART VI – STAPHYLINIDAE.

Sub-fam. OXYTELINAE.

Tribus PIESTINI.

1. SUB-TRIBUS *TRIGONURI.*

1. Apatetica, Westw.

Westw. Cab. Orient. Ent., Lond., 1848, 86.

javanica, Shp., Notes Leyd. Mus., XIV, 1892, 61; Fauv. Rev. d'Ent., XIV, 1895, 194. — Bengal. Burma.

lebioides, Westw., Cab. Orient. Ent., 1848, 86, pl. 41, f. 9; Lac. Gen. Col., pl. 16, f. 6; Fauv. R. d'Ent., XIV, 1895, 193. — Himalayas. Sikkim.

rotundicollis, Fauv., Rev. d'Ent., XXIII, 1904, 85. — Khasi Hills.

sikkimi, Fauv., *id.*, XIV, 1895, 193. — Sikkim., Darjiling.

viridipennis, Fauv., Rev. d'Ent., XIV, 1895, 192. — Sikkim., Darjiling.

2. Nodynus, Waterh.

C. O. Waterhouse, Tr. Ent., Soc., Lond., 1876, 12.

nitidus, Waterh., Tr. Ent. Soc., Lond., 1876, 13. — East Indies.

2. SUB-TRIBUS *PIESTI.*

3. Eupiestus, Kr.

Kraatz, Arch. Naturg., XXV, 1859, I, 182; Fauv., Rev. d'Ent. XXI, 1902, 29.

angulatus, Fauv., Rev. d'Ent., XXI, 1902, 29. — Darjiling., Sikkim.

feae, Fauv., Rev. d'Ent., XIV, 1895, 187 — Tenasserim.

sculpticollis, Kr., Arch. Naturg., XXV, 1859; I, 182, pl. 3, fig. 4. — Nilgiris., Sikkim. Burma., Ceylon.

sikkimi, Fauv., Rev. d'Ent., XXI, 1902, 30. — Sikkim., Darjiling.

spinifer, Fauv., *id.,* XIV, 1895, 186. Burma.

temporalis, Fauv., *id.,* XIV, 1895, 188. Burma.

(4) Siagonium, Kirby.

Kirby and Spence, Introd., Ent. I 1815, pl. 1.

indicum, Fauv., Rev. d'Ent., XXI, 1902, 21. Sikkim. Darjiling.

3. SUB-TRIBUS *ELEUSII.*

5. Eleusis, Cast.

Castelnau, Etud. Ent., I, 1835, 131.

feae, Fauv., Rev. d'Ent., XIV, 1895, 188. Burma.

fusciceps, Kr. Arch. Naturg., XXV, 1859, I, 184. Ceylon., Siwaliks.

humilis, Er., Gen. Spec. Staph., 1840, 839. *fasciata* Lec., New Spec., Col. I, 1863, 59. *indica,* Kr., Arch. Naturg., XXV, 1859, I, 183. Ceylon., Nilgiris. Burma., (Tropical cosmopolite)

pallidipennis, Fauv., Bull. Soc. Lin. Norm., IX, 1865, 40.

subtilis, Shp., Ann. Mag. Nat. Hist. (0), III, 1889, 467.

tennis, Fauv., Bull. Soc. Lin. Norm., IX, 1865. 41.

tenuis, Walk., Ann. Mag. Nat. Hist. (3), III, 1859, 52.

kraatzi, Favu., Ann. Mus. Civ. Gen. XII, 1878, 207 Ceylon.

apicipennis, Kr., Arch. Naturg., XXV, 1859, I, 183.

laticeps, Favu., Rev. d'Ent., XIV, 1895, 190. Burma.

longiceps, Favu., Ann. Mus. Civ. Gen., XII, 1878, 208. ?Ceylon.

plagiata, Fauv., Rev. d'Ent., XXIII, 1904, 83. Sikkim., Darjiling.

pusilla, Fauv., *id.,* XIV, 1895, 190 Burma.

quadridens, Fauv., *id.,* XXIII, 1904, 84. Further India.

rotundiceps, Fauv., *id.,* 83. Sikkim., Darjiling.

vermicularis, Favu., *id.,* XIV, 1895, 189. Burma.

viridans, Fauv., *id.,* 189. Burma., Tenasserim.

4. SUB-TRIBUS *LEPTOCHIRI.*

6. Leptochirus, Germ.

Germ., Ins. Spec., Nov. 1823, 35.

Sub-gen. *Strongylochirus* Bernh., D. E. Z., 1903, 120.

atkinsoni, Fauv., Rev. d'Ent. XIV, 1895, 182. Burma., Tavoy.

laevis, Cast., Hist. Nat., I, 1840, 826. Dehra Dun.

quadridens, Motsch., B. M., XXX, 1857, IV, 501. Tharrawaddy. Bengal., Assam.

7. Borolinus, Bernh.

Bernh., Deutsche Ent. Zeitschr., 1903, 116, 133.

antilope, Fauv., Rev. d'Ent., XIV, 1895, 181. Burma.

minutus, Cast., Hist. Nat. I, 1840, 186. Tharrawaddy.

semirufus, Fauv., Rev. d'Ent, XIV, 1895, 181. Burma.

sikkimensis, Bernh., D. E. Z., 1919, 359. Sikkim.

8. Priochirus, Sharp.

Sharp, B. C. A. I. (2), 1887, 740; Heller, Abhand. Mus. Dresden, VII, 9; Bernh., D. E. Z., 1903, 114, 135; Kraatz, Arch. Naturg., XXV, 1859, I, 190, pl. 3.

Sub-gen. *Triacanthus* Bernh.

Bernh., D. E. Z., 1903, 136.

apicalis, Epp., W. E. Z., XIV, 1895, 68. Poom Pa, Burma. Novlai, Himalayas.

bipunctatus, Fauv., Rev. d'Ent., XIV, 1895, 180. Burma.

ceylanensis, Fauv., *id.*, XXI, 1902, 12. Ceylon., N. Kanara.

pentagonalis, Bernh., W. E. Z., 1914, 78. *fletcheri* var. *analis,* Cam., E. M. M., LVI, 1920, 142. Shembaganur, Palnis. Iyerpadi, Kodaikanal.

tridens, Motsch., B. M., XXX, 1857, II, 502. Bengal.

Sub-gen. *Cephalomerus* Bernh.

Bernh., D. E. Z., 1903, 139.

bifoveatus, Epp., W. E. Z., XIV, 1895, 69. Sikkim.

combustus, Fauv. Rev. d'Ent., XXI, 1902, 10. Sikkim.

exaratus, Epp., D. E. Z., 1895, 406. Burma.

hoplites, Fauv., Rev. d'Ent., XIV, 1895, 182. Burma.

indicus, Bernh., Tyd. Ent., LVIII, 1915, 214. N. E. Burma.

major, Cam., E. M. M., LVI, 1920, 142. Palni Hills, South India.

micrognathus, Fauv., Rev. d'Ent., XXI, 1902, 13. Sikkim.

opacinus, Bernh. W. Z. B., LXIV, 1914, 79. Shembaganur.

pygmaeus, Kr., Arch. Naturg., XXV, 1859, I, 191. Dehra Dun.

refus, Cam., E. M. M., LVI, 1920, 142. N. Coorg.

sanguinosus, Motsch., B. M., XXX, 1857, II, 501. Nilgiris, Anamalais. Tenasserim, Burma.

Sub-gen. *Pseucodontus* Bernh.

Bernh., D. E. Z., 1903, 140.

mandibularis, Kr., Arch. Naturg., XXV, 1859, I, 189, pl. 3, f. 3a-b. Ceylon.

Sub-gen. *Leptarthrus* Bernh.

Bernh., D. E. Z., 1903, 141.

longicornis, Fauv., Bull. Soc. Norm., IX, 1865, 14. East Indies

Sub-gen. *Plastus* Bernh.

Bernh., D. E. Z., 1903, 142.

ascedens, Fauv., Rev. d'Ent., XXII, 1903, 171. Burma.

brachycerus, Kr., Arch. Naturg., XXV, 1859, I, 191. Ceylon

excavatus, Motsch., B. M., XXX, 1857, II, 502. W. Himalayas. Dehra Dun.

kuluensis, Bernh., W. Z. B., LXIV, 1914, 78. Himalayas.

Sub-gen. *Priochirus,* S. Str. Bernh.

Bernh., D. E. Z., 1903, 142.

brevicornis, Er., Gen. Spec. Staph., 1840, 827. East Indies.

9. **Thoracochirus,** Bernh.

Bernh., D. E. Z., 1903, 116, 155.

simoni, Fauv., Rev. d'Ent., XXI, 1902, 21. Ceylon.

variolosus, Fauv., *id.,* XIV, 1895, 183. Burma.

verrucifer, Fauv., *id.,* 182. Tenasserim.

5. SUB-TRIBUS *LISPINI*

10. **Paralispinus,** Bernh.

Bernh., D. E. Z., 1921, 67.

Ancacus, Fauv., Bill. Soc. Norm., IX, 1865, 60 (praeoc.).

exiguus, Er., Gen. Spec. Staph., 1840, 830.
aruensis, Fauv., Ann. Mus. Civ. Gen., XII, 1878, 200. Ceylon. (Tropical
fauveli, Shp., Tr. Ent. Soc. Lond., 1876, 392. cosmopolite).
laevigatus, Kr., Arch. Naturg., XXV, 1859, I, 188 (*Lispinus*).
pallescens, Blackb., Tr. Roy. Soc. Dub. (2), III, 1902, 126.
refescens, Lec., New Spec. Col., I, 1863, 59; Sharp, B. C. A., I (2), 1887, 718.

11. Lispinodes, Sharp.

Sharp, Tr. Ent. Soc. London, 1880, 54.

bistriatus, Fauv., Rev. d'Ent. XIV, 1895, 185.
rugosus, Fauv., *id.,* XXI, 1902, 31. Burma., Nilgiris.
seriatus, Fauv., *id.,* XXI, 1902, 31. Sikkim., Darjiling
Nilgiris.

12. Holosus, Motsch.

Burma.

Motsch., B. M., XXX, 1857, II, 496.

aberrans. Fauv., Rev. d'Ent., XXIII, 1904, 78.
ambigenus, Fauv., *id.,* XIV, 1895, 183. Ceylon
brevipennis, Fauv., *id.,* XXIII, 1904, 77. Burma
conuriformis, Motsch., B. M., XXX, 1857, II, 499. Siwaliks.
fossulatus, Motsch., *id.,* 495. East Indies.
faveolatus, Motsch., *id.,* 496. Nilgiris.
madurensis, Bernh., W. Z. B., LXIV, 1914, 83. East Indies.
mycetoporiformis, Motsch., B. M., XXX., 1857, II, 500. Shembaganur (Palnis).
olisthaeriformis, Motsch., *id.,* 500. Siwaliks.
tachiniformis, Motsch., *id.,* 498; Kr., Arch. Naturg., XXV, 1859, I, pl. 3, fig. 5. East Indies.
East Indies.
tachyporiformis,, Motsch., B. M., XXX, 1857, II, 498.

13. Lispinus, Er.

East Indies.

Er., Gen. Spec. Staph., 1840, 828.

birmanus, Fauv., Rev. d'Ent., XIV, 1895, 184.
brevicornis, Kr., Arch. Naturg., XXV, 1859, I, 187. Burma., Tenasserim.

coarcticollis, Kr., Arch. Naturg., XXV, 1859, I, 186, pl. 3, f. 6.	Dehra Dun.
coriaceus, Fauv., Rev. d'Ent., XIV, 1895, 185.	East Indies.
fulvus, Motsch, B. M., XXX, 1857, II, 495.	Burma., Tenasserim.
impressicollis, Motsch., *id.,* 495; Kr., Arch. Naturg., XXV, 1859, I, 186; Fauv., Rev. d'Ent., XXII, 1903, 150.	East Indies. Ceylon., Anamalais.
filiformis, Walk., Ann. Mag. Nat. Hist. (3), II, 1858, 205.	Dehra Dun.
quadratus, Blackb., Tr. Roy. Soc. Dub. (2), III, 1885, 125.	(Tropical cosmopolite).
iyeri, Bernh., W. Z. B., LXIV, 1914, 82.	
	Singhbhum.
laevipennis, Kr., Arch. Naturg., XXV, 1859, I, 185.	Dehra Dun.
madurensis, Bernh., Ent. Blatt., 1915 (Heft 10-12), 251.	Ceylon.
strigiventris, Kr., Arch. Naturg., XXV, 1859, I, 186.	Madura.
sublucens, Favu., Rev. d'Ent. XIV, 1895, 184.	Nilgiris.
	Nilgiris., N. Kanara.
subopacus, Kr. Arch., Naturg., XXV, 1859, I, 187.	Tenasserim., Burma.
tenuicornis, Kr., *id.,* 187.	Ceylon.
testaceus, Kr., *id.,* 188.	Ceylon.

14. Tetrapleurus, Bernh.

Ceylon.

Bernh., W. Z. B., LXIV, 1914, 84.

indicus, Bernh., *id.,* 85.	
sculptus, Kr., Arch. Naturg., XXV, 1859, I, 188 (*Lispinus*).	Madura. Ceylon.

6. SUB-TRIBUS *THORACOPHORI*

15. Bothrys, Fauv.

Fauv., Rev. d'Ent., XIV, 1895, 185.

personatus, Fauv., Rev. d'Ent., XIV, 1895, 186.	Burma.

16. Thoracophorus, Motsch.

Motsch., B. M. V., 1837, 98 (*Thoraxophorus*), B. M., XIII, 1840, 197. Etud. Ent., VII, 1859, 66.

Sub-gen. *Thoracophorus,* S. Str.

subnitidus, Motsch., Etud. Ent., VIII, 1859, 66.	East Indies.

Tribus Micropeplini.

Micropeplida, Heer, Faun. Helv., I, 169.

17. Micropeplus, Latr.

Latr., Gen. Crust. Ins., IV, 1809, 377.

sikkimi, Fauv., Rev. d'Ent., XXI, 1902, 8. Sikkim.

vulcanus, Favu., *id.,* 9. Sikkim.

Tribus pseudopsini.

18. Pseudopsis, Newm.

Newm., Ent. Mag., II, 1834, 313.

sulcata, Newm., Ent. Marg., II, 1834, 314. W. Almora.
Simla Hills., Chakrata.
_______, Europe.

Tribus proteinini.

Proteinini, Er., Käf. Mark. Brand, I, 641 (exc. *Micropeplus*); Gen. Spec. Staph., 1840, 901 (excl. *Glyptoma, Micro-peplus*).

19. Megarthrus, Steph.

Steph., III, Brit. Ent., V, 1832, 330.

basicornis, Fauv., Rev. d'Ent., XXIII, Sikkim. 1904, 87. Sikkim.

bimaculatus, Fauv., Rev. d'Ent., XXIII, 1904, 87. Ceylon.

birmanus, Fauv., *id.* XIV, 1895, 194. Burma.

rufomarginatus, Cam., Tr. Ent. Soc., Lond., 1913, 525. Nilgiris.

umbonatus, Fauv., Rev. d'Ent., XIV, 1895, 194. Burma., Sikkim.

Tribus omalini.

Omalini. Ganglb., Käf. Mitt.-Eur., II. 1895, 697.

20. Anthobium, Steph.

Steph., Ill, Brit. Ent., V, 1832, 335.

indicum, Champ., E. M. M., LVI, 1920, 243. Kumaon.

sikkimi, Fauv., Rev. d'Ent., XXIII, 1904, 88. Sikkim.

21. Omalium. Grav.
Grav., Col. Micr., 1802, III.

Sub-gen. *Omalium,* S. Str.

Luze, W. Z. B., LVI, 1906, 505.

almorense, Champ., E. M. M., LVI, 1920, 241.	Kumaon.
altivagans, Bernh., Col. Rundsch., 1915 (3), 49.	Kashmir.
angulatum, Motsch., B. M., XXXIV, 1861, I, 142.	Ceylon.
quadrifossulatum, Motsch., Etud., Ent., 1859, 68.	Ceylon.

22. Phloeonomus, Heer.

Heer, Fauna Helv. I, 184.

Sub-gen. *Phloeostiba,* Thoms.

Thoms., Skand. Col., III, 1861, 208.

pinicola, Champ., E. M. M., LVI, 1920, 242.	Kumaon.

Sub-gen. *Phloeonomus,* S. Str. Ganglb.
Ganglb., Käf. Mitt-Eur., II, 1895, 731.

obscurus, Kr., Arch. Naturg., XXV, 1859, I, 181.	Ceylon.
singulare, Kr., *id.,* 181.	Nilgiris., Ceylon. Sikkim.
discalis, Cam., Tr. Ent. Soc., Lond., 1913, 525.	Nilgiris.

23. Philorhinum, Kr.

Kr., Nat. Ins. Deutsch., II, 1856-1858, 965.

floricola, Champ., E. M. M., LVI, 1920, 242. (*nec Wdl.*)	Kumaon., Chakrata.
florivagum, n. n., *id.,* LIX, 1923, 80.	

24. Arpedium, Er.

Er., Käf. Mark., Brand, I, 618.

Pallens, Motsch. B. M., XXX, 1857, IV, 493.	Bombay.

25. Amphichroum, Kr.

Kr., Nat. Ins. Deutsch., II, 1856-1858, 947.

pindarense, Champ., E. M. M., LVI, 1920, 242.	Sunderdhunga. Kumaon.

26. Lesteva, Latr.

Latr., Précis. Car. Ins., 1796, 75.

fluviata, Champ., E.M.M., LVI, 1920, 246. Kumaon., Chakrata. Masuri.

27. Eudeliphrum, Champ.

Champ., E. M. M., LVI, 1920, 244.

gracilipalpe, Champ., *id.,* 244. Kamaon.

28. Geodromicus, Redt.

Redt., Fn. Aust., ed. II, 1858, 224.

amnicola, Champ., E. M. M., LVI, 1920, 246. Kumaon.

crassipalpis, Champ., *id.,* 245. Kumaon.

puncticollis, Wse., B. E. Z., XIX, 1875, 364. W. Himalayas.

29. Tetradelus, Fauv.

Fauv., Rev. d'Ent., XXIII, 1904, 90.

trigonuroides, Fauv., Rev. d'Ent., XXIII, 1904, 91. Sikkim.

Tribus oxytelini.

Oxytelini, Er., Käf. Mark. Brand, I, 1839, 576.

1. SUB-TRIBUS *COPRPOHILI.*

30. Coprophilus, Latr.

Latr., in Cuv. Regne anim., IV, 1829, 429.

alticola, Fauv., Rev. d'Ent., XXIII, 1904, 93. Sikkim., Himalayas.

2. SUB-TRIBUS *OXYTELI.*

31. Planeustomus, Jacq. du Val.

Jacq. du Val, Gen. Col. d'Eur., II, 1859, 58.

bengalensis, Champ, E. M. M., LV, 1919, 155. Bengal (Sarda).

indicus, Fauv., Rev. d'Ent., XXIII, 1904, 21. Burma.

longiceps, Champ., E. M. M., LV, 1919, 154. Bengal (Sarda).

pusae, Cam., E. M. M., LVI, 1920, 143. Bihar (Pusa).

subcarinatus, Champ., E. M. M. LV, 1919, 156. Bengal (Sarda).

32. Trogophloeus, Mannerh.

Mannh., Brach., 1830, 49.

Sub-gen. *Thinodromus*, Kr.

Kr., Naturg., Ins. Deutsch., II, 866.

fuscipalpis, Fauv., Rev. d'Ent., XXIII, 1904, 94.	Ceylon.
lewisi, Cam., E.M.M., LV, 1919, 224.	Ceylon.
lunatus, Motsch., B. M., XXX, 1857, H, 504, Klima, M. K. Z., II, 1904, 50 (note).	East Indies.

Sub-gen. *Trogophlocus*, S. Str.

calcuttanus, Bernh., Ent. Blatt. VII, 55.	Calcutta.
faveicollis, Kr., Arch. Nat., XXV, 1859, I, 180.	Ceylon.
indicus, Kr., *id.*, 179; Fauv., Rev. d'Ent., V, 1886, 144; *id.*, VIII, 1889, 247.	Ceylon. Burm, Pondicherry. Singhbhum.
ceylonicus, Bernh., D. E. Z., 1902, 44.	
flavipes, Motsch., B. M., XXXIV, 1861, 43; Fauv., Rev. d'Ent., XXII, 1903, 150.	
nigrita, Woll., Cat. Col. Mad., 1857, 202; Fauv., Rev., d'Ent. XXI, 1902, 61; Klima, M.K.Z., II, 1904, 57.	East Indies.
bilineatus, Woll., Cat. Can., 1864, 599.	
insularis, Kr., B. E. Z., II, 1858, 127.	
oculatus, Woll., Col. Atlant., 1865, App. 74.	
nitidipennis, Cam., E. M. M., LV, 1919, 225.	Ceylon.
planicollis, Bernh., D. E. Z., 1902, 45.	Ceylon.
scabrosus, Kr., Arch. Nat., XXV, 1859, I, 179.	Ceylon.
siamensis, Fauv., Rev. d'Ent., V, 1886, 144; Ann. Soc. Ent. Fr. (6), VI, 1886, 178.	East Indies.
simplex, Motsch., B. M., XXX, 1857, II, 505; Kr., Arch. Nat., XXV, 1859, I, 180; Fauv., Ann. Mus. Civ. Gen., XIII, 1878, 26.	Ceylon. Calcutta.

Sub-gen. *Boopinus*, Klima.

Klima, M. K. Z., 1904, 56.

peregrinus, Cam., E. M. M., LV, 1919, 225.	Ceylon.

Sub-gen., *Troginus,* Muls., and Rey. Muls., and Rey, Brevip., 1879, 317; Ganglb., Käf. Mitt.-Eur., II, 1895, 648; Klima, M. K. Z., II, 1904, 46.

exiguus, Er., Käf. Mark. Brand, 604; Gen., Spec. Staph., 809; Kr., Naturg, Ins. Deutsch., II, 877; J. Sahlb., En. Col. Brach. Fenn., 1876, 180; Fauvel. Fn. Gallo. Rhen., III, 157; Ann. Mus. Civ. Gen., XIII, 1878, 195; Muls., and Rey Brevip., 1878, 317; Klima, M. K. Z., II, 1904, 65. — Simla., Ceylon. Burma. (Cosmopolitan)

aberrans, Rosh., Thiere Andal., 1856, 85.

atomus, Saulcy, Ann. Soc. Ent. Fr. (4), IV, 1864, 658.

bledioides, Woll., Cat. Can. 601.

despectus, Ganglb., Käf. Mittel-Eur., II, 658.

discolor, Baudi, B. E. Z., XIII, 1869, 400.

glabricollis, Motsch., B. M., XXXIII, 1860, II, 552.

luteiucornis, M. and R., Brevip., 319.

minimus, Kr., Arch. Nat., XXV, 1859, I, 180.

minusculus, Motsch., B. M., XXIV, 1851, I, 144.

taprobanae, Walker, Ann. Mag. Nat. Hist., 1859, 52. — Ceylon.

33. **Xerophygus,** Kraatz.

Kraatz, Arch. Nat., XXV, 1859, I, 178.

flavipes, Motsch., B. M., XXXIV, 1861, I, 143. — Ceylon.

pallipes, Motsch., B. M., XXX, 1857, II, 505; Kraatz., Arch. Nat., XXV, 1859, I, 178, pl. 3, fig. 1. — East Indies.

34. **Apocellagria,** Cam.

Cam., E. M. M., LVI, 1920, 143.

indica, Cam., *id.,* 144. — Nilgiris.

35. **Oxytelopsis,** Fauv.

Fauv., Rev. d'Ent., XIV, 1895, 199.

apicipennis, Fauv. Rev. d'Ent., XIV, 1895, 200. — Burma.

cimicoides, Fauv., *id.,* 200. — Burma.

pseudopsina, Fauv., *id.,* 201. — Burma., Masuri. Siwaliks.

36. Delopsis, Fauv.

Fauv., Rev. d'Ent. XIV, 1895, 198.

cornuta, Favu., Rev. d'Ent., XIV, 1895, 198. Burma.

seticornis, Fauv., *id.,* 199. Burma.

37. Oxytelus, Grav.

Grav., Col. Micr., 1802, 101.

Sub-gen. *Anotylus,* Thoms, Skand. Col., III, 130.

Sub-gen. *Caccoporus,* Thoms, *id.,* 127.

Sub-gen. *Epomotylus,* Thoms, *id.,* 128.

Sub-gen. *Oxylelus,* S. Str. Thoms., *id.,* 126.

Sub-gen. *Tanycraerus,* Thoms., *id.,* 129.

armiger, Fauv., Rev. d'Ent., XIV, 1895, 202. Burma.

bellicosus, Fauv., *Id.,* 202 (*Tanycraerus*). Burma.

bengalensis, Er., Gen. Spec. Staph., 789; Kraatz, Arch. Nat., XXV, 1859, I, 171. Bengal. Ranchi.

bicolor, Walk., Ann. Mag. Nat. Hist., 1859, 52. Ceylon.

cephalotes, Epp., W. E. Z., XIV, 1895, 66. Simla Hills.

cribrum, Fauv., Rev. d'Ent., XXIV, 1905, 114. Simla., Sikkim. Bengal.

distincticollis, Cam., E.M.M., LV, 1919, 226. Ceylon.

exasperatus, Kr., Arch. Nat., XXV, 1859, I, 175. Ceylon.

ferrugineus, Kr., *id.,* 173; Fauv., Rev. d'Ent. XX, 1901, 71; Shp., Fn. Haw., III, V, 1908, 543 (*Caccoporus*). Ceylon., Belgaum. Dehra Dun., Coonoor., Calcutta. (Tropical cosmopolite)

bledioides, Blackb., Tr. E. Soc., Dub. (2), III, 1885, 125.

laevior, Shp., Tr. Ent. Soc., Lond., 1874, 92.

laxipennis, Fairm., Ann. Soc. Ent. Bleg., 1893, 527.

flavipennis, Kr., Arch. Nat., XXV, 1859, I, 172. East Indies.

hirtulus, Epp., W. E. Z., XIV, 1895, 68. Simla Hills. Kumaon.

incicus, Motsch, B. M., XXX, 1857, II, 504. W. Himalayas. Assam.

kraatzi, nom. nov.

pulcher, Kr., Arch. Nat., XXV, 1859, I, 173 (nec. Grav).

latiusculus, Kr., *id.*, 176; Fauv., Rev. d'Ent., XXI, 1902, 67.	Ceylon., Belgaum.
sulcifrons, Fauv., Fn. Gallo-Rhen., III, 1875, Cat. XI.	Simla Hills. Dehra Dun., ——— Malay Peninsula. Africa.
lividus, Motsch., B. M., XXX, 1857, II, 503; Kraatz, Arch. Nat., XXV, 1859, I, 171 (*Caccoporus*).	Ceylon., Ranchi.
megacephalus, Fauv., Rev. d'Ent., XXIII, 1904, 99.	Darjiling., Chakrata.
megacerus, Fauv., Rev. d'Ent., XIV, 1895, 201 [*Tanycraerus*].	Simla Hills., Chakrata. Masuri., ——— Malay Archipelago.
micans, Kr., Arch. Nat., XXV, 1859, I, 175. [*Tanycraerus*].	Ceylon.
puncticeps, Kr., (♀), *id.*, 176.	
monoceros, Cam., E. M. M., LV. 1919, 226.	Ceylon., Dehra Dun. Masuri, Simla Hills.
myrmecophilus, Cam., Tr. Ent. Soc., Lond., 1913, 526 (*Anotylus*).	Nilgiris.
nigriceps, Kr., Arch. Nat., XXV, 1859, I 176 (*Caccoporus*).	Dehra Dun. Belgaum., Burma Ceylon.
nitidifrons, Woll., Ann. Mag. Nat. Hist. (4), VIII, 1871, 411.	Burma. Nilgiris.
advena, Shp., Tr. Ent. Soc., Lond., 1880, 50; Tr., Roy. Soc., Dub., III, 1885, 223; Fn., Haw., III, V, 1908, 543; Fauv., Rev. d'Ent., XXI, 1902, 65.	——— Philippines. Hawaii., Africa
parasitus, Motsch., Etud., Ent., VIII, 1859, 69.	Ceylon.
punctipennis, Fauv., Rev. 'D'Ent. XXIV, 1905, 113.	Sikkim.
pygmaeus, Kr., Arch. Nat., XXV, 1859, I, 176; Shp., Fn. Haw., III, V, 1908, 543 (*Anotylus*).	Nilgiris., Burma Ceylon.
pusillimus, Kr., Arch. Nat., XXV, 1859, I, 177.	———, Aden.
tenuis, Motsch., Etud., Ent.,. VIII, 1859, 70.	Hawaii.
raffrayi Fauv., Rev. d'Ent., XXIV, 1905, 117.	East Indies., ——— Malay Peninsula.
robustus, Schub., D. E. Z., 1906, 383 (*Tanycraerus*).	Kashmir.
rufus, Kr., Arch. Nat., XXV, 1859, I, 174 (*Anotylus*).	Dehra Dun, Ceylon.
sikkimi, Fauv. Rev. d'Ent., XXIV, 1905, 115.	Sikkim.
simplex, Motsch., Etud. Ent., VIII, 1859. 70.	Ceylon.

thoracicus, Motsch., B. M., XXX, 1857, II, 504 (*Anotylus*). N. India.

Kulu

tibialis, Schub., D. E. Z., 1908, 624 (*Anotylus*). Dehra Dun.

varipennis, Kr., Arch. Nat., XXV, 1859, I, 172. Belgaum, Khandesh. Nilgiris., Burma.

38. Platystethus, Mannerh.

Mannerh., Brach., 1830, 46.

Sub-gen, *Pyctocraerus,* Thoms, Skand. Col., III, 125.

Sub-gen. *Platystethus,* Thoms., *id.,* 122.

cornutus, Grav., Col. Mic., 109, Er. Kaf. Mark. Brand, I, 586; Gen. Spec. Staph., 782; Kraatz, Nat. Ins. Deutsch., II, 841; Jacq. du Val., Gen. Col. d'Eur., II, 55; Thoms., Skand. Col. III, 122; Fauv., Fn. Gallo-Rhen., III, 180; Muls. and Rey, Brevip., 23; Ganglb., Käf. Mittel-eur., II, 631. Burma., Nilgiris. Coonoor., Theog. Simla Hills. ———— Europe.

degener, Muls, and Rey, Brevip., 26.

maxillosus, Peyron., Ann. Soc. Ent. Fr. (3), VI, 432.

scybalarius, Runde., Brach. Hal., 19.

crassicornis, Motsch., B. M., XXX, 1857, II, 506. Nilgiris., Burma. Purneah. Nilgiris.

dilutipennis, Cann., Tr. Ent. Sco., Lond., 1913, 527.

robustus, Cam., E. M. M., 1918, 102. Nilgiris

spectabilis, Kr., Arch. Nat., XXV, 1859, I, 170. East Indies.

testaceus, Motsch., B. M., XXX, 1857, II, 506. East Indies.

39. Bledius, Mannerh.

Mannh., Brach., 1830, 44.

Bledius+Hesperophilus, Steph., Ill. Brit. Ent., V, 307, 309.

Bledius+Hesperophilus+Astycops, Thoms. Skand. Coll., III, 118, 120, 121.

Bledius+Tadunus+Bargus+Astycops,+Hesperophilus, Schiödte, Nat. Tid. (3), IV, 144, 145, 164, 165.

Sub-gen. *Astycops,* Muls. and Rey, Brevip. Oxypor. and oxytel., 1879, 195.

Sub-gen. *Belidus,* Muls. and Rey, *l. c.,* 215.

Sub-gen. *Bledioides,* Muls, and Rey, *l. c.,* 134.

Sub-gen. *Bledius,* s. str. Muls. and Rey, *l. c.,* 113.

Sub-gen. *Elbidus,* Muls. and Rey, *l. c.* ,130.

Sub-gen. *Hesperophilus,* Muls. and Rey, *l. c.* 189.

Sub-gen. *Pucerus,* Muls. and Rey, *l. c.* 212.

arenicola Fauv., Rev. d'Ent., XXIII, 1904, 112.	Malabar.
bispinus, Kr., Arch. Nat., XXV, 1859, I, 169.	East Indies.
brunnipennis, F. Syst, Eleuth., II, 596; Er., Gen. Spec. Staph. 779; Kr., Arch. Nat., XXV, 1859, I, 168 (note); Fauv., Rev. d'Ent., XXII, 1903, 151 (*Bledius*)	Kanara., Ceylon. Calicut., Burma. Mahé., Coonoor. Khulna.
productus, Walk., Ann. Mag. Nat. Hist. (3). II, 1858, 205 (*Oxytelus*).	
rudis, Walk., *id.,* 205 (*Oxytelus*).	
dilutipennis, Motsch., B. M., XXX, 1857, II, 507.	East Indies.
gracilicornis, Kr., Arch. Nat., XXV, 1859, I, 169 (*Pucerus*).	Belgaum; Khanapur: Mahé; Calicut; Ceylon; Genji; Ginlon; Travancore.
helferi, Fauv., Rev. d'Ent, XXIII, 1904, 112.	Burma.
hoplites, Fauv., Rev. d'Ent., V, 1886, 146 (*Bledius,* S. Str.).	East Indies.
lucidus, Shp., Tr. Ent. Soc., Lond., 1874, 90 (*Hesperophilus*).	Karachi., Mahé. Burma., Purneah.
capitalis, Fauv., Ann. Mus. Gen., X, 1877, 204.	
fischeri, Bernh., Stett. Ent. Zeit., LXIV, 1903, 35.	
maindroni, Fauv., Rev. d'Ent. XXII, 1903, 151.	Mahé., Calicut.
minusculus, Motsch., B. M., XXXIV, 1861, I, 144.	Ceylon.
palliatus, Fauv., Rev. d'Ent. XIV, 1895, 203.	Burma.
pulchellus, Kr., Arch. Nat., XXV, 1859, I, 169.	Mahé., Calicut. Ceylon., Pondicherry.
quadricornis, Bernh., W. Z. B., 71-1915, 135	Madura.
rugosicollis, Bernh., D. E. Z., 1902, 43.	Ceylon.
taruensis, Cam., E. M. M., 56, 1920, 144 (*Elbidus*).	Taru (Peshawar).
tuberculatus. F., Ent. Syst. Supp., 1798, 181; Kraatz. Arch. Nat., XXV, 1859, I, 168; Fauv., Rev. d'Ent., XXII, 1903, 151 (*Pucerus*).	Madura; Kanara; Ceylon; Mahé; Calicut; Genji; Boligai; Puri.

Tribus Oxoriini.

Oxytelini, Subtribus *Osorini,* Er., Gen. Spec. Staph., 753.

Oxytelides, Subtribus Osoriides, Lacord., Gen., Col., II 111.

Osorii, Fauv., Fn. Gallo-Rhen., III, 213, note.

Oxytelini, Group *Osorii,* Lec., Horn. Class. Col., N. Amer., 102.

Oxytelini, I. *Osoraria,* Lynch., Bol. Ac. Nac. Cord., 1882, 344.

Oxytelinae, Group *Osoriina,* Shp., B. C. A. I., 2, 677.

1. SUB-TRIBUS *OSORII.*

40. **Paragonus,** Fauv.

Fauv. Rev. d'Ent. XIV, 1895, 197. — Burma.

birmanus, Fauv., *id.,* XIV, 1895, 197.

41. **Mimogonus,** Fauv.

Fauv., Rev. d'Ent. XXII, 1903, 261

fumator, Fauv., *id.,* VIII, 1889, 246. — Nilgiris., Ceylon. Burma., _______ (Tropical cosmopolite).

rufipennis, Motsch. B. M. XXX, 1857, II, 508. — East Indies.

42. **Holotrochus,** Er.

Er., Gen. Spec. Staph. 757.

annandalei, Bernh., Ent. Blatt., VII, 1911. 56. — Travancore.

43. **Osorius,** Latr.

Latr., Regne Anim., IV, 438.

Molosoma, Say, Trans. Am. Phil. Soc. (n. s.). IV, 452.

annandalei, Bernh., Ent. Blatt., VII, 1911, 56. — Paresnath (Bengal).

beibei, Bernh., W. Z. B.., LXIV, 1914, 86. — N. E. Burma.

calvus, Epp., D. E. Z., 1895, 406. — Sukna (Sikkim).

constrictus, Fauv. Rev. d'Ent., XIV, 1895, 196. — Burma.

indicus, Cam., Tr. Ent. Soc., Lond., 1913, 528. — Nilgiris.

kraatzi, nom. nov. — East Indies.

rugiceps, Kr., Arch. Nat. XXV, 1859, I, 166.

madurensis, Schub., D. E. Z., 1911, 5. Madura.

monticola, Cam., Tr. Ent. Soc., Lond., 1913, 527. Nilgiris.

nilgiriensis, Fauv., Rev. d'Ent., XXII, 1903, 152. Nilgiris., Coonoor.

pectinifrons. Fauv., *id.,* XIV, 1895, 196. Burma.

peguanus, Bernh., W. Z. B., LXIV, 1914, 87. Pegu.

puncticollis, Kr., Arch. Nat., XXV, 1859, I, 168. East Indies.

punctipennis, Fauv. Rev. d'Ent. XIV, 1895, 195. Burma.

punctulatus, Motsch., B. M., XXX, 1857, II, 508. East Indies.

robustus, Cam., E. M. M., LVI, 1920, 145. Sidapur.

rufipes, Motsch., B. M., XXX, 1852, II, 508. *compactus* Walk., Ann. Mag. Nat. Hist., 1859, 52. Nilgiris. Burma., Tudai.

syriacus. Fauv., Ann. Soc. Ent. Fr. (5), V. 1875; Bull., VIII; Fn. Gallo-Rhen., III, Cat. 1875, p. XIII; Bull. Soc. Linn. Norm. (3), II, 160. Ceylon., ——— (Tropical cosmopolite).

rugicollis, Kr., Arch. Nat., XXV, 1859, I, 167. Ceylon.

scolytinus, Fauv., Rev. d'Ent., XIV, 1895-195. Burma.

sikkimensis, Bernh., Arch. Naturg., 84, A-10 1918, 177. Sikkim., Darjiling

strangulatus, Fauv., *id.,* XXIII, 1904, 47. N. Kanara.

stricticollis, Fauv., *id.,* XIV, 1895, 196. Burma.

Sub-Fam. *OXYPORINAE.*

Oxyporina, Thoms, Skand. Col., III, 143.

Oxytelini, Group I *Oxypori,* Leeonte Class. Col., N. Amer., 1861, 68.

Oxytelini, Section 41 *Oxypori,* Fauv., Fn. Gall.Rhen., III, 212.

Oxyporiens, Muls. and Rey. Ann. Soc. Agr., Lyon (5), VIII, 208; Brevip. Oxyp. Oxyt., 1879, 1.

Oxytelini, Group II *Oxypori,* Lec. and Horn., Class. Col., N. Amer., 1883, 102.

Oxyporinæ, Sharp., B. C. A., 1, 2, 676; Gaglb., Käf. Mittel-Eur., II, 601; Reitter Fn. Germ., II, 162.

Tribus Oxyporini.

Ganglb., Käf. Mittel-Eur., II, 602.

44. Oxyporus, F.

F., Syst. Ent., 1775, 267.

apicalis, Cam., Tr. Ent. Soc., Lond., 1913, 529. Burma.

bucephalus, Fauv., Rev. d'Ent. XIV, 1895, 204. Burma.

flavicornis, Schub., D. E. Z., 1908, 623. Himalayas. Simla Hills.

stigma, Fauv., Rev. d'Ent., XIV, 1895, 203. Burma.

Sub-Fam. MEGALOPINAE.

Shp., B. C. A., I, 2, 668.

Oxytelini, Subtrib. *Megalopini,* Er., Gen. Spec. Staph., 750.

Oxytelides, Sous-tribe I, *Megalopides,* Lac., Gen. Col., II, 110.

Megalopides, Motsch., B. M., XXX, 1857, II, 509.

Oxytelini, Group I. *Megalopes,* Lec. and Horn., Class Col., N. Amer., 1883, 102.

Tribus Megalopini.

45. Megalops, Er.

Er., Gen. Spec. Staph., 751.

sexdentatus, Cam., Tr. Ent. Soc., London, 1913, 529. Assam., Ceylon.

subfasciatus, Champ., E. M. M., LIX, 1923, 45. Kumaon.

Sub-fam. STENINAE.

Sharp., B. C. A., I, 2, 640.

Stenini, Er., Käf. Mark. Brand, I, 526; Gen. Spec. Staph., 687 (excl. *Euaesthetus*); Kraatz, Nat. Ins. Deutsch., II, 732 (excl. *Euaesthetus*); Leconte and Horn., Class. Col., N. Amer., 1883, 97; Lynch., Bol. Ac. Nac. Cord., VII, 1884, 331; Casey, Rev. Stenini of Amer., North of Mexico, 1884, 3.

Stenides, Lac., Gen. Col., II, 106 (excl. *Euaesthetus*).

Stenites, Jacq. de Val., Gen. Col. d'Eur., II, 50 (excl. *Euaesthetus*).

Stenina, Thoms., Skand. Col., II, 210.

Stenini, Section II *Steni,* Fauv., Fn. Gallo-Rhen., III, 224.

Stenides, Muls. and Rey, Ann. Soc. Agr., Lyon (5), VIII, 1875, 205; Rey, Ann. Soc. Linn. Lyon, XXX, 1883, 175.

Tribus Stenini.

Ganglb., Käf. Mitel-Eur., II, 548.

46. **Stenus,** Latr.

Latr., Préc. Caract. Gen. Ins., 1796, 77.

Stenus, s. Str., *Nestus,* Rey, Ann. Soc. Linn., Lyon, XXX, 1883, 183; Ganglb., Käf. Mittel-Eur., II, 552.

Tesnus, Rey, Ann. Soc. Linn., Lyon, 1883, 183; *Mutinus,* Cas., Rev. Sten., North of Mexico, 1884, 146.

Hypostenus, Rey, Ann. Soc. Linn., Lyon, 1883, 183; *Areus,* Cas., *l. c.* 150 (*exparte*); *Astenus,* Lynch., Bol. Nc. Nac. Cord., VII, 1884, 341.

Hemistenus, Motsch., B. M., XXXIII, 1860, II, 557 (*ex parte*).

Stenosidotus, Lynch., *l. c., 338.*

Hemistenus, Motsch., B. M., XXXIII, 1860, II, 557 (*ex parte*); *Areus,* Cas., *l. c.,* 150 (*ex parte*).

Mesostenus, Rey, Ann. Soc. Linn., Lyon, 1883, 183.

Parastenus, Heyden, W. E. Z., XXIV, 1905, 262.

Sub-gen. *Stenus,* s. str.+*Nestus,* Rey.

annamita, Fauv., Rev. d'Ent., XIV, 1895, 207.	Burma., Khadesh. Cylon.
aurichalceus, Champ., E. M. M., LVI, 1920, 172 (*Stenus,* *s. str.*).	W. Almora. N. W. Himalayas.
braceatus, Champ., *id.,* 173 (*Stenus,* s. str.).	W. Almora. N. W. Himalayas.
carinatus, Cam., Tr. Ent. Sco., London, 1913, 532 (*Nestus*).	Ceylon.
collium. Fauv., Rev. d'Ent., XIV, 1895, 207 (*Stenus,* s. str.).	Tenasserim., Burma.
comes, Fauv., *id.,* 207 (*Stenus,* s. str.).	Burma.
confluens, Cam., E. M. M., LIV, 1918, 103 (*Stenus,* s. str.).	Lebong.
crenicollis Epp., W. E. Z., XIV, 1895, 65.	Simla.
cupreoaeneus, Champ., E. M. M., LIX, 1923, 45.	Bodiar., Kumaon.

diffidens. Cam., Tr. Ent. Soc., Lond., 1913, 530. Assam.

fuscipes, Gr., Col. Mic., 157; Er., Käf. Mark. Brand., I, 555; Gen., Spec. Staph., 716; Kraatz. Nat. Ins. Deutsch. II, 772; Thoms., Skand. Col., II, 225; Fauv., Fn. Gallo-Rhen., III, 259; Rhen., Rey, Ann. Soc. Linn., Lyon, XXX, 1883, 303; Ganglb., Käf. Mittel-Eur., II, 579 (**Nestus*). East Indies. ——— Europe.

argyrostoma, Steph., Ill. Brit. Ent., V, 288.

cariniger, Motsch., B. M., XXX, 1857, IV, 513.

curvipes, Steph., III, Brit. Ent., V, 285.

femorellus, Zett., Ins. Lapp., 71.

fulvipes, Boisd. and Lac., Fn. Ent., Paris, I, 446.

vilis, Kr., Arch. Naturg., XXV, 1859, I, 162.

kraatzi Bernh., Ent. Blatt., VII, 1911 (2-3) 59 (*Stenus,* s. str.). Fagu (Simla Hills).

bidentatus. Champ., E. M. M., LVI, 1920, 170. W. Almora.

mucronatus, Champ., E. M. M., LVI, 1920, 171 (*Stenus,* s. str.). W. Almora.

nigrovirens, Fauv. Rev. d'Ent., XIV, 1895, 206 (*Stenus,* s. str.). Burma.

rugicollis, Kr., Arch., Naturg., XXV, 1859, I, 162 (*Stenus,* a str.). East Indies.

seminiger, Champ., E. M. M., LVI, 1920, 171 (*Stenus,* s. str.). W. Almora.

venator, Fauv., Rev. d'Ent., XIV, 1895, 208 (*Stenus,* s. str.). Burma.

viriditinctus, Champ., E. M. M., LVI, 1920, 172 (*Stenus,* s. str.). W. Almora.

Sub-gen *Hypostenus,* Rey.

acuminatus, Kr., Arch. Naturg., XXV, 1859, I, 165. Ceylon.

albidicornis, Bernh., Ent. Blatt., 1915, (10-12), 252. Madura.

andrewesi, Fauv., Rev. d'Ent., XXIII, 1904, 49. Dugeli.

angusticollis, Epp., D. E. Z., 1895, 405. Pegu.

barbatus, Nietn., Ann. Mag. Nat. Hist., 1857, 381; Kr., Arch. Naturg., XXV, 1859, I. 163. Ceylon. Darjiling.

arachnipes, Bernh., D. E. Z., 1902, 42.	
basicornis, Kr., Arch., Naturg., XXV, 1859, I, 163.	Ceylon.
birmanus, Fauv., Rev. d'Ent., XIV, 1895, 213.	Tenasserim., Burma.
bispinus, Motsch., B. M., XXX, 1857, II, 514.	Niligiris., Burma.
bivulneratus, Motsch., *id.*, 514.	East Indies.
brachypterus, Kr., Arch. Nat., XXV, 1859, I, 164.	Ceylon.
ceylonicus, Bernh., D. E. Z., 1902, 41.	Ceylon.
consors, Fauv., Rev. d'Ent., XXIII, 1904, 49.	Nilgiris., Simla Hills.
cribellatus, Motsch., B. M. XXX, 1857, II, 515.	W. Himalayas.
flavovittatus, Champ., E. M. M., LVI, 1920, 174.	W. Almora.
flexuosus, Champ., *Id.*, 174	W. Almora.
fulvescens, Motsch, B. M., XXX, 1857, II, 515.	East Indies.
gastralis, Fauv. Rev. d'Ent., XIV, 1895, 214.	Burma., Rangoon.
guttalis, Fauv., Rev. d'Ent., XIV, 1895, 212.	Tharrawaddy, Burma.
himalayicus, Bernh., Col. Rundsch., 1915 (3), 50.	Himalyas.
kurseonginus, Bernh., Ent. Blatt., VII, 1911, (2-3), 58.	W. Himalayas; Kurseong.
lacertoides, Nietn., Ann. Mag. Nat. Hist., 1857, 382; Kr., Arch. Naturg., XXV, 1859, I, 165.	Ceylon.
millepunctus, Fauv., Rev. d'Ent., XXIII, 1904, 49.	Nilgiris.
monomerus, Fauv., Rev. d'Ent., XIV, 1895, 214.	Burma., Pegu.
nitidulus, Cam., Tr. Ent. Soc., Lond., 1913, 530.	Manipur.
nilgiriensis, Cam., Tr. Ent. Soc., Lond., 1913, 532.	Nilgiris.
pictus, Motsch, B. M., XXX, 1857, II, 515.	East Indies.
piliferus, Motsch., *id.*, 514.	East Indies.
planifrons, Fauv., Rev. d'Ent., VIII, 1889, 253.	Puri., Dehra Dun. N. W. Himalayas.
pulcher, Motsch., Etud., Ent., 1859, 71.	Pegu., Kanara. Nilgiris., Ceylon.
pustulatus, Bernh., W. Z. B. 1914, 95.	Pegu.
rugosissimus, Champ., E. M. M., LVI, 1920, 173.	W. Almora. W. Himalayas.
signifer, Fauv., Rev. d'Ent., 1895, 213.	Burma.

tricarinatus, Kr., Arch. Naturg., XXV, 1859, I, 164.	Calcutta., Dugeli. Kanara., Burma.
ventricosus, Fauv., Rev. d'Ent., XXIII, 1904, 48.	Dehra Dun. Singhbhum. Dugeli., Khandesh.
wasmanni, Fauv., *id.,* XIV, 1895, 214.	Burma., Sikkim. Darjiling.

Sub-gen. *Hemistenus+Mesostenus,* Rey.

abdominalis, Fauv., Rev. d'Ent., XIV, 1895, 210.	Burma.
aceris, Steph., III, Brit. Ent., V, 292 (*Mesostenus*).	W. Himalayan. ————, Europe.
belli, (Fauv., Rev. d'Ent., XXIII, 1904, 47.	Nilgiris.
circumflexus, Fauv., *id.,* XIV, 1895, 211 (*Mesostenus*).	Burma.
feae, Fauv., *id.,* 211.	Tharrawaddy.
gestroi, Fauv., *id.,* 212.	Burma., Masuri.
immsi, Bernh., Col. Rundesch., 1915 (3), 49 (*Mesostenus*).	Kumaon.
peraffinis, Fauv., Rev. d'Ent., XIV, 209.	Burma.
pilicornis, Fauv., *id.,* 209.	Burma.
posticus, Fauv., *id.,* 209.	Burma., Masuri. Chakrata.
stigmaticus, Fauv., *id.,* 210.	Darjiling., Sikkim. Burma.
virgula, Fauv., *id.,* 210.	Bengal., Kurseong. Burma., Sikkim. Darjiling., Matiana.

47. Dianous, Sam.

Sam., Ent. Comp., 1819, 173.

aereus, Champ., E.M.M., 1919, 53.	Chakrata., W. Almora.
andrewesi, Cam., Tr. Ent., Soc., Lond., 1913, 534.	W. Almora., Lebong
annandalei, Bernh., Ent. Blatt., VII, 1911, 57.	W. Himalayas. Kurseong.,
azureus, Champ., E. M. M., 1919, 97.	Tanakpur.
bifoveifrons, Champ., E. M. M. LVII, 1921, 201.	Kumaon.
bisignatus, Cam., *id.,* LVII, 1921, 270.	Lebong.
caeruleonotatus, Champ., E. M. M., 1919, 45.	W. Almora., Siwaliks.

cameroni, Champ., *id.,* 50. W. Almora.

championi, Cam., E. M. M., LVI, 1920, 145. Lebong.

cribrarius, Champ., E. M. M., 1919, 98. Tanakpur.

cyanogaster, Champ., *id.,* 1919, 47. W. Almora.

distigma, Champ., *id.,* 1919, 49. W. Almora.

gracilipes, Champ., *id.,* 1921, 183. Ranikhet.

inaequalis, Champ., *id.,* 1919, 45. W. Almora.

lobigerus, Champ., *id.,* 1919, 48. W. Almora., Shillong.

luteoguttatus, Champ., *id.,* 1919, 98. W. Almora.

minor, Champ., *id.,* 1919, 53. W. Almora.

obliquenotatus, Champ., E M. M., LVII, 1921, 183. Sikkim.

punctiventris, Champ., E. M. M., 1919, 48. W. Almora.

radiatus, Champ., *id.,* 51. Chakrata., W. Almora. Masuri.

scabricollis, Champ., *id.,* 1919, 46. W. Almora., Kumaon.

subvorticosus, Champ., *id.,* 1919, 52. W. Almora., Chakrata.

subtortuosus, Champ., E. M. M. LVII, 1921, 201. Kumaon.

tortuosus, Champ., E. M. M., 1919, 52. W. Almora.

versicolor, Cam., Tr. Ent. Soc., Lond., 1913, 534. Lebong.

verticosus, Epp., D. E. Z. 1895, 404. Pegu.

Sub-fam. EUAESTHETINAE.

Sharp., B. C. A., I, 2, 639.

Tribus Stenaesthetini.

48. Stenaesthetus, Shp.

Sharp, Tr. Ent. Soc., Lond., 1874, 79.

sunioides, Sharp, Tr. Ent. Soc., Lond., 1874, 80. Derha Dun. Colombo.

———

China., Japan.

Tribus Euaesthetini.

49. Edaphus, J. Lec.

Lec., Class. Col. N. Amer., I, 1861, 67.

major, Fauv., Rev. d'Ent., XIV, 1895, 205. Burma.

punctatus, *id.,* 206. Burma.

Sub-fam. PAEDERINAE.

Ganglb. Käf. Mittel-Eur., II, 1895, 493.

Tribus Pinophilini.

Er., Gen. Spec. Staph., 1840, 668.

1. SUB-TRIBUS PINOPHILI.

50. Pinophilus, Gr.

Gr., Col. Mic., 1802, 201.

aegyptius, Er., Gen. Spec. Staph., 673. Belgaum., Ceylon. Pegu., Sampgaon. Sahibgunge.

rufipes, Kr. (*ruficeps* err. typ.), Arch., Nat, XXIV, 1859, I, 155.

——— Egypt., Java.

beccarii, Fauv., Rev. d'Ent., 1895, XIV, 220. Burma.

bomfordi, Epp., W. E. Z., IX, 1890, 279. Burma., Sampgaon.

bornensis, Fauv., Rev., d'Ent., 1895, 221. Burma., ——— Burneo.

brachypterus, Kr., Arch. Naturg., XXV, 1859, I, 161. Ceylon.

brevis, Kr., *id.,* 159. Khandesh., Ceylon. Bengal.

complanatus, Er., Gen., Spec., Staph., 672; Kr., Arch. Naturg., XXV, 1859, 155. Pegu., Burma.

morio, Motsch., B. M., XXX, 1859, I, 517.

depressus, Kr., Arch. Nat., XXV, 1859, I, 157. Burma., Pegu.

verticalis, Fauv., Rev. d'Ent., XIv, 1895, 219. Bagewadi.

ferrugineus, Bernh., D. E. Z., 1902, 40. Ceylon.

funestus, Bernch., *id.,* 40. Ceylon.

genalis, Fauv., Rev d'Ent., XIV, 1895, 219. Burma.

javanus, Er., Gen. Spec. Staph., 672; Kr., Arch. Naturg., XXV, 1859, I, 159. Ceylon. Pegu.

pallipes, Kr., *id.,* 156. Burma.,

insignis, Shp., Tr. Ent. Soc., Lond., 1874, 77. Ranchi., Calcutta. Purneah., ——— Java., China., Japan.

leucopus, Kr., Arch., Naturg., XXV, 1859, I, 158. Tranquebar.

melanocephalus, Motsch., B. M., XXX, 1857, II, 516; Kr., Arch. Naturg., XXV, 1859, I. 159. East Indies.

melanomerus, Kir., *id.,* 160. East Indies.

mixtus, Cam., Tr. Ent. Soc., Lond., 1913, 534. Daccca.

nigripes, Cam., *id.,* 536. North India.

piceus, Fauv., Rev. d'Ent., XXIII, 1904, 51. Belgaum.

picticornis, Kr., Arch. Naturg., XXV, 1859, I, 158. Ceylon.

pilicollis, Motsch., B. M., XXX, 1857, II, 516. Toungoo.

planus, Kr., Arch. Naturg., XXV, 1859, I, 157. East Indies.

rotundicollis, Kr., *id.,* 160. Ceylon.

speculifrons, Cam., Tr. Ent. Soc., Lond., 1913, 535. Nilgiris.

thoracicus, Fauv., Rev. d'Ent., XIV, 1895, 220. Burma.

variipes, Fauv., *id.,* 218. Burma.

2. SUB-TRIBUS *PROCIRRI.*

51. Procirrus, Latr.

Latr., Regne Anim., IV, 1829, 436.

feae, Fauv. Rev. d'Ent., XIV, 1895, 215. Burma.

fusculus, Shp., Ann. Mag. Nat. Hist. (6), III, 1889, 324. Dacca.

52. Palaminus, Er.

Er., Gen. Spec. Staph., 681.

ceylanensis, Kr., Arch. Naturg., XXV 1859, I, 153. Ceylon.

indicus, Kr., Arch. Naturg., XXV, 1859, I, 153. East Indies.

simoni, Fauv., Rev. d'Ent., XXIII. 1904, 50. Ceylon., Nilgiris.

truncatus, Fauv., *id.,* XXII, 1903, 152. Coonoor.

53. Oedichirus; Er.

Er., Gen. Spec. Staph., 684.

alatus, Nietn., Ann. Mag. Nat. Hist. (2), XIX, 1857, 247; Kr., Arch. Naturg., XXV, 1859, I, 154.	Ceylon.
birma us, Fauv., Rev. d'Ent., XIV, 1895, 217.	Tharrawaddy.
dimidiatus, Epp., W. E. Z., IX, 1890, 280.	Nilgiris., Nagpur.
longipennis, Kr., Arch. Naturg., XXV, 1859, I, 155.	N. India., Nilgiris. Ceylon.
minor, Cam., Tr. Entom. Soc., Lond., 1913, 537.	Ceylon.
niger, Cam., *id.,* 536.	Nilgiris.
ruficeps, Kr., Arch., Naturg., XXV, 1859, 1, 155.	East Indies.
rufotestaceus, Bernh., D. E. Z., 1902, 39.	Ceylon.
schultheissi, Fauv., Rev. d'Ent., XIV, 1895, 217.	Burma.

Tribus Paederini.

Er., Gen. Spec. Staph., 560.

54. Paederus, F.

F., Syst., Ent., 1775, 268.

aliiceps, Cam., Tr. Ent. Soc., Lond., 1913, 538.	Manipur.
alternans, Walk., Ann. Mag. Nat. Hist. (3), II. 1858, 205.	Kawkareik., Ceylon.
ruficoxis, Kr., Arch. Naturg., XXV, 1859, I, 151.	Burma., Himalayas. Kanara., Pegu.
amplicollis, Kr., *id.,* 150.	Ceylon.
andrewesi, Fauv., Rev. d'Ent., XXIII, 1904, 53.	Nilgiris.
argentatus, Cam., Tr. Ent., Soc., Lond., 1913, 540.	Karachi.
basalis, Bernh., W. Z. B., LXIV, 1914, 98.	Siwaliks., Dehra Dun.
birmanus, fauv., Rev. d'Ent., XIV, 1895, 233.	Burma.
capillaris, Fauv., *id.,* 231.	Burma.
ceylonicus, Bernh., D. E. Z., 1902, 37.	Ceylon.
conicollis, Motsch., Etud., Entr., VIII, 1859, 73.	Belgaum., Khandesh. Pondicherry., Genji. Madura.
coxalis, Fauv. Rev. d'Ent., XI, 1895, 232.	Burma.
cyanocephalus, Er., Gen. Spec. Staph. 662; Kr., Arch. Naturg., XXV, 1859, I, 152.	East Indies.

indicus, Motsch., B. B. M., XXXI, 1858, II, 634.	
extraneus, Wied., Zool., Mag. II, I, 1818, 133; Er., Gen. Spec., Staph., 661; Kr., Arch. Naturg., XXV, 1859, I, 152.	Bengal.
pubescens, Cam., Tr. Ent. Soc., Lond., 1913, 539.	
feae, Fauv., Rev. d'Ent., XIV, 1895, 232.	Burma.
fuscipes, Curt., Ent. Brit., III, 108.	Karachi.
aestuans, Er., Gen. Spec., Staph., 635.	Dugeli.
angolensis, Er., Arch. Naturg., IX. 1843, I, 222.	Khanapur., Madura.
corsicus, Gaut., Ann. Sco. Ent. Fr. (4), I, 1861, 393.	Nilgiris., Dehra Dun.
erichsoni, Woll., Coll. Hesp. 247.	Coonoor.
fennicus, J. Sahlb., En. Col. Brach. Fenn., 38.	Genji., Calcutta.
idae, Shp., Tr. Ent. Soc., Lond., 1874, 75.	Puri., Bangalore.
longipennis, Er., Kaf. Mark. Brand, 517.	Ranchi.
riparius, Gr., Col. Mic., 62.	Siliguri District.
var. *peregrinus,* Er., Gen. Spec. Staph., 656; Kr., Arch. Naturg., XXV, 1859, I, 151.	Purneah., Tenasserim.
Breviceps, Bernh, D. E. Z., 1902, 37.	Burmah., Lahore. Jhelum Valley. ————, Europe. New Guinea. Australia.
gratiosus, Fauv., Rev. d'Ent., XXIII, 1904, 54.	Nagargali., Belgaum.
horni, Bernh., D. E. Z., 1902, 38.	Ceylon.
himalayicus, Bernh. W. Z. B., LXIV, 1914, 98 (n. n.).	Himalayas.
indicus, Bernh., Ent. Blatt., VII, 1911, 61 (praeoc.)	Matiana.
melampus, Er., Gen., Spec. Staph., 660; Kr., Arch. Naturg., XXV, 1859, I, 153.	East Indies.
nepalensis, Bernh., Ent. Blatt., VII, 1911, 60.	Nepal., Soondrigal
nigerimus, Bernh., W. Z. B., LXV, 1915, 137 (*Pseudopaederus*).	Madura.
nigricornis, Bernh., Ent. Blatt., VII, 1911, 61.	Nepal., Soondrigal. Fagu.
pallidus, Bernh., W. Z. B., LXV, 1915, 138 (*Pseudopaederus*).	Madura.

pilifer, Motsch., Etud., Ent., VIII, 1859, 74. East Indies.

puberulus, Motsch., *id.,* 74. East Indies.

kuluensis, Bernh., W. Z. B., LXIV, 1914, 99. Himalayas., Kulu

setifer, Cam., Tr. Ent. Soc., Lond., 1913, 538. Madulsima (Ceylon)

sharpi, Cam., *id.,* 530. Dacca.

sondiacus, Fauv., Rev. d'Ent., XIV, 1895, 232. Nagargali., Belgaum.

javanus, Er., Gen. Spec. Staph., 654 (nec. Cast); Kr., Arch. Naturg., XXV, 1859, I, 150. Khanapur. Sampgaon., Ceylon. Burma., Khasi Hills. Nilgiris.

amulus, Er., Gen. Spec. Staph., 661 ; Kr., Arch. Naturg., XXV, 1859, I, 153. Nagargali., Gunji.

dubius, Kr., *id.,* 151. Ceylon.

mixtus, Shp., Tr. Ent. Soc., Lond., 1874, 75. Ghizeh., Belgaum.

rugipennis, Motsch., Etud., Ent., VIII, 1859, 74. Pondicherry., Genji. Nilgiris., Kumaon. Lucknow., Travancore.

variceps, Kr., Arch. Naturg., XXV, 1859, I, 152. Ceylon.

variicornis, Fauv., Rev. d'Ent., XXII, 1903, 154. Ghozeh., Madura. Kanara., Belgaum. Khandesh., Nilgiris. Anamalai., Coonoor.

55. Astenus, Steph.

Steph., Ill. Brit. Ent., V, 275.

Sunius, Er., Käf. Mark. Brand, 523.

Sub-gen. *Astenognathus,* Reitt., Fn. Germ., II, 150.

Sub-gen. *Neognathus,* Shp., Tr. Ent. Soc., Lond., 1874, 69.

biplagiatus, Motsch., B. M., XXXI, 1858, II, 636. East Indies.

birmanus, Fauv., Rev. d'Ent., XIV, 1895, 222. Burma.

bispinus, Motsch., B. M. XXXI, 1858, II, 636. East Indies. Ceylon.

major, Kr., Arch. Naturg., XXV, 1859, I, 146.

concolor, Kr., *id.,* 149. East Indies.

gracilentus, Fauv. Ann. Mus. Civ. Gen., XV, 1879-80, 83. East Indies.

grucilis, Kr., Arch. Naturg., XXV, 1859, I, 147.

gratellus, Fauv., Ann. Mus. Civ. Gen., XV, 1879-80, 83 (*Neognathus*). Nilgiris., Bombay.

pulchellus, Kr., Arch. Naturg., XXV, 1859, I, 147. Pondicherry. Ceylon., Burma.

hindostanus, Cam., E. M. M., LV, 1919, 227. Ceylon.

h-signatus, Cam., Tr. Ent., Soc., Lond., 1913, 540. Assam.

indicus, Kr., Arch. *Naturg.*, XXV, 1859, I, 148 (*Astenognathus*). Bombay. Ceylon.

oculatus, Shp., Tr. Ent. Soc., Lond., 1874, 72. Burma., ——

pallidulus, Woll., Cat. Canar., 591. Europe.

parviceps, Ragusa Nat. Sic., X, 1891, 239.

kraatzi, Bernh., D. E. Z., 1902, 36. Ceylon.

leptocerus, Epp., W. E. Z., XIV, 1895, 64. Nagpur.

maculatus, Cam., E. M., LVI, 1920, 146. Nilgiris.

maculipennis, Kr., Arch. Naturg., XXV, 1859, I, 148. Ceylon.

melanurus, Kust., Käf Eur., XXVI, 1853, 76. Dehra Dun. Nagargali. Dugeli., Sampgaon.

var. *subnotatus,* Fauv., Rev. d'Ent., XXIII, 1904, 51. ——— Europe (type). Burma.

ophis, Fauv., Rev. d'Ent., XIV, 1895, 222. Ceylon.

taprobanus, Cam., E. M. M., LV, 1919, 227.

56. Stilicopsis, Sachse.

Sachse, Stett. Ent. Zeit., XIII, 1852, 144.

plagiata, Fauv., Rev. d'Ent. XIV, 1895, 223. Burma.

strigella, Fauv., *id.,* 223. Burma.

trinotata, Kr., Arch. Naturg., XXV, 1859, I, 149. Ceylon.

umbilicata, Fauv. Rev. d'Ent., XIV, 1895, 223. Burma.

57. Solerochiton, Kr.

Kr., Arch. Naturg., XXV, 1859, I, 133. *Saurellus,* Motsch., Etud., Ent., VIII, 1859, 71.

andrewesi, Cam., Tr. Ent. Soc., Lond., 1913, 541. Nilgiris.

indicus, Motsch., B. M., XXXI, 1858, II, 637. East Indies.

ochraceus, Kr., Arch., Naturg., XXV, 1859, I, 134.

58. **Stiliderus,** Motsch.

Motsch., B. M., XXXI, 1859, II, 639.

Styliderus, Gem., and Har., Cat., II. 1868, 623.

cicatricosus, Motsch., B. M., XXXI, 1858, II, 639.	East Indies.

59. **Stilicus,** Latr.

Latr. Enc. Meth., X, 1825, 495.

ceylanensis, Kr., Arch. Naturg., XXV, 1859, I, 126.	Ceylon., Chakrata., Masuri.
gracilis, Epp., W. E. Z., XIV, 1895, 63.	East Indies.
indicus, Cam., Tr. Ent. Soc., Lond., 1913, 542.	Assam.
luteipennis, Kr., Arch. Naturg., XXV, 1859, I, 126.	Ceylon.
melancholicus, Schub., D. E. Z., 1911, 10.	Shembaganur (Palnis).
ocularis, Fauv., Rev. d'Ent., XIV, 1895, 226.	Burma.
pygmaeus, Kr., Arch. Naturg., XXV, 1859, I, 126.	Ceylon.
rufescens, Shp., Tr. Ent. Soc., Lond., 1874, 61.	Nilgiris., Ceylon.
rubellus, Epp., D. E. Z., XXIX, 1885, 121.	Burma., ________ Japan., W. Africa.
sericeus, Motsch., B. M., XXXI, 1858, II, 640.	East Indies
velutinus, Fauv., Rev. d'Ent., XIV. 1895, 226.	Burma., Chakrata. Masuri.

60. **Stilicoderus,** Sharp.

Sharp., Ann., Mag. Nat. Hist. (6), III, 1889, 320.

discalis, Fauv., Rev. d'Ent., XIV, 1895, 225.	Pegu.
subseriatus, Epp., D. E. Z., 1895, 403 (*Stilicus*).	
feae, Fauv., Rev. d'Ent., XIV, 1895, 224.	Burma., Siwaliks. Masuri,
fenestratus, Fauv., *id.,* 225.	Burma., Chakrata., Masuri.
umbratus, Fauv., *id.,* XXIII, 1904, 52.	Nilgiris.

61. **Psilotrachelus,** Kr.

Kr., Arch. Naturg., XXV, 1859, I, 124.

crassus, Kr., *id.,* 124. Ceylon.

sculptipennis, Kr., *id.,* 125. Dehra Dun., Masuri.

62. Thinocharis, Kr.

Kraatz, Arch. Naturg., XXV, 1859, I, 142.

carinicollis, Kr., *id.,* 143. Ceylon

pygmaea, Kr., *id.,* 143. Ceylon.

63. Acanthoglossa, Kr.

Kraatz., Arch., Naturg., XXV, 1859, I, 144.

brachycera, Kr., *id.,* 145, pl. 2, f. 10. Ceylon.

hirta, Kr., *id.,* 144. Parleh., Ceylon.

rufa, Kr., *id.,* 146. Ceylon.

testaceipennis, Kr., id., 145. East Indies.

64. Medon, Steph.

Steph., Ill. Brit. Ent., V, 273.

Sub-gen. *Medon,* s. str.

Muls., and Rey Ann. Soc. Linn., Lyon, XXIV, 1877, 122.

aspericollis, Fauv., Rev. d'Ent., XIV, 1895, 229. Burma.

auripilis, Fauv., Rev. d'Ent., XIV, 1895, 228. Burma.

basalis, Cam., E. M. M., 1919, 252. Ceylon.

birmanus, Fauv., Rev. d'Ent., XIV, 1895, 228. Burma.

championi, Cam., E. M. M., 1919, 251. Ceylon.

discipennis, Fauv., Rev. d'Ent., XIV, 1895, 230. Burma.

fauveli, nom. nov. Burma.

luctuosus, Fauv., *id.,* 229.

fortepunctatus, Bernh., D. E. Z., 1902, 34. Ceylon.

immsi, Bernh., W. Z. B., LXIV, 1914, 99. Dehra Dun.

opacellus, Fauv., Rev. d'Ent., XIV, 1895, 231. Burma.

picipennis, Fauv., *id.,* 230. Burma.

praeustus, Fauv., *id.,* XXIV, 1905, 138. Burma.

propinquus, Bris., Harold Col., Hefte II, 1867, 116. W. Himalayas.

Europe.

Sub-gen. *Cryptoporus,* Motsch.

Motsch., B. M., XXXI, 1858, II, 654.

flavipes, Motsch., *id.,* 655. East Indies.

Sub-gen. *Hypomedon,* Cas.

Casey, Tr. Ac., St. Louis, XV, 1905, 152.

latecinctus, Fauv., Rev. d'Ent., XXIV, 1905, 139. Tenasserim.

Sub-gen. *Chloecharis,* Lynch.

Lynch. Bol. Ac. Cord., VII, 1884, 259.

debilicornis, Woll. Cat. Col. Mad., 194. Ceylon.

aegyptiacus, Motsch., B. M., XXXI, 1858, II, 644. Dehra Dun.

---

brevicornis, Allard. Ann. Soc. Ent. Fr. (3), V, 1857, 747. (Tropical cosmopolite)

occultus, Waterh., Ann. Mag. Nat. Hist. (4), XVIII, 1876, 108.

rufulus, Lynch., Boll. Ac. Nac. Cord. VII, 1884, 259.

minutissimus, Bernh., D. E. Z., 1902, 35. Ceylon.

nigriventris, Cam., Tr. Ent. Soc., Lond., 1913, 542. Ceylon.

Sub-gen. *Lithocharis,* Boisd. et Lacord.

Boisd. et Lacord., Fn. Ent., Paris, 1835, 431.

breviusculus, Kr., Arch. Naturg., XXV, 1859, I, 137. Ceylon.

cinnamopterus, Kr., Arch. Naturg., XXV, 1859, I, 141. East Indies.

curtus, Kr., Arch. Naturg., XXV, 1859, I, 136. Ceylon.

flavescens, Motsch., B. M., XXXI, 1858, II, 643. East Indies.

fuscipennis, Kr., Arch. Naturg., XXV, 1859, I, 138. Ceylon.

humeralis, Motsch., B. M., XXXI, 1858, II, 644. East Indies.

nigerrimus, Kr., Arch., Naturg., XXV, 1859, I, 141. Ceylon.

ochraceus, Gr., Col. Mic., 1802, 59. Pondicherry.

alutacea, Cas., Bull. Calif., II, 1880, 29. Ceylon.

brunniceps, Fairm., Rev. Zool., 1849, 290. Dehra Dun.

fastidiosus, Fairm., et Germ. Ann. Soc. Ent. Fr. (4), II, 1861, 438. ____________

(Cosmopolite).

nigriceps, Kr., Arch. Naturg., XXV, 1859, I, 139.

quadricollis, Cas., Bull. Calif., II, 1886. 30.

rubricollis, Gr., Col. Mic., 138.

pallidus, Motsch., B. M., XXXI, 1858, II, 644. East Indies.

planus, Kr., Arch. Naturg., XXV, 1859, I, 136. East Indies.

rufulus, Kr., *id.,* 140. East Indies.

uvidus, Kr., *id.,* 138. Ceylon.

vicinus, Kr., *id.,* 139. East Indies.
137. Nilgiris.

erythropterus, Gem., and Har., Cat., II. 621. Sikkim., Ceylon.

fuscipennis, Shp., Trans. Dub. Soc. (2), III, 1885, 223 (nec. Kr.). Burma.
Dehra Dun.

sororculus, Kr., Arch. Naturg., XXV, 1859, I, 140.

Sub-gen. *Isocheilus,* Sharp.

Sharp, Ann. Mag. Nat. Hist. (6), II, 1889, 263.

staphylinoides Kr., Arch., Naturg., XXV, 1859, I, 134. Ceylon., East Indies.

____

Japan.

Sub-gen. *Charichirus,* Sharp.

Sharp, Ann. Mag. Nat. Hist. (6), II, 1889, 262.

chinensis, Boh., Resa. Eugen., 1858, 32. Ceylon.

dimidiatus, Motsch., B. M., XXXI, 1859, II, 645. Nilgiris., Ranchi.

obliquus, Walk., Ann. Mag. Nat. Hist. (3), III, 1859, 52. ————, China.

spectabilis, Kr., Arch. Nat., XXV, 1859, I, 135. Japan., Java, etc.

ceylonicus, Cam., E. M. M., LV, 1919, 228. Ceylon.

robustus, Bernh., D. E. Z., 1902, 33. Ceylon.

Sub-gen. *Aderocharis,* Shp.

Sharp, B. C. A., I, 2. 552.

princeps, Fauv., Rev. d'Ent., XIV, 1895, 227. Burma.

65. **Scopaeus.** Er.

Er., Gen. Spec. Staph., 604.

bicuspis, Kr., Arch. Naturg., XXV, 1859, I, 129.	East Indies.
dilutus, Motsch., B. M., XXXI, 1858, II, 642.	Mahé., Ceylon.
suturalis, Kr., Arch. Naturg., XXV, 1859, I, 130.	Burma.
felvescens., Motsch., B. M., XXXI, 1858, II, 642.	East Indies.
fusculus, Motsch., *id.,* 641.	East Indies.
limbatus, Kr., Arch. Naturg., XXV, 1859, I, 130.	East Indies.
micros, Kr., *id.,* 132.	Ceylon.
nitidulus, Motsch., B. M., XXXI, 1858, II, 643.	East Indies.
subfasciatus., Kr., Arch. Naturg., XXV, 1859, I, 129.	
pallidulus, Kr., *id.,* 131.	Calcutta., Ceylon.
planiusculus, Kr., *id.,* 132.	East Indies.
procerus, Kr., *id.,* 127.	East Indies.
puberulus., Kr., *id.,* 128.	East Indies.
puncticeps, Kr., *id.,* 132. ♂	East Indies.
testaceus, Motsch., B. M., XXXI, 1858. II. 642.	Burma., Ceylon.
rubrotestaceus, Kr., Arch. Naturg., XXV, 1859, I, 128 ( ).	Nagargali.
semifuseus, Kr., *id.,* 128 (♀) .	
thoracicus, Motsch., B. M., XXXI, 1858, II, 641.	East Indies.
velutinus, Motsch., *id.,* 642.	Burma.
decipiens., Kr., Arch. Naturg., XXV, 1859, I, 131.	Ceylon., Nagargali.

66. **Lathrobium,** Grav.

Grav., Col. Micr., 1802, 51.

cafrum, Boh. Ins. Caff., I, 1848, 285.	Ceylon.
pulchellum, Kr., Arch. Naturg., XXV, 1859, I, 116.	Pegu., Mahé., Calicut., ——— Africa., Philippines.
gracilentum, Kr., ***id.,*** 115.	Ceylon.
guttula, Fauv., Rev. d'Ent., XIV, 1895, 234.	Burma.
semicaeruleum, Cam., E. M. M., LVII, 1921, 271.	Masuri., Sikkim.

var. *ochreonotatum,* Champ., *id.,* 58, 1922, 34. Kumaon., Fagu.

seminigrum, Kr., Arch. Naturg., XXV, 1859, I. 117. Ceylon.

sublaeve, Motsch., B. M., XXXI, 1858, II, 647. East Indies.

unicolor, Kr., Arch. Naturg., XXV, 1859, I. 117. South India., Pegu.

pallens, Gem. and Har., Cat. Col., II, 1868, 611.

seriatum. Shp., Ann. Mag. Nat. Hist. (6), III, 1889, 259 (veris).

testaceum, Motsch., B. M. XXXI, 1858, II, 646.

67. **Pseudobium,** Muls., & Rey.

Muls., & Rey, Ann. Soc. Linn., Lyon, XXIV, 1877, 104.

laeviventre, Champ., E. M. M., LVIII, 1922, 34. E. & W. Almora; Dehra Dun.

68. **Scimbalium,** Er.

Er. Gen. Spec. Staph., 1840, 579.

asperiventre, Fauv., Rev. d'Ent., XIV, 1895, 235. Burma.

badium, Motsch, B. M., XXXI, 1858, II, 646. South India; Pegu.

indicum, Epp., W. E. Z., IX 1890, 277. Calcutta.

69. **Achenium,** Curt.

Curt. Brit. Ent., III., 1836, 115.

humerale, Motsch., B. M., XXXI, 1858, II, 648. East Indies.

70. **Dolicaon,** Cast.

Cast., Etud. Ent., I, 1835, 119.

indicus, Kr., Arch. Naturg., XXV, 1859, I, 118. Tharrawaddy; Toungoo; Ceylon, ________ China., Africa.

angolensis, Qued., B. E. Z., XXV, 1881, 294.

eastaneus, Epp., D. E. Z., XXIX, 1885, 114.

vicinus, Kr., Arch., Naturg., XXV, 1859, I, 118.

parviceps, Fauv., Rev. d'Ent., XIV, 1895, 234. Burma., Pegu.

robustus, Kr., Arch. Naturg., XXV, 1859, I. 117. East Indies.

71. **Calliderma,** Motsch.

Motsch., B. M., XXXI, 1858, I, 653.

Cephalochetus, Kr., Arch. Naturg., XXV, 1859, I, 122.	
aspericolle, Fauv., Rev. d'Ent., XIV, 1895, 236.	Burma.
brunneum, Motsch., B. M., XXXI, 1858, II, 653.	East Indies.
elegans, Kr., Arch. Naturg., XXV, 1859, I, 123.	Ceylon.
indicus, Kr., *id.,* 124.	East Indies.
kraatzi, Fauv., Rev. d'Ent., XIV, 1895, 235.	Burma.

72. Cryptobium, Mannerh.

Mannerh. Brach., 1830, 38.

abdominale, Motsch., B. M., XXXI, 1858, I, 651.	Ceylon, Burma.
pygiale, Kr., Arch. Naturg., XXV, 1859, I, 121.	Dehra Dun, Pusa.
var. *discoideum,* Epp., D. E. Z., 1895, 402.	
var *indicum,* Kr., Arch. Naturg., XXV, 1859, I, 121.	
var. *rufipenne,* Motsch., B. M., XXXI, 1858, I, 658.	Purneah.
suturale, Motsch., *id.,* 652.	
brunnipes, Cam., Tr. Ent. Soc., Lond., 1913, 543.	Dacca.
ceylanense, Kr., Arch. Naturg., XXV, 1859, I, 119; Epp., D. E. Z., 1895, 401 (♂).	Bhim Tal, Kumaon. Ceylon.
elephas, Fauv., Rev. d'Ent., XXIII, 1904, 54.	Nilgiris.
extraneum, Fauv., *id.,* 55.	Tudia, Bengal. Satara.
filum, Kr., Arch. Naturg., XXV, 1859, I, 119.	Ceylon, Burma.
rugiceps, Epp., D. E. Z., XXIX, 1885, 120.	
fluviatile, Champ., E. M. M., LVII, 1921, 181.	W. Almora.
fossigerum, Kr., Arch., Naturg., XXV, 1859, I, 120.	Ceylon, Pondicherry.
fulvescens, Motsch., B. M., XXXI, 1858, 207 [*Homoeotarsus*].	Burma.
testaceum, Kr., Arch., Naturg., XXV, 1859, I, 120.	Burma.
gastrale, Fauv., Rev. d'Ent., XIV, 1895, 237.	Siwaliks, W. Almora.
kumaonense, Champ., E. M. LVII, 1921, 181.	Tudia, Pegu.
marginatum, Motsch., B. M., XXXI, 1858, II, 652.	Belgaum, Kanara, Burma.
limbaeum, Kr., Arch. Naturg., XXV, 1859, I, 121.	Mahé, Tenasserim. Genji, Pondicherry.

mutator, Fauv., Rev. d'Ent., XIV, 1895, 236. Burma.

nigrobadium, Epp., W. E. Z., IX, 1890, 278. East Indies.

nilgiriensis, Cam., Tr. Ent., Soc., Lond., 1913, 543. Nilgiris.

nisus, Fauv., Rev. d'Ent., XIV, 1895, 239. Paresnath, Burma. Ceylon.

paricolor, Epp., D. E. Z., 1895, 401. Ten Malai. W. Ghats.

rosti, Schub., D. E. Z., 1908, 622. Himalayas., Kulu.

sanguinolentum, Motsch., B. M., XXXI, 1858, II, 650. Khandesh, Burma. Tenasserim, Pegu.

semiopacum. Epp., D. E. Z., 1895, 402. Tenasserim.

biforme, Fauv., Rev. d'Ent., XIV, 1895, 237. Pegu.

semirufum, Bernh., D. E. Z., 1902, 32. Cylon.

sharpi, Fauv., Rev. d'Ent., XIV, 1895, 238. Paresnath, Burma.

spectabile, Kr., Arch. Naturg., XXV, 1859, I, 118. Himalayas.

waageni, Bernh., Ent. Blatt., 1915 (10-12), 252. Pegu.

Sub-fam. STAPHYLININAE.
Sharp, B. C. A., I, 2, 320.

Tribus Xantholinini.

Ganglb. Käf. Mittel-Eur., II, 466.

73. **Oligolinus,** Cas.

Casey, Tr. Ac., St. Loius, XVI, 1906.

excellens, Bernh., Ent. Blatt, 1915 (10-12), 253. Madura.

74. **Metolinus,** Cam.

Cam., E. M. M., LVI, 1920, 147.

basalis, Cam., *id.,* 147. Anamalai.

leucocnemis, Kr., Arch., Naturg., XXV, 1859, I, 107 (*Metoponcus*). Kanara; Sikkim;

exiguus, Kr., *id.,* 108. Ceylon; Tenasserim.

75. **Leptacinus,** Er.

Er., Käf. Mark. Brand, 429.

Xanthophius, Motsch., Et. Ent., VIII, 1859.

aethiops, Bernh., Arch. Naturg., 84 A. 10, 1918, 180.	Sikkim; Darjiling.
filum, Kr., Arch. Naturg., XXV, 1859, I, 111.	Ceylon.
fusciceps, Epp., D. E. Z., 1895, 399.	N. India.
gracilis, Fauv., Rev. d'Ent., XIV, 1895, 240.	Nilgiris, Burma.
parumpunctatus, Gyll., Ins. Suec., IV, 481, ? var. *tricolor,* Kr., Arch. Naturg., XXV, 1859, I, 110.	Belgaum, Ghozeh. Khandesh.
flavipennis, Kr., *id.,* 111.	Nilgiris., Dehra Dun. Pondicherry. Ceylon., _______ (Cosmoplite)
pusillimus, Kr., *id.,* 112.	Ceylon.
serpentarius, Motsch., Etud., Ent., VIII, 1859, 76.	Ceylon.
trigonocephalus, Kr., Arch. Naturg., XXV, 1859, I, 109.	Ceylon;, Nilgiris.

76. Pachycorynus, Motsch.

Motsch., B. M., XXI, 1858, 204.

Holisomorphus, Kr., Arch. Naturg., XXV, 1859, I, 100.	
dimidiatus, Motsch., B. M., XXXI, 1858, 205.	Nilgiris; Calcutta;
ceylanensis, Kr., Arch. Naturg., XXV, 1859, I, 101.	Ceylon.
cinctus, Walk., Ann. Mag. Nat. Hist. (3), II, 1858, 205	

77. Mitomorphus, Kr.

Kr., Arch. Naturg., XXV, 1859, I, 105.

cupreolus, Fauv., Rev. d'Ent., XIV, 1895, 240.	Burma.
indicus, Kr., Arch. Naturg., XXI, 1859, I, 107.	East Indies.
nigroaeneus, Kr., *id.,* 106.	Ceylon.
nigerrimus, Bernh., Ent. Blatt., 1915 (10-12), 253.	Madura.
obsoletus, Fauv., Rev. d'Ent., XXIII, 1904, 56.	Nilgiris.
ovaliceps, Cam., E. M. M., LVI, 1920, 147.	Nilgiris.

78. Xantholinus, Serv.

Serv., Encycl. Meth., X, 1825, 475.

cinctus, Fauv., Rev. d'Ent., XXIII, 1904, 56.	Nilgiris.
dispilus, Er., Gen. Spec. Staph., 317; Kr., Arch. Naturg. XXV, 1859, I, 104.	East Indies.

eppelsheimi, Nom. nov.	East Indies.
rudis, Epp., W. E. Z., XIV, 1895, 62.	
feae, Fauv., Rev. d'Ent., XIV, 1895, 242.	Burma
hypsibathus, Bernh., Col. Rundsch., 1915, (3), 51.	Kulu.
metallicus, Fauv., Rev. d'Ent., 1895, 243.	Burma.
nigerrimus, Kr., Arch. Naturg., XXV, 1859, I, 103.	East Indies.
parcus, Epp., D. E. Z., 1895, 397.	Himalayas.

79. **Gauropterus,** Thoms.

Thoms., Skand. Col., II. 167.

Eulissus, Epp., Cat. Eur., 1891, 1905. [nec. Mannerh].	
fulgidus, F., Mant. Ins., I, 220.	Ceylon.
angulatus, Kust., Käf. Eur., XIII, No. 7.	———
intermedius, Kust., *id.,* No. 8.	Europe;
pyroplerus, Gr., Col. Mic., 102.	North America.
nigroaeneus., Cam., E. M. M., LVI, 1920.	Pusa.

80. **Eulissus,** Mannerh.

Mannern, Brach., 35.

anachoreta. Er., Gen. Spec., Staph., 316.	Belgaum.;
hongkongensis, Redt., Reise, Novara, II, 28.	Masuri; Ceylon;
morio, Mosch., B. M., XXXI, 1858, 207.	Nagargali; Dehra Dun; Mahé; Calicut; Coonoor; Burma; Calcutta; ———, China; Madagascar; Mauritius.

81. **Thyreocephalus,** Guer.

Guer., Rev. Zool., 1844, 10.

Indoscitalinus, Heller, Abh. Mus., Dresd., IX, 5.	
annulatus, Fauv., Rev. d'Ent., XIV, 1895, 241.	Burma.
gestroi, Fauv., *id.,* 243.	Burma.
indicus, Fauv., *id.,* 242.	Burma.
lorquini, Fauv., Ann. Mus. Civ. Gen., X, 1877, 211.	?India.

82. Tesba, Sharp.

Sharp. Tr. Soc., Lond., 1876, 194.

indica, Epp., D. E. Z., 1895, 398. Burma.

83. Diochus, Er.

Er., Gen. Spec., Staph., 300.

Rhegmatocerus, Motsch., B. M., XXXI, 1858, II, 657.

antennatus, Motsch., B. M., XXXI, 1858, II, 659. East Indies.

indicus, Kr., Arch., Naturg., XXV, 1859, I, 113.

conicollis, Motsch., B. M., XXXI, 1858, II, 658. East Indies.

major, Kr., Arch. Naturg., XXV, 1859, I, 113.

punctipennis, Motsch., B. M., XXXI, 1858, II, 659. East Indies.

Purneah.

84. Platyprosopus, Mannerh.
Mannerh, Brach., 36.

aequalis, nom. nov.

parallelus, Bernh., Ent. Blatt., VII, 1911, 86. Tharrawaddy.

bilineatus, Fauv., Rev. d'Ent., XIV, 1895, 239. Calcutta;

biseriatus, Epp., D. E. Z., 1895, 400. Tharrawaddy; Toungoo; Pegu.

Tharrawaddy; Burma; Pegu.

consularis, Fauv., Rev. d'Ent., XIV, 1895, 239.

Burma; Pegu.

elatus, Fauv., Rev. d'Ent., XXIII, 1904, 70. Bengal;

indicus, Kr., Arch. Naturg., XXV, 1859, I, 117.

fuliginosus, Er. Gen. Spec., Staph., 923. Burma.

orientalis, Motsch., B. M., XXXI, 1858, II, 212. Purneah.

var. *tamulus,* Er. Gen. Spec., Staph., 923.

fulvicollis, Motsch., B. M. XXXI, 1858, II, 212. Calcutta;

linearis, Kr., Arch. Naturg., XXV. 1859, I, 115.

indicus Motsch., B. M., 1858, II, 212.

laevicollis, Kr., Arch. Naturg., XXV, 1859, I, 114. Berhampur; Sahebgunj; Dacca.

Tribus Staphylinini.
Ganglb. Käf Mittel-Eur., II, 412.

1. SUB-TRIBUS *STAPHYLINI.*

85. Neobisnius, Ganglb.

Ganglb., Käf, Mittel-Eur., II, 464.

flavicornis, Fauv., Rev. d'Ent., XIV, 1895, 261.	Burma.
praelongus, Gem., et. Har., Cat., II, 591.	Ceylon.
longulus, Kr., Arch. Naturg., XXV, 1859, I, 99.	Burma.
refipennis, Cam., E. M. M., 1919, 252.	Ceylon.

86. Actobius, Fauv.

Fauv., Fn. Gallo-Rhen., III, Supp. 1874, 72.

basalis, Motsch., B. M., XXXI, 1858, II, 664.	Paresnath; (W. Bengal).
castaneipennis, Kr., Arch. Naturg., XXV, 1859, I, 96.	
humeralis, Cam., E. M. M., LVI, 1920, p. 214.	Nilgiris.

87. Philonthus, Curt.

Curt., Brit. Ent., XIII, 610.

Thorax with dorsal series of two punctures.

bipunctatus, Bernh., Ent., Blatt., VII, 1911, 90.	Nepal; Masuri; Chittagong.

Thorax with dorsal series of three punctures.

cinctulus, Gr., Coll. Mic., 167.	Calcutta; Ranchi; Khandesh; Bengal; Burma; Ceylon; Genji; Dehra Dun.
industanus, Fauv. Rev. d'Ent. XXII, 1903, 158.	Coonoor; Kurseong; Darjiling.
notabilis, Kr., Arch. Naturg., XXV, 1859, I, 79.	Dehra Dun; Ceylon.
tripunctatus, Cam., E. M. M., 1919, 253.	Ceylon.
pubipennis, Cam., *id.,* 253.	Ceylon.

Thorax with dorsal series of three or four punctures.

ruficollis, Schub., D. E. Z., 1908, 615.	Himalayas; Kulu.

Thorax with dorsal series of four punctures.

aeger, Epp., W. E. Z., 1895, 59.	Simla Hills.
aeneus, Rossi, Faun., Etr., I, 249.	Himalayas. ———, Europe; North America; Tasmania; New Zealand.
adversus, nom. nov.	Burma.
misellus, Fauv., Rev. d'Ent., XIV, 1895, 262.	
andrewesi, Cam., E. M. M., LVI, 1920, 214.	Nilgiris.
bengalensis, Bernh., Ent. Blatt., IV, 1911, 91.	Paresnath; W. Bengal; Kangra Valley; Punjab; Siwaliks.
birmanus, Fauv., Rev. d'Ent., XIV, 1895, 261.	Burma.
castaneus, Gemm., and Har., Cat., II, 586.	East Indies.
badius, Motsch., B. M., XXXI, 1858, I, 661.	
chalceus, Steph., III, Brit. Ent., 227.	N. India.
carbonarius, Er., Käf. Mark. Brand, 448.	———
nigritus, Runde. Brach. Hal., 7	Europe.
proximus, Kr., Arch. Naturg., XXV, 1859, I, 80.	
succicola, Thoms, Skand. Col., II, 157.	
concolor, Kr., Arch. Naturg., XXV, 1859, I, 80.	Ceylon.
cyaneliytrius, Kr., *id.,* 80.	Simla Hills. Murree; Tahe-Pum Mt.; Chinese Frontier.
delicatulus, Boh., Eng. Resa, 29.	Ceylon;
subirideus, Kr., Arch. Nat., XXV, 1859, I, 81.	Belgaum; Ranchi; Nilgiris; Burma; Dehra Dun; ——— China; Java, etc.
diversus, Schub., D. E. Z., 1906, 382.	Kashmir., (Sintan).
donckieri, Bernh., Ent. Blatt., 1915 (10-12), 254.	Madura.
fimetarius, Gr., Col. Mic., 175.	East Indies.
interpunctatus, Motsch., B. M., XXXIII, 1860, II, 566.	———, Europe.

longulus, Motsch., *id.*, XXXI, 1858, II, 662.	
rigidicornis, Gr., Col. Mic., 23.	
stenoderus, Reiche, Ann. Soc. Ent. Fr. (4) I, 204.	
xanthomerus, Kr., Arch. Naturg., XXV, 1859, I, 83.	
hindostanus, Schub., D. E. Z., 1911, 23.	Shembaganur (Palnis).
kempi, Cam., Rec., Ind., Mus., XXVI, 1924, 181.	Assam.
maindroni, Fauv., Rev. d'Ent., XXII, 1903, 158.	Coonoor.
annandalei, Cam., Rec., Ind., Mus. XXVI, 1924, 118.	Masuri.
nitens, Kr., Arch. Naturg., XXV, 1859, I, 82.	Ceylon.
peliomerus, Kr., Arch. Naturg., XXV, 1859, I, 82.	Khandesh; Ceylon; Burma;
flavoterminatus, Woll, Col., St. Helenae, 26.	———, St. Helena; Java, etc.
proximatus, Schub., D.E.Z., 1908, 616.	Bhim Tal; Kumaon; Kulu.
protenus, Schub., D.E.Z., 1906, 382.	Kashmir; Sangam.
rotundicollis,* Menet., Cat. Rais., 145.	Pamir;
duplopunctatus, Motsch., B.M. XXXIII, 1860, II, 567.	Yarkand; ———, Europe
semiaenescens, Bernh., Arch. Naturg., 84, A. 10, 1918, 182.	Murree.
simlaensis, Bernh., Ent. Blatt., VII, 1911, 90.	Himalayas; Fagu.
sordidus, Gr., Col. Mic., 176.	Pamir; ——— Europe; America; Australia.
tractatus, Epp., W.E.Z., 1895, 61.	Simla Hills.

Thorax with dorsal series of five punctures.

acroleucus, Kr., Arch. Naturg., XXV, 1859, I, 91.	East Indies.
aeneipennis, Boh., Eugen., Resa, 30.	Belgaum; Dehra Dun;
erythropus, Kr., Arch. Naturg., XXV, 1859, I, 88.	Ceylon; India generally. ———
lewisius, Shp., Tr. Ent. Soc., Lond., 1894, 42.	China; Japan; Java, etc.

* Occasionally 5 or 6 punctures are present in the dorsal row.

var. *cliens,* Epp., W. E. Z., IX, 1890, 277. Nagpur., Belgaum. Khandesh., Kanara. Karachi., Masuri.

aliquatenus, Schub., D. E. Z., 1908, 618. Himalayas., Kulu.

asemus, Kr., Arch. Naturg., XXV, 1859, I, 86. Himalayas.

cinctipennis, Fauv. Fn. Gallo-Rhen. Cat., 1875, XXX. Burma., Khandesh.

circumductus, Fauv., Rev. d'Ent., XIV, 1895, 263. Burma.

convalescens, Epp., W. E. Z., IX, 1890, 276. Simla., Fagu.

crassicornis, Fauv., Rev d'Ent., XIV, 1895, 264. Tenasserim. Burma., Ceylon. Nilgiris.

curtipennis, Bernh., D. E. Z., 1902, 26. Ceylon.

discoideus, Gr., Col. Mie. 1802, 38. East Indies.

conformis, Boisd., & Lac., Fn. Ent., Paris, I, 398 Ceylon. ——— (Cosmopolitan).

lepidulus, Steph., Ill, Brit. Ent., V, 223.

ruficornis, Melsh., Proc. Ac. Phil., II, 1846, 38.

suturalis, Steph., III, Brit. Ent., V, 224.

testaceus, Gr., Col. Mic., 60.

disjunctus, nom. nov. Burma.

distans, Fauv., Rev. d'Ent., XIV, 1895, 265.

distinctus, Gemm., & Har., Cat., II, 587. East Indies.

antennatus, Motsch., B. M., XXXI, 1858, II, 664.

duplex, Bernh., Arch. Naturg., 84, A. 10, 1918, 182. E. Indies.

fauvelianus, nom. nov. Burma.

dispersus, Fauv., Rev., d'Ent., XIV, 1895, 264.

flavipes, Kr., Arch. Naturg., XXV, 1859, I, 88. Belgaum., Ceylon. Burma.

flavocinctus, Motsch., B. M., XXXI, 1858, II, 663. Ceylon., Burma.

rufomarginatus, Kr., Arch. Naturg., XXV, 1859, I, 90. Belgaum.

frontalis, Schub., D. E. Z., 1908, 616. Himalayas., Kulu.

gemellus, Kr., Arch. Naturg., XXV, 1859, I, 91. Belgaum., Ceylon., Nilgiris.

geminus, Kr., Arch. Naturg., XXV, 1859, I, 87. — Nilgiris., Burma. Ceylon., Pegu.

var, *inornatus,* Cam., E. M. M., LVI, 1920, 215. — Nilgiris.

indicus, Cam., E. M. M., LVI, 1920, 216. — Nilgiris.

idiocerus, Kr., Arch., Naturg., XXV, 1859, I, 85. — Ceylon., Ranchi. Dehra Dun.

kashmiranus, Bernh., Col. Rundsch., 1915, (3), p. 52. — Kashmir.

kuluensis, Schub., D. E. Z., 1908, 617. — Himalayas., Kulu.

leucopygus, Kr., Arch. Naturg., XXV, 1859, I, 99. — East Indies.

longicornis, Steph., Ill, Brit. Ent., V, 237. — Nilgiris.

algiricus, Motsch., B. M., XXI, 1858, II, 663. — Ceylon., Coonoor.

fumosus, Solsky, Hor. Soc. Ent. Ross., 1867-1868, 134. — Dehra Dun., Ranchi.

fuscicornis, Nord., Symb., 96. — Bangalore.

laetabilis, Oliff., Pr. Lin. Soc., N. S. Wales, 1887, 501. — ———— (Cosmopolitan).

pedestris, Walk., Ann. Mag. Nat. Hist. (3), III, 1859, 51.

perplexus, Fairm., Ann. Soc. Ent. Fr. (4), I, 1861, 431.

promptus, Er., Gen. Spec. Staph., 929.

scybalarius, Nord., Symb., 94.

varians, Fairm., Rev. Zool., 1819, 290.

varians, var B. Er., Kaf. Mark. Brand, 466.

maculatus, Cam., E. M. M. LVI, 1920, 217. — Mysore.

madurensis, Bernh., Ent. Blatt., 1915 (10-12), 255. — Madura.

minutus, Boh., Ins. Caff. I. 279. — Kanara.

longiceps, Fauv., Ann. Mus. Civ. Gen., XV, 104. — Ceylon., Nilgiris.

mutans, Shp., Tr. Ent. Soc. London, 1874, 43. — Burma., Pegu.

parviceps, Kr., Arch. Naturg., XXV, 1859, I, 86. — ————, Africa.

rufocinctus, Fauv., Bull. Soc. Lin. Norm. (3), II, 126. — S. Europe., China. Japan. E. Indies generally. N. Guinea.

nilgiriensis, Cam., E. M. M., LVI, 1920, 215. — Nilgiris.

obsoletus, Epp., W. E. Z., XIV, 1895, 61. — Himalayas.

paederoides, Motsch., B. M. XXXI, 1858, II, 662. — Nilgiris.

bellus, Kr., Arch. Naturg., XXV, 1859, I, 83.	Katihar., Calcutta. Ceylon., Belgaum. Sikkim., Burma. Pegu.
pamirensis, Shp., Col. Soc. Yark. Miss., 1891, 41.	Pamir.
peregrinus, Fauv., Ann. Soc. Ent. Fr. (4), VI, 1866, 315.	Kashmir., Chamba.
piceicoxis, Epp., D. E. Z., 1895, 127.	Sikkim.
var. *scutatus*, Epp. *id.*, 127.	Chakrata., ______ Africa., Seychelles.
productus, Kr., Arch. Naturg., XXV, 1859, I, 89.	Ceylon.
punctatissmus, Schub., D. E. Z., 1908, 619.	Pathankot. Himalayas.
quisquiliarius, Gyll., Ins. Suec., II, 1810, 335.	Bhogaon.
var. *inquinatus*, Steph., Ill, Brit. Ent. V., 223.	Purneah District. North Bengal.
dimidiatus, Boisd., & Lac., Fn. Ent., Paris, I, 1825, 402.	Burma., Belgaum.
variipennis, Kr., Arch. Naturg., XXV, 1859, I, 85.	Dehra Dun., Ranchi.
stoliczkae, Sharp, Col. Second., Yarkand, Miss, 1891, 40.	India generally. ________ (Cosmopolitan.)
remotus, Fauv., Rev. d'Ent., XIV, 1895, 264.	Burma.
refiventris, Cam., E. M. M., 1918, 170.	Kandy., Ceylon.
signifrons, Schub, D. E. Z., 1908, 618.	Himalayas., Kulu.
sericoilius, Cam., E. M. M., LVI, 1920, 216.	Nilgiris.
tardus, Kr., Arch. Naturg., XXV, 1859, I, 84.	Ceylon.
tricoloris, Schub., D. E. Z., 1908, 620.	Palumbus. Himalayas.
vagus, Fauv. Rev. d'Ent. XIV, 1895, 263.	Burma.
ventralis, Gr., Col. Micr., 1802, 174.	Dehra Dun.
anthrax, Gr., *id.*, 176.	________
fulvipes, Steph, Ill. Brit. Ent. V, 219.	(Cosmopolitan).
immundus, Gr., Col. Micr., 1802, 66.	
picicollis, Steph., III. Brit. Ent., V, 224.	
rotundiceps, Steph., *id.*, 248.	

Thorax with dorsal series of six punctures.

acuminatus, Kr., Arch. Naturg., XXV. 1859, I, 93. Ceylon.

asiaticus, Bernh., D. E. Z., 1902, 27. Ceylon.

basipilosus, Schub., D. E. Z., 1908, 621. Himalayas., Kulu.

christiei, Bernh., Arch. Naturg., 81, A-10, 1918, 181. Sikkim., Darjeeling.

fuscatus, Kr., Arch. Naturg., XXV, 1859, I, 95. Ceylon., Burma.

himalayicus, Bernh., Col. Rundsch., 1915 (3), 62. Kulu.

imitator, Fauv., Rev. d'Ent., XIV 1895, 265. Burma.

oculatus, Bernh., D. E. Z., 1902, 28. Ceylon.

perniger, Bernh, Arch. Naturg., 84, A-10, 1918, 181. Madura.

rubricollis, Motsch., B. M. XXXI, 1858, II. 665. Katihar.

erythrosticus, Kr., Arch. Naturg., XXV, 1859, I, 94.

turbidus Er., Gen. Spec. Staph., 484. Assam.

Thorax with dorsal series of more than six punctures.

amabilis, Kr., Arch. Naturg., XXV, 1859, I, 97. N. India.

eustilbus, *id.,* 98. Ceylon.

leptoceros, Fanv., Rev. d'Ent. XIV, 1895, 266. Burma.

leucotus, Er. Gen. Spec. Staph, 507; Kr., Arch. Naturg., XXV, 1859, I, 98. Nepal

versicolor, Cam., E. M. M., 1919, 254. Ceylon.

88. **Gabrius,** Steph.

Steph., Ill., Brit., Ent. V. 249.

nigritulus, Gr., Col. Micr., 41. Himalayas., Ceylon.

(Cosmopolitan)

pulchellus, Kr., Arch. Nat. XV, 1859, I, 92. East Indies., Ceylon.

thermarum, Aubé., Ann. Soc. Ent. Fr. (2), VIII, 316. Ceylon., Dehra Dun.

angustatus Kr. Arch. Naturg. XXV, 1259, I. 92. W. Himalayas.

breviventer, Sperk, B. M., VIII, 1835, 153. (Cosmopolitan).

exilis, Kr., Stett. Ent. Zeit., XII, 1851, 293.

fuscolaterus, Motsch., Et. Ent., VIII, 1859, 76.

inclinans, Walk., Ann. Mag. Nat. Hist. (3), III, 1859, 51.

pygmaeus, Kr., Arch. Naturg., XXV, 1859, I, 93.

sulcifrons, Sharp, Ann. Mag. Nat. Hist. (6), III, 1889, 39.

89. **Cafius,** Steph.

Steph., Ill, Brit. Ent., V, 245.

ceylonicus, Bernh., D. E. Z., 1902, 29. — Ceylon.

nauticus, Fairm., Rev. Zool., 1849, 288. — Ceylon.

densiventris, Fauv., Ann. Mus. Civ. Gen., X, 1877, 258. — ————

longipennis, Walk, Ann. Mag. Nat. Hist. (3), II, 1858, 205; Kr., Arch. Naturg., XXV, 1859, I, 99. — E. Indies generally. China., Japan. Polynesia.

parallelus, Kr, *id.,* 99. — E. Africa.

puncticollis, Bohem, Eugen. Resa, 31.

90. **Hesperus,** Fauv.

Fauv., Fn. Gallo-Rhen., III, 426.

Feae, Fauv, Rev. d'Ent., XIV, 1895, 260. — Burma.

inaequalis, Fauv., *id.,* 258. — Burma

laevigatus, Fauv., *id.,* 259. — Burma.

trimerus, Fauv. *id.,* 261. — Burma.

wasmani, Fauv., *id.,* 260. — Burma., Dehra Dun.

91. **Anisolinus,** Sharp.

Sharp, Ann. Mag. Nat. Hist. (6), III, 113.

auronotatus, Fauv., Rev. d'Ent. XIV, 1895, 270. — Burma.

fasciiventris, Fauv., *id.,* 270. — Burma., Pegu. Bomanhalli.

92. **Tolmerus,** Bernh.

Bernh., Ent. Blatt., VII, 1911, 89.

rufipennis, Fauv., Rev. d'Ent., XIV, 1895, 271. — Burma.

93. **Amichrotus,** Sharp.

Sharp, Ann. Mag. Nat. Hist. (6), III, 1889, 114.

costatus, Fauv., Rev. d'Ent., XIV, 1895, 269. — Burma.

elegans, Cam., E. M. M., LVI, 1920, 218. — Sidapur.

semicyaneus, Fauv., Rev. d'Ent., XIV, 1895, 269. Tenasserim.

94. Belonuchus, Nord.

Nord., Symbol., 1837, 129.

aeneipennis, Fauv., Rev. d'Ent., XIV, 1895, 268. Nilgiris., Burma.

grandiceps, Kr., Arch. Naturg., XXV, 1859, I, 95. Ceylon., Kandy.

bicolor, Motsch., Et. Ent., VIII, 1859, 79.

maculipennis, Fauv., Rev. d'Ent., XIV, 1895, 267. Burma.

nalandensis, Bernh., D. E. Z., 1902, 30. Ceylon.

quadratus, Kr., Arch., Natrug., XXV, 1859; I, 96. Calcutta.

rufoniger, Fauv., Rev. d'Ent., XIV, 1895, 266. Burma.

tripunctatus, Fauv., *id.,* 267. Burma.

95. Saniderus, Fauv.

Fauv., Rev., d'Ent., XIV, 1895, 257.

ruficollis, Fauv., Rev. d'Ent. XIV, 1895, 257. Burma.

96. Phucobius, Sharp.

Sharp, Tr. Ent. Soc. Lond., 1874, 35; Cam., E. M. M., LV, 1919, 207.

Orthidus, Muls. & Rey, Ann. Soc. Agric. Lyon (4), VIII, 1875, 339.

congruus, Walk., Ann. Mag. Nat. Hist. (3), II, 1858, 205. Ceylon.

punctilinea, Walk., *id.,* 205.

horni, Bernh., D. E. Z., 1902, 28.

97. Staphylinus, L.

L., **Syst. Nat., 1758, 421.**

aeneicollis, Bernh., Ent. Blatt., VII, 1911, 87. Masuri.

asemus, Kr., Arch. Naturg., XXV, 1859, I, 177. N. India.

auronotatus, Fauv., Rev. d'Ent., XIV, 1895, 256. Burma.

basicornis Fauv. *id.,* 256. Burma.

bengalensis, Bernh., W. Z. B., 1914, 100. Bengal.

birmanus, Fauv., Rev. d'Ent., XIV, 1895, 255. Burma.

brachypterus, Kr., Arch. Naturg., XXV, 1859, I, 76.	N. India.
bryanti, Cam., E. M. M., 1918, 104.	Ceylon.
chrysostigma, Fauv., Rev. d'Ent., XIV, 1895, 252.	Burma.
circumcinctus, Bernh., W. Z. B. 1914, 102.	East Indies.
curticornis, Fauv., Rev. d'Ent., XIV, 1895, 256.	Burma.
decipiens, Kr., Arch., Naturg., XXV, 1859, I, 75.	Ceylon., Kandy.
gemmatus, Fauv., Rev. d'Ent. XIV, 1895, 251.	Burma.
goryi, Casteln., Et. Ent., I, 1835, 113; Er., Gen. Spec. Staph., 595.	Deccan., N. India.
auripennis, Kraatz, Arch. Naturg., XXV, 1859, I, 72.	
indicus, Kr., *id.,* 75.	Nilgiris.
lineatus, Walk., Ann. Mag. Nat. Hist. (3), III, 51.	Anamalai., Ranchi. Ceylon., Dehra Dun.
maculicollis Fauv., Rev. d'Ent., XIV, 1895, 253.	Soondrijal., Nepal. Burma.
maculipennis, Kr., Arch. Naturg., XXV, 1859, I, 77.	Birbhum. Dehra Dun.
marmorellus, Fauv., Rev. d'Ent., XIV, 1895, 253.	Burma.
purpurescens, Cam., E. M. M., LVI, 1920, 217.	S. Coorg.
semipurpureus, Kr., Arch. Naturg., XXV, N. India 1859, I, 73.	N. India.
subirideus. Kr., *id.,* 78.	East Indies.
submarmorellus, Schub., D. E. Z., 1908, 613.	Himalayas., Kulu.
suspectus, Fauv., Rev d'Ent., XXIII, 1904, 57.	Nilgiris., Mahé. Matheran.
trimaculatus, Fauv., *id.,* XIV, 1895, 252.	Burma., Kurseong.
virgulatus, Fauv., *id.,* 254.	Burma.
xanthocephalus, Kr., Arch. Naturg., XXV, 1859, I, 74.	Kashmir., Chamba.
bomfordi, Epp., W. E. Z., XIV, 1895, 56.	Sikkim., Deoban. Fagu., Masuri.

98. **Ocypus,** Steph.

Steph., Ill, Brit. Ent., V. I., 1832, 211.

pendjabensis, Bernh., Col. Rundsch., 1915 (3) 53.	Shahpur.
wasmanni, Bernh., Arch. Naturg., 84, A-10, 1918, 184 (*Tasgius*).	Sikkim.

sikkimensis, Bernh, *id.*, 185 (*Tasgius*). Sikkim.

Sub-gen. *Parocypus,* Bernh., Col.

Rundsch., 1915 (3), 52.

dehradunensis, Bernh., Col. Rundsch., 1915 (3), 52. Dehra Dun.

99. **Ontholestes,** Ganglb.

Ganglb., Käf. Mittel-Eur., II, 417.

Leistotrophus, Kr., Naturg, Ins. Deutsch. II, 532 (nec. Perty.)

aurosparsus, Fauv., Rev. d'Ent. XIV, 1895, 245. Burma.

birmanus, Fauv., *id.*, 246. Burma.

marmoratus, Er., Gen. Spec. Staph., 826. Masuri., Ceylon.
————, Java.

tenuicornis, Kr., Arch. Naturg., XXV, 1859, I, 68. Ghoom., Chakrata.

100. **Eucibdelus,** Kr.

Kr., Arch. Naturg., XXV, 1859, I, 70.

angusticeps, Bernh., Arch. Naturg., 84, A-10, 1918, 186. Sikkim.

bicolor, Fauv., Rev d'Ent., XIV, 1895, 245. Burma.

feae, Fauv., *id.*, 247. Burma.

gracilis, Kr., Arch. Naturg., XXV, 1859, I, 71. N. India.

varius, Fauv., Rev. d'Ent., XIV, 1895, 24. Burma.

101. **Rhyncocheilus,** Sharp.

Sharp, Ann. Mag. Nat. Hist. (6), III, 1889, 120.

andrewesi, Cam., E. M. M., 1918, 105. Nilgiris.

argenteus, Fauv., Rev. d'Ent., XIV, 1895, 249. Burma.

aureus, F., Mant. Ins., I, 219; Er., Gen. Spec., Staph., 344; Kr., Arch. Naturg., XXV, 1859, I, 74 (note). Burma., Kanara. Sibsagar., Assam. Ceylon., Tenasserim.

chrysites, Fauv., Rev. d'Ent., XIV, 1895, 248. Burma.

kraatzi, Epp., D. E. Z., 1895, 392. Pegu.

limbatus, Fauv., Rev. d'Ent., XIV, 1895, 250. Burma.

102. **Trichocosmetes,** Kr.

Kr., Arc. Naturg., XXV, 1859, I, 69.

eucomus, Er., Gen. Spec. Staph., 362; Kr., Archiv. Naturg., XXV, 1859, I. 70. — Sikkim., Nepal.

prainae, Epp., W. E. Z., XIV, 1895, 58. — Sikkim.

103. **Seleucus,** Fauv.

Fauv. Rev. d'Ent., XXII, 1903, 157.

mutillarius, Er., Gen. Spec. Staph., 924. — Bengal.

waageni, Epp., D. E. Z., 1895, 396. — Calcutta.

104. Palaestrinus, Er.

Er. Gen. Spec. Staph., 343.

sykesi, Er., Gen. Spec. Staph., 344; Kr., Arch. Naturg., XXV, 1859, I, 71. — East Indies.

var. *Kraatzi,* Fauv., Rev. d'Ent., XXIII, 1903, 156. — Ceylon.

105. **Menoedius,** Fauv.

Fauv., Rev. d'Ent., XXII, 1903, 155.

andrewesi, Fauv., *id.,* 156. — Nilgiris., Coonoor.

106. **Naddia,** Fauv.

Fauv., Col. Heft II, 1867, 117.

Caranistes, Er., Gen. Spec. Staph., 925 (nec. Schönh.).

miniata, Fauv., Rev. d'Ent., XIV, 1895, 250. — Tenasserim., Thagata.

rufipennis, Bernh., Col. Rundsch., 1915 (3), 54. — Upper Burma.

westermanni Er., Gen. Spec. Staph., 925. — Bengal., Kurseong.

107. **Physetops,** Mann.

Mann., Brach., 1830, 32.

transcaspius, Bernh., Soc. Ent., XXV, 1910, 74. — Nushki., Baluchistan.
———, Transcaspia.

108. **Wasmannellus,** Bernh.

Bernh., Arch. Naturg., 84, A-10, 1918, 186.

tristis, Bernh., *id.,* 187. — Sikkim.

109. **Barygnathus,** Bernh.

Bernh., D. E. Z., 1902, 31.

opacus, Bernh., *id.,* 31. — Ceylon.

2. SUB-TRIBUS *XANTHOPYGI.*

Xanthopigina, Sharp., B. C. A. I., 2, 342 (+*Creophilus, Mannerh.*)

110. **Tympanophorus,** Nord.

Nord., Symb. 9. (*Diaphoetes* Waterh. Proc. Lin. Soc., Lond., 1884, 214).

rugosus, Waterh., Proc. Lin. Soc., London, 1884, 214. — Burma.

longiceps, Bernh., Tydsch., Ent. LVIII, 1915, 235. — Murree.

111. **Pammegus,** Fauv.

Fauv., Rev. d'Ent., XIV, 1895, 271.

andrewesi, Cam., E. M. M., LVII, 1921, 271. — Nilgiris.

flavipes, Fauv., Notes Leyden Mus., VI, 1884, 246. — Himalayas.

var. *ruficollis,* Fauv., Rev. d'Ent., XIV, 1895, 272. — Burma.

112. **Agacerus,** Fauv.

Fauv., Rev. d'Ent., XIV, 1895, 245.

Eurycerus, Fauv., *id.,* 244.

pectinatus, Fauv., *id.,* 245. — Burma.

113. **Creophilopsis,** Cam.

Cam., E. M. M., LVII, 1921, 272.

semiaeneus, Cam., *id.,* 273. — Burma.

114. **Creophilus,** Mannerh.

Mannerh., Brachel., 16.

maxillosus, L., Syst. Nat., ed. X, 421. — Kogyar, Simla Hills. (Cosmopolitan.)

villipennis, Kr., Arch. Naturg., XXV, 1859, I, 67. — N. India., Deccan.

115. **Craspedomerus,** Bernh.

Bernh., Ent. Blätt., VII, 1911, 88.

glenoides, Schub., D. E. Z., 1908, 622 (*Philonthus*). — Himalayas, Kulu.

Tribus Quediini.

Ganglb. Käf. Mittel-Eur., II, 378.

116. **Heterothops,** Steph.

Steph., Ill, Brit., Ent., V, 256.

flavicollis, Motsch., B. M., XXXI, 1858, II. 660.	East Indies.
oculata, Fauv., Rev. d'Ent., XIV, 1895, 275.	Tenasserim.

117. **Algon,** Sharp.

Sharp., Tr. Ent. Soc., Lond., 1874, 22.

Securipalpus, Schub., D. E. Z. 1908, 613 (verisim).

mmsi, Bernh, Col. Rund., IV, 1915. 54.	Central Himalayas.
psittacus, Fauv., Rev. d'Ent., XIV, 1895, 273.	India.
rudepunctatus, Schub., D. E. Z., 1908, 613 (*Securipalpus*).	Himalayas., Kulu.
semiaureus, Fauv., Rev. d'Ent., XIV, 1895, 273.	Bengal., Kerseong. Kombir., Sikkim. Darjiling.
tricolor, Fauv., *id.,* 272.	Burma, Teinzo.

118. **Quedius,** Steph.

Steph., Ill. Brit. Ent., V. 214.

Sub-gen. *Quedionuchus,* Shp., B. C. A. I., 2, 1884, 136.

Sub-gen. *Microsaurus,* Muls., and Rey, Ann. Soc. d'Agric., Lyon, (4), VIII, 1875, 685.

Sub-gen. *Raphirus,* Steph., Ill, Brit. Ent., V, 241.

Sub-gen *Sauridus,* Muls., and Rey, Ann. Soc. d'Agric., Lyon, (4), VIII, 1875, 700.

apicicornis, Epp., D. E. Z., 1895, 391, (*Microsaurus*).	Sikkim.
aureiventris, Bernh., Col. Rund., IV, 1915, 57, (*Raphirus*).	Central Himalayas.
bipunctatus, Epp., D. E. Z., 1895, 390, (*Raphirus*).	Paresnath.
chlorophanus, Er., Gen. Spec. Staph., 532.	Bengal.
conicus, Champ., E. M. M., LVIII, 1922, 33, (*Riphirus*).	Kumaon.
cruentus, Ol., Ent. III, 42, (*Microsaurus*) *fulgidus,* Marsh., Ent. Brit., 503.	East Indies,______ Europe.
filicornis, Epp., W. E. Z., XIV, 1895, 55.	Sikkim.
fulgidus, F., Mant. Ins., I, 220, (*Microsaurus*).	N. India, ________ Europe., North America. New Zealand.

himalayicus, Bernh., Col. Rundsch., IV, 1915, 55, (*Sauridus*).	Central Himalayas. Masuri., Chaktrata.
intricatus, Fauv., Rev. d'Ent., XIV, 1895, 274.	Burma.
nilgiriensis, Cam., E. M. M., LVI, 1920, 219, (*Quedionuchus*).	Nilgiris.
ochripennis, Men., Cat. Rais., 145.	Narkanda.
oculatus, Fauv., Rev. d'Ent., XIV, 1895, 274.	Burma.
rugosus, Cam, E. M. M., LVII. 1921, 273, (*Raphirus*).	Lebong.
spectabilis, Kr., Arch. Naturg., XXV 1859, I, 66.	N. India
taprobanus, Cam., E. M. M., 1918, 171, (*Raphirus*).	Ceylon.

119. Acylophorus, Nordm.

Nordm., Symb., 127.

Rhygmacera, Motsch., B. M., XVIII, 1845, I, 40.

bipunctatus, Cam., E. M. M., LVI, 1920, 219.	Nilgiris.
flavipes, Motsch., B. M., XXXI, 1858, I, 657.	East Indies.
furcatus, Motsch., B. M., XXXI, 1858, I, 658.	East Indies.
microceros, Fauv., Rev. d'Ent., XV, 1895, 276.	Burma.
puncticeps, Fauv., *id.,* 275.	Burma.
ruficollis, Motsch., B. M., XXXI, 1858, II, 657.	East Indies. Ceylon.

120. Anchocerus, Fauv.

Fauv., Rev. d'Ent., XXIV, 1905, 141.

birmanus, Fauv., *id.,* 141.	Burma. Tenasserim.

121. Atanygnathus, Jacobson.

Jacobson., Käf., VII, 521.

Tanygnathus, Er., Käf. Mark. Brand., I, 417.

brevicollis, Fauv., Rev. d'Ent., XIV, 1895, 276.	Burma, Siwaliks.
fuscus, Kr., Arch. Naturg., XXV, 1859, I, 65.	East Indies.
piceus, Motsch., B. M., XXVI, 1853, III, 213.	East Indies.
terminalis, Er., Käf. Mark. Brand., 418.	Khandesh.
var. *pictus,* Motsch., B. M., XXVI, 1853, III.	Dehra Dun. Pondicherry.

var. *ruficollis,* Kr. Arch. Naturg., XXV, 1859, I, 64.. Genji., Ceylon.
Calcutta., ———
Europe., N. Africa.
Malay Peninsula.

Sub-fam. TERMITODISCINAE

Wasm., Ziet. für. Wiss Zool., C. I., 1912, 91.

122. **Discoxenus,** Wasm.

Wasm., Zool, Jahrb., Supp. VII, 1904, 655.

acuticornis, Wasm., Zool. Jahrb., XXXIX, 1916, 179. Ceylon.

assmuthi, Wasm., *id.,* Supp. VII, 1904, 656. Khandala, Bombay.

crassicornis, Wasm., *id.,* XXXIX, 1916, 179. Ceylon.

lepisma, Wasm., *id.,* Supp. VII, 1904, 656. Ahmednagar.

123. **Termitodiscus,** Wasm.

Wasm., D. E. Z., 1899, 147.

butteli, Wasm., Zool. Jahrb. XXXIX, 1916, 182. Ceylon.

escherichi, Wasm. in Escherich Term. Leb., Ceylon, 1911, 231. Ceylon.

var. *picea,* Wasm., Zool. Jahrb. XXXIX, 1916, 182. Ceylon.

heimi, Wasm., D. E. Z., 1899, 147. Further India.
Chilka Lake.

Sub-fam. PYGOSTENINAE.

Wasm., Verh. Deutsch. Zool. Ges., 1902, 91-93.

Tribus Pygostenini.
Fauv., Rev. d'Ent., XVIII, 1899. 5.

124. **Doryloxenus,** Wasm.

Wasm., W E. Z., XIX, 1898, 101.

Mimocoete, Fauv., Rev. d'Ent., XVIII, 1899, 7.

Phocasoma, Kr., D. E. Z., 1899, 363.

ceylonicus, Wasm., Zool. Jahrb. XXXIX, 1916, 175. Ceylon.

butteli, Wasm., *id.,* 175. Ceylon.

eutermitis, Was., *id.,* 176. Khandala, Bombay.

phocus, Fauv., Rev. d'Ent., XVIII, 1899, 10. Ceylon.

peradenyiae, Wasm., Zool. Jahrb. XXXIX, 1916, 175. Ceylon.

splendidus, Wasm., *id.,* 175. Ceylon.

termitophilus, Wasm., Zool. Jahrb., Supp. VII, 1904, 655. Khandala, Bombay.

transfuga, Wasm., *id.,* 655. Khandala, Bombay.

Sub-fam. TACHYPORINAE.
Ganglb., Käf. Mittel-Eur., II. 329.

Tribus Bolitobiini.
Ganglb. Käf. Mittel-Eur., II. 358.

125. **Mycetoporus,** Mannerh.
Mannerh., Brach., 62.

braminus, Motsch., B. M., XXXI, 1858, III, 215. East Indies.

fasciatocollis, Champ., E. M. M., LVIII, 1922, 32. W. Almora.

quadriguttatus, Champ., *id.,* LIX, 1923, 47. Kumaon.

splendidus, Gr., Col. Mic., 24. East Indies.

testaceus, Kr., Arch. Naturg., XXV, 1859, I, 64. ————, Europe. N. America.

126. **Bryoporus,** Kr.

Kr., Naturg. Ins. Deutsch., II, 452.

lineipennis, Fauv., Rev. d'Ent., XXIII, 1904, 60, (*Megacronus*). Himalayas., Simla. Nilgiris. Kashmir.

127. **Bolitobius,** Mannerh.
Mannerh., Brach., 11.

distinctus, Schub., D. E. Z., 1906, 381. Simla Hills.

indicus, Bernh., Col. Rundsch., 1917, (7-9), 1. Madura.

lambda, Fauv., Rev. d'Ent., XIV, 1895, 277, (*Megacronus*). Burma.

nitidus, Motsch., B. M., XXXI, 1858, III, 215. Nilgiris., Ceylon.

bimaculatus, Kr., Arch. Naturg., XXV, 1859, I, 63.

septemseriatus, Champ., E. M. M., LIX, 1923, 46. Lansdowne, U.P.

spinipes, Champ., E. M. M., LVIII, 1922, 31. W. Almora.

vittula, Fauv., Rev. d'Ent., XIV, 1895, 277, (*Megacronus*). Burma.

xanthopterus, Champ., E. M. M., LVIII, 1922, 32. Kumaon.

Tribus Tachyporini.
Ganglb. Käf Mittel-Eur., II. 335.

128. Conosoma, Kr.

Kr., Naturg. Ins. Deutsch., II, 431.

Conurus, Steph., Ill. Birt. Ent., V, 188.

anale, Fauv., Rev. d'Ent., XIV, 1895, 283.	Tenasserim.
antennarium, Bernh., D. E. Z., 1902, 25.	Ceylon.
biguttatum, Motsch., B. M., XXXI, 1858, III, 222.	East Indies.
distigma, Kr., Arch. Naturg., XXV, 1859, I. 61.	
birmanum, Fauv., Rev. d'Ent., XIV, 1895, 284.	Nilgiris., Burma. Tenasserim.
breve, Fauv. *id.,* 284.	Anamalai., Burma.
brevipenne, Motsch., Et. Ent., VIII, 859, 82.	Ceylon.
ceylanense, Kr., Arch. Naturg., XXV, 1859 I, 62.	Ceylon., Burma. Tenasserim.
dimerum, Fauv., Rev. d'Ent., XIV, 1895, 285.	Burma.
fascipenne, Epp., D. E. Z., 1895, 389.	Pegu.
flavofasciatum, Bernh., Ent. Blätt., XI, 1915, 255.	S. India.
gracile, Kr., Arch. Naturg., XXV, 1859, I, 62.	Ceylon.
hemisphaericum, Bernh., Col. Rundsch., IV, 1915, 57.	E. Himalayas.
immaculatum, Steph., Ill, Brit. Ent., V, 190.	East Indies.
cinctum Motsch., B. M., XXXI, 1858, III 223.	Europe.
kaschmiricum., Bernh., Col. Rundsch, IV, 1915, 58.	W. Himalayas.
montanum, Cam., E. M. M., LV, 1919, 255.	Ceylon.
notatum, Fauv., Rev. d'Ent., XIV, 1895, 283.	Burma.
pictipenne, Kr., Arch. Naturg., XXV, 1859, 1, 63.	Ceylon.
plagiatum, Favu., Rev. d'Ent., XIV, 1895, 282.	Burma.
rufum, Kr., Arch. Naturg., XXV, 1859, I, 63.	N. India.
suave, Fauv. Rev. d'Ent., XIV, 1895, 284.	Tenasserim.
virgula, Fauv., *id.,* 282.	Burma.

129. **Tachyporus,** Gr.

Gr., Col. Mic. 1.

dslutus, Motsch., Et. Ent., VIII, 1859, 80. — Ceylon.

flavopictus, Fauv., Rev. d'Ent., XIV, 1895, 282. — Darjiling., Burma.

hypnorum, F., Syst. Ent., 1775, 266. — Kashmir., Sind. Burma., Dehra Dun. ————, Europe.

montanus, Bernh., Col. Rundsch., IV, 1915, 58. — W. Himalayas.

himalayicus, Bernh., Arch. Naturg., 87, 1918 (1920), 187. — Himalayas.

130. **Tachinus,** Gr.

Gr., Col. Mic., 134.

birmanus, Fauv., Rev. d'Ent., XIV, 1895, 278. — Burma.

scorteus, Schub., D. E. Z., 1908, 612. — Jalaori, Himalayas.

stoliczkae, Shp., Jl. As. Soc., Bengal, (2), XLVII, 170. — Pamir.

131. **Tachinomorphus,** Kr.

Kr. Arch. Naturg., XXV, 1859, I, 54.

angularis, Fauv., Rev. d'Ent., XIV, 1895, 278. — Burma.

ceylonicus, Bernh., D. E. Z., 1902, 23. — Ceylon, Kandy.

fulvipes, Er., Gen. Spec. Staph., 921; Kr., Arch. Naturg., XXV, 1859, I, 55. — East Indies.

sanguinolentus, Motsch., B. M., XXXI, 1858, III, 216.

vitattus, Kr., Arch Naturg., XXV, 1859, I, 55.

132. **Tachinoderus,** Motsch.

Motsch., B. M., XXXI, 1858, II, 217.

longicornis, Motsch., *id.,* 218. — East Indies.

133. **Olophrinus,** Fauv.

Fauv., Rev. d'Ent., XIV, 1895, 280.

striatus, Fauv., *id.,* 281. — Burma.

134. **Coproporus,** Kr.

Kr., Naturg. Ins. Deutsch. II, 1856-1858, 399 (note)

Erchomus, Motsch., B. M., XXXI, 1858, II, 218.	
atomus, Kr., Arch. Naturg., XXV, 1859, I, 58.	Ceylon.
brunneicollis, Motsch., B. M., XXXI, 1858, II, 220.	Nilgiris., Sikkim.
punctipennis, Kr., Arch. Naturg., XXV, 1859, I, 57.	Burma., Pegu.
castaneipennis, Kr., *id.,* 60.	Ceylon., Dehra Dun.
cuprinicollis, Fauv., Rev. d'Ent., XIV, 1895, 279.	Tenasserim.
fasciipennis, Kr., Arch. Naturg., XXV, 1859, I, 59.	Ceylon.
flavicornis, Kr., *id.,* 56.	East Indies.
holisinus, Fauv., Rev. d'Ent., XIV, 1895, 279.	Burma.
laevigatus, Motsch., B. M., XXXI, 1858, II. 219.	East Indies.
latus, Motsch., *id.,* 219.	
spectabilis, Kr., Arch. Naturg., XXV, 1859, I, 55.	East Indies.
melanarius, Er., Gen. Spec. Staph., 252.	Nilgiris.
subdepressus, Kr., Arch. Naturg., XXV, 1859, I, 57.	
subpunctulatus, Motsch., Et. Ent., VIII, 1859, 81.	
minimus, Motsch., B. M., XXXI, 1858, II, 220.	Nilgiris., Ceylon.
granulum, Motsch., B. M., XXXI, 1858, II, 221.	Burma., Pondicherry.
orientalis, Solsky, B. M., XXXVI, 1864, II, 439.	
pygmaeus, Kr., Arch. Naturg., XXV, 1859, I, 58.	
rubiginosus, Motsch., B. M., XXXI, 1858, II, 221.	
tantillus, Motsch., *id.,* 221.	
planus, Fauv., Rev. d'Ent., XXIV, 1905, 142.	Burma.
rufus, Kr., Archiv. Naturg., XXV, 1859, I, 59.	East Indies.
fulvus, Motsch., B. M., XXXI, 1858, II, 220.	
sanguinolentus, Motsch., *id.,* 220.	East Indies.
tachyporoides, Kr., Arch. Naturg., XXV, 1859, I, 60.	Ceylon.
triangulus, Cam., E. M. M., LVI, 1920, 49.	Ceylon.

135. Leucoparyphus, Kr.

Kr., Arch. Naturg. Ins. Deutsch. II, 393.

silphoides, L., Syst. Nat., I, 2, 684.	Ceylon.

var. *limbifer,* Motsch., B. M., XXXI, 1858, II, 222. — Belgaum., Burma. Dehra Dun.

Tribus Symmixini.
Bernh., Col. Rundsch, IV, 1915, 56. — (Cosmopolitan).

136. Symmixus, Bernh.

Bernh., Col. Rundsch., IV, 1915, 56.

sikkimensis, Bernh., Col. Rundsch., IV, 1915, 56. — E. Himalayas.

Tribus Hypocyptini.
Ganglb. Käf Mittel-Eur., II, 330.

137. Hypocyptus, Mannerh.

Mannerh., Brach., II.

glaber, Kr., Arch. Naturg., XXV, 1859, I, 53. — Ceylon.

Sub-fam. ALEOCHARINAE.

Tribus Deinopsini.
Sharp, B. C. A., 1 (2), 294.

138. Deinopsis, Matth.

Matth., Ent. Mag., V, 193.

cinnamomea, Kr., Arch. Naturg., XXV, 1859, I, 51. — Ceylon.

Tribus Oligotini.

139. Oligota, Mannerh.

Mannerh., Mem. Ac. Sc., St. Peters., I, 486.

Sub-gen. *Holobus,* Sol. in Gay. Nat. Hist., Chil. Zool., IV, 335.

chrysopyga, Kr., Arch. Naturg., XXV, 1859, I, 45, (*Holobus*). — Ceylon., ——— New Caledonia.

pusillima, Gr., Col. Mic., 17, (*Oligota,* s. str.). — Calcutta., ———

indica, Motsch., B. M., XXXI (3), 236. — Europe.

indica, Kr., Arch. Naturg., XXV, 1859, I, 44. — N. & S. America.

testacea, Kr., Arch. Naturg., XXV, 1859, I, 44, (*Oligota,* s. str.). — Ceylon.

semibrunnea, Kr., *id.,* 45, (*Oligota,* s. str.). — Ceylon.

Tribus Termitoptochini.

140. Termitoptochus, Silv.

Silv., Boll. Lab. Zool., Port., V, 37.

indicus, Silv., Boll. Lab. Zool., Port., V, 39. — Burma.

141. Termitotima, Wasm.

Wasm., Zool. Jahrb., XXXIX, 1916, 188 (Note).

assmuthi, Wasm., *id.,* 188 (note). — Khandala.

Tribus, Bolitocharini.
Ganglb., Käf Mittel-Eur., 2, 258.

142. Diestota, Rey.

Rey, Op. Ent., 14, 194.

subopaca, Cam., E. M. M. LVI, 1920, 50. — Ceylon.

testacea, Kr., Arch. Naturg., XXV, 1859, I, 7, (*Bolitochara*). — Ceylon., ————

mayeti, Rey, Op. Ent., 14, 196. — Hawaii., France.

143. Linoglossa, Kr.

Kr., Arch. Naturg., XXV, 1859, I, 10.

bifoveolata, Kr., *id.,* 14. — East Indies.

144. Plagiusa, Bernh.

Bernh., Col. Rundsch., 1915 (2), 28.

Neosilusa, Cam., Tr. Ent. Soc., Lond., 1920, 232.

ceylonica, Kr., Linn. Ent., XI, 8, (*Stenusa*). — Nilgiris., Burma. Ceylon., ———— Malay Peninsula.

145. Eccoptogenia, Kr.,

Kr., Arch. Naturg., XXV, 1859, I, 8.

rufa, Kr., *id.,* 9. — Ceylon.

146. Coenonica, Kr.

Kr., Linn. Ent., XI, 45; Cam., E. M. M., LV, 1919, 230.

puncticollis, Kr., Linn. Ent., XI, 46. — Ceylon.

indica, Motsch., B. M., 1858, III, 258. — Burma., Dugeli., Nagargali.

————

Malay Archipelago W. Indies

147. Homalota, Mannerh.

Mannerh., Mem. Ac. Sci., St. Peters., I, 487.

Epipeda, M. and R. Hist. Nat. Col., Fr., V, 136.

crenulata, Fauv., Rev. d'Ent., XIII, 1904, 65. — Nilgiris.

nigrescens, Fauv., *id.,* XIV, 1905, 147. — Pusa.

fuscipennis, Cam., Tr. Ent. Soc. Lond. 1920, 24. — ————, Java, etc.

platygaster, Kr., Arch. Naturg., XXV, 1859, I, 33. — Ceylon., Pusa.

tuberculicollis, Kr., Arch. Naturg., XXV, 1859, I, 33. — Calcutta., Ceylon.

variventris, Kr., *id.,* 34. — Ceylon.

poetinalis, Fauv., Ann. Mus. Civ. Gen., XII, 1878, 298.

148. Peliusa, Er.

Er., Gen. Spec. Staph., 192.

pallescens, Motsch., B. M., XXXI, 3, 226. — East Indies

149. Placusa, Er.

Er., Gen. Spec. Staph., 194; Ganglb., Käf. Mittel-Eur., II, 295.

acuminata, Kr., Arch. Naturg., XXV, 1859, I, 43. — Ceylon., Chakrata. N. W. Himalayas.

intermedia, Cam., E. M. M., LVII, 1921, 274. — Nilgiris.

pygmaea, Kr., Arch. Naturg., XXV, 1859, I, 44. — Ceylon.

spinigera, Kr., *id.,* 43. — Ceylon.

150. Brachida, Muls. & Rey.

Muls. & Rey, Brevip., V, 4

crassiuscula, Kr, Arch. Naturg., XXV, 1859, I, 41, (*Homalota*). — Ceylon.

151. Neobrachida, Cam.

Cam., E. M. M., LVI, 1920, 51.

castanea, Cam., E. M. M., LVI, 1920, 51. — Ceylon.

152. **Pseudobrachida,** Cam.

Cam., E. M. M., LVI, 1920, 51.

nigriventris, Cam., E. M. M., LVI, 1920, 52.	Ceylon.

153. **Gyrophaena,** Mannerh.

Mannerh., Mem. Ac. Sci., St. Peters., I, 488. Sub-gen. *Orphnebividea,* Schub., D. E. Z., 1908, 611.

appendiculata, Motsch., B. M., XXXI (3), 228.	East Indies.
laminata, Kr., Arch. Naturg., XXV, 1859, I, 45.	
rosti, Schub., D. E. Z., 1908, 611.	Simla Hills. Himalaya., Kulu.

Sub-gen. *Gyrophaena,* s. str.

Ganglb., Käf Mitt., 2, 300.

cicatricosa, Motsch., B. M., XXXI, (3) 1858, 231.	W. Himalayas.
granulifera, Kr., Arch. Natrug., XXV, 1859, I, 48.	Ceylon.
furcata, Motsch., B. M., XXXI, (3) 1858, 228.	East Indies.
indica, Motsch., *id.,* 230.	
humeralis, Kr., Arch. Naturg., XXV, 1859, I, 46.	Ceylon.
immatura, Kr., *id.,* 47.	Ceylon.
kashmirensis, Bernh., Col. Rundsch., X, 1922, 127.	Uri.
livida, Motsch., B. M., XXXI, (3) 1858, 231.	East Indies.
pygmaea, Kr., Arch. Naturg., XXV, 1859, I, 49.	Ceylon.
oxyteloides, Motsch., Et. Ent., VIII. 85.	Ceylon.
pallipes, Kr., Arch. Naturg., XXV, 1859, I, 47.	Ceylon.
permutaria, Schub., D. E. Z., 1906, 380.	Kashmir.
puncticollis, Schub., *id.,* 381.	
rigida, Motsch., B. M., XXXI, (3) 1858, 230.	East Indies.
rufiventris, Cam., E. M. M., LVI, 1920, 50.	Ceylon.
soror, Bernh., Col. Rundsch., X, 1922, 127.	Uri.
trifida, Motsch., Et. Ent., VIII, 84.	Ceylon.
tuberculiventris, Bernh., Col. Rundsch., 1915, (3), p. 58.	Kashmir., Simla Hills.

Sub-gen. *Phaenogyra.*
M. & R., Brevip., V, 76.

curtula, Motsch., Et. Ent., VIII, 85. Ceylon.

nigra, Kr., Arch. Naturg., XXV, 1859, I, 49.

nigra, Motsch, Et. Ent., VIII, 85. Ceylon.

154. Hygroptera, Motsch.

Motsch., Et. Ent., VIII, 86.

castanea, Motsch., B. M., XXXIV, 1860, 150. Ceylon.

termitis, Motsch., *id.,* 87. Ceylon.

155. Autalia, Mannerh.

Mannerh., Mem. Ac. Sci., St. Peters., I, 43?

angustata, Motsch., B. M., XXXI, (3) 1858, 261. East Indies.

riparia, Motsch., Et. Ent., VIII, 93. India.

156. Bolitochara, Mannerh.

Mannerh., Brach., 75.

amabilis, Motsch., B. M. 1861, I, 154. Ceylon.

semiaspera, Fauv., Rev. d'Ent., 1904, 65. Nilgiris.

157. Eustenia, Fauv.

Fauv., Rev. d'Ent., 1905, 145.

opacicollis, Fauv. Rev. d'Ent., 1905, 146. Burma.

158. Leptusa, Kr.

Kr., Naturg., Ins. Deutsch., II, 1857, 60.

annuliventris, Kr., Arch. Nat., 1859, I, 12. East Indies.

varicornis, Kr., *id.,* 13. Ceylon.

159. Tachychara, Cam.

Cam., E. M. M., LVI, 1920, 52.

discipennis, Cam., E. M. M., LVI, 1920, 53. Ceylon., ———— Boreno.

Tribus Myllaenini.

160. Camacopselaphus Gemm. & Har.

Gemm. & Har., Cat., II, 549.

Camacopalpus, Motsch., B. M., 1858, III, 231.

bituberculatus, Motsch., B. M., 1858, III, 233, T. I, figr. East Indies.

flavicornis, Motsch., B. M., 1858, III, 233. East Indies.

fulvus, Motsch. *id.*, 234. East Indies.

161. Myllaena, Er.

Er., Käf. Mark. Brand., I, 382.

apicalis, Kr., Arch. Nat., XXV, 1859, I, 51. Ceylon.

lateritia, Kr., *id.*, 50. Pondicherry, Genji. Ceylon.

nitidula, Kr., *id.*, 50. Ceylon.

picticornis, Fauv., Rev. d'Ent., XIV, 286. Burma.

regina, Fauv., *id.*, 286. Burma.

Tribus Pronomaeini.

162. Pronomaea, Er.

Er., Käf. Mark. Brand., I, 378.

bramina, Motsch., B. M., 1858, III, 224. East Indies.

subrufa, Motsch., *id.*, 1861, I, 149. Ceylon.

Tribus MYRMEDONIINI

163. Schistogenia, Kr.

Kr., Linn. Ent., XI, 1857, 39.

crenicollis, Kr., Linn. Ent., 1857, 40. Khandesh., Ceylon. Dehra Dun.

Malay Peninsula,

164. Thamiaraea, Thoms.

Thoms., Ofv. Vet. Ac. Forh., 1858, 35.

armata, Fauv., Rev. d'Ent. XXII, 161. Coonoor.

basiventris, Bernh., Col. Rundsch., 1916, (1-3) 22. Tenasserim.

insigniventris, Fauv., Ann. Mus. Civ., Gen., XII, 299. Ceylon.

---

miriventris, Cam., Tr. Ent. Soc., Lond., 1920, 259, (*Atheta*). New Guinea, etc. Malay Peninsula.

165. **Amaurodera,** Fauv.

Fauv., Rev. d'Ent. XXIV, 142.

bomfordi, Epp., W. E. Z., IX, 1890, 273, (*Drusilla*), Schub., D. E. Z., 1908, 611. — Simla.

opacicollis, Epp., D. E. Z., 1895, 385, (*Astilbus*). — Sikkim.

veluticollis, Motsch., B. M., 1858, III, 261. — East Indies.

166. **Cardiola,** Muls, & Rey.

Muls. & Rey. Brevip., IV, 452.

obscura, Gr., Col. Micr., 74. — E. India., ——— Europe. N. America.

vestita, Boh., Eugen. Res., 1858, Ins. 25, (*Falagria*). — Dehra Dun. ———, China. Malay Peninsula.

167. **Falagria,** Mannerh.

Mannerh., Brach., 14.

Sub-Gen. *Anaulacaspis,* Ganglb., Käf. Mittel-Eur., II, 255.

asperata, Cam., E. M. M., LVI, 1920, 95. — Ceylon.

ceylonicus, Cam., E. M. M., LVI, 1920, 94. — Ceylon.

dimidiata, Motsch., B. M., 1858, III, 260. — Sikkim.

flavocincta, Kr., Arch. Nat., XXV, 1859, I, 4. — Nilgiris., Tenasserim. Ceylon., Burma.

gracilis, Motsch., B. M., 1858, III, 260, (*Anaulacaspis*).

madurensis, Bernh., Ent. Blätt., 1915, (10-12), 256. — Madura.

nilgiriensis, Fauv., Rev. d'Ent., XXIII, 62. — Nilgiris.

opacicollis, Kr., Arch. Nat., XXV, 1859, I, 4. — Ceylon.

parva, Kr., *id.*, 6, (*Anaulacaspis*). — Ceylon

pygmaea, Kr., *id.*, 7, (*Anaulacaspis*). — Ceylon

subrugosa, Kr., *id.*, 5, (*Anaulacaspis*). — Ceylon

vilis, Kr., *id.*, 5, (*Anaulacaspis*). — Ceylon

168. **Tachyusa,** Er.

Er., Käf. Mark. Brand, I, 307.

bicolor, Schub., D. E. Z., 1906, 379. Kashmir.

cordicollis, Bernh., *id.*, 1902, 23. Ceylon.

lewisiana, Cam., E. M. M., LVI, 1920, 95. Ceylon.

169. **Gnypeta,** Thoms.

Thoms., Skand Col., III. 6.

elegans, Bornh., D. E. Z., 1902, 22. Ceylon.

170. **Aloconota,** Thoms.

Thoms., Skand. Col., III, 7.

peguana, Bernh., Col. Rundsch., 1915, (2) 30. Pegu.

suspiciosa, Motsch., Et. Ent., 1859, 90. Ceylon.

171. **Amischa,** Thoms.

Thoms., Oefv. Svens. Vet. Acad. Forh., 1858, 33.

lineola, Kr., Arch. Naturg., 1859, I, 27, (*Oxypoda*). Ceylon.

exigua, Kr., *id.*, 36.

172. **Atheta,** Thoms.

Thoms., Skand. Col., I, 39, III, 61.

Sub-gen. *Acrotona,* Thoms., Skand. Col., III. 35.

Sub-gen. *Atheta,* s. str., Ganglb. Käf. Mittel-Eur., II, 150.

Sub-gen. *Bessobia,* Thoms., Ocfv. Sv. Vet. Akad., Forh., 1858, 35.

Sub-gen. *Chaetida,* Muls. & Rey, Brevip., III. 212.

Sub-gen. *Coprothassa,,* Thoms., Skand. Col., III. 33.

Sub-gen. *Dimetrota,* Muls. & Rey Brevip., III, 401.

Sub-gen. *Dralica,* id., IV, 186.

Sub.-gen. *Metaxya, id.*, IV, 187.

Sub-gen. *Microdota, id.*, III, 295.

Sub-gen. *Philhygra, id.*, III, 308.

Sub-gen. *Taxicera, id.*, IV, 289.

acuticollis, Kr., Arch. Naturg., 1859, I, 28. tab. I, f. 10, (*Bessobia*). Ceylon.

anepsia, Cam., E. M., LVI, 1920, 95, (*Coprothassa*).	Ceylon.
annuliventris, Kr., Arch. Naturg., 1859, I, 40, (*Acrotona*).	Ceylon.
aterrima, Gr., Col. Micr., 83, (*Acrotona*).	Dehra Dun. ———, Europe.
atramentaria, Gyll., Ins. Suec. (II), I, 408, (*Dimetrota*).	Dehra Dun. ———, Europe. E. Africa.
binidens, Bernh., Ent., Blätt., 1915, (10-12), 258, (*Atheta*, s. str.).	Madura.
dilutipennis, Motsch., B. M., 1858, III, 252, (*Atheta*, s. str.).	Pondicherry. Bombay.
mucronata, Kr., Arch., Naturg., 1859, I, 29.	Coonoor., Ceylon. ———, West Indies. Canaries.
dubia, Kr., Arch. Naturg., 1859, I, 37, (*Dimetrota*).	Ceylon.
exasperata, Kr., *id.*, 32.	Thago, Simla Hills. Ceylon.
floralis, Bernh., Col. Rundsch., 1916, (1-3), 21 (*Taxicera*).	Tenasserim.
fungi, Gr., Col. Micr., 157, (*Acrotona*).	East Indies.
lugens, Motsch., B. M., 1858, III, 257.	———
rhyssoptera, Kr., Arch. Naturg., 1859, I, 38.	Europe
horni, Bernh., D. E. Z., 1902, 21.	Ceylon.
inornata, Kr., Arch. Naturg., 1859, I, 39, (*Acrotona*).	Ceylon.
inutilis, Kr., *id.*, 35, (*Microdota*).	Ceylon.
longicornis, Gr., Col. Micr., 87, (*Chaetida*).	? East Indies.
tropica, Motsch., B. M., 1858, III, 256.	———, Europe.
marcida, Er., Käf. Mark. Brand., I, 328, (*Dimetrota*).	? East Indies. ———
tenuicornis, Motsch., B. M., 1858, III, 250.	Europe.
microcephala, Motsch, *id.*, 256.	East Indies.
nana, Kr., Arch. Nat., 1859, I, 36.	Ceylon.
nilgiriensis, Fauv., Rev. d'Ent., 1904, XXIII, 62, (*Dimetrota*).	Nilgiris.
parva, Sahlb., Ins. Fenn., I, 380, (*Acrotona*)	Dehra Dun. ———, Europe.
pelioptera, Kr., Arch. Naturg., XXV, 1859, I, 30.	East Indies.

peregrina, Kr., *id.,* 39, (*Acrotona*).	Belgaum., Ceylon.
piceata, Schub., D. E. Z., 1906, 378, (*Metaxya*).	Kashmir.
picticollis, Bernh. Ent. Blätt., 1915, (10-12, 257, (*Atheta,* s. str.).	Madura.
platystethoides, Motsch., B. M., 1861, I, 152.	Ceylon.
prona, Motsch., *id.,* 1858, III, 252.	East Indies.
putridula, Kr., Arch. Naturg., XXV, 1859, 1, 35, (*Microdota*).	Ceylon.
rugatipennis, Kr, *id.,* 40, (*Acrotona*).	Ceylon.
scrobicollis, Kr., *id.,* 31. (*Dralica*).	East Indies.
cava, Fauv., Fn. Gall. Rhen., III, 738.	————
indica, Kr., Arc. Naturg., 1859, I, 42.	S. Europe.
sinuatocrenata, Bernh., Col. Rundsch., 1916, (1-3), 22, (*Philhygra*).	Himalayas.
sordida, Marsh. Ent., Brit. 514, (*Coprothassa*).	Nilgiris, Himalayas. Ceylone., ____ Europe, N. America. S. Africa.
splendida, Kr., Arch. Naturg., XXV, 1859, I, 38.	Ceylon.
termitophila, Motsh., Et. Ent., 1859, 91.	Ceylon.
testaceipennis, Motsch., B. M., 1858, III, 251.	East Indies.
tridentata, Kr. Arch. Naturg., XXV, 1859, I, 31.	Ceylon. Dehra Dun.
tuberculata, Kr., *id.,* 32.	E. Indies.
vicaria, Kr., *id.,* 38, (*Acrotona*).	Simla Hills. Ceylon.

173. **Sipalia,** Muls., and Rey.

Muls., and Rey, Opusc. Ent., II, 45.

circellaris, Gr., Mon., 155.	East Indies.
cursor, Motsch., B. M., 1858, III, 251.	

174. **Astilbus** Steph.

Steph. Ill. Brit. Ent., V, 106.

incola, Fauv., Rev. d'Ent., 1904, XXIII, 63.	Nilgiris., Burma.
migratorius, Fauv., *id.,* 1903, XXII, 162.	Coonoor., Ceylon.

obliquus, Bernh., W. Z. B., LXVI, 1916, 427. Burma.

planaticollis, Kr., Lin. Ent., II, 1857, 51. Belgaum., Ceylon.

punctatissimus, Kr., *id.,* 52. Nilgiris., Ceylon.

tuberculatus, Bernh., W. Z. B., LXVI, 1916, 427. Burma.

175. Doratoporus, Wasm.

Wasm., D. E. Z., 1893, 207.

mendax, Wasm., D. E. Z., 1893, 207. Sikkim.

176. Jacobsonella, Silv.

Silv. Boll. Lab. Zool. Sc. Agric., Portici, V, 1910, 59.

termitobia, Silv., *id.,* 61. Burma

177. Orphnebius, Motsch.

Motsch., B. M., XXXI, 1858, 263.

bryanti, Cam., E. M. M., LVI, 1920, 97. Ceylon.

cingulatus, Cam., E. M. M., LVI, 1920, 96. Ceylon.

flaviventris, Champ., E. M. M., LVII, 1921, 179. Himalayas.

hauseri, Epp., W. E. Z., 1895, 54. Simla.

laticornis, Wasm., Ann. Mus. Civ. Gen., XXXVI, 624. Sikkim

178. Orphnebiota, Cam.

Cam., E. M. M., LVI, 1920, 97.

rufocastanea, Cam., *id.,* 98. Ceylon.

179. Pelioptera. Kr.

Kr., Linn. Ent., XI, 55.

Termitopora, Motsch., Et. Ent., VIII, 1859, 91.

micans, Kr., Linn. Ent., 1857, 56. Ceylon.

adjustipennis, Motsch., Et. Ent., 1859, 93, T. I, f. 7.

opaca, Kr., Linn. Ent., 1857, 56. Ceylon.

pernitida, Gam., E. M. M., LVI, 1920, 96. Ceylon.

180. Porus, Westw.

Westw. in Royle, III. Bot. Himal., I, Introd. 55.

ochraceus, Westw., *id.*, 55. Nepal.

181. Pseudoporus, Wasm.

Wasm., D. E. Z., 1893, 206.

furcifer, Wasm., D. E. Z., 1893, 206. Sikkim.

182. Wroughtonilla, Wasm.

Wasm, D. E. Z., 1899, 157.

lobopeltae, Wasm., D. E. Z., 1899, 157. India.

183. Myrmedonia, Er.

Er., Käf. Mark. Brand., I, 286.

Sub-gen., *Glossacantha,* Gem., and Har., Cat., II, 519.

Sub-gen. *Myrmedonia,* s. str.

Sub-gen. *Pella,* Steph., Ill, Brit. Ent., V, 434.

Sub-gen. *Rhynchodonia,* Wasm., Ann. Mus. Civ. Gen., XVI, 1896, 620.

Sub-gen., *Zyras,* Steph., Ill. Brit. Ent., V, 430.

afinis, Kr., Arch. Nat., XXV, 1859, I, 23, (*Pella*). Ceylon.

armifrons, Epp., W. E. Z. IX, 274, (*Pella*). Belgaum., Nagpur. Khandesh., Nilgiris.

badia, Motsch., Et. Ent., 1859, VIII, 89, t. I, f. 5, (*Glossacantha*). Ceylon.

bicolor, Epp., W. E. Z., 1895, 388, (*Rhynchodonia*). Calcutta.

bidentatus, Bernh., W. Z. B., LXIV, 1914, 108, (*Rhynchodonia*). Pegu.

clavicornis, Kr., Arch. Nat., XXV, 1859, I, 23. Ceylon.

exasperatus, Schub., D. E. Z., 1908, 610. Himalayas., Kulu.

excisa, Kr., Lin., Ent., 1857, 50, (*Pella*). Cylon.

feae, Wasm., Ann. Mus. Civ. Gen. (2), XVI, 1896, 620 (*Rhynchodonia*). Burma.

forticornis, Fauv., Rev. d'Ent., XXIII, 1904, 64, (*Pella*). Nilgiris.

gemina, Kr., Arch. Nat., XXV, 1859, I, 27, (*Zyras*). Ceylon., Assam.

gestroi, Wasm., Ann. Mus. Civ. Gen. (2), XVI, 1896, 622, (*Rhynchodonia*). Burma.

gracilicornis, Bernh., D. E. Z., 1902, 20.	Ceylon.
hastatus, Fauv., Rev. d'Ent., XXIII, 1904, 64, (*Zyras*).	Belgaum.
heimi, Wasm., D. E. Z., 1899, 150, (*Glossocantha*).	Gunji., Belgaum. Maypore., Mahé
hirta, Kr., Arch. Nat., XXV, 1859, I, 25, t. I, f. 8.	Ceylon.
hoplonota, Kr., Linn. Ent., 1857, 49, (*Rhynchodonia*).	Ceylon.
humerosus, Motsch., Et. Ent., 1859, VIII, 90, (*Glossacantha*).	Ceylon.
impressicollis, Kr., Linn., Ent., II, 1857, 53, (*Myrmedonia*).	Ceylon.
indorum, Fauv., Rev. d'Ent., XXII, 102, (*Rhynchodonia*).	Coonoor., Kanara. Nilgiris.
kraatzi, Schub., D. E. Z., 1908, 609.	Kulu., Himalayas.
laevigata, Kr., Arch. Naturg., XXV, 1859, I, 26, t. 8, f. 9.	Ceylon.
laminata, Motsch., Et. Ent., 1859, 87, (*Rhynchodonia*).	Ceylon.
lineatocollis, Kr., Arch. Naturg., XXV, 1859, I, 26.	Ceylon.
nietneri, Kr., Linn. Ent., 1857, 52.	Ceylon.
nigrescens, Motsch. B. M., 1858, III, 262, (*Zyras*).	East Indies.
nigriceps, Kr., Arch. Naturg., XXV, 1859, I, 22, (*Rhnychodonia*).	East Indies.
nigripennis, Bernh., Col. Rundsch., 1915, (3), 59.	Dehra Dun.
obscura, F., Syst. Eleuth., II, 595, (*Rhynchodonia*).	Genji., Mahé, Madura.
tricuspidata, Kr., Linn. Ent., 1857, 49.	Khandesh.
hauseri, Epp., W. E. Z., IX, 275, 1890, (♂)1895, 54.	Ranchi., Bijnor. Ceylon., Dhampur. Travancore. Shencottah. Western Ghats.
opacicollis, Cam., E. M. M., LVI, 1920, 98.	Ceylon.
opacipennis, Bernh., D. E. Z., 1902, 19.	Ceylon.
perforata, Champ., E. M. M., LVII, 1921, 178, (*Zyras*).	Himalayas.
pindarae, Champ., *id.*, 179, (*Zyras*).	Himalayas.
procera, Kr., Arch. Nat., XXV, 1859, I, 20, (*Rhynchodonia*).	Ceylon.
rubricollis, Kr., Linn. Ent., 1857, 51, (*Pella*).	Ceylon.

rufiventris, Epp., D. E. Z., 1895, 386. Pegu.

sculpticollis, Wasm., D. E. Z., 1899, 151, (*Myrmedonia*). Ahmednagar.

semirufa, Bernh., D. E. Z., 1902, 20, (*Pella*). Ceylon.

serraticornis, Kr., Linn. Ent., 1857, 48, Arch. Nat., XXV, 1859, I, 20, (*Rhyn-chodonia*). Ceylon.

sordida, Kr., Arch. Nat., XXV, 1859, I, 23, (*Pella*). Ceylon.

termitaria, Wasm., Ann. Mus. Civ. Gen. (2), XVIII, 30, (*Rhynchodonia*). Burma.

ermiticola, Wasm., Ann. Mus. Civ. Gen. (2), VI, 1888, 110, (*Rhynchodonia*). Burma.

termitobia, Wasm., *id.,* (2), XVI, 1896, 622.

termitophila, Wasm., *id.,* (2) XVIII, 1897, 28, (*Rhynchodonia*). Burma.

tridens, Wasm., D. E. Z., 1899, 149, (*Glossocantha*). Ahmednagar.

trisulcata, Wasm., Ann. Mus. Civ. Gen. (2), XVI, 1896, 623. Burma.

waageni, Epp., D. E. Z., 1895, 386, (*Rhynchodonia*). Pegu., Sedon. Upper Burma.

Tribus Gymnusini. Ceylon.

184. Leucocraspedum, Kr.

Kr., Arch. Nat., XXV, 1859, I, 51.

Euryglossa, Motsch., Et. Ent., VIII, 82.

horni, Bernh., Ent. Blätter, 1915, (10-12), 256. Ceylon.

limbatum, Cam., E. M. M., LVI, 1920, 49. Ceylon.

nigromaculatum, Cam., E. M. M., LIV, 1918, 218. Ceylon.

var. *obsoletum,* Cam, E. M. M., LVI, 1920, 50. Ceylon., Coonoor.

plumbeum, Fauv., d'Ent., XXII, 160. Ceylon.

pulchellum, Kr., Arch. Nat. XXV, 1859, I, 53, t. I, f. 12. Ceylon.

flavocincta, Motsch., Et. Ent., 1859, 84, t. I.

rufum, Favu., Rev. d'Ent., XIV, 285. Burma.

Tribus Oxypodini.

185 Apimela, Muls., and Rey.

Muls., and Rey, Brevip. Aleochar., IV, 48.

indica, Cam., E. M. M., LVI, 1920, 99. Ceylon.

186. Phloeopora.

Er., Gen. Spec. Staph., 76.

ceylanensis, Kr., Arch. Naturg., XXV, 1859, I, 41. Ceylon.

impressicollis, Motsch., B. M., 1858, III, 257. East Indies.

indica, Kr., Arch. Naturg., XXV, 1859, I, 42. East Indies

187. Ocalea, Er.

Er., Käf. Mar. Brand., I, 1837, 298.

indica, Kr., Arch. Naturg., XXV, 1859, I, 12. W. Himalayas.

188. Oxypoda, Mannerh.

Mannerh., Mem., Ac. Sci., St. Petersb., I, 483; Er., Gen. Spec. Staph., 141.

atriceps, G. and H., Cat., II, 526. East Indies.

atricapilla, Mostch., B. M., 1858, III, 244.

brunnescens, Motsch., *id.,* 243. East Indies.

lineola, Kr., Arch. Naturg., XXV, 1859, I, 87. Ceylon.

nigricauda, Motsch., B. M., 1861, I, 153. Ceylon.

palleola, Motsch., *id.,* 1858, III, 245. East Indies.

pallidipennis, Motsch., *id.,* 243. East Indies

plagiata, Motsch., B. M. 1858, III, 242. East Indies.

vilis, Kr., Arch. Naturg., XXV 1859, I, 28. East Indies.

189. Dinarda, Mannerh.

serricornis, Walk., Ann. Mag. Nat. Hist., 1859, 52. Ceylon.

190. Dinusa, Saulcy.

Saulcy, Ann. Soc. Fr. (4), Vol. 4, 433, 1864.

heimi, Wasm., Zeit. Wiss. Zool., CI, 1912, 97. Further India. Further India.

Tribus Aleocharini.

191. Pseudoplandria, Fenyes.

Fenyes, Bull. Mus. Comp. Zool., Harvard, LXV, 2, 1921, 30.

fuscipennis, Kr., Arch. Naturg., XXV, 1859, I, 9, (*Hoplandria*).	Ceylon.

192. Aleochara, Gr.

Gr., Col. Micr., 67 ; Bernh., W. Z. B., LI, 1901, 436.

Sub-gen. *Aleochara,* s. str. Bernh., W. Z. B., LI, 1901, 441.

Sub-gen. *Baryodma,* Thoms., Oefv. Sv. Vet. Akad. Forh., 1858, 31.

Sub-gen. *Ceranota,* Steph., Man. Brit. Col., 1839, 351.

Sub-gen. *Coprochara,* Muls., and Rey, Col. Fr. Aleoch., II, 1874, 146.

Sub-gen. *Dyschara,* Muls., and Rey, Col. Fr. Aleoch., II, 1874, 141.

Sub-gen. *Euryodma,* Reitt., Fn. Germ., II, 1910, 23.

Sub-gen. *Heterochara,* Muls., and Rey, Col. Fr. Aleoch., II. 1874, 15.

Sub-gen. *Polychara,* Muls., and Rey, Col. Fr. Aleoch., II, 1874, 64.

Sub-gen. *Xenochara,* Muls., and Rey, Col. Fr. Aleoch., II, 1874, 60.

andrewesì. Fauv., Rev. d'Ent., XXIII, 1895, 66, (*Aleochara,* s. str.).	Nilgiris., Coonoor.
asiatica, Kr., Arch. Naturg., XXV, 1859, I, 15, (*Euryodma*).	Ceylon., Coonoor. ———— China., Japan.
badia, Motsch. B. M. 1858, III, 237.	East Indies.
bipustulata. L., Faun. Suec. (2) 1761, p. 232. (*Coprochara*).	Fagu, Simla Hills. Dehra Dun.
nitida, Gr., Col. Micr., 97.	———— Europe. N. America.
brunneiventris, Kr., Arch. Nat., XXV, 1859, I, 14 (*Polychara*).	Nigiris. Burma.
castanea, Motsch., B. M., 1858, III, 239.	East Indies.
clavicornis, Redt., Fn. Austr., 1848, 822. (*Heterochara*).	Pondicherry. ———, S. Europe.
croceipennis, Motsch., B. M., 1858, 238, (*Eurydoma*).	N. Africa.

sanguinipennis, Kr., Arch. Naturg., XXV, 1859, I, 17.	Nilgiris., Burma.
v. *maculipennis*, Kr., *id.*, 17.	Ceylon., Dehra Dun.
denticulata, Motsch., B. M., 1858, III, 238, (*Coprochara*).	East Indies.
haematopyga, Kr., Arch. Naturg., XXV, 1859, I, 15.	Ceylon.
hindustana, Motsch., B. M., 1858, III, 237.	East Indies.
incerta, Epp., W. E. Z., 1895, 53, (*Dyschara*).	Himalayas.
minutissima, Kr., Arch. Naturg., XXV, 1859, I, 19.	Ceylon.
mutata, G. and H., Cat., II. p. 514, 1858, (*Baryodma*).	East Indies.
tenuicornis, Motsch., B. M., 1858, III, 240.	
nigra, Kr., Arch. Naturg., 1859, I, 13, (*Aleochara,* s. str.).	Ceylon., ——— Malay Peninsula. Java, etc.
pernigra, Schub., D. E. Z., 1906 378, (*Alcochara,* s. str.).	Kashmir.
postica, Walk., Ann. Mag. Nat. Hist. III, 2, 1858, 205, (*Aleochara,* s. str.).	Ceylon.
puberula, Klug., Abh. Ak. Wiss. Berl., 1833, 139, (*Xenochara*).	Mahé, Coonoor. Belgaum., Nilgiris. Ranchi. Chota-Nagpur. ——— Cosmopolitan.
punctiventris, Kr., Arch., Naturg., XXV, 1859, I, 19.	Ceylon.
rutilipennis, Kr., Arch. Naturg., XXV, 1859, I, 17 (*Polychara*).	Ceylon.
subjecta, Walk., Ann. Mag. Nat. Hist., 1859, 52.	Ceylon.
translata, Walk., *id.*, 52.	Ceylon.
tristis, Gr., Mon., 170, (*Isochara*).	W. Himalayas. ———
trivialis, Kr., Arch. Naturg., XXV, 1859, I, 18, (*Polychara*).	Europe., Nilgiris. Ceylon., Dehra Dun. ——— E. Africa., Arabia.
tuberiventris, Kr., *id.*, 14, (*Ceranota*).	East Indies.
viatica, Fauv., Rev. d'Ent., 1905, XXIV, 67, (*Euryodma*).	Burma. Nilgiris.

193. Maseochara, Sharp.

Sharp, B. C. A., I. (2), 154.

indica, Fauv., Rev. d'Ent., XXIII, 66. Nilgiris.

194. Tetrasticta, Kr.

Kr., Linn. Ent., 1857, 54.

polita, Kr., Linn. Ent., 1857, 55. Ceylon.

APPENDIX.

Sub-fam. OXYTELINAE.

13. Lispinus Er.

beesoni Cam., Tr. Ent., Soc. Lond., 1924, 161. Siwaliks. Dehra Dun.

quadricollis, Cam., *id.,* 160. Almora. Kanarichina.

14. Tetrapleurus Bernh.

himalayicus, Cam., Tr. Ent. Soc. Lond., 1924, 161. Masuri.

Proteinus Latr.

Latr., Préc. Caract. Ins., 1796, 9.

fuscotestaceus Cam., Tr. Ent. Soc. Lond., 1924, 162. Dehra Dun.

ruficollis Cam., *id.,* 162. Masuri., Simla Hills.

19. Megarthrus, Steph.

alticola Cam., Tr. Ent. Soc. Lond., 1924, 164. Simla Hills. Narkanda.

chatterjeei Cam., *id.,* 162. Chakrata.

flavolimbatus Cam., *id.,* 164. Simla Hills., Matiana. Kotgarh., Narkanda.

trisinuatus Cam., *id.,* 163. Chakrata.

Phyllodrepa Thoms.

Thoms., Skand., Col. I., 1860, 52.

himalayica Cam., Tr. Ent., Soc. Lond., 1924, 166 (s. str.). Chakrata. Simla.

21. Omalium Grav. (s. str.).

beesoni Cam., Tr., Ent., Soc. Lond., 1924, 166.	Chakrata.
cariosum Cam., *id.*, 167.	Dehra Dun., Masuri.
humerale Cam., *id.*, 167.	Chakrata.
secretum Cam, *id.*, 167.	Deoban.

22. Phloeonomus, (s. str.)

morosus Cam., Tr. Ent., Soc. Lond., 1924, 169.	Chakrata.
opacus Cam., *id.*, 168.	Chakrata., Simla.
subopacus Cam., *id.*, 168.	Simla Hills.

Lathrimaeum. Er.

Er., Käf. Mark. Brand., I., 624.

monticola Cam., Tr. Ent. Soc. Lond., 1924, 169.	Chakrata.
nigrum Cam., *id.*, 170.	Simla Hills.
	Simla Hills.

26. Lesteva Latr.

torrentum Cam., Tr. Ent., Soc. Lond., 1924, 171.	Masuri Dist.

28. Geodromicus Redt.

championi Cam., Tr. Ent., Soc. Lond., 1924, 171.	Chakrata Dist.
coriaceus Cam., *id.*, 172.	Chakrata Dist.
curvipes Cam., *id.*, 173.	Simla Hills.
recticollis Cam., *id.*, 172.	Charkata. Dist.

Hygrogoeus Rey.

Rey, Ann. Soc. Linn. Lyon, XXVII, 1880, 56.

cyanipennis Cam., Tr. Ent. Soc. Lond., 1924, 174.	Chakrata Dist.

Ancyrophorus Kr.

Kr., Ins. Deutsch. II, 886.

monticola Cam., Tr. Ent. Soc. Lond., 1924, 174.	Simla Hills.
nitidus Cam., *id.*, 175.	Smila Hills.

36. Delopsis Fam.

tenuicornis, Cam., Tr. Ent. Soc. Lond., 1924, 175.	Masuri., Chakrata.

Thinobius Kiesw.

Kiesw., Stett. Ent., Ziel., V, 1844, 355.

antennarius Cam., Tr. Ent., Soc. Lond., 1924, 177.	Dehra Dun.
himalayicus Cam., *id.*, 176.	Masuri. Dist.
pruinosus Cam., *id.*, 177 (s. str.).	Lachiwala.

39. Bledius Mannerh.

quadricornis Bernh., W. Z. B. LXXI, 1915, 135.	Madura.

43. Osorius Latr.

granulifrons Cam., Tr. Ent. Soc. Lond., 1924, 178.	Simla Hills.

Sub-fam. OXYPORINAE.

44. Oxyporus, F.

v. **nigriceps** Cam., Tr. Ent. Soc. Lond., 1924, 177.	Simla Hills.

46. Stenus Latr.

coelogaster Champ., E.M.M., LX, 1924, 159 (*Hypostenus*).	Ranikhet.
paederinus Champ, *id.*, 160 (*Hypostenus*).	Kumaon.
rufoplagiatus Champ., *id.*, 160 (*Hypostenus*).	Kumaon.

47. Dianous Sam.

robustus Cam., Tr. Ent. Soc. Lond., 1924, 179.	Kaligad.

Sub-fam. PAEDERINAE.

52. Palaminus Er.

morosus Cam., Tr. Ent. Soc. Lond., 1924, 179.	Chakrata Dist.

53. Oedichirus Er.

elegans Cam., Tr. Ent. Soc. Lond., 1924, 180.	Masuri.

54. Paederus F.

nigripennis Cam., Tr. Ent. Soc. Lond., 1924, 180.	Masuri.

56. Stilicopsis Sachse.

dorsolineata Cam., Tr. Ent. Soc. Lond., 1924, 181.	Dehra Dun., Masuri.

59. **Stilicus** Latr.

lithochoroides Cam., Tr. Ent., Soc. Lond., 1924, 183.	Masuri.
plagiatus Cam., *id.,* 181.	Dehra Dun.
praelongus Cam., *id.,* 182.	Simla Hills.

64. **Medon** s. str.

asperulus Cam. Tr. Ent. Soc. Lond., 1924, 186.	Chakrata Dist.
beesoni Cam., *id.,* 185.	Singhbhum.
castaneipennis Cam., *id.,* 184.	Siwaliks.
chatterjeei Cam., *id.,* 188.	Dehra Dun.
dolorosus Cam., *id.,* 187.	Masuri.
exasperatus Cam., *id.,* 187.	Masuri., Chakrata.
niger Cam., *id.,* 185.	Simla.,
parkeri Cam., *id.,* 183.	Almora., Zodh.
rufoferrugineus, Cam., *id.,* 186.	Masuri.
umbilicatus Cam., *id.,* 184.	Masuri., Chakrata. Simla.

Neosclerus Cam.,

Cam., Tr. Ent. Soc. Lond., 1924, 188.

fortepunctatus Cam., *id.,* 189.	Masuri., Chakrata
frater Cam., *id.,* 190.	Masuri.
granulicollis Cam., *id.,* 190.	Masuri.

Domene Fauv.

Fauv., Fn. Gall. Rhen., iii, 1872, 305.

Sub-gen *Ennalagium* Bernh., W. Z. B., 1915, 139.

aeneipennis Cam., Tr. Ent. Soc. Lond., 1924, 191.	Masuri.
castaneipenne Cam., *id.,* 192 (*Ennalagium*).	Siwaliks.
indicum Cam., *id.,* 191 (*Ennalagium*).	Masuri.

66. **Lathrobium** Grav.

badium Cam., Tr. Ent. Soc. Lond., 1924, 193 (*Lobrathium*).	Dehra Dun., Masuri.

cylindricolle Cam., *id.,* 194 (*Lobrathium*). Dehra Dun.

nigerrimum Cam., *id.,* 192 (*Lobrathium*). Masuri., Siwaliks.

triste Cam., *id.,* 193 (*Lobrathium*).

67. Pseudobium Muls. and Rey. Chakrata.

assimile Cam., Tr. Ent., Soc. Lond., 1924, 195. Dehra Dun.

semiflavum Cam., *id.,* 195.

72. Cryptobium Mannerh. Dehra Dun.

bernhaueri, Cam., Tr. Ent. Soc. Lond., 1924, 197. Siwaliks.

var. **distinctum** Cam., *id.,* 197. Masuri.

capitalis Cam., *id.,* 196. Siwaliks.

humerale Cam., *id.,* 197. Masuri.

[While this catalogue has been passing through the Press the following papers have appeared:—

Champion, G. C., 1925, Ent., Mo. Mag., pp. 101—102, (11 *nov. spp., 2 nov. var.* from Kumaon).

Wasmann, E., 1925, Deutsche Ent., Zeitschr., pp. 1—10, t. I, f. 1— 4*d.* (1 *nov. sp.* from Bihar).

C. F. C. B.]

PART VII
LASIOCAMPIDAE

1. **Bhima, Moore** (1888).

Bhima, Mo., P. Z. S., 1888, 403-404; Hmpsn., F. I. Moths I 404 (1892) [redescr.]; Grünberg, Seitz Lep. Pal. II, 177 (1911) [redescr.]	Type:—*undulosa, Wlk.*

undulosa, Walker.

Pœcilocampa undulosa, Wlk., Cat. VI, 1477 (1855).[1] ♀	Kangra.[2 3 6 8] Nepal.[1 3]
Bhima undulosa, Mo., P. Z. S. 1888. 404 (1888)[2]; Hmpsn., F. I. Moths. I 404, f. 279 (1892)[3]; Swinh., T. E. S., 1895. 27 (1895)[4]; Dudgeon, B. J., XIII, 406 (1901)[5]; Grünberg, Seitz Lep. Pal., II, 178 (1911)[6]; Turner, T. E. S., 1918, p. 180, f. 46 [neuration][7]; Grünberg, Seitz Lep. Indo-austral, II, 393, t. 34*a* (1923) [redescr].[8]	Sikkim.[5] Darjiling. Khasis.[3] Cherrapunji.[2 4] Manipur.[3,] ——— South Asia.[8]
Larva, shortly descr.[2]	Sunda Isds.[8]

2. **Taragama, Moore** (1859).

Taragama, Mo., Cat. Lep. E. I. Co., p. 427 (1859); Mo., Lep. Ceylon II 147 (1883); Hmpsn., F. I. Moths I, 404 (1892) [redescr.]; Grünberg, Seitz Lep. Pal. II 175 (1911) [redescr]. *id.,* Seitz Lep. Indo-austral. II 393 (1923).	Type:–*siva,* Lef.

dorsalis, Walker.

Megasoma dorsalis, Wlk., Cat. XXXV 1947-1948 (1866).[1]♂	N.-W. Himalaya.[1 13] Kangra.[5 6 11]
Taragama igniflua, Mo., Lep. Ceylon II 147, t. 142 ff. 2, 2*a* (1883).[2] ♂ ♀	Khatmandu (Nepal). Sikkim (1800 ft.).[9]
Taragama hyperantheræ Mo., P. Z. S. 1883, 18, t. 5 f. 5 (1883)[3] ♂	Calcutta.[3 7] Bengal.[4]
Taragama intensa, Mo., P. Z. S. 1883. 18, t. 5. f. 4 (1883).[4] ♂	South India.[7] Ceylon.[2 7 12 13]
Taragama castanoptera, Mo., P. Z. S. 1888. 404-405 (1888)[5]; Butl. Ill. Het. VII 38, t. 124 f. 1 (1889)[6]	Kandy.[8] Pundaluoya.[8] ———

Taragama dorsalis, Hmpsn., F. I. Moths. I, 405, f. 280 (1892)[7]; Hmpsn., Ill. Het. IX 9 (1893)[8]; Dudgeon, B. J. XIII 406 (1901)[9]; Hmpsn., B. J. XVI 197 (1905)[10]; Grünberg, Seitz Lep. Pal. II 177, t. 30*e* (1911)[11]; Rutherford, Trop. Agric. XLIII 129-130 (Aug. 1914)[12]; Grünberg, Seitz. Lep. Indo-Austral. II 393 (1923).[13]

Java.[7]
Sunda Isds.[13]
Borneo.[1 7]
Philippines.[7 13]

Larva described[10], on *Acacia* spp.[10], destr. to *Erythrina* in Ceylon[12], on *Hyperanthera.*[3]

siva, Lefebvre.

Bombyx siva, Lef., Zool. Journ. III 210 (1827).[1]

Throughout India.[2]
Burma.[10]

Bombyx ganesa, Lef., Zool. Journ. III 211 (1827) [male][2]; Mo., Cat. Lep. E. I. Co. II, t. 22 ff. 4, 4[a] (1859).[3]

Megasoma venustum, Wlk., Cat. VI 1449 (1855).[4] ♀

Megasoma pallidum, Wlk., Cat. VI 1450 (1855).[5]

Megasoma albicans, Wlk., Cat. VI 1450 (1855).[6]

Taragama siva, Hmpsn., F. I. Moths. I 405-406 (1892)[7]; Grünberg, Seitz Lep. Pal. II 177 (1911)[8]; Fletcher, Pusa Bull. 89, pp. 68-71, f. 40 (1919)[9]; Grünberg, Setiz Lep. Indo-austral. II 394, t. 34 ff. *a* [male, female] (1923) [redescr.][10]; Lefroy, Ind. Ins. Life, p. 497, t. 41 ff. 8-11 (1909).[11]

Larva descr.[3 7 8 9 11], on *Tamarix gallica*[9], *Murraya exotica*[9], *Polyalthia longifolia*[9], rose[9], *Acacia arabica*[9], *Zizyphus jujuba.*[8]

3. Suana, Walker (1855).

Suana, Wlk., Cat. VI 1502 (1855); Mo., Lep. Ceylon II 151 (1883); Hmpsn., F. I. Moths I. 406 (1892) [redescr.]; Grünberg, Seitz Lep. Pal. II 178 (1911) [redescr.].

Type :– *concolor,* Wlk

concolor, Walker.

Lebeda concolor, Wlk., Cat. VI 1463 (1855).[1]

Throughout India.[9]
Kashmir.[13]

Lebeda bimaculata, Wlk., Cat. VI 1463 (1855).[2]

Sikkim.[12], Bhutan.[12]

Suana ampla, Wlk., Cat. VI 1502 (1855).[3]

Sylhet.[1 2 3]

Suana carvina, Mo., P. Z. S. 1879 410 (1879)[4]; Mo., Lep Ceylon II 152, t. 139 ff. 2, 2[a] (1883).[5]

Shillong.[11]
Cherrapunji.[11]
Pollibetta (Coorg)·

Suana bimaculata, Mo., Cat. Lep. E. I. Co. II 428, t. 13[a], ff., 2, 2[a] (1859)[6]; Mo., Lep. Ceylon II 152, t. 140 ff. 1, 1[a], 1[b] [larva] (1883)[7]; Butl., Ill. Het. VI 2, t. 101 ff. 3, 4 (1886).[8] — Shembaganur (Palnis; 6,000 ft.) Ceylon.[4 5 7 9 13 14] Kandy.[10], Matale[14]

Suana concolor, Hmpsn., F. I. Moths I 406-407, f. 281 [male], f. 278 [larva] (1892) [redescr.][9]; Hmpsn. Ill. Het. IX 9 (1893)[10]; Swinh., T. E. S. 1895 27 (1895)[11]; Dudgeon. B. J. XIII 406 (1901)[12]; Grünberg, Seitz Lep. Pal. II 178, t. 29*c* (1911)[13]; White, Spol. Zeyl. XI 299-302, tab. (1920)[14]; Grünberg, Seitz Lep. Indo-austral. II 393 (1923).[15] — Nawalapitiya.[10] Ramboda.[10]

———————

Sumatra.[15] Java.[1 9 15 3] Philippines.[3 9 15]

Larva descr.[7 9 13 14], on *Careya arborea*[7], *Psidium guyava*[14], *Cajanus indicus*[14], *Hibiscus rosa-sinensis*[14], *Shorea robusta*[14], *"Daminiya" (?=Grewia tiliæfolia)*[14], occasionally on tea.[14] Biology.[14]

4. Paralebeda, Aurivillius (1894).

Paralebeda, Auriv., Deuts. ent. Zeitschr. Lep. VII 178 (1894); Grünberg, Seitz Lep. Pal. II 174-175 (1911) [redescr.]; *id.*, Seitz Lep. Indo-austral II 393 (1923). — Type:–*plagifera*, Wlk.

[*nec Paralebeda*, Hmp., A. M. N. H. (8) VI 119 (1910) (Teragridæ).]

piagifera, Walker.

Lebeda plagifera, Wlk., Cat. VI 1459-1460 (1855).[1] — N.W. Himalaya.

Opsirhina crinodes, Felder, Reise Novara, Lep. IV, t. 84 f. 3 (1874).[2] — Simla.[4] Nepal.[1 4 8]

Odonestis plagifera, Butl., Ill. Het. V 33, t. 99 f. 5 (1881)[3]; Hmpsn., F. I. Moths I 427 (1892) [redescr][4]; Swinh., T. E. S. 1895 28 (1895)[5]; Dudgeon., B. J. XIII 411 (1901).[6] — Khatmandu.[2] Sikkim (2,000-3,000 ft.)[6 8], Darjiling. Cherrapunji.[5]

———————

Java.[4. 8]

Paralebeda plagifera, Grünberg, Seitz Lep. Pal. II 175, t. 30*d* (1911)[7]; *id.*, Seitz Lep. Indo-austral. II 393 (1923).[8] — Guiana[2] [error loci.]

5. Lebeda, Walker (1855s).

Type:–*nobilis*, Wlk.

Kashmir.[6], Kangra.[6]

Lebeda, Wlk., Cat. VI 1453 (1855); Mo., Lep. Ceylon II 150 (1883); Hmpsn., F. I. Moths I 407 (1892) [redescr.]; Grünberg. Seitz Lep. Pal. II 174 (1911) [redescr.]; *id.*, Lep. Indo-Austral. II 392 (1923) [redescr.].

nobilis, Walker.

Lebeda nobilis, Wlk., Cat. VI 1456 (1855)[1]; Butl., Ill. Het. V 74, t. 100 ff. 5, 6 (1881)[2]; Hmpsn., F. I. Moths I 407-408, f. 282 (1892) [redescr.][3]; Swinh., T. E. S. 1895 27 (1895)[4]; Dudgeon, B. J. XIII 406 (1901)[5]; Grünberg, Seitz Lep. Pal. II 174, t. 30*c* (1911) [redescr.][6]; *id.,* Lep. Indo-austral. II 392 (1923).[7]

Simla., Masuri. Nepal.[1 6] E. Himalaya.[3] Sikkim.[5], Bhutan.[5] Sylhet.[1 3 6], Shillong. Cherrapunji.[4] Nagas.[3 6]

6. Dendrolimus, Germar (1812).

Dendrolimus, Germ., Syst. Gloss, Prodr., p. 48 (1812); Curt., Brit. Ent. I, t. 9 (1824); Kirby, Cat. Het. I 813 (1892); Leech, T. E. S. 1899 109; Grünberg, Seitz Lep. Pal. II 171 (1911) [redescr.]; *id.*, Lep. Indo-austral. II 405 (1923) [sep. from *Metanastria*]. — Type:–*pini,* Linn. [Europe; N. Asia].

Metanastria, Hübner, Verz. p. 186 (1822); Hmpsn., F. I. Moths I 408 (1892) [redescr.]; Grünberg, Seitz Lep. Indo-austral. II 394 (1923) [redescr.]. — Type: *hyrtaca,* Cr.

Eutricha, Stephens, Ill, Brit. Ent., Haust. II 50 (1828); Wlk., Cat. VI 1405 (1855). — Type: *pani,* Linn.

Oeona, Walker, Cat. VI 1417 (1855). — Type: *punctata,* Wlk. [China].

Chatra, Moore, Lep. Atk., p. 80 (1879). — Type: [*latipennis,* Wlk. = *grisea,* Mo.]

aconyta, Cramer.

Bombyx aconyta, Cram., Pap. Exot. II, t. 131A (1777).[1]

Bombyx quadricincta, Fabr., Mantissa II 111 (1787).[2]

Dirphia? aconyta, Wlk., Cat. VI 1370 (1855).[3]

Lasiocampa trifascia, Wlk., Cat. VI 1439 (1855).[4]

Lasiocampa substrigosa, Wlk., Cat. VI 1441 (1855).[5]

Lebeda opponens, Wlk., Cat. VI 1462 (1855).[6]

Lasiocampa subparalletla, Wlk., Cat. XXXII 562 (1865).[7]

Metanastria aconyta, Hmpsn., F. I. Moths I 408-409 (1892) [redescr.][8]; Swinh., T. E. S. 1895, 27 (1895)[9]; Dudgeon, B. J. XIII 407 (1901) [not found in Sikkim][10]; Grünberg, Seitz Lep. Indo-austr. II 395, t. 34 f. *c.* [female] (1923) [redescr.][11]

Sikkim.[8], Derjiling.[11] Bengal.[3 11 4] Cherrapunji.[9] Kanara.[8 11] Coromandel.[3]

ampla, Walker.

Odenestis ampla, Wlk., Cat. VI 1412 (1855).[1]	Kangra.[6]
Lebeda ferruginea, Wlk., Cat. VI 1458 (1855).[2]	Sikkim.[4,6], Bhutan.[6]
Lebeda ampla, Butl., Ill. Het. V 72, t. 99 ff. 3, 4 (1881).[3]	Khasis.[5], Shillong.
Metanastria ampla, Hmpsn., F. I. Moths I, 410 (1892) [redescr.][4]; Swinh., T. E. S 1895 27 (1895)[5]; Dudgeon, B. J. XIII 407 (1901)[6]; Grünberg, Seitz Lep. Indo-austr. II 395, t. 34 f. *c.* [male] (1923).[7] *Larva on Artemisia.*[6]	Sylhet.[1,2,4], Burma.[4] ——— Sumatra.[7], Java.[4,7] Philippines.[4,7]

fia, Swinhoe.

Lenodora fia, Swinh., A. M. N. H. (7) III 113 (1899)[1]; Hmpsn., B. J. XIII 233, t. B. ff. 17, 18 (1900) [redescr.][2]; Grünberg, Seitz Lep. Indo-austr. II 404, t. 33 f. *g* [female], t. 46 f. *b* [male] (1923) [redescr.].	Kashmir.[1,2,3]

fulgens, Moore.

Lebeda fulgens, Mo., Lep. Atk. p. 81 (1879).[1]	Darjiling.[1,3] Sikkim.[2–4]
Metanastria fulgens, Hmpsn., F. I. Moths I, 409 (1892) [redescr.][2]; Dudgeon, B. J. XIII 407 (1901)[3]; Grünberg, Seitz Lep. Indo-austr. II 195, t. 34 f. *b* (1923) [redescr.].	Bhutan (6,400 ft).[3]

hyrtaca, Cramer.

Bombyx hyrtaca, Cram., Pap. Exot. III, t. 249 F (1780).[1]	Dharmsala.[9] E. Himalaya.[11]
Bombyx lusca, Fabr., Mantissa II 112 (1787).[2]	Kalimpong.[13], Pusa.
Bombyx laricis, Hübner, Beytr. Schm. II (*ii*) p. 48, t. 3 f. O, II (*iii*) p. 67, t. 2 f. M (1791).[3]	Calcutta., Assam.[11] Bombay.[11]
Bombyx brahma, Lef., Zool. Journ. III 208 (1827).[4]	Mhow.[7,11] Balaghat (C. P.)
Bombyx buddha Lef. Zool. Journ. III 209 (1827)[5]; Mo., Cat. Lep. E. I. Co., t. 12*a* ff. 2, 2*a*[6]; Forsayeth, T. E. S., 1884 407, t. 15 ff 3, 3*a*, 3*b* [larva].[7]	Ganjam.[11], Madras.[11] Trivandrum.[15] Ceylon.[8,10,11,16]
Lebeda plagiata, Wlk., Cat. VI 1464 (1855).[8]	Colombo.[12], Kandy.[12]
Lasiocampa bhira, Mo., P. Z. S. 1879 410, t. 34 f. 2 (1879).[9]	Rangoon., ———
Metanastria hyrtaca, Mo., Lep. Ceylon II 148, t. 141 ff. 1, 1*a*, 1*b* (1883)[10]; Hmpsn., F. I. Moths I 410-411, f. 283 (1892) [redescr.][11]; Hmpsn., Ill. Het. IX 9 (1893)[12];	Sunda Isds.[16] China.[16]

Elwes, B. J. XIII 407 (1901)[13]; Grünberg, Seitz Lep. Pal. II 177, t. 29 f. *a* (1911) [*Taragama*][14]; Fletcher, Pusa Bull. 59, p. 25, f. 14 (1919) [gynandrous specimen][15]; Grünberg, Seitz Lep. Indo-austr. II 395, t. 34 ff. *e* (1923)[16]; Lefroy, Ind. Ins. Life, p. 497, ff. 332, 333 (1909)[17]; Fletcher, S. Ind. Ins. pp. 409-410, f. 282 (1914)[18]; *id.*, Rept. Second Entl., Meeting, pp. 38, 249 (1917)[19]; *id.*, Rept. Third Entl. Meeting, p. 103 (1919).[20]

Larva descr.[10 11 14 18], on *Schima wallichü*[13], *Eugenia jambolana*[15], *Acacia arabica*[20], *Mimusops elengi*[18], *Terminalia catappa*[18], *Nyctanthes arbor-tristis*[18], *Albizzia stipulata.*[20]

latipennis, Walker.

Lebeda latipennis, Wlk., Cat. VI 1457 (1855).[1]

Lebeda variegata, Mo., A. M. N. H. (4) XX 347 (1877)[2]; Mo., Lep. Ceylon II 151, t. 139 f. 1 (1883).[3]

Chatra grisea, Mo., Lep. Atk. p. 80. (1879).[4]

Lebeda placida, Mo., Lep. Atk. pp. 80-81 (1879).[5]

Lebeda vulpina, Mo., Lep. Atk. p. 81 (1879).[6]

Eutricha cheela, Mo., P. Z. S. 1879 408 (1879).[7]

Lebeda fasciata, Mo., T. E. S. 1884 375 (1884).[8]

Odonestis latipennis, Butl., Ill. Het. V 72, t. 99, ff. 1, 2 (1881).[9]

Metanastria latipennis Hmpsn., F. I. Moths I. 409 (1892) [redescr.][10]; Hmpsn., Ill. Het. IX 9 (1893)[11]; Swinh., T. E. S. 1895 27 (1895)[12]; Dudgeon, B. J. XIII 407 (1901)[13]; Grünberg, Seitz Lep. Indo-austr. II 395, t. 34. ff. *b, c* (1923) [redescr.][14]
Larva descr.[14]

Dharmsala.[7], Darjiling.[5 6]
Sikkim.[10 13]
Khasis.[8], Shillong.
Cherrapunji. [4 12]
Calcutta.[10] [?error loci]
Cylon.[2 10 14]
Dikoya. [11]
Burma.[10 14], ______
Java.[14], Borneo.[10 14]

lidderdalii, Butler.

Lebeda lidderdalii Butl., Ill. Het. V 73, t. 100 ff. 1, 2 (1881).[1]

Lebeda stigmata, Butl., Ill. Het. V 73, t. 100 ff. 3, 4 (1881).[2]

Metanastria lidderdalii, Hmpsn., F. I. Moths I 410 (1892) [redescr.][3]; Dudgeon, B. J. XIII 407 (1901)[4]; Grünberg. Seitz Lep. Indo-austr. II 395-396, t. 34 f. *d* (1923) [redescr.].[5]

Darjiling.[2 5]
Bhutan.[1 3 4 5]

lineata, Moore.

Lebeda lineata Mo., Lep. Atk. p. 81 (1879).[1] — Darjiling.[1,6], Sikkim.[2,5]

Odonestis lineata, Hmpsn., F. I. Moths I 427-428 (1892) [redescr.].[2] — Kurseong. Cherrapunji.[3]

Metanastria fuscomarginata, Swinh., T. E. S. 1895 27-28, t. 1 f. 7 (1895)[3]; Hmpsn., F. I. Moths IV 487 (1896) [syn. of *lineata*].[4]

Metanastria lineata, Dudgeon, B. J. XIII 408 (1901)[5]; Grünberg, Seitz Lep Indo-austr. II 396, t. 35 f. *c.* (1923) [redescr.][6]

nanda, Moore.

Lebeda nanda, Mo., Cat. Lep. E. I. Co. II 423, t. 12*a* f. 3 (1859).[1] — Darjiling.[1] Sikkim.[2–4]

Metanastria nanda, Hmpsn., F. I. Moths I 409 (1892) [redescr.][2]; Dudgeon, B. J. XIII 407 (1901)[3]; Grünberg, Seitz Lep. Indo-austr. II 395, t. 34 f. *b* (1923). [redescr.][4]

obliquifascia, Swinhoe.

Metanastria obliquifascia, Swinh., Cat. Het. Mus. Oxon. I 263-264 (1892)[1]; Hmpsn., F. I Moths I 412 (1982) [redescr.][2]; Swinh., T. E. S. 1895 27 (1895)[3]; Grünberg, Seitz Lep. Indo-austr. II 396 (1923) [redescr.].[4] — Darjiling.[1,4] Sikkim.[2] Shillong.[3]

recta, Walker.

Lebeda recta, Wlk., Cat. VI 1461 (1855).[1] — N. India.[1,2,4,5]

Metanastria recta Hmpsn., F. I. Moths I 411 (1892) [redescr.][2]; Dudgeon, B. J. XIII 408 (1901) [female descr., probably of *repanda*][3]; Grünberg, Seitz Lep. Pal. II 177 1911 [*Taragama*][4]; *id.,* Lep. Indo-austr. II 396, t. 34 f. *e.* (1923) [redescr.].[5] — Sikkim.[3]

repanda, Walker.

Lebeda repanda, Wlk., Cat. VI 1460 (1855)[1]; Butl., Ill. Het. VI 2, t. 101 ff. 1, 2 (1886).[2] — Nepal.[3] Darjiling distr. Sikkim (1,800 ft.).[5]

Metanastria repanda, Hmpsn., F. I. Moths I 411 (1892) [redescr.][3]; Swinh., T. E. S. 1895 27 (1895)[4]; Dudgeon, B. J. XIII 407-408 (1901) [male descr. ?*repanda*][5]; Grünberg, Seitz Lep. Indo-austr. II 396, t. 33 f. *b* (1923) [redescr.].[6] — Assam.[3,6] Khasis.[4] Sylhet.[1] Bassein.[3]

subpurpurea, Butler.

Pœcilocampa subpurpurea, Butl., T. E. S. 1881 18 (1881).[1] ♀	Khasis.[3], _______
Odonestis dieckmanni Gräser, Berl. Ent. Zeit. XXXII 128 (??).[2]	Japan.[1 5 6] Amur.[2 5 6]
Metanastria rubra, Hmpsn., F. I. Moths IV 486-487 (1896)[3]; *id.,* B. J. XI 294 (1897) [syn. of *dieckmanni*].[4]	
Metanastria subpurpurea, Grünberg, Seitz Lep. Pal. II 161, t. 26 f. *c* (1911)[5]; *id.,* Lep. Indo-austr. II 396 (1923).[6]	

undans, Walker.

Lebeda undans, Wlk., Cat. VI, 1458-1459 (1855).[1]	Khasis. [10 13]
Bombyx fasciatella, Ménetriès, Bull. Acad. Petersb. XVII 218 (1858)[2]; Schrank, Amur Reisen, p. 55, t. 4, f. 8 (1859)[3]; Bremer, Lep. Ost.-Siber., p. 43 (1864)[4]; Fixsen, Rom. Mem. III 344 (1877).[5]	Sylhet.[1 9] Masuri. _______
Odonestis excellens, Butl., A. M. N. H. (4) XX, 481 (1877)[6]; *id.,* Ill. Het. II 19, *t,* 26, ff. 4, 5 (1878).[7]	Japan.[6 7 9 12] Korea.[12], Amur.[2 12]
Odonestis excellens var. *unicolor,* Obth., Et. Ent. V 38 (1881).[8]	Askold Isd.[8]
Metanastria undans, Hmpsn., F. I. Moths I 410 (1892) [redescr.][9]; Swinh. T. E. S. 1895 27 (1895)[10]; Grünberg, Seitz Lep. Indo-austr. II 396, t. 34 f. *d* (1923) [redescr.].[11]	
Dendrolimus undans, Leech, T. E. S. 1899 110-111 (1899)[12]; Grünberg, Seitz Lep. Pal II 173 (1911).[13]	

vitta, Moore.

Odonestis vitta, Mo., Cat. Lep. E. I. Co. Mus. II 424, t. 12*a* f. 4. (1859)[1]; Kirby, Cat. Het. I. 811 (1892) [*vita–laps.*].[2]	Calcutta.[3] _______, Sumatra.[4]
Metanastria vitta, Hmpsn., B. J. XX 110 (1910) [redescr.].[3]	Java.[1–4]
Arguda vitta, Grünberg, Seitz Lep. Indo-austr. II 399, t. 35 f. *a* (1923) [redescr.][4]	Borneo.[3 4]

7. Arguda, Moore (1879).

Arguda, Moore, Lep. Atk., p. 79 (1879); Hmpsn., F. I. Moths I 412 (1892) [redescr.]; Grünberg, Seitz Lep. Pal. II 178-179 (1911) [redescr.]; *id.,* Lep. Indo-austr. II 398-399 (1923) [redescr].	Type: *decurtata,* Mo.
Radhica, Mo., Lep. Atk., p. 79 (1879).	Type: *flavovittata,* Mo.

albigutta, Walker.

Opsirhina albigutta, Wlk., Cat. VI 1419-1420 (1855).[1]	Sabathu.[2]
Arguda albigutta, Hmpsn., F. I. Moths I 414 (1892) [redescr.][2]; Grünberg, Seitz Lep. Indo-austr. II 400, t. 35 f. *b* (1923) [redescr.].[3]	______ Australia.[2 3] Tasmania.[1 3]

bheroba, Moore.

Odonestis bheroba, Mo., Cat. Lep. E. I. Co. Mus. II 424, t. 12*a* f. 5 (1859).[1]	Darjiling.[2] Sikkim.[2 3]
Arguda bheroba, Hmpsn., F. I. Moths I 413 (1892) [redescr.][2]; Dudgeon, B. J. XIII 408 (1901) [variation][3]; Grünberg, Seitz Lep. Indo-austr. II 399, t. 32 f. *d* (1923) [redescr.].[4]	Cylon.[2], ______ Singapore.[2] Sunda Isds.[4], Amoy.[2], Philippines.[2]

bipartita, Leech.

Arguda bipartita, Leech., T. E. S. 1899 116 (1899)[1]; Grünberg, Seitz Lep. Indo-austr. II 400, t. 33 f. *c* (1923).[2]	Pu-tsu-fong.[1 2] [E.Tibet border.]

decurtata, Moore.

Arguda decurtata, Mo., Lep. Atk. p. 79 (1879)[1]; Hmpsn., F. I. Moths I 414 (1892) [redescr.][2]; Swinh., T. E. S. 1895 28 (1895)[3]; Dudgeon, B. J. XIII 408 (1901)[4]; Grünberg, Seitz Lep. Indo-austr. II 400, t. 35 ff. *b* (1923) [redescr.].[5]	Darjiling.[1 5] Sikkim (1,800 ft.).[4] Cherrapunji.[3]

erectilinea, Swinhoe.

Arguda erectilinea, Swinh., T. E. S. 1904, 152 (1904)[1]; Grünberg, Seitz Lep., Indo-austr. II 399, t. 35 f. *a* [as *bheroba, ex errore*] (1923) [redescr.].[2]	India.[2], Ceylon.[2] _____, Sunda Isds.[2] Philippines.[2]

flavovittata, Moore.

Radhica flavovittata, Mo., Lep. Atk., pp. 79-80 (1879).[1]	Kashmir.[3] Dharmsala.[1 2]
Arguda flavovittata, Hmpsn., F. I. Moths I 412-413, f. 284 (1892) [redescr.][2]; Grünberg, Seitz Lep. Pal. I 179, t. 29 f. *d* (1911)[3]; *id.,* Lep. Indo-austr. II 399, t. 34 f. *e* (1923) [redescr.].[4]	Masuri. Naini Tal.[1 2] Bhutan.[2]

rectilinea, Hampson.

Arguda rectilinea, Hmpsn., F. I. Moths I. 414 (1892)[1]; Grünberg, Seitz Lep. Indo-austr. II 400 (1923) [redescr.].[2]	Bassein (Burma).[1]

rosea, Hampson.

Radhica rosea, Hmpsn., Ill. Het. VIII 65, t. 144 f. 18 (1891).[1] — Nilgiris (6,000ft.)[1 2]

Arguda, rosea, Hmpsn., F. I Moths I 413 (1892) [redescr.][2]; Grünberg, Seitz Lep. Indo-austr. II 399, t. 35 f. *a* (1923) [redescr.].[3]

vinata, Moore.

Lebeda vinata, Mo., P. Z. S. 1865, 820 (1865).[1] — Darjiling.[1 3 4]

Arguda vinata. Hmpsn. F. I. Moths I 413-414 (1892) [redescr.][2]; Elwes, B. J. XIII 408 (1901)[3]; Grünberg, Seitz Lep. Indo-austr. II 399, t. 35 f. *b* (1923) [redescr.].[4]

8. Syrastrena, Moore (1884).

Syrastrena, Mo., T. E. S. 1884, 373 (1884); Hmpsn., F. I. Moths I 414-415 (1892) [redescr.]; Grünberg, Seitz Lep. Indo-austr. II 401 (1923) redescr.]. — Type: *minor,* Mo.

minor, Moore.

Metanastria minor, Mo., Lep. Atk. pp. 78-79 (1879).[1] — Darjiling.[1], Sikkim. (?low elevation).[2]

Syrastrena minor, Hmpsn., F. I. Months I 415, f. 285 (1892) [redescr.][2]; Elwes, B. J. XIII 408 (1901)[3]; Grünberg, Seitz Lep. Indo-austr. II 401, t. 35 f. *b* (1923) [redescr.].[4]

9. Bharetta, Moore (1865).

Bharetta, Mo., P. Z. S. 1865, 820 (1865); Hmpsn., F. I. Moths I 415 (1892) [redescr.]; Grünberg, Seitz Lep. Indo-austr. II 401 (1923) [redescr.]. — Type : *cinnamomea,* Mo.

cinnamomea, Moore.

Bharetta cinnamomea, Mo., P. Z. S. 1865, 820, t. 43 f. 6 (1865)[1]; Hmpsn., F. I. Moths I 415-416, f. 286 (1892) [redescr.][2]; Dudgeon, B. J. XIII 409 (1901)[3]; Grünberg, Seitz Lep. Indo-asutr. II 401, t. 35 f. *c* (1923) [redescr.].[4] — Darjiling.[1 3 4]

flammans, Hampson.

Bharetta flammans, Hmpsn., F. I. Moths I 416 (1892)[1]; Grünberg, Seitz Lep. Indo-austr. II 401 (1923) [redescr.].[2] — Manipur (6,000 ft.).[1]

10. Odontocraspis, Swinhoe (1894).

Odontocraspis, Swinh., A. M. N. H. (6) XIV 439 (1894); Hmpsn., F. I. Moths IV 487, f. 258 (1896) [redescr.]; Grünberg, Seitz Lep. Indo-austr. II 401 (1923) [redescr.]. — Type: *hasora,* Swinh.

hasora, Swinhoe.

Odontocraspis hasora, Swinh., A. M. N. H. (6) XIV 439 (1894)[1]; *id.,* T. E. S. 1895, 28, t. 1 ff. 11, 11*a* (1895)[2]; Hmpsn., F. I. Moths IV 487. f. 258 (1896) [redescr.][3]; Grünbeg, Seitz Lep. Indo-austr. II 401, t. 35 f. *c* (1923) [redescr.][4]. — Cherrajpunji.[1 2] ——— Borneo.[4]

11. Chilena, Walker, (1855).

Chilena, Wlk., Cat. V 1070-1071 (1855) Mo., Lep. Ceylon III 543 (1887); Hmpsn., F. I Moths I 416 (1892) [redescr.]; Grünberg, Seitz Lep. Indo-austr. II 402 (1923) [redescr.]. — Type : *similis,* Wlk.

similis, Walker.

Chilena similis, Wlk., Cat. V 1071 (1855)[1]; Hmpsn., F. I. Moths I 416, f. 287 (1892) [redescr.][2]; Grünberg, Seitz Lep. Indo-austr. II 402, t. 33 f. *a* (1923) [redescr.].[3] — Peshawar. Punjab.[2 3] Nepal.[1 2 3], Saran.

strigula, Walker.

Lasiocampa strigula, Wlk., Cat. XXXII 563 (1865)[1]; Forsayeth, T. E. S. 1884, t. 14 ff. 6, 6*a* 6*b* (1884).[2] — Peshawar. N. India.[1], Nepal.[5]

Chilena strigula, Hmpsn., F. I. Moths I 417 (1892) [redescr.][3]; *id.,* Ill. Het. IX 9 (1893)[4]; Grünberg, Seitz Lep. Indo-austr. II 402 (1923) [redescr.].[5] — Saran., Mhow.[2 3] Surat. Campbellpur.[3]

[?] *Chilena similis* [*nec* Wlk.], Mo., Lep. Ceylon II 543, t. 212 f. 5 (1887).[6] — Ceylon.[3 5] Trincomali.[4]

Larva descr.[2 3 5], on *Acacia.*[5]

12. Clisiocampa, Curtis, (1828).

Clisiocampa, Curt., Birt. Ent. V, t. 229 (1828); Steph., Ill. Brit. Ent. Haust. II 48 (1829); Wlk., Cat. VI 1485 (1855); Kirby, Cat. Het. I 838 (1892); Hmpsn., F. I. Moths I 417 (1892) [redescr.]. — Type: *castrensis,* Linn. [Europe].

Malacosoma [*nec* Hb.] Grünberg, Seitz Lep. Pal. II 150 (1911) [redescr.]; *id.,* Lep. Indo-austr. II 402 (1923).

[Note — *Malacosoma, Hübner,* 1822, equivalent to *Diplura,* Rambur 1866, both having *loti,* Ochs., as genotype, is apparently distinct from *Clisiocampa*].

flavomarginata, Poujade.

Bombyx flavomarginata, Pouj., Bull. S. E. Fr. (6) VI, p. Xcii (1886).[1] — Moupin.[1]

Malacosoma ? flavomarginata, Kirby, Cat. Het. I 819 (1892)[2]; Leech, T. E. S. 1899, 117 (1899).[3]

indica, Walker.

Clisiocampa indica, Walk., Cat. VI 1489 (1855)[1]; Hmpsn., F. I. Moths I 417, f. 288 (1892) [redescr.].[2] — Simla.[2], Dehra Dun.[2] Muktesar (Kumaon; 7,000 ft.)·

Malacosama indica, Grünberg, Seitz Lep. Indo-austr. II 402, t. 33 f. *a* (1923) [redescr.].[3]

Testacea, Motschulsky.

Clisiocampa testacea, Mots., Et. Ent. IX, 32 (1860).[1] — Pu-tsu-fong.[2], ——

Clisiocampa neustria, [*nec* Linn.], Leech, T. E. S. 1899, 111 (1899)[2]; Grünberg, Seitz Lep. Pal. II 150-151, t. 24 f. *b* (1911).[3] — N., C. and W. China.[2] Japan.[1 2], Korea.[2] Amur.[2]

vulpes, Hampson.

Clisiocampa vulpes, Hmpsn., B. J. XIII 233, t. B f. 26 (1900).[1] — Chitral (10,000 ft.).[1]

Malacosoma vulpes, Grünberg, Seitz Lep. Indo-austr. II 402 (1923) [redescr.].[2]

13. Alompra, Moore (1872).

Type: *ferruginea,* Mo.

Alompra, Mo., P. Z. S. 1872, 579 (1872); Hmpsn., F. I. Moths I, 417-418 (1892) [redescr.]; Grünberg, Setiz Lep. Indo-austr. II 402 (1923) [redescr.].

ferruginea, Moore.

Alompra ferruginea, Mo., P. Z. S. 1872, 580, t. 33 f. 8 (1872)[1]; Hmpsn, F. I. Moths I 418, f. 289 (female, not male as stated] (1892) [redescr.][2]; Dudgeon, B. J. XIII 409 (1901) [female descr.][3]; Hmpsn., B. J. XX 111 (1910) [male descr.][4] ; Grünberg, Setiz Lep. Indo-austr. II 402, t. 35 f. *d* (1923) [redescr.].[5] — Darjiling.[1] Sikkim. (2,000 ft.).[3] Bhutan (2,000 ft).[3] Assam.[8]

Taragama, indicus, Conte, Rapp. Lab. Et. Soie XIII 24, t. 6 f. 1 (1909)[6]; Hmpsn., B. J. XX 111 (1910) [syn. of *ferruginea*][7]; Grünberg, Seitz Lep. Indo-austr. II 394 (1923) [redescr.].[8]

14. Kosala, Moore (1879).

Kosala, Mo., P. Z. S. 1879, 407 (1879); Hmpsn., F. I. Moths I 418 (1892) [redescr.]; Grünberg, Seitz Lep. Indo-austr. II 402 (1923) [redescr.].

Type: *sanguinea*, Mo.

flavosignata, Moore.

Eutricha flavosignata, Mo., Lep. Atk. pp. 77-78, t. 3 f. 17 (1879).[1]

Darjiling.[1]

Kosala flavosignata, Hmpsn., F. I. Moths I 419 (1892) [redescr.][2]; Grünberg, Seitz Lep. Indo-austr. II 403, t. 35 f. *d* (1923) [redescr.].[3]

modulata, Swinhoe

Gastropacha modulata, Swinh., T. E. S. 1890, 197, t. 6 f. 18 (1890)[1].

Bhamo.[1 2 3]

Kosala modulata, Hmpsn., F. I. Moths I 419, f. 290 (1892) [redescr.][2]; Grünberg, Seitz Lep. Indo-austr. II 403, t. 35 f. *d* [as *undulata*] (1923) [redescr.].[3]

rufa, Hampson.

Kosala rufa, Hmpsn., F. I. Moths I 419 (1892)[1]; Grünberg, Seitz Lep. Indo-austr. II 403 (1923) [redescr].[2]

Nagas.[1 2]

sanguinea, Moore.

Kosala sanguinea, Mo., P. Z. S. 1879, 408, t. 33 f. 8 (1879)[1]; Hmpsn., F. I. Moths I 418 (1892) [redescr.][2]; Swinh., T. E. S. 1895, 28 (1895)[3]; Leech., T. E. S. 1899, 112 (1899)[4]; Elwes, B. J. XIII 409 (1901)[5]; Grünberg, Seitz Lep. Indo-austr. II 403, t. 35 f. *d* (1923) [redescr.].[6]

Khasis.[1 2 5 6]
Cherrapunji.[3]
Sikkim.[2 6]
Darjiling.[5]
Omei-shan.[4]

15. Trichiura, Stephens (1828).

Trichiura, Steph., Ill. Brit. Ent., Haust. II 42 (1828); Wlk., Cat. VI 1479 (1855); Rambur, Cat. Lép. Andal., p. 360 (1866); Hmpsn., F. I. Moths I 420 (1892) [redescr.]; Grünberg, Seitz Lep. Pal. II 152 (1911) [redescr.]; *id.*, Lep. Indo-austr. II 403 (1923) [redescr.].

Type: *crataegi*, Linn. [Europe].

khasiana, Moore.

Trichiura khasiana, Mo., Lep. Atk., p. 82, t. 3 f. 21 (1879)[1]; Hmpsn., F. I. Moths I. 429, f. 291 (1892) [redescr.][2]. Swinh., T. E. S. 1895, 28; (1895)[3]; Elwes, B. J. XIII 409 (1901)[4]; Grünberg, Seitz Lep. Indo-austr. II 403, t. 35 f. *e* (1923) [redescr.].[5]

Khasis.[1 2 5]
Cherrapunji.[3]
Sikkim.[2 5]
Darjiling.[4]

16. Crinocraspeda, Hampson (1892).

Crinocraspeda, Hmpsn., F. I. Moths I 420, f. 292 (1892); Grünberg, Seitz Lep. Indo-austr. II 403 (1923) [redescr.].	Type: *torrida,* Mo.

torrida, Moore.

Gastropacha torrida, Mo., Lep. Atk., p. 76, t. 3 f. 19 (1879)[1]; *Crinocraspeda torrida,* Hmpsn., F. I. Moths I 420-421, f. 292 (1892) [redescr.][2]; Swinh., T. E. S. 1895, 28 (1895)[3]; Dudgeon, B. J. XIII 409 (1901)[4]; Grünberg, Seitz Lep. Indo-austr. II 403, t. 35 f. *d* (1923) [redescr.].[5]	Darjiling.[1] Shillong.[3] Cherrapunji.[3] Manipur.[2 5] Bhutan.[4]

17. Trabala, Walker (1856).

Amydona, Wlk., Cat. VI 1413 (1855) [*prœocc.*].

Trabala. Wlk. Cat. VII 1785 (1856); Mo., Lep. Ceylon II 145 (1883); Hmpsn., F. I. Moths I 421 (1892) [redescr.]; Grünberg, Seitz Lep. Pal. II 165-166 (1911) [redescr.]; *id.,* Lep. Indo-austr. II 403 (1923).	Type : *vischnou,* Lef.

irrorata, Moore.

Trabala irrorata, Mo., T. E. S. 1884, 375 (1884)[1]; *id.,* J. A. S. B. LIII 205 (1885)[2]; *id.,* J. Linn. Soc. XXI 55 (1886)[3]; Hmpsn., F. I. Moths I 442 (1892) [redescr.][4]; Grünberg, Seitz Lep. Indo-austr. II 403, t. 33 f. *d* (1923) [redescr.].[5]	Tavoy.[4] Mergui.[4] ——— Java.[4 5] Borneo.[4 5]

vischnou, Lefebvre.

Gastropacha vischnou, Lef., Zool. Journ. III 207 (1827).[1]	Throughout India, Burma and Ceylon.
[?*Bombyx hibisci,* Fabr., Syst. Ent. p. 564 (1775); Kiry, Cat. Het. I 825 (1892) (East Indies)].	——— Java. [3 7]
Gastropacha sulphurea, Kollar, Hügel's Kaschmir IV 471 (1848).[2]	China.[3 4 7 10]

Amydona basalis, Wlk., Cat. VI 1415 (1855).[3]

Amydona prasina, Wlk., Cat. VI 1417 (1855).[4]

Amydona pallida, Wlk., Cat. VI 1417 (1855).[5]

Trabala mahananda, Mo., P. Z. S. 1865, 821 (1865).[6]

Trabala vishnu, Hmpsn., F. I Moths I 421-422, f. 293 (1892) [redescr.][7]; Mo., Lep. Ceylon II 146, t. 138 ff. 1, 1*a,* 1*b,*

2, *2a 2b* (1883)[8]; Hmpsn., Ill. Het. IX 9 (1893)[9]; Swinh., T. E. S. 1895, 28 (1895)[9] ; Leech, T. E. S. 1899, 113 (1899)[10]; Dudgeon, B. J. XIII 410 (1901)[11]; Lefroy. Ind. Ins. Life. p. 498, tab. 66 (1909)[12]; Fletcher, Rept. Second Entl. Meeting, pp. 87, 246 (1917)[13]; *id.*, Rept. Third Entl. Meeting, I 103 (1919).[14]

Trabala vishnou, Kirby, Cat. Het. I 825 (1892) [*vishnon,* err. typogr.][15]; Grünberg, Seitz Lep. Pal. II 166, t. 27 f. *a* (1911) [larva descr.][16]; *id.,* Lep. Indo-austr. II 403, t. 35 ff. *e* (1923).[17]

Larva descr.[7 8 12], on *Polygonum orientale*[11], *Ricinus communis* [12 13 14], *Eugenia jambolana*[12], *Shorea robusta*[12], *Terminalia catappa*[12], cultivated Geranium[14] and rose[14], *Quisqualis*[14], etc.

18. Lenodora, Moore (1883).

Lenodora, Mo., Lep. Ceylon II 144-145 (1883); Hmpsn., F. I. Moths I 422-423 (1892) [redescr.]; Grünberg, Seitz Lep. Indo-austr. II 404 (1923) [redescr.] — Type: *vittata,* Wlk.

crenata, Hampson.

Lenodora crenata, Hmpsn., B. J. XIII 233 (1900)[1]; Hmpsn., B. J. XVI 197 (1905) [male descr.][2]; Grünberg, Seitz Lep. Indo-austr. II 404, t. 33 f. *g* (1923) [redescr.].[3] — Horton Plains (Ceylon).[1 2]

hyalomelæna, Hampson.

Lenodora, hyalomelæna, Hmpsn., B. J. XVI 197, t. D, f. 9 (1905)[1]; Grünberg, Seitz Lep. Indo-austr. II 404, t. 46 f. *b* (1923) [redescr.].[2] — Palnis (7,000 ft).

pallidochrea, Hampson.

Lenodora pallidochrea, Hmpsn., F. I. Moths IV 487 (1896)[1]; Grünberg, Seitz Lep. Indo-austr. II 404, t. 46 f. *a* (1923) [redescr.].[2] — Khasis.[1 2]

semihyalina, Swinhoe.

Lenodora semihyalina, Swinh., T. E. S. 1890, 196, t. 6 ff. 10, 16 (1890)[1]; Hmpsn., F. I Moths I 424 (1892) [redescr.][2]; Grünberg, Seitz Lep. Indo-austr. II 404, t. 35 [as *semihyalinata*] (1923) [redescr.].[3] — Nagas.[2 3] Rangoon.[2] Bassein.[2] E. Pegu.[2]

signata, Moore.

Lenodora, signata, Mo., T. E. S. 1884, 374 (1884)[1]; Hmpsn., F. I. Moths I 423-424 (1892) [redescr.][2]; Grünberg, Seitz Lep. Indo-austr. II 404, t. 35 f. *f* [form *fusca*] (1923) [redescr.].[3] — Bengal.[1] Dehra Dun.[2] Bombay.[2 6] Nilgiris.[2 5]

Lenodora fasciata, Mo., T. E. S. 1884, 374 (1884)[4]; Hmpsn., Ill. Het. VIII 65, t 142 f. 6 (1891).[5]	
Lenodora fusca, Mo., T. E. S. 1884, 374 (1884).[6]	
vittata, Walker.	
Lasiocampa vittata, Wlk., Cat. VI 1440 (1855) [N. India].[1]	Almora.[4] Raipur (C. P.).
Miresa ? subcostalis, Wlk., Cat. XXXII 476 (1865).[2]	Throughout South India.[4], Ceylon. [2 3 4]
Lenodora subcostalis, Mo., Lep. Ceylon II 145, t. 137, ff. 3, 3*a,* 3*b* (1883).[3]	Colombo.[5] Kandy.[5]
Lenodora vittata, Hmpsn., F. I. Moths I 423, f. 294 (1892) [redescr.][4]; *id.,* Ill. Het. IX 73, t. 161 ff. 4, 11 (1893) [redescr.][5]; Grünberg, Seitz Lep. Indo-austr. II 404, t. 35 f. *f* (1923) [redescr.].[6] *Larva* descr.[3 4 6], on Gramineæ. [3 6]	Pundaluoya.[5] Peradeniya.

19. Estigena, Moore (1859).

Estigena, Mo., Cat. Lep. E. I. Co. Mus. II 426 (1859); Mo., Lep. Ceylon II 148-149 (1883) Hmpsn., F. I. Moths I. 424 (1892) [redescr.]; Grünberg, Seitz Lep. Pal. II 169-170 (1911); *id.,* Lep. Indo-austr. II 405 (1923) [redescr.].	Type: *pardalis,* Wlk.
pardalis, Walker.	
Megasoma pardale, Wlk., Cat. VI 1453 (1855).[1]	Throughout India and Ceylon.[7]
Estigena nandina, Mo., Cat. Lep. E. I. Co. Mus. II 427 (1859)[2]; Mo., Lep. Ceylon II 149, t. 142 ff. 1, l*a* (1883).[3]	Masuri., Khatmandu. Sikkim.[8]
Gastropacha abstracta, Wlk. Cat. XXXII 551 (1865).[4]	Bhutan.[8], Saran.
Lebeda scriptiplaga, Wlk., Cat. XXXII 569 (1865).[5]	Pusa., Nagas.[7]
Estigena pardalis, Mo., Cat. Lep. E. I. Co. Mus. 11 426, t. 13*a,* ff. 1, 1*a* (1859)[6]; Hmpsn., F. I. Moths I 424-425, f. 295 (1892) [redescr.][7]; Dudgeon, B. J. XIII 410 (1901)[8]; Grünberg, Seitz Lep. Pal. II 170, t. 27 f. *c* (1911)[9]; *id.,* Lep. Indo-austr. II 405, t. 35 f. *f* (1923) [redescr.].[10]	Burma.[10], Mergui.[7] Andamans.[7] Pundaluoya.[11] Ceylon.[4 5], ______ Java. [1 7 9], Arabia.
Estigena pardale, Hmpsn., Ill. Het. IX 9 (1893).[11]	

20. Odonestis, Germar (1811).

Odonestis, Germar, Prodr. p. 49 (1811); Hmpsn., F. I. Moths I 425 (1892) [redescr.]; Grünberg, Seitz Lep. Indo-austr. II 405 (1923).	Type : *potatoria,* Linn. [Europe].

Cosmotriche, Hübner, Verz. p. 188 (1822); Grünberg, Seitz Lep. Pal. II 163-164 (1911) [redescr.]; *id.,* Lep. Indo-austr. II 406 (1923) [part.]. — Type: *potatoria,* Linn.

Chrostogastria, Hübner, Verz. p. 189 (1822); Hmpsn., B. J. XX 111 (1910) [redescr.]. — Type: *pruni,* Linn.

Odenestis, Wlk., Cat. VI 1409 (1855). — Type: *potatoria,* Linn.

castanea, Hampson.

Odonestis castanea, Hmpsn., F. I. Moths I 426 (1892)[1]; *id.,* B. J. XIII 410 (1901)[2]; Grünberg, Seitz Lep. Indo-austr. II 407, t. 32 f. *b* (1923) [redescr.] [*Cosmotriche*].[3]

Darjiling.[2]
Sikkim.[1 3]

divisa, Moore.

Odonestis divisa, Mo., P. Z. S. 1879, 408 (1879)[1]; *id.,* Lep. Ceylon II 150, t 141 f. 2 (1883)[2]; Hmpsn., F. I. Moths I 426 (1901)[4]; [part.][3]; Elwes, B. J. XIII 410 (1901)[4]; Grünberg, Seitz Lep. Indo-austr. II 407, t. 32 f. *a* (1923) [redescr.].[5]
..........*sulphurea,* Auriv.,

Sikkim.[3 4]
Ceylon.[1 2 5]
_____, Sunda Isds.[5]
China.[5], Japan.[5]
Amur.[5]
Jaintia Hills.[1-3]

fossa, Swinhoe.

Odonestis fossa, Swinh. A. M. N. H. (6) XIX 410 (1897)[1]; Hmpsn., B. J. XVI 198, t. D. f. 14 (1905) [redescr.][2]; Grünberg, Seitz Lep. Indo-austr. II 408, t 33 f. *c* (1923) [redescr.] [*Cosmotriche*].[3]

Pu-tsu-fong.[1 2]

hampsoni, Leech.

Odonestis hampsoni, Leech, T. E. S. 1899, 115 (1899)[1]; Grünberg, Seitz Lep. Indo-austr. II 407 (1923) [redescr.] [*Cosmotriche*].[2]

Nagas (6,000 ft.).[1 2]

isocyma, Hampson.

Odonestis isooyma, Hmpsn., F. I. Moths I 427 (1892)[1]; Grünberg, Seitz Lep. Indo-austr. II 407 (1923) [redescr.] [*Cosmotriche*].[2]

N.-W. Himalaya.[3]

læta, Walker.

Amydona læta, Walk., Cat. VI 1416 (1855)[1]; Mo., Cat. Lep. E. I. Co. Mus. II, t. 12*a* ff. 7, 7*a* (1859).[2]

Odonestis læta, Hmpsn., F. I. Moths I 425-426, f. 296 (1892) [redescr.][3]; Leech, T. E. S. 1899, 114 (1899)[4]; Dudgeon, B. J. XIII 410 (1901)[5].

Sikkim.[3], Sylhet.[1 3]
Burma.[3]
Ta-tsien-lu.[11]
Chia-ting-fu.[11]
______

Cosmotriche læta, Grünberg, Seitz Lep. Pal. II 165, t. 26, f. *g* (1911)[6]; *id.*, Lep. Indo-austr. II 406-407, t. 32 f. *c* (1923) [redescr.].[7]

China.[4], Korea.[4,6] Japan.[4,6] Siberia.[4,6] Sumatra.[3,7] Java.[3,7]

Lasiocampa decisa, Wlk., Cat. VI 1441 (1855) [N. India][8]; Elwes, B. J. XIII 410 (1901) [*Odonestis;* distinct from *læta*].[9]

Lasiocampa inobtrusa, Wlk., T. E. S. (3) I 85 (1862)[10]; Leech., T. E. S. 1899, 114-115 (1899) [*Odonestis;* distinct from *læta*].[11]

Odonestis ossa, Swinh., A. M. N. H. (6) XIX 410 (1897)[12]; Hmpsn., B. J. XI 294 (1897) [syn. of *læta*].[13]

Larva descr.[3,6,7], on grasses.[5]

lidderdalii, Druce.

Odonestis lidderdalii, Druce, A. M. N. H. (7) III 471 (1899)[1]; Hmpsn. B. J. XIII 233 (1900) [male redescr.][2]; Grünberg, Seitz Lep. Indo-austr. II 408, t. 33 f. *d* (1923) [redescr.] [*Cosmotriche*].[3]

Darjiling.[1,3]

pruni, Linnæus.

Bombyx pruni, Linn., Syst. Nat. (ed. X) I 498 (1758)[1]; Hübner, Bomb. f. 186 (1804 ?).[2]

Khasis.[5], Jaintias.[5] Pu-tsu-fong.[4]

Odonestis pruni, Alpha., Rom., Mém. IX 133 (1897)[3]; Leech. T. E. S. 1899, 115 (1899)[4]; Hmpsn., B. J. XX 111 (1910) [redescr.] [*Chrostogastria*][5]; Grünberg, Seitz Lep. Pal. 11 170, t. 27 f. F. (1911) [redescr.].[6]

Larva descr.[6]

_____, W. China.[3,5] Japan.[4,5] Amur.[4,5] Bithynia.[5] C. Europe.[1,5]

purpurascens, Hampson.

Estigena purpurascens, Hmpsn. Ill. Het. IX 74, t. 160 f. 10 (1893).[1]

Kelani Valley[1,2] (Ceylon).

Odonestis divisa, Hmpsn. F. I. Moths I 426 (1892) [partim].[2]

Cosmotriche purpurascens, Grünberg, Seitz Lep. Indo-austr. II 407, t. 33 f. *d* (1923) redescr.].[3]

pyriformis, Moore.

Odonestis, pyriformis, Mo., P. Z. S. 1879, 408, t. 34 f. 7 (1879)[1]; Hmpn., F. I. Moths I 427 (1892) [redescr.][2]; Dudgeon, B. J. XIII 411 (1901)[3]; Elwes, *l. c*[4]; Grünberg, Seitz Lep. Indo-austr. II 407, t. 32 f. B (1923) [redescr.] [*Cosmotriche*].[5]

Larva descr.[2]

Masuri.[1], Tonglu (Sikkim, 10,000 ft.).[4] Darjiling.[5] Sikkim. (5,000 ft.)[3]

signata, Moore.

Odonestis signata, Mo., Lep. Atk. p. 76 (1879)[1]; Hmpsn., F. I. Moths I 426 (1892) [redescr.][2]; Dudgeon, B. J. XIII 411 (1901)[3]; Grünberg, Seitz Lep. Indo-austr. II 407, t. 32 f. B (1923) [redescr.] [*Cosmotriche*].[4]	Darjiling.[1 4] Sikkim. (1,000 ft.).[2] Bhutan.[3]

urda, Swinhoe.

Odonestis urda, Swinh., A. M. N. H. (8) XVI 178 (1915).[1]	Khasis.[1]

21. Lasiocampa, Schrank (1802).

Lasiocampa, Schrank, Fauna Boica II (ii) 147, 154 (1802).	Type: *quercifolia,* Linn. [Europe].
Gastropacha, Ochsenheimer, Schmett. III 239 (1810); Hmpsn., F. I. Moths I 428 (1892) [redescr.]; Grünberg, Seitz Lep. Pal. II 168 (1911) [redescr.]; *id.,* Lep. Indo-austr. II 406 (1923).	Type: *quercifolia,* L.
Phyllodesma, Hübner, Verz., p. 190 (1822).	Type: *quercifolia,* L.

divaricata, Moore.

Gastropacha divaricata, Mo., T. E. S. 1884, 374 (1884)[1]; Hmpsn., F. I. Moths I 429 (1892) [redescr.][2]; Grünberg, Seitz Lep. Indo-austr. II 406, t. 33 f. c. (1923) [redescr.].[3]	Darjiling.[1 3]

khasiana, Swinhoe.

Gastropacha khasiana, Swinh., A. M. N. H. (7) XV 499 (1905)[1]; Hmpsn., B. J. XX 111-112 (1910) [male redescr.].[2]	Khasis.[1 2]

undulifera, Walker.

Gastropacha undulifera, Wlk., Cat. VI 1395 (1855)[1]; Hmpsn., F. I. Moths I 428, f. 297 (1892) [redescr.][2]; Grünberg, Seitz Lep. Indo-austr. II 406, t. 33 f. *c* (1923) [redescr.].[3]	N. India.[1] Dharmsala.[2 3]

vitellius, Oberthur.

Lasiocampa vitellius, Obth. Et. Lep. Comp. V 337, t. 83 f. 785 (1911) ♂	Tchang-kou (E. Tibet border.)

22. Stenophylloides, Hampson (1892).

Stenophylloides, Hmpsn., F. I. Moths I 429, f. 298 (1892); Grünberg, Seitz Lep. Indo-austr. II 406 (1923) [redescr.].	Type: *sikkima,* Mo.

encausta, Hampson.

Stenophylloides encausta, Hmpsn., B. J. XIII 233-234 (1900)[1]; Grünberg, Seitz Lep. Indo-austr. II 406, t. 33 f. *G* [as *encrusta*] (1923) [redescr.].[2] — Simla.[1 2]

sikkima, Moore.

Gastropacha sikkima, Mo., Lep. Atk. pp. 75-76 (1879)[1]. — Darjiling.[1 3 4]

Stenophylloides sikkima, Hmpsn., F. I. Moths I 429-430, f. 298 (1892) [redescr.][2]; *id.,* B. J. XIII 411 (1901)[3] Grünberg, Seitz Lep. Indo-austr. II 406, t. 32 f. *d* (1923) [redescr.].[4]

sinuata, Moore.

Gastropacha sinuata, Mo., Lep. Atk. p. 76 (1879)[1]; Hmpsn., F. I. Moths I 428-429 (1892) [redescr.].[2] — Darjiling.[1 3]

Stenophylloides, sinuata, Grünberg, Seitz Lep. Indo-austr. II 406, t. 35 f. F. (1923) [redescr.].[3]

75 Species.

PART VIII
AMATIDAE (SYNTOMIDAE).

Syntomidæ : Hampson, F. I. Moths I 209-228 (1892) [key to genera]; *id.*, Cat. Lep. Phal. I (1898) [revision ; keys to genera and species]; Seitz, Lep. Pal. II 37-42 (1910); *id.*, Lep. Indo-Austr. II 61-92 (1912-1913).

Amatidæ : Hampson, Cat. Phal. Suppl. I, pp. 1-388 (1914).

1. Psichotoë, Boisduval (1829).

Psichotoë, Boisd., Mon. Zyg., p. 129 (1829); Hampson, F. I. Moths I 225, f. 141 (1892) [redescr.]; *id.*, Cat. Phal. I 32, f. 9 (1898) [redescr.]; Seitz., Lep. Indo-Austr., II 63 (1912).Type: *duvauceli,* Boisd.

Cacosoma, Felder, Reise Novara, Lep. IV, p. 2 (1874). Type *gnatula,* Boisd. [S. Africa].

duvauceli, Boisduval—

Psichotoë duvauceli, Boisd., Mon. Zyg., p. 129, t. 8, f. 5 (1829)[1]; Hmpsn., F. I. Moths I 226, f. 141 (1892) [redescr.][2]; *id.*, Cat. Phal. I 32, f. 9 (1898) [redescr.][3]; Seitz, Lep. Indo-Austr. II 63, t. 10, f. A (1912).[4]

Karachi.[2, 4], Lahore. Pusa., Calcutta.[2, 3] Gauhati., Jorhat. Assam.[3, 4], Burma.[3, 4] Rangoon., Minbu. Mandalay., Hsipaw. Pyinmana., Pegu. Mergui.

2. Callitomis, Butler (1876).

Callitomis, Butl., J. Linn. Soc. XII 351, t. 27, f. 1 (1876); Hmpsn., F. I. Moths I 224, f. 139 (1892) [redescr.] *id.*, Cat. Phal. I 33 (1898) [redescr.]; *id.*, B. J. XIII 45 (1900) [key to spp.]; Seitz, Lep. Pal. II 41 (1910) [redescr.]; *id.*, Lep. Indo-Austr. II 63 (1912) [redescr.]. Type: *syntomoides,* Butl.

leucosoma, Butler—

Callitomis leucosoma, Butl., J. Linn. Soc. XII 351 (1876)[1]; Hmpsn., F. I. Moths I 225 (1892) [redescr.][2]; *id.*, Cat. Phal. I 34, t. 1, f. 3 (1898) [redescr.][3]; Seitz, Lep. Pal. II 41, t. 9, f. H (1910).[4]

Kashmir.[1-4]

multifasciata, Hampson.

Callitomis multifasciata, Hmpsn., F. I. Moths I 225 (1892)[1]; *id.,* Cat. Phal. I 33, t. 1, f. 1 (1898) [redescr.][2]; Seitz, Lep. Indo-Austr. II 63, t. 10, f. A (1912)[3].

Nagas.[1-3]
Sikkim.[2, 3]

syntomoides, Butler.

Callitomis syntomoides, Butl., J. Linn. Soc. XII 351 (1876)[1]; Hmpsn., F. I. Moths I 224, f. 139 (1892) [redescr.][2]; *id.,* Cat. Phal. I 34, f. 12 (1898) [redescr.][3]; Seitz, Lep. Pal. II 41 (1910)[4]; *id.,* Lep. Indo-Austr. II 63 (1912).[5]

Kashmir.[1-4]
Murree.[3]
Simla.[2]

3. Ceryx, Wallengren (1863).

Ceryx, Wlgn., Wien. Ent. Mon. VII 140 (1863); Hmpsn., Cat. Phal I 35 (1898) [redescr.]; *id.,* B. J. XIII 45-46 (1900) [key to spp.]; Seitz, Lep. Pal. II 41 (1910) [redescr.; habits]; *id.,* Lep. Indo-Austr. II 63 (1912) [redescr.]. Type: *anthraciformis,* Wlgn. [S. Africa].

Agaphthora Meyrick, Proc. Linn. Soc. N. S. Wales (2) I 774 (1886). Type: *guttulosa,* Wlk. [Papuan subregion].

Syntomoides, Hmpsn., F. I. Moths I 209-210, f. 137 (1892). Type: *imaon* Cram.

cherra, Moore.

Syntomis cherra, Mo., Lep. Atk., pp. 12-13 (1879)[1]; Hmpsn., F. I. Moths I 220 (1892) [redescr.][2].

Khasis.[1-8]
Shillong.[3, 4]
Cherrapunji.[1]

Syntomoides quisqualis, Swinh., A. M. N. H. (6) XIV 441 (1894)[3]; *id.,* T. E. S. 1895, 30 (1895)[4]; Hmpsn., F. I. Moths IV 465 (1896).[5]

Ceryx cherra, Hmpsn., Cat. Phal. I 41, t. 1, f. 13 (1898) [redescr.][6]; *id.,* B. J. XIII 46 (1900)[7]; Seitz, Lep. Indo-Austr. II 65, t. 10, f. D (1912).[8]

diptera, Fabricus.

Zygaena diptera Fb., Syst. Ent. p. 555 (1775).[1]

S. India[4, 11]
Tranquebar.[2]

Sphinx atercus, Stoll, Pap. Exot. IV 240, t. 400, f. A. (1782)[2].

Coromandel Coast.[2, 7]
Coimbatore.[3, 8]

Syntomis incipiens, Wlk., Cat. XXXI 68-69 (1864).[3]

Syntomoides incipiens, Hmpsn., F. I. Moths I 211 (1892)[4].

Syntomoides brachypecten, Hmpsn., F. I. Moths I 211 (1892)[5].

Ceylon.[8, 11]

Psychotoë brachypecten, Hmpsn., Ill. Het. IX 61, t. 157, f. 23 (1893).[6]	Kanthalai.[4] Anuradhapura.[5]
Syntomis aterea, Hmpsn., F. I. Moths I 223 (1892).[7]	
Ceryx diptera, Hmpsn., Cat. Phal. I 36-37, f. 15 (1898) [redescr.][8]. *id.,* B. J. XIII 45 (1900)[9]; Seitz, Lep. Indo-Ausr. II 64, t. 9, f. F (1912)[10].	
Ceryx diptera f. *brachypecten,* Seitz, Lep. Indo-Austr. II 64 (1912).[11]	

ginorea, Swinhoe.

Eressa ginorea, Swinh., A. M. N. H. (6) XIV 441 (1894)[1]; *id.,* T. E. S.. 1895, 32, t. 1, f. 8 (1895)[2]; Hmpsn., F. I. Moths IV 465 (1896).[3]	Khasis.[1, 5] Cherrapunji.[1, 2]
Ceryx ginorea, Hmpsn., Cat. Phal. I 37 (1898) [redescr.][4]; Seitz., Lep. Indo-Austr. II 64, t. 10, f. B (1912).[5]	

codarti, Boisduval.

Syntomis godarti, Boisd., Mon. Zyg., p. 115, t. 7, f. 3 (1829)[1].	Khatmandu (Nepal). Sikkim.[6]
Syntomis libera, Wlk., Cat. XXXI 78-79 (1864).[2]	Gauhati., Shillong.
Syntomis fytchei, Mo, P.Z.S. 1871, 246, t. 18, f. 3 (1871).[3]	Cherrapunji.[9], Yunnan.[3, 6]
Syntomis cupreipennis, Butl., J. Linn. Soc. XII 347-348 (1876).[4]	Burma.[6] Lashio (U. Burma).
Syntomoides godartii, Hmpsn., F. I. Moths I 210 (1892) [*partim*].[5]	Decca., Pusa.
Ceryx godarti, Hmpsn., Cat. Phal. I 43-44 (1898) [redescr.][6]; *id.,* B. J. XIII 46 (1900)[7]; Seitz, Lep. Indo-Austr. II 88, t. 9, f. F. (1913)[8].	Calcutta.[9, 6] S. India.[6, 8] Ceylon.[6, 8]
Syntomoides cupreipennis, Swinh., T. E. S. 1895 30 (1895).[9]	S. China.[8,] Malacca.[2, 6] Sumatra.[6]

hyalina, Moore.

Syntomis hyalina, Mo., Lep. Atk. p. 13 (1879).[1]	Sikkim.[3, 6, 7] Darjiling.[1]
Syntomis volans, Swinh. T. E. S. 1890, 173, t. 6, f. 6 (1890).[2]	
Syntomoides hyalina, Hmpsn., F. I. Moths I 211 (1892).[3]	Assam.[7]
Syntomoides volans, Hmpsn., F. I. Moths I 211 (1892).[4]	Nagas.[3, 6] Cherrapunji.

Ceryx hyalina, Hmpsn., B. J. XIII 46 (1900)[5]; *id.,* Cat. Phal. I 42 (1898) [redescr.][6]; Seitz, Lep. Indo-Austr. II 66, t. 9, f. G. (1912).[7] — Shillong. Burma.[7] Karen Hills.[2, 4, 6]

imaon, Cramer.

Sphinx imaon, Cram., Pap. Exot. III 94, t. 248, f. E (1780).[1] — Kashmir.[12, 13] Kangra.[9]

Syntomis imaon, Wlk., Cat. I 127 (1854).[2] — Sikkim.[9], Darjiling.

Syntomis fusiformis, Wlk. Cat. VII 1595 (1856).[3] [Hindostan]. — Khasis.[9], Shillong.[14, 15]

Syntomis approximata, Wlk., Cat. XXXI 79 (1864).[4] — Cachar.[6, 9]

Syntomis artina, Butl. J. Linn. Soc. XII 347 (1876)[5]; Mo., Lep Ceylon II 35, t. 94, f. 4, (1883).[5a]

Syntomis sargania, Butl., T. E. S. 1879 4 (1879).[6] — Calcutta.[5, 9]

Syntomis mota, Swinh., T. E. S. 1891. 134 (1891).[7] — Bombay.[2]

Syntomoides imaon, Hmpsn., F. I. Moths I 210, f. 137 (1892) [redescr.].[8] — Nilgiris.[7, 9, 13]

Ceryx imaon, Hmpsn., Cat. Phal. I 44. f. 18 (1898) [redescr.][9]; *id.,* B. J. XIII 46 (1900)[10]; Piepers, Tijds. Ent. XLVII 49[11]; Seitz, Lep. Pal. II 41, t. 9, f. H (1910)[12]; *id.,* Lep. Indo-Austr. II 65, t. 10, f. E. (1912).[13] — Ceylon.[5, 9], Kandy.[13] Lashio (U. Burma) Karen Hills.[9] Maymyo., Tenasserim.[9]

Syntomoides sargania, Swinh., T. E. S. 1895. 30 (1895).[14] — ______________

Syntomoides fusiformis, Swinh. T. E. S. 1895. 30 (1895) [*approximata* is synon.].[15] — Hongkong.[9, 13] Sumatra.[4]

Larva descr.[11]

pleurasticta, Hampson.

Ceryx pleurasticta, Hmpsn., A. M. N. H. (7) VIII 165 (1901) [*nec pleurosticta,* Hmp. 1898][1]; *id.,* B. J. XVI 151, t. D, f. 2 (1904)[2]; *id.,* Cat. Phal. Suppl. I 6, t. 1, f. 7 (1914).[3] — N. Kanara.[1, 5]

Ceryx laterimacula, Zerny, Junk's Cat. Lep., pt. 7, p 8[4]; Seitz, Lep. Indo-Asutr. II 88 (1913).[5]

emicincta, Hampson.

Syntomoides semicincta, Hmpsn., T. E. S. 1895 282 (1895)[1]; *id.,* F. I. Moths IV 465 (1896) [redescr.].[2] — Tenasserim.[1, 2, 4] Dawna Hills.[3]

Ceryx semicincta, Hmpsn., Cat. Phal. I 43, t. 1, f. 16 (1898) [redescr.][3]; Seitz, Lep. Indo-Austr. II 66, t. 10, f. E. (1912)[4]. — __________ Malacca.[3, 4] Borneo.[3, 4]

4. Trichaeta, Swinhoe (1892).

Trichaeta, Swinh., Cat. Het. Mus. Oxon. I. 52 (1892); Hmpsn., F. I. Moths I 225, f. 140 (1892) [redescr.]; *id.,* Cat. Phal. I 53 (1898) [redescr.]; *id.,* B. J. XIII 46 (1900) [key to spp.]; Seitz, Lep. Indo-Austr. II 66 (1912) [redescr.]. Type: *teneiformis,* Wlk.

teneiformis, Walker.

Syntomis teneiformis, Wlk., Cat. VII 1595-1596 (1856).[1] [Hindostan]. — Simla.[3, 4, 5, 7] Sabathu.[9]

Trichæta teneiformis, Swinh., Cat. Het, Mus. Oxon. I 52 (1892)[2]; Hmpsn., F. I. Moths I 225, f. 140 (1892) [redescr.][3]; *id.,* Cat. Phal. I 58, f. 29 (1898) [redscr.[4]]; Seitz, Lep. Indo-Austr. II 89 (1913)[5]; Hmpsn., Cat. Phal. Suppl. I, p. 12 (1914).[6] — Bhutan.[4, 5]

Eressa simplex, Rothschild, Nov. Zool. XVII 439 (1910)[7]; *id.,* Nov. Zool. XIX 376 t. 5, f. 20 (1912)[8]; Seitz, Lep. Indo-Austr. II 85 (1913)♂[9].

5. Amata, Fabricius (1807).

Amata, Fabr., Illiger's Mag. VI 289 (1807); Hmpsn., Cat. Phal. Suppl. I 13 (1914). Type: *passalis,* Fb.

Syntomis, Ochsenheimer, Eur., Schmett. II 103 (1808); Hübner, Verz. p. 121 (1820); Hmpsn., F. I. Moths I 212 (1892) [redescr.]; *id.,* Cat. Phal. I 59 (1898) [redescr.]; *id.,* B. J. XIII 46-49 (1900) [key to spp.]; Seitz, Lep. Pal. II 38 (1910) [redescr.]; *id.,* Lep. Indo-Austr. II 67-68 (1912) (redescr.). Type; *phegea,* Linn. (Europe.]

Cænochromia, Hübner, Verz. p. 121 (1820). Type: *passalis,* Fb. *Hydrusa,*Walker, Cat. I 255 (1854). Type: *bicolor,* Wlk. [Queens-land].

Asinusca, Wallengren, Œfv. Vet. Akad. Förh. XIX 197 (1862). Type: *atricornis,* Wlgn. [S. Africa]. — Type: *atricornis*

Asinutea, Wlgn., Wien. Ent. Mon. VII 139 (1863). Wlgn. [S. Africa].

Buthysia, Wlgn., Wien. Ent. Mon. VII 139 (1863). Boisd. — Type: *huebneri*

actea, Swinhoe (Two subspecies).

(1) actea actea, Swinhoe.

Hydrusa actea, Swinh., T. E. S. 1891. 474, t. 19, f. 7 (1891).[1] — Khasis.[1, 2, 5, 6]

Syntomis baioaea [*nec* Swinh.], Hmpsn., F. I. Moths I 222 (1892) [*partim*].[2] — Shillong.[3]

Eressa actea, Swinh., T. E. S. 1895. 32 (1895).[3]

Syntomis actea, Hmpsn., B. J. XI 284 (1897) [baiaea is distinct][4]; *id.,* Cat. Phal. I 64 (1898) [redeser.][5]; Seitz, Lep. Indo-Austr. II 74 (1913).[6]

(2) actea swinhoei, Leech.

Syntomis, swinhoei, Leech. Entom. XXXI, 152 (1898)[1]; *id.,* T. E. S. 1898. 322 (1898).[2] — Moupin.[2, 3] Chia-ting-fu.[2]

Syntomis actea subsp. I, Hmpsn., Cat. Phal. I 64 (1898).[3] — W. China.[2, 5, 6] Chusan.[3]

Syntomis actea swinhoei, Hmpsn., Cat. Phal. I 537 (1898)[4]; Seitz, Lep. Pal II 40 (1910)[5]; *id.,* Lep. Indo-Austr. II 74 (1913).[6] — Pundaluoya.[3, 4]

Syntomis swinhoei var. *obsoleta,* Leech, Entom. XXXI, 152 (1898).[7]

albapex, Hampson.

Syntomis albapex, Hmpsn., F. I. Moths I 214 (1892)[1]; *id.,* Ill. Het. IX 62, t. 157, f. 1 (1893)[2]; *id.,* Cat. Phal. I 105-106 (1898) [redescr.][3]; Seitz, Lep. Indo-Austr. II 69, t. 9, f. G. (1913).[4] — Nawalapitiya.[3, 4]

albifrons, Moore.

Syntomis albifrons, Mo., P. Z. S. 1878. 845, t. 53, f. 6 (1878)[1]; Hmpsn., F. I. Moths I 216 (1892) [redescr.][2]; *id.,* Cat. Phal. I 100 (1898) [redescr.][3]; Seitz, Lep. Indo-Austr. II 73, t. 9 f. D (1913).[4] — Tenasserim.[3] Momeit.[2] Dawna Hills.[3]

assamica, Rothschild.

Amata khasiana [*nec* Butl.], Roths., Nov Zool. XVII 435 (1910).[1] — Khasis.[1, 4]

Amata assamica, Roths., Nov. Zool. XIX 375, t. 3, f. 21 (1912)[2]; Hmpsn., Cat. Phal. Suppl. I 31-32 (1914).[3]

Syntomis lucina assamensis, Zerny; Seitz, Lep. Indo-Austr. II 69 (1913).[4]

aurea, Swinhoe.

Syntomis aurea, Swinh., A. M. N. H. (8) I. 62-63 (1908) ♂♀[1]; Seitz, Lep. Indo-Austr. II 71 (1913).[2] — Khasis.[1, 2]

aureola, Swinhoe.

Syntomis aureola, Swinh., A. M. N. H. (7) IX. 78-79 (1902) ♂ ♀[1]; Seitz, Lep. Indo-Austr. II 71, t. 9, f. G. (1913).[2]

Khsis.[1, 2]

baiæa, Swinhoe.

Hydrusa baiæa, Swinh., T. E. S 1891. 473, t. 19 f. 10 (1891).[1]

Syntomis baiæa, Hmpsn., F. I. Moths I 222 (1892) [*partim*][2]; *id.,* B. J. XI 284 (1897)[3]; *id.,* Cat. Phal. I. 67-68 (1898) [redescr.][4]; Seitz, Lep. Indo-Austr. II 74, t. i, f. F (1913).[5]

Eressa baiæa, Swinh. T. E. S. 1895. 32 (1895).[6]

Khasis.[1, 2, 4, 5]
Shillong.[6]
Cherrapunji.[6]
____
Java. [4, 5]
Borneo.[4, 5]

berinda, Moore.

Syntomis berinda, Mo., P. Z. S. 1878. 845, t 53, f. 8 (1878)[1]; Hmpsn., F. I. Moths I 219 (1892) [redescr.][2]; *id.,* Cat. Phal. I 98 (1898) [redescr.][3]; Seitz, Lep. Indo-Austr. II 70 t. 10, f. H (1913).[4]

Dacca.
Tenasserim.[1, 4]
Moulmein.[3, 4]

bicincta, Kollar.

Syntomis bicincta, Kollar, Hügel's Kaschmir IV 460, t. 19, f. 8 (1844)[1]; Hmpsn. F. I. Moths I 213 (1892) [redescr.][2]; Hmpsn., Cat. Phal. I 106-107 (1898) [redescr.][3]; Seitz, Lep. Pal. II 38, t. 9, f. B (1910)[4]; *id.,* Lep. Indo-Austr. II 68 (1912).[5]

Kashmir.[4, 5], Simla.
Kangra.[2, 3], Kulu.[2]
Masuri., Dehra Dun.[3]
Sikkim.[2, 3] Calcutta.[3, 4]
Khasis.[2, 3], ______
Fuchau (China).[3]

chlorocera, Hampson.

Syntomis chlorocera, Hmpsn., F. I. Moths I 218 (1892)[1]; *id.,* Cat. Phal. I 93, t. 3, f. 22 (1898) [redescr.][2]; Seitz, Lep. Indo-Austr. II 74, t. 11, f. C (1913).[3]

Nagas.[1-3]

compta, Walker.

Syntomis compta, Wlk., Char. Undescr. Het. p. 89 (1869)[1]; Hmpsn., Cat. Phal. I 99, t. 5, f. 1 (1898) [redescr.][2]; Dudgeon. B. J. XII 293, t. 1, f. 24 (1899)[3]; Hmpsn., B. J. XIII 50 (1900) [redescr.][4]; Seitz, lep. Indo-Austr. II 70, t. 10, f K (1913).[5]

* Benares.[1]
Khasis [2, 4, 5]
Sikkim.[4]

* Walker described this and many other species as "collected near Benares by the late Colonel John Graham, of Her Majesty's Indian Army." It is, however, obvious that the locality given is incorrect for S. *compta* and many of the other species described at the same time. T. B. F.

confluens, Leech.

Syntomis acrospila [*nec* Feld.], Leech, T. E. S. 1898. 322 (1898)[1]; Hmpsn., Cat. Phal. I 91-92 (1898) [*partim*].[2] — Ta-tsien-lu.[1, 2] Nitou.[1, 2] Washan.[1, 2, 7]

Syntomis acrospila var. *confluens,* Leech, T. E. S. 1898. 322 (1898).[3] — Chia-kou-ho.[1]

Amata confluens, Hmpsn., Cat Phal. Suppl. I 33 (1914).[4] — Huang-mu-chang.[2]

Amata leechi, Roths., Novit. Zool. XVII 433 (1910)[5]; *id.,* Novit. Zool XIX 375, t. 3, f. 23 (1912).[6]

Syntomis leechi, Seitz, Lep. Pal. II 445 (1912).[7]

congenita, Hampson.

Amata congenita, Hmp., Nov. Zool. XXV, 93 (1918) ♂♀[1]. — Nilgiris.[1], Shevaroys.[1]

connectens, Rothschild.

Amata connectens, Roths., Novit. Zool. XVII 433 (1910)[1]; *id.,* Novit. Zool. XIX 375, t. 3, f. 37 (1912)[2]; Hmpsn., Cat. Phal. Suppl. I 33 (1914)[3]; Seitz, Lep. Indo-Austr. II 71, t. 10, f. L (1913) [*Syntomis*].[4] — Bernardmyo.[1]

Syntomis sladeni [*nec* Moore] Hmpsn., F. I. Moths I 217 (1892) [*partim*].[5]

consequa, Leech.

Syntomis consequa, Leech. Entom., XXXI 153 (1898)[1]; Hmpsn., Cat. Phal. I 96-97, t. 5, f. 2 (1898) [redescr.][2]; Leech, T. E. S. 1898. 324 (1898)[3]; Seitz, Lep. Pal. II 40, t. 9, f. E (1910).[4] — Moupin.[1-3] Tibet.[4] W. China.

cyssea, Stoll.

Sphinx cysseus, Stoll, Pap. Exot. IV, t. 355 f. B (1782).[1] — Cherat (N.W.F.P.). Kashmir.[15, 16]

Zygæna collaris, Fabr., Ent. Syst. III (*i*) p. 388 (1793)[2]. — Kangra.[12, 14]

Syntomis schoenerrhi, Boisd., Mon. Zyg. p. 112, t. 7, f. 1 (1829).[3] — Nepal.[14] Darjiling.

Syntomis cyssea, Boisd., Mon. Zyg. p. 114, t. 7, t. 2 (1829).[4] — Sikkim.[12, 14], Kurseong.

Syntomis cuprea, Prittw., Stett. ent. Ztg. XXVIII, 277 (1867).[5] — Shillong.[13], Pusa.

Syntomis georgina, Butl., J. Linn. Soc. XII 345 (1876)[6]; Mo., Lep. Ceylon II 34-35, t. 95, f. 2 (1882)[7]; Hmpsn., F. I Moths I 213 (1892) [redescr.][8]; *id.,* F. I. Moths IV 465 (1896).[9] — Calcutta.[14, 16] Karachi.[14, 16] Surat.[2], Mhow.[14] Bombay.

Syntomis cysseoides, Butl., J. Linn. Soc. XII 346 (1876)[10]; Mo., Lep. Ceylon II 35, t. 95 f. 1 (1882).[11]

Nilgiris.[14]
Coimbatore., Coorg.

Syntomis cyssea, Hmpsn., F. I. Moths I 213, f. 138 (1892) [redescr.][12]; Swinh., T. E. S. 1895. 30 (1895)[13]; Hmpsn., Cat. Phal. I 105, f. 36 (1898) [redescr.][14]; Seitz, Lep. Pal. II 39, t. 1, f. E (1910)[15]; *id.,* Lep. Indo-Austr. II 68 (1912).[16]

Ceylon.[6, 7, 10, 11, 14, 16]
Peradeniya.

diaphana, Kollar.

Syntomis diaphana, Kollar, Hügel's Kaschmir IV 460, t. 19, f. 7 (1844).[1]

Kashmir.[4, 6, 7]
Kasauli.[6]

Syntomis vitreata, Herr. Schäff., Aussereur. Schmett. f. 267 (1855).[2]

Simla.
Dalhousie.[6]

Syntomis ænone, Butl., J. Linn. Soc. XII 344 (1876).[3]

Almora.[6], Jeolikote.

Syntomis diaphana, Hmpsn., F. I. Moths I 216 (1892)[4]; Swinh., T. E. S. 1895. 31 (1895)[5]; Hmpsn., Cat. Phal. I. 67 (1898) [redescr.][6]; Seitz. Lep Pal. II 40 (1910)[7]; *id.,* Lep. Indo-Austr. II 74, t. 9, f. G (1913).[8]

Masuri.[1, 6]
Darjiling district.
Assam.[8], Khasis.[6]
Shillong.[5]
Cherrapunji.[5]
Sibsagar.[4, 6]
Burma.[4, 6, 7, 8]

dichotoma, Leech.

Syntomis dichotoma, Leech, Entom. XXXI 153 (1898)[1]; *id.,* T. E. S. 1898. 323 (1898)[2]; Hmpsn., Cat. Phal. I 100, t. 4, f.1 (1898) [redescr.][3]; Seitz, Lep. Pal. II 39, t. 9, f. E (1910).[4]

Moupin (Tibet).[1-3]

W. China[2, 4]

Syntomis dichotoma var. *concurrens,* Leech, Entom. XXXI, 153 (1898).[5]

divisa, Walker.

Syntomis divisa, Wlk., Cat. I 131 (1854)[1]; Butl., Ill. Het. I 16, t. 6, f. 14 (1877)[2]; Hmpsn., F. I. Moths I 215 (1892) [redescr.][3]; Swinh., T. E. S. 1895. 30 (1895)[4]; Hmpsn., Cat. Phal. I 98-99 (1898) [redescr.][5]; Seitz., Lep. Indo-Austr. II 70, t. 9. f. D (1913).[6]

Sikkim.[3, 5, 6], Gauhati.
Sylhet.[1, 3, 5]
Khasis.[3, 4, 5]
Shillong.
Chin Hills.[5], Maymo.[2]
Upper Tenasserim.[5, 7]

Syntomis disrupta, Mo. P. Z. S. 1878. 845, t. *53,* f. 5 (1878).[7]

Dawna Hills.[5]

elongata, Hampson.

Syntomis elongata, Hmpsn., B. J. XIII 50-51 (1900)[1]; *id.,* Cat. Phal. Suppl. I 38, t. 2, f. 29 (1914) [rdescr.][2]; Seitz, Lep. Indo-Austr. II 78 (1913).[3]

Belgaum.[2]
Kanara.[1-3]
Pirmâd.[1, 2]

elwesi, Rothschild.

Syntomis tigrina [*nec* Wlk.], Hmpsn., F. I. Moths I 218 (1892) ♂.[1] — Burma.[1, 4, 5]

Amata elwesi, Roths., Novit. Zool. XVII 432 (1910)[2]; *id.,* Novit. Zool XIX 375, t. 3, f. 34 (1912).[3]

Amata elvesi, Hmpsn., Cat. Phal. Suppl. I 29 (1914) [possibly tigrina ♂][4]

Syntomis elwesi, Seitz, Lep. Indo-Austr. II 71 (1913).[5]

era, Swinhoe.

Hydrusa, era, Swinh., T. E. S. 1891, 473, t. 19, f. 15 (1891) ♂.[1] — Khasis.[1-4]

Syntomis era, Hmpsn., F. I. Moths I 222 (1892) [redescr.] ♀[2]; *id.,* Cat. Phal. I 68 (1898) ♂[3]; Seitz, Lep. Indo-Austr. II 80 (1913)[4]. — Shillong.[5]

Eressa era, Swinh., T. E. S. 1895. 32 (1895).[5]

euryzona, Leech.

Syntomis euryzona, Leech, Entom. XXXI 153 (1898) ♂[1]; Hmpsn., Cat. Phal. I 94, t. 3, f. 19 (1898) ♂ [redescr.][2]; Seitz, Lep. Pal. II 40, t. 9, f. G (1910).[3]

Moupin.[1, 2]
Tibet.[3], ————
W. China.[3]

extensa, Walker.

Syntomis extensa, Wlk., Cat. XXXV 1863 (1866)[1]; Hmpsn., Ill. Het. VIII 3, t. 143, f. 14 (1890)[2]; *id.,* F. I. Moths I 214 (1892) [redescr.][3]; *id.,* Cat. Phal. I 103 (1898) [redescr.][4]; Seitz, Lep. Indo-Austr. II 72, t. 11, f. A (1913).[5]

Mahableshwar.[3, 4]
Matheran.[3-5]
Nilgiris.[2-5]
Ootacamund.

fervida, Walker.

Syntomis fervida, Wlk., Cat. I 131 (1854)[1]; Butl., Ill. Het. I 18, t. 6, f. 16 (1877)[2]; Hmpsn., F. I. Moths I 217 (1892)♀[3]; *id.,* Cat. Phal. I 93 (1898) ♀[4]; Seitz., Lep. Indo-Austr. II 71, t. 9, f. D (1913).[5]

Moulmein.[1, 3, 4]
E. Pegu.[3, 4]

flavifrons, Hampson.

Syntomis flavifrons, Hmpsn., F. I. Moths I 215 (1892)[1]; *id.,* Cat. Phal. I 101, t. 4, f. 3 (1898) [redescr.][2]; Seitz. Lep. Indo-Austr. II 69, t. 10, f. H (1913)[3]; Swinh., T. E. S. 1895, 30 (1895).[4]

Khasis.[2]
Shillong.[4]
Momeit (Burma).[1, 2]
————
Indo-China.[3]

flaviguttata, Hampson.

Syntomis flaviguttata, Hmpsn., B. J. XIII 50, t. B. f. 2 (1900)[1]; Seitz., Lep. Indo-Austr. II 77 (1913)[2]; Hmpsn., Cat. Phal. Suppl. I 32 (1914) [redescr.][3]	Chitral.[1-3]

flavolavata, Rothschild.

Amata flavolavata, Roths., Novit. Zool. XVII 434 (1910)[1]; *id.,* Novit. Zool. XIX 375, t. 3, f. 24 (1912)[2]; Hmpsn., Cat. Phal. Suppl. I 33 (1914) [redescr.; may be a var. of *melas*][3]; Seitz., Lep. Indo-Austr. II 73 (1913) [*Syntomis.*][4]	Khasis.[3, 4]

formosæ, Butler.

Syntomis formosæ, Butl. J. Linn. Soc. XII 346 (1876)[1]; Hmpsn., Cat. Phal. I 98, t. 3, f. 26 (1898) [redescr.][2]; Seitz, Lep. Pal. II 39, t. 9, f. E. (1910)[3]; *id.,* Lep. Indo-Austr. II 69 (1913).[4]	Margarita.[2], E. Pegu.[2] _____, Fuchau.[2, 3, 5] Chusan.[2], Formosa.[1-4]
Syntomis emma, Butl., J. Linn. Soc. XII 350-351 (1876).[5]	

gelatina, Hampson.

Syntomis, gelatina, Hmpsn., Ill. Het. VIII 43, t. 139, f. 1 (1890)[1]; *id.,* F. I. Moths I 214 (1892) [redescr.][2]; *id.,* Cat. Phal. I 103 (1898) [redescr.][3]; Seitz, Lep. Indo-Austr. II 72, t. 10, f. L (1913).[4]	Nilgiris.[1-4]

grotei, Moore.

Syntomis grotei, Mo., I. Z. S. 1871. 245, t. 18, f. 4 (1871)[1]; Hmpsn., F. I. Moths I 218 (1892) [redescr.][2]; *id.,* Cat. Phal. I 92 (1898) [redescr.][3]; Seitz., Lep. Indo-Austr. II 71, t. 10, ff. K [♂♀] (1913).[4]	Upper Tenasserim.[2, 4] Yannan.[1-4] ______ Hongkong.[3, 4]

huebneri, Boisduval.

Syntomis huebneri, Boisd., Mon. Zyg. p 127, t. 8, f. 4 (1829).[1]	Bombay,[10, 11] Throughout India.[12]
Syntomis marsdeni, Mo., Cat. Lep. E. I. Co. Mus. II 323 (1859)[2]; Mo., P. Z. S. 1859. 157, t. 60, f. 3 (1859).[3]	Dhamai (E. Bengal.)
Syntomis xanthomela, Wlk., J. Linn. Soc. III 184 (1860).[4]	
Naclia cingulata, Wlgn., Wien. Ent. Mon. IV 39 (1860).[5]	________
Buthisia sangaris, Wlgn., Wien. Ent. Mon. VII 139 (1863).[6]	Singapore.[4, 7, 10, 11]
Syntomis contermina, Wlk., Cat. XXXI 78 (1864).[7]	

Hydrusa pyrrhodera Meyr., Proc. Linn. Soc. N. S. W. (2) I 777 (1886).[8]
Java.[2, 6, 10, 11]

Syntomis frustulenta, Swinh., Cat. Het. Mus. Oxon., I 44 (1892).[9]
Manila.[5]

Syntomis huebneri, Hmpsn., F. I. Moths I 219 (1892) [redescr.][10]; *id.,* Cat. Phal. I 69 (1898) [redescr.][11]; Seitz, Lep. Indo-Austr. II 74 (1913).[12]
N. Australia.[8, 9, 10, 11]

hydatina, Butler.

Syntomis hydatina, Butl., J. Linn. Soc. XII 346 (1876)[1]; Hmpsn., F. I. Moths I 210-211 (1892) [redescr.] [*Syntomoides*][2]; *id.,* Cat. Phal. I 106; t. 4, f. 14 (1898) [redescr][3]; Setiz; Lep. Indo-Austr. II 68, t. 10, f. G. (1912).[4]
Sikkim.[5]
Calcutta.[1-3]
Central India.[2-4]

insueta, Swinhoe.

Syntomis insueta, Swinh., Cat. Het. Mus. Oxon. I 41, t. 2, f. 16 (1892)[1]; Hmspn. F. I. Moths I 213 (1892) ♀[2]; *id.,* Cat. Phal. I 105 (1898)♀[3]; Seitz, Lep. Indo-Austr. II 69 (1913).[4]
Barrackpur.[1-4]

khasiana, Butler.

Syntomis khasiana, Butl., J. Linn. Soc. XII 345-346 (1876)[1]; Hmpsn., F. I. Moths I 214 (1892) [redescr.][2]; *id.,* Cat. Phal. I 103, t. 4, f. 9 (1898) [redescr.][3]; Seitz., Lep. Indo-Austr. II 69, t. 10, f. H (1913)[4]; Swinh., T.E.S. 1895. 30 (1895).[5]
Khasis.[1-4]
Shillong.[5]

lucina, Butler.

Syntomis lucina, Butl., J. Linn. Soc. XII 345 (1876)[1]; Hmpsn., F. I. Moths I 214 (1892) [redescr.][2]; *id.,* Cat. Phal. I 104 t. 4, f. 10 (1898) [redescr.][3]; Seitz, Lep. Indo-Austr. II 69, t. 10, f. H. (1913).[4]
Khatmandu.
Nepal.[1-4], Sikkim.[2-4]
Calcutta.[3-4]
[?error loci].

luteifascia, Hampson.

Syntomis luteifascia, Hmpsn., F. I. Moths I 218-219 (1892) ♂[1]; *id.,* Cat. Phal. I 97-98, t. 3, f. 25 (1898)♂[2]; Seitz, Lep. Indo-Austr. II 70, t. 10, f. I (1913).[3]
Nagas.[1-3]

madurensis, Hampson.

Syntomis madurensis, Hmpsn., A.M.N.H. (7) VIII 166 (1901) ♂[1]; *id.,* B. J. XVI 151, t. D, f. 27 (1904) [redescr.][2]; Seitz., Lep. Indo-Austr. II 69, t. 9, f. F (1913)[3]; Hmpsn., Cat. Phal. Suppl. I 15, t. 1, f. 23 (1914)♂ [*Amata*].[4]
Kodaikanal.[4]
Madura District.[1 4]

masoni, Moore.

Syntomis masoni, Mo., P.Z.S. 1878. 845, t. 53, f. 4 (1878)[1]; Hmpsn., F. I. Moths I 215-216 (1892) [redescr.][2]; *id.,* Cat. Phal. I 95-96 (1898) [rdescr.][3]; Seitz., Lep. Pal. II 39, t. 9, f. F (1910)[4]; *id.,* Lep. Indo-Austr. II 71 (1913).[5]	Upper Tenasserim.[1, 2, 3, 5] Tibet.[2, 4] Moupin.[3, 6], ______
Syntomis davidi, Poujade, Bull. S.E.Fr. (6) IV 136 (1885)[6]; Hmpsn., F. I. Moths I 218 (1892)[7]; Leech, T.E.S. 1898. 323 (1898).[8]	W. China[4, 8] S. China.[5] Central China.[8]

melas, Walker.

Syntomis, melas, Wlk., Cat. I 133-134 (1854)[1]; *Butl.,* Ill. Het. I 17, t. 6, f. 10 (1877)[2]; Seitz. Lep. Indo-Austr. II 73, t. 9, f. G. (1913).[3]	Nepal. [1, 3, 6, 8] Sikkim.[6, 8], Assam.[3] Khasis.[6, 8]
Syntomis andersoni, Mo., P.Z.S. 1871. 244, t. 18, f. 1 (1871)[4]; *id.,* Anderson's Researches in W. Yunnan, p. 926, t. 81, f. 4 (1878).[5]	Shillong.[6, 7] Chin Hills.[8]
Syntomis melæna; Hmpsn., F. I. Moths I 216 (1892) [emend. pro *melas*] [redescr.][5]; Swinh., T.E.S. 1895. 31 (1895)[7]; Hmpsn., Cat. Phal. I 96 (1898) [redescr.].[8]	Yunnan.[3, 4, 5, 8]

minor, Warren.

Syntomis minor, Warr., P.Z.S. 1888, 294 (1888)[1]; Hmpsn., F. I. Moths I 213 (1892) [redescr.][2]; *id.,* Cat. Phal. I 107, t. 4, f. 15 (1898) [redescr.][3]; Seitz, Lep. Indo-Austr. II 69, t. 10, f. H (1913).[4]	Kala Pani (Punjab).[1, 3] Abbottabad.

muirheadi, Felder.

Syntomis muirheadi, Feld., Wien. Ent. Mon. VI 37 (1862)[1]; Leech, T.E.S. 1898, 322 (1898)[2]; Seitz, Lep. Pal. II 40, t. 9. f. G (1910).[3]	Chia-ting-fu [E.Tibet border.][2] ________ Kwei-chow Prov.[2, 4]
Syntomis aucta, Leech, T.E.S. 1898. 323 (1898).[4]	S. China.[3] Central China.[2]
Syntomis muirheadi ab. *aucta,* Seitz, Lep. Pal. II 40 (1910).[5]	N. E. China.[2]

newara, Moore.

Syntomis newara, Mo., Lep. Atk., p. 12 (1879)[1]; Hmpsn., F. I. Moths I 217 (1892) ♀[2]; *id.,* Cat. Phal. I 94, t. 3, f. 18 (1898) ♀[3]; Seitz., Lep, Indo-Austr. II 70, t. 10 f. i (1913).[4]	Darjiling.[1] Sikkim.[2-4] Bhutan.[2-4]

ochreipuncta, Hampson.

Syntomis ochreipuncta, Hmpsn., F. I. Moths I 222 (1892) ♂[1]; *id.,* Cat. Phal. I 64, t. 2, f. 17 (1898) ♂[2]; Seitz, Lep. Indo-Austr. II 74, t. 11, t. B (1913).[3]	Ganjam.[1-3]

owstoni, Rothschild.

Amata owstoni, Roths., Novit. Zool XVIII 155 (1911)[1]; *id.,* Novit. Zool. XIX 377, t. 5, f. 6 (1912)[2]; Seitz, Lep. Indo-Austr. II 71 (1913) [*Syntomis*].[3]	Yunnan.[4] ________, Annam.[3] Hainan.[1-3]
Amata oustoni, Hmpsn., Cat. Phal. Suppl. I 36 (1914) ♂.[4]	

pasca, Leech.

Syntomis pascus, Leech, T.E.S. 1889. 124, t. 9, f. 1 (1889)[1]; *id.,* T.E.S. 1898, 320-321 (1898).[2]	Tibet.[4], Moupin.[2, 3] ________
Syntomis pasca, Hmpsn., Cat. Phal. I 94-95 (1898) [redescr.][3]; Seitz, Lep. Pal. II 40, t. i, f. H (1910).[4]	W. China.[2, 4] Central China.[2, 4] Kinkiang.[1, 3]

passalis, Fabricius.

Zygœna cerbera [*nec* Linn.] Sulzer, Gesch. Ins. t. 20, f. 8 (1776).[1]	Assam.[10, 11] Calcutta.[11]
Zygœna passalis, Fabr., Spec. Ins. II 159 (1781).[2]	Pusa.
Sphinx creusa [nec Linn.], Cram., Pap. Exot. III, t. 248, f. F (1788)[3]; Boisd., Mon. Zyg. p. 121, t. 7, f. 7 (1829) [*Syntomis*][4].	Kirkee. Nagpur. Mysore City.
Syntomis latreillii, Boisd., Mon. Zyg. p. 117, t. 7, f. 5 (1829).[5]	Bangalore.[11]
Syntomis montana, Butl., J. Linn. Soc. XII 349 (1876)[6]; Mo., Lep. Ceylon II 36, t. 94, f. 5 (1882).[7]	Coimbatore,[11] Tanjore., Travancore.[14]
Syntomis passalis, Mo., Lep. Ceylon II 35 (1882).[8]	Ceylon.[7, 8, 11]
Zygœna passalis, Kirby, Cat. Het. I 92-93 (1892).[9]	
Syntomis passalis, Hmpsn., F. I. Moths I 219-220 (1892) [redescr.][10]; *id.,* Cat. Phal. I 86 (1898) [redescr.][11]; *id.,* B. J. XX 95 (1910)[12]; Seitz., Lep. Indo-Austr. II 72, t. 11, ff. A [♂♀] (1913)[13]; Fletcher, Proc. Third Entl. Meeting I 52 (1919).[14]	Colombo.[8, 13] Kandy.[6, 13, 15] Peradeniya.
Syntomis passalis ab. *montana,* Seitz., Lep. Indo-Austr. II 72, t. 11, f. A [♂] (1913).[15]	

Larva descr.[13], on *I pomœa batatas.*[14]

pectoralis, Walker.

Syntomis pectoralis, Wlk., Cat. I 133 (1854)[1]; Butl., Ill. Het. I 17, t. 6, f. 5 (1877)[2]; Hmpsn., F. I. Moths I 224 (1892) [redescr.][3]; *id.,* Cat. Phal. I 62-63 (1898) ♀[4]; Seitz, Lep. Indo-Austr. II 73, t. 9, f. E (1913).[5]	Burma.[5], E. Pegu.[3] Moulmein.[1, 3] Dawna Hills.[4] Andamans.[4]
Amata pectoralis, Hmpsn., Cat. Phal. Suppl. I 383 (1914).[6]	
Hydrusa basirufa, Swinh., Cat. Het. Mus. Oxon. I 51 (1892).[7]	——, Cambodia.[3-5]
Eressa rubribasis, Joannis, Bull. S. E. France 1912. 287 (1912).[8]	Cochin China.[8]

pentazonata, Hampson.

Syntomis pentazonata, Hmpsn., Cat. Phal. I 99, t 3, f. 27 (1898)[1]; *id.,* B. J. XIII 50 (1900) [redescr.][2]; Seitz., Lep. Indo-Austr. II 72, t. 10, f. L (1913).[3]	N. Chin. Hills.[1, 2]

perixanthia, Hampson.

Syntomis perixanthia, Leech, Entom. XXXI 152 (1898) [non descr.][1]; Hmpsn., Cat. Phal. I 97, t. 3, f. 17 (1898)[2]; Leech, T.E.S. 1898. 321 (1898)[3]; Seitz, Lep. Indo-Austr. II 70 (1913).[4]	Moupin.[2, 3] Ni-tou.[6] Wa-shan.[2, 3] Omei-shan.[2, 3, 6]
Syntomis persimilis, Leech, Entom., XXXI 152 (1898)[5]; Leech, T.E.S. 1898. 322 (1898) ♀.[6]	——— W. China.[3, 10, 11]
Syntomis perixanthia persimilis, Seitz., Lep. Pal. II 39 (1910).[7]	Formosa,[2, 3, 4, 10]
Amata perixanthia sinensis, Roths., Novit. Zool. XVII 435 (1910)[8]; *id.,* Novit. Zool. XIX 375, t. 3, f. 17 (1912)[9]; Seitz, Lep. Pal. II 39, 445, t. 9, f. F (1910) [*Syntomis*][10]; Hmpsn., Cat. Phal. Suppl. I 36 (1914).[11]	

phœnicozona, Hampson.

Syntomis phœnicozona, Hmpsn., B. J. XI 284 (1897) ♂[1]; *id.,* Cat. Phal. I 63, t. 2, f. 13 (1898)♂[2]; Seitz, Lep. Indo-Austr. II, 73, t. 11 f. C (1913).[3]	Andamans.[1-3]

quadrifascia, Hampson.

Syntomis quadrifascia Hmpsn., F. I. Moths I 215 (1892) ♂[1]; *id.,* Cat. Phal. I 100, t. 4 f. 2 (1898)♂[2]; Seitz., Lep. Indo-Austr. II 70, t. 10, f. K (1913).[3]	Bernardmyo (Burma)[1-3]

rubrozonata, Poujade.

Syntomis rubrozonata, Poujade, Bull. S. E. Frnace (6) VI 117 (1886)[1]; Hmpsn., Cat. Phal. I 85, t. 3 f. 7 (1898) [redescr.][2]; Leech, T. E. S. 1898, 324 (1898)[3]; Seitz, Lep. Pal. II 40, t. 9, f. F (1910).[4]	Moupin.[1, 3, 5] Omei-Shan.[2, 3, 7] Tibet.[4]

Syntomis rubrizonata, Kirby, Cat Het. I 93 (1892).[5]	
Syntomis leucoma, Leech, Entom., XXXI, 154 (1898)[6]; *id.,*T. E. S. 1898, 324 (1898).[7]	W. China.[4]

serrata, Hampson.

Syntomis serrata, Hmpsn., F. I. Moths I 221 (1892)[1]; *id.,* Cat. Phal. I 70, t. 2, f. 24 (1898)[2]; Seitz., Lep. Indo-Austr. II 70, t. 10, f. i (1913).[3]	Khasis.[1-4]
Syntomis grotei [*nec* Moore], Swinh., T. E. S. 1895, 31 (1895).[4]	

sladeni, Moore.

Syntomis sladeni, Mo., P.Z.S. 1871. 245, t. 18, f. 5 (1871)[1]; *id.,* Anderson's Researches in W. Yunnan, p. 927, t. 81, f. 8 (1878)[2]; Hmpsn., F. I. Moths I 217 (1892) [redescr.][3]; *id.,* Cat. Phal. I 91 (1898) [redescr.][4]; Seitz, Lep. Indo-Austr. II 71, t. 10, f. L (1913).[5]	Yunnan.[1-5] E. Pegu.[3-5] Bernardmyo.[3, 4] Chin Hills.[4] Upper Tenasserim.[4, 5]

sperbius, Fabricius.

Zygœna sperbius, Fabr., Mantissa II 103 (1787).[1]	Pusa., Sikkim.[6, 8]
Syntomis atkinsoni, Mo., P.Z.S. 1871, 245, t. 18, f. 2 (1871)[2]; Butl., J. Linn. Soc. XII 347 (1876)[3]; Anderson's Researches in W. Yunnan, p. 927, t. 81, f. 5 (1878)[4]; Hmpsn., F. I. Moths I 212 (1892) [redescr.].[5]	Margherita., Cachar.[6] Shillong. Upper Burma.[6] Lower Burma.[6] Moulmein.[2, 6]
Syntomis sperbius, Hmpsn., Cat. Phal. I 106 (1898) [redescr.][6]; *id.,* B. J. XIII 51 (1900)[7]; Seitz., Lep. Indo-Austr. II 68-69, t. 10, ff. G [♂], H [♀] (1912).[8]	Tenasserim.[6] Yunnan.[4, 5], ______ Tonkin.[6], Hainan.[8] China.[1]

stellaris, Snellen.

Syntomis stellaris, Snell., Deuts. Ent. Ziet., Lep. VIII 130 (1895)[1]; Hmpsn., Cat. Phal. I 112, t. 4, f. 22 (1898) ♂.[2]	Tenasserim.?, ______ Siam.[2], Perak.[2] Sumatra.[2], Borneo.[2]

submarginalis, Walker.

Syntomis submarginalis, Wlk., Char. Undescr. Het. p. 89 (1869)[1]; Hmpsn., F. I. Moths I 218 (1892) [redescr.][2]; *id.,* Cat. Phal. I 91, t. 3 f. 15 (1898) [redescr.][3]; Swinh., T. E. S. 1898. 31 (1898)[4]; Seitz., Lep. Indo-Austr. II 71, t. 10, f. i (1913).[5]	Sikkim.[2, 3, 5] Khasis.[3, 4] Toungoo (Burma).[3] Bernardmyo.[2, 3]

thoracica, Moore.

Phacusa thoracica, Mo., A.M.N.H. (4) XX (343) (1877)[1]; *id.,* Lep. Ceylon II 38, t. 95, f. 7 (1882).[2]	Ceylon.[1-4], Kandy. Peradeniya.[5]

Syntomis thoracica, Hmpsn., F. I. Moths I 224 (1892) [redescr.][3]; *id.,* Cat. Phal. I 63, f. 32 (1898) [redescr.][4]; Seitz., Lep. Indo-Austr. II 73, t. 11, f. B (1913).[5]

tigrina, Walker.

Syntomis tigrina, Wlk., Cat. XXXI, 71 (1864) ♀.[1]

Trichœta tigrina, Hmpsn., Cat. Phal. I 54, f. 26 (1898) ♀[2] *id.,* B. J. XIII 46 (1900).[3] — Cambodia.[1, 2]

Amata tigrina, Hmpsn., Cat. Phal. Suppl. I 29 (1914).[4]

[Note—This may be opposite sex of *elwesi* (q.v.)].

unifascia, Hampson.

Syntomis unifascia, Hmpsn., F. I. Moths I 212 (1892) ♀[1]; *id.,* Cat. Phal. I 112, t. 4, f. 20 (1898) ♀[2]; Seitz., Lep. Indo-Austr. II 79, t. 11, f. H (1913).[3] — Sikkim.[1 3]

verecunda, Swinhoe.

Syntomis verecunda, Swinh., A.M.N.H. (7) IX 78 (1902) ♂[1]; Seitz., Lep. Indo-Austr. II 71 (1913).[2] — Jaintia Hills.[1, 2]

vitrea, Walker.

Syntomis vitrea, Wlk., Cat. VII 1594 (1856)[1]; Hmpsn., F. I. Moths I 218 (1892) ♀[2]; *id.,* Cat. Phal. I 90-91, t. 3, f. 14 (1898) ♂♀[3]; Seitz., Lep. Indo-Austr. II 70, t. 10, f. K (1913).[4] — Assam.[3, 5] Sylhet.[1, 2] Chin Hills.[3, 6, 7]

Syntomis austeni, Mo., P.Z.S. 1879. 389 (1879).[5]

Syntomis polyzonata, Hmpsn., T.E.S. 1895. 283 (1895) ♀[6]; *id.,* F. I. Moths IV 466 (1896) ♀.[7]

wimberleyi, Swinhoe.

Syntomis wimberleyi, Swinh., P.Z.S. 1889. 400, t. 43, f. 11 (1889)[1]; Hmpsn., F. I. Moths I 219 (1892) ♀[2]; Hmpsn., Cat. Phal. I 88-89 (1898) ♂♀[3]; Seitz., Lep. Indo-Austr. II 78, t. 11, f. F. (1913).[4] — Andamans.[1-4]

xanthograpta, Hampson.

Amata xanthograpta, Hmpsn., Cat. Phal. Suppl. I 32, t. 3, f. 6 (1914) ♂.[1] — Kashmir.[1]

yunnanensis, Rothschild.

Amata yunnanensis, Roths., Novit. Zool. XVIII 155 (1911) ♂[1]; *id.,* Novit. Zool. XIX 377, t. 5, f. 25 (1912)[2]; Hmpsn., Cat. Phal. Suppl. I 35 (1914) ♂[3]; Seitz, Lep. Indo-Austr. II 70 (1913) [*Syntomis*].[4] — Yunnan.[1-4]

6. Eressa, Walker (1854).

Eressa, Wlk., Cat. I 149 (1854); Hmpsn., Cat. Phal. I 115, f. 38 (1898) [redescr.]; *id.,* B. J. XIII 223 (1901) [key to spp.]; Seitz, Lep. Pal. II 445 (1912); *id.,* Lep. Indo-Austr. II 82 (1913) [redescr.]. Type: *confinis,* Wlk.

Ctenandra, Felder, Seitz, Akad. Wien. LXIII 38 (1861). Type: *naclioides,* Felder [Amboina].

Trianeura, Butler, J. Linn. Soc., XII 353-354, t. 28, f. 3 (1876). Type: *subaurata,* Wlk.

Chloromeles, Meyr., Proc. Linn. Soc. N.S.W. (2) I 785 (1886). Type: *geographica,* Meyr. [Queensland].

Eressades, Beth.-Baker, A.M.N.H. (8) VII 531 (1911). Type: *flavipunctata,* B.-B. [N. Angola.]

affinis, Moore.

Eressa affinis, Mo., P.Z.S. 1877. 596, t. 59, f. 3 (1877)[1]; Hmpsn., F. I. Moths I 223 (1892) [*partim; Syntomis*][2]; *id.,* Cat. Phal. I 117 (1898) ♂ ♀[3]; Seitz. Lep. Indo-Austr. II 83 (1913).[4] — Andamans.[1 4]

Eressa politula, Swinh., Cat. Het. Mus. Oxon. I 52 (1892).[5]

annosa, Walker.

Syntomis annosa, Wlk., J. Linn. Soc. III 183 (1859).[1] — Dawna Hills.[5], ____

Syntomis marcescens, Felder, Reise Novara, Lep. IV, t. 102, f. 12 (1875).[2] — Malay Penins.[7]

Syntomis ? lasara, Pagenst., Jahrb. Nass. Verh. XXXVIII 13, t. 2, f. 8 (1885).[3] — Singapore.[1, 4, 5] Java.[2, 5, 7]

Syntomis annosa, Hmpsn., F. I. Moths IV 466 (1896) ♂.[4] — Nias.[3]

Eressa annosa, Hmpsou, Cat. Phal. I 120 (1898) ♂ ♀[5]; *id.,* B. J. XIII 224 (1900) [synon.][6]; Seitz, Lep. Indo-Austr. II 82, t. 12, f. B (1913).[7]

aperiens, Walker.

Syntomis aperiens, Wlk., Cat. XXXI 68 (1864).[1] — Khasis.[7, 10]

Trianeura moorei, Butl., J. Linn. Soc. XII 354 (1876).[2]

Syntomis lydia, Swinh., T.E.S. 1891. 134, t. 8, f. 7 (1891) [aberr.].[3] — Bombay.[2 5]

Syntomis aperiens, Hmpsn., F. I. Moths I 221-222 (1892)♂♀.[4] — Nilgiris.[1, 3, 4, 5]

Eressa aperiens, Hmpsn., Cat. Phal. I 119-120 (1898) ♂♀[5]; Seitz, Lep. Indo-Austr. II 83 (1913).[6]	Ceylon.[6] ————
Eressa aperiens khasiana, Roths., Novit. Zool. XVII 438 (1910)[7]; *id.*, Novit. Zool. XIX 376, t. 4, f. 8 (1912)[8]; Hmpsn., Cat. Phal. Suppl. I 48 (1914)[9]; Seitz, Lep. Indo-Austr. II 83, t. 12, f. C (1913).[10]	Sumatra.[6]

confinis, Walker.

Glaucopis confinis, Wlk., Cat. I 149-150 (1854)[1]; *id.*, Cat. VII 1592 (1856) [*Syntomis*].[2]	Simla.[8], Kangra.[5, 8] Kulu.[6]
Eressa confinis, Mo., Lep. Ceylon II 36, t. 95, f. 6 (1882).[3]	Almora.[6]
Eressa musa, Swinh., P.Z.S. 1885. 290, t. 20, f. 1 (1885)[4]; Hmpsn., F. I. Moths I 222-223 (1892) [redescr.] [*Syntomis*].[5]	Sikkim.[8]
Syntomis confinis, Hmpsn., F. I. Moths I 223 (1892) ♂.[6]	Bhutan.[8], Shillong.
Eressa confinis, Swinh., T.E.S. 1895. 32 (1895)[7]; Hmspn., Cat. Phal. I 116, f. 38 (1898) [redescr.][8]; Seitz., Lep. Indo-Austr. II 83 (1913)[9]; Hmpsn., Cat. Phal. Suppl. I 46-47 (1914) [Synon.].[10]	Cherrapunji.[7] Jaintias.[11], Calcutta.[8] Hoshangabad. Bombay.[4, 5, 8]
Eressa catoria, Swinh., A.M.N.H. (7) VI 305 (1900).[11] ♂.	Balgaum.[15]
Syntomoides finitima, Wileman, Entom. XLIII 220 (1910).[12]	Nilgiris.[5, 8] S. India.[17]
Eressa confinis malaccensis, Roths., Novit. Zool. XVII 437 (1910)[13]; *id.*, Novit. Zool. XIX 376, t. 4, f. 6 (1912).[14]	Ceylon.[3, 5, 8, 9] Burma.[5 8, 9] Myitkyina (U.Burma.)
Eressa confinis intensa, Roths., Novit. Zool. XVII 438 (1910)[15]; *id.*, Novit. Zool, XIX 376, t. 4, f. 7 (1912)[16]; Seitz, Lep. Indo-Austr. II 83, t. 12, f. B (1913).[17]	————, Penang.[13] Tonkin.[13] Formosa.[12]

discinota, Moore.

Syntomis discinota, Mo., Lep. Atk. P. 13 (1879)[1]; Hmpsn., F. I. Moths I 216-217 (1892) ♂[2]; Swinh., T.E.S. 1895. 31 (1895).[3]	Khasis.[1-5]
Eressa discinota, Hmpsn. Cat. Phal. I 123 f. 44 (1898) [redescr.][4]; Seitz., Lep. Indo-Austr. II 84, t. 9, f. H (1913).[5]	Tenasserim.[2, 4]

eressoides, Hampson.

Syntomis eressoides, Hmpsn., F. I. Moths I 220 (1892) ♀.[1]	Momeit (Burma).[1, 2]

Eressa eressoides, Hmpsn., Cat. Phal. I 118, t. 4, f. 26 (1898) ♀[2]; *id.,* B. J. XIII 223 (1900)[3]; Seitz, Lep. Indo-Austr. II 84, t. 12, f. B (1913).[4]

erythrosoma, Hampson.

Syntomis erythrosoma, Hmpsn., F. I. Moths I 223-224 (1892) ♂.[1] — Matelei (Burma)[1, 2]

Eressa erythrosoma, Hmpsn., Cat. Phal. I 119, t. 5, f. 23 (1898) ♂[2]; Seitz, Lep. Indo-Austr. II 84, t. 12, f. C (1913).[3]

ichneumoniformis, Rothschild.

Eressa ichneumoniformis, Roths., Novit. Zool. XVII 438 (1910) ♂[1]; *id.,* Nov. Zool. XIX 376, t. 4, f. 10 (1912)[2]; Hmpsn., Cat. Phal. Suppl. I 48 (1914) ♂[3]; Seitz, Lep. Indo-Austr. II 84, t. 12, f. C (1913).[4] — Khasis.[1, 3, 4]

lepcha, Moore.

Syntomis lepcha, Mo., Lep. Atk., p. 12 (1879)[1]; Hmpsn., F. I. Moths I 217 (1892)[2]; *id.,* Cat. Phal. I 121, t. 5, f. 7 (1898) [redescr.][3]; Seitz., Lep. Indo-Austr. II 84, t. 12, f. C (1913).[4] — Darjiling.[1] Sikkim.[2-4]

microchilus, Hampson.

Syntomoides microchilus, Hmpsn., F. I. Moths I 212 (1892).[1] — Tenasserim Valley.[1, 3]

Syntomis plumalis, Hmpsn., F. I. Moths IV 466, (1896) ♂.[2] — Toungoo (Burma).[2, 3]

Eressa microchilus, Hmpsn., Cat. Phal. I 119, t. 5, f. 5 (1898) ♂[3]; *id.,* B. J. XIII 224 (1900) [synon.][4]; Seitz, Lep. Indo-Austr. II 84, t. 12, f. C. (1913).[5]

multigutta, Walker.

Syntomis multigutta, Wlk., Cat. I 134 (1854)[1]; Butl. Ill. Het. I 19, t. 7, f. 3 (1877) [*Hydrusa*][2]; Hmpsn., F. I. Moths I 220 (1892) [redescr.][3]; *id.,* Cat. Phal. I 120-121, f. 39 (1898) [redescr.][4]; *id.,* B. J. XIII 224 (1900) [synon][5]; Leech, T.E.S. 1898. 324 (1898)[6]; Hmpsn., P.Z.S. 1906. 486 (1906)[7]; Seitz, Lep. Pal. II 445 (1912)[8]; *id.,* Lep. Indo-Austr. II 84 (1913).[9]* — Nepal.[1, 3, 4] Sikkim.[3, 4, 11] Gantok.[7], Tibet.[8] Moupin.[3, 4, 10, 1] Omei-Shan.[12] Wa-shan.[12], Ni-tou.[6] Shillong., Burma.[3, 4]

---

* Seitz[9] states that three are two forms [subspecies], (1) *multigutta,* occurring in Burma, and (2) *blanchardi,* ranging from Sikkim to Tibet; but *multigutta,* was described from Nepal as that this name cannot be applied to the Burmese form which has as yet no name if sufficiently distinct to deserve one. T. B. F.

Syntomis blanchardi, Poujade, Bull. S. E. France (6) IV 136 (1884)[10]; Hmpsn., F. I. Moths I 221 (1892) ♂[11]; Leech, T.E.S. 1898. 323-324 (1898) [variation][12]; Seitz, Lep. Pal. II, t. 9, f. H (1912).[13] — China.[4, 11]

nigra, Hampson. — Simla.[1-3]

Syntomis nigra, Hmpsn., F. I. Moths I 221 (1892) ♂[1]; *id.,* Cat. Phal. I 121, t. 5, f. 8 (1898) ♂[2]; Seitz, Lep. Indo-Austr. II 84, t. 12, f. C (1913).[3]

subaurata, Walker.

Glaucopis subaurata, Wlk., Cat. I 149 (1854).[1] — Ceylon.[1-4]

Eressa subaurata, Mo., Lep. Ceylon II 36-37, t. 95, f. 4 (1882)[2]; Hmpsn., F. I. Moths I 223 (1892) [redescr.] [*Syntomis*][3]; *id.,* Cat. Phal. I 117 (1898) [redescr.][4]; Seitz, Lep. Indo-Austr. II 83, t. 12, f. B (1913).[5] — Kandy.[5] Peradeniya.

vespa, Hampson.

Eressa vespa, Hmpsn., Cat. Phal. I 118, t. 4, f. 27 (1898) ♂♀[1]; *id.,* B. J. XIII 224 (1900)[2]; Seitz, Lep. Indo-Austr., II 84, t. 12, f. C (1913).[3] — Karen Hills.[1, 2]

vespoides, Rothschild.

Eressa vespoides, Roths., Novit. Zool. XVII 438 (1910)[1]; *id.,* Nov. Zool. XIX 376, t. 4. f. 9 (1912)[2]; Hmpsn., Cat. Phal. Suppl. I 51 (1914) ♂[3]; Seitz, Lep. Indo-Austr. II 84, t. 12, f. D (1913)[4] — Khasis.[1-4]

7. Dysauxes, Hübner (1822).

Dysauxes, Hb., Verz., p. 171 (1822); Hmpsn., Cat. Phal. I 132 (1898) [redescr.]; *id.,* B. J. XIII 224 (1900); Seitz, Lep. Pal. II 41 (1910) [redescr.; habits.]. Type; *ancilla,* Linn. [Europe.]

Naclia, Boisd., Index method, p. 60 (1840); Hmpsn., F. I. Moths I 226 (1892). Type: *ancilla,* Linn.

punctata, Fabricius.

Bombyx punctata, Fb., Spec. Ins. II 207 (1781).[1] — Kashmir.[7, 8, 15, 17]

Noctua ancilla, var., Esper. Eur. Schmett. p. 51, t. 85, f. 3 (1786).[2]

Bombyx serva, Hb., Eur. Schmett. III, ff. 115, 126, 127 (1804?).[3]

Bombyx confamula, Hb., Eur. Schmett. III, ff. 226, 227 (1804?).[4]

Call. serva, Godart, Lep. France IV 381, t. 39, f. 4 (1822).[5] — Europe.[1-7, 8, 14]

Naclia punctata, Mill., Ann. Soc. Linn. Lyon (2) VIII 234, t. 4, ff. 10, 11 (1861) [larga][6]; Hmpsn., F. I. Moths I 226, f 142 (1892) [redescr.].[7] — W. Asia.[8, 14]

Dysauxes punctata, Hmpsn., Cat. Phal. I 133-134, f. 53 (1898) [redescr.][8]; *id.,* Cat. Phal. Suppl. I 59-60 (1914) [Synon][9]; *id., t. c.* p. 383 (1914) [larva][10]; *id.,* B. J. XIII 224 (1900)[11]; Kirby, Butt. Moths Europe, p. 97[12]; Hofmann, Raupen, p. 38, 10, f. 12 [*larva* descr.][13]; Seitz, Lep. Pal II 41-42, t. 9, f. i. (1910).[14]

Dysauxes punctata kashmirensis, Roths., Novit. Zool. XVII 441 (1910)[15]; *id.,* Nov. Zool. XIX 376, t. 4, f. 11 (1912)[16]; Seitz, Lep. Pal. II 445 (1912).[17] — Srinagar. Achhabal (Kashmir).

Larva descr.[6, 10, 12, 13, 14,] on lichens. [6, 10, 12, 13, 14].

8. Euchromia, Hübner (1820).

Euchromia, Hb., Verz. p. 121 (1820); Wlk. Cat. I 201 (1854); Butl., J. Linn. Soc. XII 363, t. 28, f. 20 (1876); *id.,* T.E.S. 1888. 109 (1888); Meyr., Proc. Linn. Soc. N.S.W. (2) I 786 (1886); Kirby, Cat. Het. I 117 (1892); Hmpsn., F. I. Moths I 226, f. 143 (1892) [redescr.]; *id.,* Cat. Phal. I 293 (1898); *id.,* B.J. XIII 224 (1900) [key to spp.]; Seitz, Lep. Indo-Austr. II 85 (1913) [redescr.]. Type: *guineensis,* Fabr. 1775 (=*sperchius,* Cram. 1777) [W. Africa.]

Phalanna, Wlk., Cat. I 218 (1854). Type: *polymena,* Linn. *Hira,* Wlk., Cat. I 220 (1854). Type: *guineensis,* Fb.

elegantissima, Wallengren.

Sphinx polymena [*nec* Linn.], Drury, Ill. Exot. Ent. I, t. 26, f. 1 (1773).[1] — Moulmein.[6, 11, 19]

Charida elegantissima, Wlgn., Eugenie's Resa, p. 360 (1861).[2] — Tenasserim.[11] "N. India."[3] [? error loci.]

Euchromia orientalis, Butl., J. Linn. Soc. XII 364 (1876)[3]; *id.,* T.E.S. 1888. 114, t. 4, f. 6 (1888)[4]; Hmpsn., F. I. Moths I 227-228 (1892).[5] — ———— Siam.[9, 11, 20]

Euchromia fraterna, Butl., J. Linn. Soc. XII 364 (1876).[6] — China.[1], Formosa,[11, 18] Hainan.[17]

Euchromia laura, Butl., J. Linn. Soc. XII 364 (1876)[7]; *id.,* T.E.S. 1888. 114, t. 4, f. 8 (1888).[8] — Philippines.[11, 13, 21]

Euchromia siamensis, Butl., J. Linn. Soc. XII 365 (1876).[9]

Celebes.[22]

Euchromia formosana, Butl., T.E.S. 1888. 114, t. 4, f. 7 (1888).[10]

Euchromia polymena, orientalis, Hmpsn., Cat. Phal. I 298 (1898)[11]; *id.,* B.J. XIII 224 (1900).[12]

Euchromia elegantissima var. *diffusihelvola,* Schultz, Philipp. Journ. Gen. Sci. 1908, p. 29, t. 1, f. 3. (1908).[13]

Euchromia polymena elegantissima, Hmpsn., Cat. Phal. Suppl. I 197 (1914) [probably a distinct sp.].[14]

Euchromia polymena ab. *laura,* Seitz, Lep. Indo-Austr. II 85, t. 12, f. D. (1913).[15]

Euchromia polymena orientalis, Seitz, Lep. Indo-Austr. II 85 (1913).[16]

Euchromia polymena ab. *hainana,* Seitz, Lep. Indo-Austr. II 85 (1913).[17]

Euchromia polymena formosana, Seitz, *l.c.*[18]

Euchromia polymena fraterna Seitz, *l.c.*[19]

Euchromia polymena siamensis, Sietz, *l.c.*[20]

Euchromia polymena elegantissima, Seitz, *l.c.*[21]

Euchromia polymena celebensis, Butl., Seitz, *l.c.*[22]

magna, Swinhoe.

Syntomis magna, Swinh., T.E.S. 1891. 133, t. 8, f. 1 (1891).[1]

N. Kanara.[1-3]

Euchromia magna, Hmpsn. F. I. Moths I 227 (1892)[2]; *id.,* Cat. Phal. I 305 (1898)[2]; Seitz, Lep. Indo-Austr. II 86 (1913).[4]

Trincomali.[2-4]

polymena, Linnæus.

Sphinx polymena, Linn. Syst. Nat. (ed. X) I 494 (1758)[1]; Cram., Pap. Exot. I, t. 13, f. D (1775).[2]

Bombay.[6] Bassein Fort (Bombay)

Euchromia polymena, Wlk., Cat. I 219 (1854)[3]; Mo., Lep. Ceylon II 39, t. 94, ff. 6, 6[a] (1882)[4]; Hmpsn., F. I. Moths I 227, f. 143 (1892)[5]; *id.,* Cat. Phal. I 297-298, f. 137 (1898)[6]; Seitz., Lep. Indo-Austr. II 85, t. 12, f. D (1913)[7]; Fletcher, Spol. Zeylan. V 63 (1907)[8]; *id.,* Rept. Second Entl. Meeting, p. 291 (1917)[9]; *id.,* Rept. Third Entl. Meeting I 52 (1919)[10]; Turner, Proc. R. Soc. Victoria (n.s.) XXXV 29 (1922).[11]

Larva descr.[4, 5] on *Ipomœa batatas,* [9, 10] Convolvulaceæ.[4]

Bengal.[6], Laheria Serai. Dacca., Coimbatore. Nilgiris.[6] Travancore.[9, 10] Ceylon.[6, 8], Peradeniya. Upper Burma.[6] Andamans.[6], _____ Sangir.[6] N. W. Australia.[11]

98 Species.

PART-IX

ZYGÆNIDÆ.

Zygænidæ : Hmpsn., F. I. Moths I 228-289 (1892): Piepers and Snellen. Tijds. voor Ent. XLV 151-166 (1902) [general remarks on Javanese spp., with key to general]; Jordan, Seitz Lep. Pal. II 3-34 (1910) [palæarctic genera and spp.]; *id.,* Lep. Ind. II 5-56 (1908-1911) [Indo-australian genera and spp.].

ZYGÆNINÆ.

Zygæninæ : Hmpsn., F. I. Moths I 229-230 (1892) [key to genera]; Jordan, Seitz Lep. Pal. II 14 (1910) [redescr.]; *id.,* Lep. Ind. II 42 (1908) [redescr.].

Anthrocerinæ : Kirby, Cat. Het. I 62 (1892).

1. **Artona, Walker** (1854).

Artona, Wlk., Cat. II 439-440 (1854); Butl., J. Linn. Soc. XII 356, t. 28, f. 14 (1876); Hmpsn., F. I. Moths I 234, f. 150 (1892) [redescr.]; Snellen, Tijds. Ent. XLV 228-229 (1902) [redescr.]; Jordan, Seitz Lep. Pal. II 14 (1910) [redescr.]; *id.,* Ind. II 43-43 (1908) [redescr.]. Type:–*discivitta,* Wlk.

Rhaphidognatha, Felder, Wien. Ent. Mon. VI 31 (1862). Type:–*sesiaeformis,* Feld. [China].

Balataea, Wlk., Cat. XXXI 110-111 (1864). Type:–*sesiaeformis,* Feld.

Bintha, Wlk., Cat XXXI 127 (1864). Type:–*gracilis,* Wlk. [Japan].

Brachartona, Hmpsn., Ill, Het. VIII 44 (1891); *id.,* F. I. Moths I 233, f. 148 (1892) [redescr.]; Snellen, Tijds. Ent. XLV 232-233 (1902) [redescr.]. Type:–*quadrimaculata,* Mo.

albofascia, Leech

Arbudas albofascia, Leech, T. E. S. 1898, 340 (1898)♂ ♀.[1]

Artona albofascia, Jordan, Seitz Lep. Pal. II 15, t. 3 f. G (1910).[2]

Ta-tsien-lu[1, 2] [E. Tibet border].

aurulenta, Poujade

Bintha aurulenta, Pouj., Bull. S. E. Fr. (6) VI 116 (1886)♂ ♀.[1]

Artona aurulenta, Leech, T. E. S. 1998, 329 (1898)[2]; Jordan, Seitz Lep. Pal. II 15, t. 3 f. C (1910).[3]

Moupin.[1, 3]

catoxantha, Hampson

Brachartona catoxantha, Hmp., F. I. Moths I 233 (1892) ♂[1]; Piepers and Snellen, Tijds. Ent. XLV 237 (1902)[2]; Pratt, Bull. No. 4, Dept. Agric., F. M. S. (1909)[3]; Susainathan, Rept. Fifth Entl. Meeting, p. 39 (1924).[4]

Artona catoxantha, Jordan, Seitz Lep. Ind. II 44-45 (1908).[5]

Larva descr.[2 3], *on Cocos nucifera.*[2 3 4]

Tenasserim Valley.[1]

Fed. Malay States,[3, 4] Singapore.[5] Java,[2, 5]

charista, Jordan

Artona charista, Jord., Seitz Lep. Ind. II 44, t. 8 f. C (1908) ♂.[1]

Khasis.[1]

clathrata, Poujade

Bintha clathrata, Pouj., Bull S. E. Fr. (6) VI 117 (1886) ♀.[1]

Bintha ? clathrata, Leech, T. E. S. 1898, 327 (1898).[2]

Artona clathrata, Jordan, Seitz Lep. Pal. II 14, t. 3 f. D (1910).[3]

Moupin.[1, 3]

cuneonotata, Leech

Artona cuneonotata, Leech, T. E. S. 1898, 328 (1898) ♂[1]; Jordan, Seitz Lep. Pal. II 15, t. 3 f. D (1910).[2]

Wa-shan.[1] Pu-tsu-fong.[1]

W. China.[2]

delavayi, Oberthür

Artona delavayi, Obth., Et. d'Ent. XIX 29, t. 5 f. 39 (1894)[1]; Leech, T. E. S. 1989, 329 (1898)[2]; Jordan, Seitz Lep. Pal. II 15, t. 3 f. D (1910).[3]

Yunnan.[1, 3] Moupin.[2] Ta-tsien-lu.[2]

W. China. [2, 3]

digitata, Hampson

Artona digitata, Hmp., Novit. Zool. XXVI 273-274 (1920).[1] — Dawna Hills.[1]

d'scivitta, Walker

Artona discivitta, Wlk., Cat. II 440 (1854) [N. India][1]; Butl. Ill, Het. 26, t. f. 8 (1881)[2]; Hmp., F. I. Moths I 234-235 (1892) ♂[3]; Jordan, Seitz Lep. Ind. II 43, t. 8 f. C (1908) ♂.[4] — Almora.[3] Nilgiris.[3, 4] Shevaroys.[3] Mercara. Virajpet (Coorg). Gudalur (Wynaad).

flaviciliata, Hampson

Artona flaviciliata, Hmp., Novt. Zool. XXVI 273 (1920).[1] — Raitdong (Sikkim).[1]

flavigula, Hampson

Chrysartona flavigula, Hmp., F. I. Moths IV 467 (1896) ♂.[1] — Khasis.[1, 2]

Artona flavigula, Jordan, Seitz Lep. Ind. II 44, t. 8 f. C (1908).[2]

flavipuncta, Hampson

Artona flavipuncta, Hmp., B. J. XIII 225, t. B. f. 22 (1900) ♂[1]; Jordan, Seitz Lep. Ind. II 44, t. 8 f. C. (1908).[2] — Khasis.[1] Shillong.

fulvida, Butler

Artona fulvida, Butl., J. Linn. Soc. XII 356 (1876)[1]; Jordan, Seitz Lep. Ind. II 43, t. 8 f. B (1908).[2] — Burma.[2, 3] Moulmein.[1] N. India.[2]

Artona confusa var *diffusa,* Obth., Et. Ent. XIX 30, t. 5 f. 34 (1894) [ab.].[3]

hypomelas, Jordan

Artona hypomelas, Jordan, Seitz Lep. Ind. II 43 (1908) ♂ ♀.[1] — Mandi (N.-W. India).[1] Chumbi Valley (Sikkim).[1]

lugubris, Jordan

Artona lugubris, Jordan, Seitz Lep. Ind. II 44, t. 8 f. C (908) ♂ ♀.[1] — Khasis.[1]

microstigma, Jordan

Artona microstigma, Jord., Seitz Lep. Ind. II 44, t. 8 f. C (1908) ♂ ♀.[1] — Khasis.[1]

phæoxantha, Hampson

Artona phaeoxantha, Hmp., Novit. Zool. XXVI 274 (1920).[1] — Shevaroys.[1]

postalba, Elwes

Artona postalba, Elwes, P. Z. S. 1890, 379, t. 32 f. 16 (1890)[1]; Hmp., F. I. Moths I 234 (1892), ♂[2], Jordan, Seitz Lep. Ind. II 44 (1908).[3] — Sikkim (12,000 ft.).[1]

posthyalina, Hampson

Araeocera posthyalina, Hmp., F. I. Moths I 244 (1892) ♀.[1] — Sikkim.[1, 2]

Artona posthyalina, Jordan, Seitz Lep. Ind. II 45 (1908) ♀.[2]

postvitta, Moore

Artona postvitta, Mo., Lep. Atk., p. 13 (1879)[1]; Hmp., F. I. Moths I 234 (1892) [redescr.][2]; Jordan, Seitz Lep. Ind. II 44, t. 8 f. C (1908).[3] — Darjiling.[1] Sikkim.[2]

quadrimaculata, Moore

Artona quadrimaculata., Mo., P. Z. S. 1870, 390 (1879)[1]; Jordan, Seitz Lep. Ind. II 44, t. 8 f. C (1908).[2]

Brachartona quadrimaculata, Hmp., Ill. Het. VIII 44 (1891)[3]; *id.,* F. I. Moths I 233, f. 148 (1892) [redescr.].[4]

Masuri.[1,4] Kangra.[4] Dharmsala.[4] Ramgarh (Kumaon). Muktesar (Kumaon; 7,000 ft.).

refulgens, Hampson

Chrysartona refulgens, Hmp., F. I. Moths I 232 (1892) ♀.[1] — Matelei (Burma).[1]

Artona refulgens, Jordan, Seitz Lep. Ind. II 43, t. 8 f. B (1908).[2]

sieversi, Alpheraky

Artona sieversi, Alph., Rom. Mem. Lep. VI 5, t. 1 f. 4 (1892)[1]; Leech, T. E. S. 1898, 328 (1898)[2]; Jordan, Seitz Lep. Pal. II 15, t. 3 f. D (1910).[3]

Artona dejeani, Obth., Et. Ent. XIX 29, t. 6 f. 51 (1894).[4]

E. Tibet. ——— Kuku-nor.[3] W. China.[2, 3] N. China.[2]

sikkimensis, Elwes

Artona sikkimensis, Elwes, P. Z. S. 1890, 379, t. 32 f. 17 (1890)[1]; Hmp., F. I. Moths I 235 (1892) ♂[2]; Jordan Seitz Lep. Ind. II 44 (1908).[3] — Sikkim (12,000 ft.)[1-3]

walkeri, Moore (Four subspecies)

Artona walkeri, Jordan, Seitz, Lep. Indo. II 43 (1908).

(1) **walkeri nigrescens, Butler**

Artona nigrescens, Butl., J. Linn. Soc. XII 350 (1876).[1] — N.-W. Himalaya.[5] Punjab Hills.[1] Sikkim.[5] Raitdong (Sikkim).[6]

Artona walkeri nigrescens, Jord., Seitz Lep. Ind. II 43, t. 8 f. B (1908).[2]

[*ab.*] *Artona confusa,* Butl., J. Linn. Soc. XII 357 (1876)[3]; *id.,* Ill, Het. V 27, t. 84 f. 10 (1886)[4]; Hmp., F. I. Moths I 236 (1892)♂ ♀.[5]; *id.,* P. Z. S. 1906, 493 (1906).[6] — ———

(2) **walkeri walkeri, Moore**

Syntomis walkeri, Mo., P. Z. S., 1859, 199, t. 60 f. 9 (1859).[7] — Java.[7,8]

Artona confusa [*nec* Butl.], Snell., Tijds. Ent. XLV 229-230 (1902).[8]

(3) **walkeri hainana, Butler**

Artona hainana, Butl., J. Linn. Soc. XII 357 (1876).[2] — Hainan.[9,10]

Artona walkeri hainana, Jordan, Seitz Lep. Ind. II 43 (1908).[10] — Tonkin.[10]

(4) **walkeri baliensis, Jordan**

Artona walkeri baliensis, Jord., Seitz Lep. Ind. II 43 (1908).[11] — Bali.[11]

zebra, Elwes

Artona zebra, Elwes, P. Z. S. 1890, 379, t. 32 f. 11 (1890)[1]; Hmp., F. I. Moths I 235 (1892) ♂[2]; Jordan, Seitz Lep. Ind. II 43 (1908).[3] — Sikkim (11,000 ft.).[1–3]

zebraica, Butler (Three subspecies).

(1) **zebraica zebraica, Butler**

Artona zebraica, Butl., J. Linn. Soc. XII 356 (1876)[1]; Butl. Ill. Het. V 27, t. 84 f. 9 (1881)[2]; Hmp., F. I. Moths I 235, f. 150 (1892) [*partim*].[3] — Almora.[1], Masuri.[1] N.-W. India.[4] Kulu.[2]

Artona zebraica zebraica, Jordan Seitz Lep. Ind. II 43 (1908).[4] — Mandi.[3], Kangra.[3] Kumaon., Sikkim.[3]

(2) **zebraica khasiana, Jordan**

Artona zebraica, Hmp., F. I. Moths I 235 (1892) [*partim*][5]; Swinh., T. E. S. 1895, 32 (1895).[6] — Khasis.[6,7], Nagas.[5] Shillong.

Artona zebraica khasiana, Jord., Seitz Lep. Ind. II 43, t. 8 f. B (1908).[7] — Cherrapunji.

(3) **zebraica guttata, Snellen**

Artona guttata, Snell.,........ — Sumatra.[8, 10] Perak.[9]

Artona zebraica, Hmp., F. I. Moths I 235 (1892) [*partim*].[9]

Artona zebraica guttata, Jord., Seitz Lep. Ind. II 43 (1908).[10]

2. **Clelea, Walker** (1854).

Clelea, Wlk., Cat. II 465 (1858); Hmp., F. I. Moths I 239, f. 156 (1892) [redescr.]; Jordan, Seitz Lep. Pal. II 17 (1910) [redescr.]; *id.*, Lep. Ind. II 45 (1908) [redescr.]. — Type:–*sapphirina,* Wlk.m

Chrysartona, Swinhoe, Cat. Het. Mus. Oxon. I 57 (1892); Hmp., F. I. Moths I 232, f. 147 (1892) [redescr.]; Snellen, Tijds. Ent. XLV 230 (1902) [redescr.]. — Type:–*stipata,* Wlk.

albomacula, Leech

Artona albomacula, Leech, T. E. S. 1898, 329 (1898).[1] — Chow-pin-sa.[1] [E. Tibet border].

Clelea albomacula, Jordan, Seitz Lep. Pal. II 17, t. 3 f. H (1910).[2] — W. China.[2]

cyanicornis, Poujade

Bintha cyanicornis, Pouj., Bull. S. E. Fr. (6) VI 116 (1886).[1] — Moupin.[1] Omei-shan.[2]

Artona cyanicornis, Leech, T. E. S. 1898, 329 (1898).[2]

Clelea cyanicornis, Jordan, Seitz Lep. Pal. II 17, t. 3 f. H (1910).[3] — W. China.[3] Centr. China.[3]

discriminis, Swinhoe

Clelea discriminis, Swinh., T. E. S. 1891, 475 (1891)[1]; Hmp., F. I. Moths I 240 (1892)[2]; Swinh., T. E. S. 1895, 33 (1895)[3]; Jordan, Seitz Lep. Ind. II 45, t. 8 f. D (1908).[4] — N. India.[4] Sikkim.[2], Nagas.[2] Khasis.[1], Shillong.[3] Cherrapunji.[3]

melaleuca, Jodan

Clelea melaleuca, Jord., Seitz Lep. Pal. II 17, t. 8 f. i (1910) ♀.[1] — Tas-tsien-lu.[1] [E. Tibet border.]

metacyanea, Hampson

Clelea metacyanea, Hmp., F. I. Moths IV 467-468 (1896) ♂[1]; Jordan, Seitz Lep. Ind. II 45, t. 8 f. D (1908).[2] — Khasis.[1, 2m]

nigroviridis, Elwes

Clelea nigroviridis, Elwes, P. Z. S. 1890, 380 (1890)[1]; Hmp., F. I. Moths I 240 (1892) ♂[2]; Jord., Seitz Lep. Ind. II 45 (1908).[3] — Nagas.[1, 2] Bernardmyo (Burma).[2]

plumbeola, Hampson

Clelea plumbeola, Hmp., F. I. Moths I. 240 (1892)[1]; Dudgeon, B. J. XII 295 (1899)[2]; Jordan, Seitz Lep. Ind. II 45, t. 8 f. D (1908).[3]

Bernardmyo.[1]
Khasis.[3]
Bhutan.[2]

refulgens, Hampson

Clelea refulgens, Hmp., B. J. XVI 193, t. D. f. 3 (1904)[1]; Jordan, Seiz Lep. Ind. II 45, t. 8 f. D (1908).[2]

Manipur (6,000 ft.).[1]
Khasis.[2]
Burma.[2]

apphirina, Walker

Clelea sapphirina, Wlk., Cat. II 465-466 (1854)[1]; Hmp., F. I. Moths I 239-240, f. 156 (1892)♂ ♀.[2]; Swinh., T. E. S. 1895, 33 (1895)[3]; Leech, T.E. S. 1898, 330 (1898)[4]; Jordan, Seitz Lep. Ind. II 45 (1908)[5]; *id.,* Lep. Pal. II 17, t. 3 f. H (1910).[6]

N. India.[5], Sikkim.[2]
Shillong.[3]
Momeit (Burma).[2]

W. China.[4, 6]
S. China.[4, 6]
Hongkong.[1, 2]
Centr. China.[4, 6]

simplex, Joda

Clelea simplex, Jord., Seitz Lep. Ind. II 45 (1908) ♀.[1]

Khasis.[1]

stipata, Walker

Procris stipata, Wlk., Cat. I 114 (1854) [N. India][1]; Butl., Ill. Het. I 13-14, t. 7 f. 9 (1877) [N. India].[2]

Chrysartona stipata, Hmp., F. I. Moths I 232, f 147 (1892)♂ ♀.[3]; Swinh., T. E. S. 1895, 32 (1895)[4]; Alph., Rom. Mém. Lép. IX 121 (1897)[5]; Leech, T. E. S. 1989, 327 (1898)[6]; Snellen, Tijds. Ent. XLV 231-232 (1902).[7]

Clelea stipata, Jordan, Seitz Lep. Ind. II 46, t. 8, f. E (1908).[8]

Kangra.[3], Sikkim.[3]
Shillong.[4]
Cherrapunji.[4]
Bernardmyo (Burma).[3]

W. Chian.[5, 6]
Java.[7, 8]

syfanica, Oberthür

Laurion syfanicum, Obth., Et. Ent. XIX 25, t. 6 f. 45 (1894).[1]

Clelea syfanica, Leech, T. E. S. 1898, 331 (1898).[2]

Clelea syfanicum, Jordan, Seitz Lep. Pal. II 17, t. 3 f. H (1910).[3]

Moupin.[2, 3]

Tongho (W. China).[1, 3]

3. **Dendrocera, Hampson** (1892).

Dendrocera, Hmp., F. I. Moths I 231, f. 146 (1892); *id.,* Ill. Het. IX 62 (1893); Jordan, Seitz Lep. Ind. II 46 (1908) [redescr.].

Type:– *quadripunctata,* Hmp.

bipuncta, Hampson

Clelea bipuncta, Hmp., T. E. S. 1895, 283 (1895) ♂[1]; *id.,* F.I. Moths IV 467 (1896). ♂[2] — Coimbatore.[1,2]

Dendrocera bipuncta, Jord., Seitz Lep. Ind. II 46, t. 8 f. E (1908).[3]

quadripunctata, Hampson

Dendrocera quadripunctata, Hmp., F. I. Moths I 231-232, f. 146 (1892) ♂[1] ; *id.,* Ill. Het. IX 62, t. 157 f. 2 (1893)[2]; Jordan, Seitz Lep. Ind. II 46, t. 8 f. D (1908).[3] — Ceylon.[1,3] Madulsima. Peradeniya. Pundaluoya.[2]

4. **Homophylotis, Turner** (1904).

Homophylotis, Turner, Tr. R. Soc. S. Austr. XXVIII 243 (1904); Jordan, Seitz Lep. Ind. 46 (1908) [redescr.]. — Type:–*thridota,* Turner [Queensland].

albicilia, Hampson

Artona albicilia, Hmp., B. J. XIII 225 (1900)♂[1]. — Nagas.[1]

———

Homophylotis albicilia, Jordan, Seitz Lep. Ind. II 46, t. 8 f. E (1908).[2] — Pulo Laut.[1,2] Amboina.[2] Tenimber.[2] Fergusson Isd.[2] Rou Isd.[2]

nigra, Hampson

Tasema nigra, Hmp., F. I. Moths I 237 (1892)♂ [*nec* ♀].[1] — Bernardmyo (Burma).[1]

Homophylotis nigra, Jord., Seitz Lep. Ind. II 47 (1908) ♀.[2]

5. **Tasema, Walker** (1856).

Tasema, Wlk., Cat. VII 1597 (1856); Hmp., F. I. Moths I 236, f. 151 (1892) [redescr.]; Jordan, Seitz Lep. Ind. II 47-48 (1908) [redescr.]. — Type:–*bipars,* Wlk.

bipars, Walker

Tasema bipars, Wlk., Cat. VII 1597 (1856) [Hindostan][1]; Hmp., F. I. Moths I 236, f. 151 (1892) ♂[2]; Jordan, Seitz Lep. Ind. II 48, t. 8 f. F (1908)♂ ♀.[3] — Kangra.[2,3] Simla.[3], Nagas.[2,3] Khasis.[3]

longipennis, Hampson

Tasema longipennis, Hmp., F. I. Moths I 236; (1892) ♂[1]; Jord., Seitz Lep. Ind. II 48 (1908).♂[2]. — Bernardmyo (Burma).[1,2]

6. **Aræocera, Hampson** (1892).

Araeocera, Hmp., F. I. Moths I 244, f. 162 (1892); Jordan, Seitz Lep. Ind. II 48 (1908) [redescr.].	Type:–*cuanescens,* Hmp.

cyanescens, Hampson

Araeocera cyanescens, Hmp., F. I. Moths I 244, f. 162 (1892) ♀.[1]; Jord., Seitz Lep. Ind. II 48, t. 8 f. F. (1908) ♀.[2]	Momeit (Burma).[1,2]

7. **Illiberis, Walker** (1854).

Illiberis, Wlk., Cat. II 280 (1854); Kirby, Cat. Het. I 88 (1892); Jordan, Seitz Lep. Ind. II 48 (1908) [redescr.]; *id.,* Lep. Pal. II 15 (1910) [redescr.].	Type:–*sinensis,* Wlk.
Goë, Hmp., F. I. Moths I 242, f. 158 (1892).	Type:–*diaphana,* Hmp.

diaphana, Hampson

Goë diaphana, Hmp., F. I. Moths I 242, f. 158 (1892) ♂[1]; Jordan, Seitz Lep. Ind. II 48 (1908)♂ [*Illiberis*].[2]	Bernardmyo (Burma).[1,2]

endocyanea, Hampson

Illiberis endocyanea, Hmp., Novit. Zool. XXVI 273 (1920).[1]	Assam.[1]

fuliginosa, Moore

Artona fuliginosa, Mo., Lep. Atk., p. 14 (1879).[1]	Darjiling.[1,3]
Tasema fuliginosa, Hmp., F. I. Moths I 236 (1892)♂ ♀.[2]	Calcutta.[2] [?error loci].
Illiberis fuliginosa, Jord., Seitz Lep. Ind. II 48 (1908) ♀.[3]	

ochracea, Leech

Illiberis ochracea, Leech, T. E. S. 1898, 335 (1898)♂ ♀.[1]; Jordan, Seitz Lep. Pal. II 16, t. 3 f. E (1910).[2]	Moupin.[1] ——— W. China.[1,2] Centr. China.[1,2]

sinensis, Walker

Illiberis sinensis, Wlk., Cat. II 280 (1854) [China].[1]	Moupin.[5] ———
Northia sinensis, Stdgr., Rom. Mém. Lép. III 170 (1887)[2]; Leech, P. Z. S. 1888, 595, t. 30 f. 20 (1889).[3]	W. China.[5,6] Centr. China.[5,6] N. China.[3]
Illiberis sinensis, Kirby, Cat. Het. I 88 (1892) [*partim*][4]; Leech, T. E. S. 1898, 334 (1898)[5]; Jordan, Seitz Lep. Pal. II 15, t. 3 f. E (1910).[6]	Korea.[3] Japan.[3]

tenuis, Butler

Northia tenuis, Butl., A. M. N. H. (4) XX 394 (1877)[1]; *id.,* Ill. Het. II 9, t. 21 f. 7 (1878)[2]; Leech, P. Z. S. 1888, 596, t. 30 f. 19 (1889).[2a]

Illiberis tenuis, Leech, T. E. S. 1898, 333 (1898)[3]; Jordan, Seitz Lep. Pal. II 16, t. 3 f. F (1910).[4]

Moupin.[3]
———
Kiushin.[3]
Korea.[3, 4]
Amurland.[3, 4]
Japan.[3, 4, 2a]

8. **Phacusa, Walker** (1854).

Phacusa, Wlk., Cat. I 150 (1854); Hmp., F. I. Moths I 240, f. 157 (1892) [redescr.]; Snellen, Tijds. Ent. XLV 223-224 (1902) [redescr.]; Jordan, Seitz Lep. Ind. II 48 (1908) [redescr.]; *id.,* Lep. Pal. II 16 (1910) [redescr.]. — Type:–*tenebrosa,* Wlk.

Notioptera, Butl., J. Linn. Soc. XII 355, t. 28 f. 2 (1876). — Type:–*dolosa,* Wlk.

birmana, Oberthür

Northia birmana, Obth., Et. Ent. XIX 31, t. 5 ff. 22, 37 (1894).[1]

Phacusa birmana, Jord., Seitz Lep. Ind. II 49, t. 8 f. H (1908).[2]

Upper Burma.[2]

cyanocera, Hampson

Phacusa cyanocera, Hmp., F. I. Moths I 241 (1892) ♂[1]; Hmp., B. J. XIII 225 (1900) [synon.][2]; Jordan, Seitz Lep. Ind. II 49, t. 8 f. G (1908).[3]

Northia ignea, Obth., Et. Ent. XIX 29, t. 5 f. 35 (1894).[4]

Bernardmyo (Burma).[1]

dolosa, Walker

Syntomis dolosa, Wlk., Cat. VII 1594 (1856).[1]

Notioptera dolosa, Butl., Ill. Het. V 27, t. 84 f. 11 (1881).[2]

Phacusa dolosa, Hmp., F. I. Moths I, 240 (1892) ♂[3]; Jordan, Seitz Lep. Ind. II 49, t. 8 f. H (1908).[4]

N. India.[1, 4]
? Assam.[3]
Upper Burma.[4]

djreuma, Oberthür

Phacusa djreuma, Obth., Et. Ent. XVIII 21, t. 2 f. 31 (1893)[1]; Leech, T. E. S. 1898, 336 (1898)[2]; Jordan, Seitz Lep. Pal. II 31, t. 3 f. i (1910).[3]

Tsc-kou.[1, 3] (E. Tibet border).

khasiana, Moore

Northia khasiana, Mo., Lep. Atk., p. 12 (1879).[1]

Phacusa khasiana, Hmp., F. I. Moths I 241 (1892) ♀.[2]; Jordan, Seitz Lep. Ind. II 49, t. 8 f. G (1908) ♀.[3]

Khasis.[1-3]
Nagas.[2, 3]
Sillong.

nicobarica, Hampson

Phacusa nicobarica, Hmp., Novit. Zool. XXVI 272-273 (1920).[1] — Nicobars.[1]

nigrigemma, Walker (Two subspecies).

(1) **nigrigemma nigrigemma, Walker**

Glaucopis nigrigemma, Wlk., Cat. I 141 (1854).[1] — Hongkong.[3, 4] N.China.[1, 5, 3]

Zama cyanecula, Herr.-Schaff., Aussereur. Schmett. I, f. 224 (1855).[2]

Northia cyanecula, Butl., Ill. Het. I 13, t. 7 f. 8 (1877).[3]

Phacusa nigrigemma nigrigemma, Jord., Seitz Lep. Ind. II 48-49 (1908).[4]

Phacusa nigrigemma, Jord., Seitz Lep. Pal. II 16, t. 3 f. G (1910)♂ ♀.[5]

(2) **nigrigemma vitrea, Jordan**

Phacusa nigrigemma vitrea, Jord., Seitz Lep. Ind. II 49, t. 8 f. F (1908).[6] — Masuri.[6]

properta, Swinhoe

Notioptera properta, Swinh., P. Z. S. 1889, 400, t. 43 f. 6 (1890).[1] — Rangoon.[1, 2] Upper Burma.[4] Nagas.[2] N. India.[4] Nicobars.[2] [?]

Phacusa properta, Hmp., F. I. Moths I 241 (1892) [redescr.][2]; *id.,* B. J. XIII 225 (1900) [synon.][3]; Jordan, Seitz Lep. Ind. II 49, t. 8 f. H (1908).[4]

Northia dohertyi, Oberthür, Et. Ent. XIX 31, t. 5 f. 36 (1894).[5]

tenebrosa, Walker (Three subspecies).

(1) **tenebrosa tenebrosa, Walker**

Glaucopis tenebrosa, Wlk., Cat. I 150 (1854).[1] — Almora.[3, 5] N. India.[1, 5, 2]

Phacusa tenebrosa, Butl., Ill. Het. I 20, t. 12 f. 1 (1877)[2]; Hmp., F. I. Moths I 241, f. 157 (1892) [redescr.] [*partim*][3]; *id.,* B. J. XIII 225 (1900) [synon.].[4]

Phacusa crawfurdi tenebrosa, Jordan, Seitz Lep. Ind. II 49, t. 8 f. G (1908).[5]

(2) **tenebrosa crawfurdi, Moore**

Syntomis crawfurdi, Mo., P. Z. S. 1859, 199, t. 60 f. 11 (1859)[6]; Mo., Cat. Lep. E. I. Co. Mus. II 327 (1859).[7] — Momeit (Burma).[8] Nicobars.[10]

Phacusa tenebrosa, Hmp., F. I. Moths I 241 (1892) [*partim*][8]; Snellen, Tijds. Ent. XLV 224-225 (1902).[9] — ——————

Phacusa erawfurdi crawfurdi, Jordan, Seitz Lep. Ind. II 49, t. 8 f. G (1908).[10] — Malacca.[10] Java.[6, 9, 10] Celebes.[9, 10] Amboina.[10] Philippines.[10]

(3) **tenebrosa siamensis, Oberthür**

Phacusa siamensis, Obth., Et. Ent. XIX 31, t. 5 f. 24 (1894).[11] — Siam.[12] Renong.[11]

Phacusa crawfurdi siamensis, Jord., Seitz Lep. Ind. II 49 (1908).[12]

translucida, Poujade

Procris translucida, Pouj., Bull. S. E. Fr. (6) IV 136 (1885).[1] — Moupin.[1, 3] Wa-sen-kow.[3] [E. Tibet border.].

Northia translucida, Obth., Et. Ent. XIX, t. 6 f. 66 (1894).[2] — ——————

Illiberis translucida, Leech, T. E. S. 1898, 333 (1898).[3] — W. China.[3, 4]

Phacusa translucida, Jordan, Seitz Lep. Pal. II 16, t. 3 f. F (1910).[4]

9. Ephemeroidea, Hampson (1892).

Ephemeroidea, Hmp., F. I. Moths I 242, f. 159 (1892); Snellen, Tijds. Ent. XLV 221-222 (1902) [redescr.]; Jordan, Seitz Lep. Ind. II 49 (1908) [redescr.]. — Type:–*ariel,* Hmp.

ariel, Hampson

Ephemeroidea ariel, Hmp., F. I. Moths I 242, f. 159 (1892) ♂[1]; Jordan, Seitz Lep. Ind. II 49, t. 8 f. H (1908)♂ ♀.[2] — Bernardmyo (Burma).[1] Assam.[2]

cyanea, Jordan

Ephemeroidea cyanea, Jord., Seitz Lep. Ind. II 50, t. 8 f. i (1908) ♀.[1] — Khasis.[1]

flavocincta, Hampson

Ephenteroidea flavoeincta, Hmp., F. I. Moths I 242-243 (1892) ♀.[1]; Jord., Seitz Lep. Ind. II 49, t. 8 f. H (1908)♀.[2] — Matelei (Burma).[1]

10. Lophosoma, Swinhoe (1892).

Lophosoma, Swinh., Cat. Het. Mus. Oxon. I 58 (1892); Hmp., F. I. Moths I 237, f. 152 (1892) [redescr.]; Jordan, Seitz Lep. Ind. II 50 (1908) [redescr.]. — Type:–*cuprea,* Wlk.

cuprea, Walker

Syntomis cuprea, Wlk., Cat. VII 1596 (1856) [Hindostan].[1] — N. India.[4], Kulu.[2]

Lophosoma cupreum, Hmp., F. I. Moths I 237, f. 152 (1892) [redescr.][2]; Swinh., T. E. S. 1895, 33 (1895).[3] — Sikkim.[2] Shillong.[3] Andamans.[2]

Lophosoma cuprea, Jordan, Seitz Lep. Ind. II 50, t. 8 f. i (1908).[4]

quadricolor, Walker

Syntomis quadricolor, Wlk., Cat. VII 1596 (1856) [Hindostan].[1] — N. India.[3] Bengal.[2]

Lophosoma quadricolor, Hmp., F. I. Moths I 237 (1892) ♂[2]; Jordan, Seitz Lep. Ind. II 50 (1908).[3] — Ganjam.[2]

11. Arachotia, Moore (1879).

Arachotia, Mo., Lep. Atk., p. 14 (1879); Hmp., F. I. Moths I 245, f. 163 (1892) [redescr.]; Jordan, Seitz Lep. Ind. II 50 (1908) [redescr.]. — Type:–*flaviplaga,* Mo.

englenia, Jordan

Arachotia euglenia, Jord., Seitz Lep. Ind. II 50, t. 8 f. K (1908)♂ ♀.[1] — Khasis.[1]

flaviplaza, Moore

Arachotia flaviplaga, Mo., Lep. Atk., p. 14 (1879)[1]; Hmp., F. I. Moths I 245, f. 163♀ (1892)♂ ♀.[2]; Swinh., T. E. S. 1895, 33 (1895)[3]; Jordan, Seitz Lep. Ind. II 50, t. 8 f. i (1908).[4] — Darjiling.[1] Sikkim.[2] Cherrapunji.[3] Nagas.[2]

vespoides, Moore

Arachotia vespoides, Mo., P. Z. S., 1879, 390 (1879)[1]; Hmp., F. I. Moths I 245 (1892) ♂[2]; Jordan, Seitz Lep. Ind. II 50, t. 8 f. i (1908).[3] — N. India.[1, 3]

12. Piarosoma, Hampson (1892).

Piarosoma, Hmp., F. I. Moths I 243, f. 160 (1892); Jordan, Seitz Lep. Ind. II 50 (1908) [redescr.]; *id.,* Lep. Pal. II 17 (1910) [redescr.]. — Type:–*albicincta,* Hmp.

albicincta, Hampson

Piarosoma albicinctum, Hmp., F. I. Moths I 243, f. 160 (1892) ♂[1]; Jord., Seitz Lep. Ind. II 51, t. 8 f. K (1908) ♀.[2] — Bernardmye.[1] Burma.[2]

hyalina, Leech (Two subspecies)

(1) hyalina hyalina, Leech

Arachotia hyalina, Leech, T. E. S. 1889, 123, t. 7 f. 6 (1889)[1]; Leech, T. E. S. 1898, 336 (1898) [*Piarosoma*].[2] — Kiukiang.[1,3]

Piarosoma hyalina hyalina, Jord., Seitz Lep. Pal. II 17, t. 3 f. i. (1910).[3]

(2) hyalina thibetana, Oberthür

Phacusa thibetana, Obth., Et. Ent. XIX 31, t. 5 f. 23 (1894).[4] — Moupin.[4] Ta-tsien-lu.[4]

Piarosoma hyalina thibetana, Jordan, Seitz Lep. Pal. II 17, t. 3 f. i (1910).[5] — W. China.[5]

sizala, Swinhoe

Phacusa sizala, Swinh., A. M. N. H. (6) XIV 441-442 (1894)[1]; *id.,* T. E. S. 1895, 33, t. 1 f. 3 (1895).[2] — Khasis.[3,4] Shillong.[1,2]

Goë sizala, Hmp., F. I. Moths IV 468 (1896) ♂.[3]

Piarosoma ? sizala, Jordan, Seitz Lep. Ind. II 51, t. 8 f. K (1908).[4]

13. Thyrassia, Butler (1876).

Thyrassia, Butler, J. Linn. Soc. XII 355, t. 28 f. 6 (1876); Kirby, Cat. Het. I 103 (1892); Hmp., F. I. Moths I 238, f. 154 (1892); Piepers and Snellen, Tijds. Ent. XLV 226-227 (1902) [redescr.]; Jordan, Seitz Lep. Ind. II 51 (1908) [redescr.]; *id.,* Lep. Pal. II 17 (1910) [redescr.]. — Type:–*subcordata,* Wlk.

Monoschalis, Hmp., F. I. Moths I 238, f. 153 (1892); *id.,* Ill. Het. IX 63 (1893). — Type:–*virescens,* Hmp.

penangæ, Moore (Two subspecies)

(1) penangæ penangæ, Moore

Syntomis penangæ, Mo., P. Z. S. 1859, 198, t. 60 f. 7 (1859)[1]; Mo., Cat. Lep. E. I. Co. Mus. II 325 (1859).[2] — Tenasserim.[4] Singapore.[8] Penang.[1,4] W. China.[6,7]

Hydrusa penangœ Kirby, Cat. Het. I 102 (1892).[3]

Syntomis penanga, Hmp., F. I. Moths IV 466 (1896).[4]

Thyrassia penanga, Hmp., B. J. XIII 224 (1900).[5]

Thyrassia penanga, Jordan, Seitz Lep. Pal. II 18, t. 3 f. i (1910)[6]; Leech, T. E. S. 1898, 330 (1898).[7]

Thyrassia penangœ penangœ, Jord., Seitz Lep. Ind. II 51 (1908).[8]

(2) penangæ rafflesi, Moore

Syntomis rafflesi, Mo., P. Z. S. 1859, 198-199, t. 60 f. 8 (1859).[9]

Thyrassia penangœ rafflesi, Jord., Seitz Lep. Ind. II 51 (1908).[10]

Java.9, 10

subcordata, Walker (Two subspecies).

(1) subcordata subcordata, Walker

Syntomis subcordata, Wlk., Cat. I 132 (1954) [N. India].[1]

Thyrassia subcordata, Butl., J. Linn. Soc. XII 355 (1876)[2]; *id.,* Ill. Het. I 19, t. 7 f. 5 (1877)[3]; Hmp., F. I. Moths I 238, f. 154 (1892)♂♀ [*partim*][4]; Fletcher, Pusa Bull. 89, Entl. Note 182, f. 48 (1919).[5]

Thyrassia subcordata subcordata, Jordan, Seitz Lep. Ind. II 51, t. 8 f. K (1908).[6]

N. India (Plains).4, 6
Pusa.[5]
Chapra.
Akra (24 Pargs.).
Surat.

(2) subcordata aurodisca, Hampson

Thyrassia aurodisca, Hmp., Ill. Het. VIII 44, t. 139 f. 10 (1891).[7]

Thyrassia subcordata, Hmp., F. I. Moths I 238 (1892) [*partim*].[8]

Thyrassia subcordata aurodisca, Jord., Seitz Lep. Ind. 51 (1908).[9]

Larva descr.[5], on *Vitis trifolia.*[5]

S. India (Plains).[8]
Nilgiris.[7]
Ceylon.[8, 9]

virescens, Hampson

Monoscholis virescens, Hmp., F. I. Moths I 238, f. 153 (1892) ♂[1]; *id.,* Ill. Het. IX 63, t. 157 f. 11 (1893).[2]; *Ramachandra Rao, Pusa Bull. 89, Entl.* Note 181 (1919).[3]

Thyrassia virescens, Jordan, Seitz Lep. Ind. II 52, t. 8 f. L (1908).[4]

Larva descr.[3], on *Vitis quadran-gularis.*[3]

Ceylon.[4]
Colombo.[1]
South India.[4]
Coimbatore.[3]

14. Platyzygæna, Swinhoe (1892).

Platyzygaena, Swinh., Cat. Het. Mus. Oxon. I 57, fig. (1892); Hmp., F. I. Moths I 243, f. 161 (1892) [redescr.]; Jordan, Seitz Lep. Ind. II 52 (1908) [redescr.]. — Type:–*mölleri*, Elwes.

molleri, Elwes

Soritia ? mölleri, Elwes, P. Z. S. 1890, 385, t. 32 f. 13 (1890).[1]

Platyzygaena mölleri, Swinh., Cat. Het. Mus. Oxon. I 59 (1892)[2], Hmp., F. I. Moths I 243-244, f. 161 (1892) ♀[a]; *id.*, P. Z. S. 1906, 493 (1906)[4]; Jordan, Seitz Lep. Ind. II 52, t. 8 f. L (1908) [*moelleri*].[5]

Sikkim.[1, 3, 5] Chengtong (Sikkim).[4] Assam.[2, 3, 5]

15. Zygæna, Fabricius (1775).

Zygaena, Fb., Syst. Ent. p. 550 (1775); Hmp., F. I. Moths I 230, f. 145 (1892) [redescr.]; Jordan, Seitz Lep. Ind. II 52 (1908) [redescr.]; *id.*, Lep. Pal. II 18 (1910) [redescr.]. — Type:–*filipendulæ*, Linn. [Europe].

Anthrocera, Scopoli, Introd. Hist. Nat., p. 454 (1777). — Type:–*filipendula*, Linn.

rubricoll's, Hampson

Zygaena rubricollis, Hmp., B. J. XIII 224-225, t. B. f. 9 (1900) ♂[1]; Jordan, Seitz Lep. Ind. II 52 (1908).[2] — Chitral.[1, 2]

16. Epizygæna, Jordan (1908).

Epizygaena, Jord., Seitz Lep. Ind. II 52 (1908); *id.*, Lep. Pal. II 31 (1910) [redescr.]. — Type:–*afghana*, Mo.

afghana, Moore

Zygaena afghana, Mo., Cat. Lep. E. I. Co. Mus. II 286, t. 7[a] f. 1 (1859)[1]; Hmp., F. I. Moths I 231 (1892) ♀.[2]

Anthrocera afghana, Kirby, Cat. Het. I 79 (1892).[3]

Epizygaena afghana, Jordan, Seitz Lep. Ind. II 52 (1908)[4]; Seitz Lep. Pal. II 31, t. 8 f. F (1910).[5]

Afghanistan.[1-4] Quetta.[2, 4]

Persia.[5]

asoka, Moore

Zygaena asoka, Mo., P. Z. S. 1879, 389 (1879).[1]

Zutulba ? asoka, Kirby, Cat. Het. I 80 (1892).[2]

Epizygaena caschmirensis ab. *asoka*, Jordan, Seitz Lep. Ind. II 52 (1908).[3]

Kashmir.[4] N.-W. India.[1, 2]

Epizygaena asoka, Seitz. Lep. Pal. II 31, t. 8 f. K (1910).[4]

caschmirensis, Kollar

Zygaena caschmirensis, Kollar, Hügel's Kaschmir IV (ii) 459, t. 19 f. 6 (1844)[1]; Wlk., Cat. I 102 (1854)[2].

Zutulba ? caschmirensis, Kirby, Cat. Het. I 80 (1892).[3]

Zygaena caschmirensis, Hmp., F. I. Moths I 230-231, f. 145 (1892).[4]

Epizygaena caschmirensis, Jordan, Seitz Lep. Ind. II 52 (1908)[5]; Seitz Lep. Pal. II 31, t. 8 f. K (1910).[6]

Afghanistan.[5] Kohat. Kashmir.[1, 4, 6] Murree., Kangra.[4] Punjab.[1] Masuri. Bhim Tal (Kumaon). Jeolikote (Kumaon). Nepal.[4, 5]

erythrosoma, Hampson

Zygaena erythrosoma, Hmp., F. I. Moths I 231 (1892) ♀.[1]

Epizygaena erythrosoma, Jordan, Seitz Lep. Ind. II 52, t. 8 f. L (1908).[2]

Almora.[1]

17. Chalcosiopsis, Swinhoe (1894).

Chalcosiopsis, Swinh., A. M. N. H. (6) XIV 442, fig. (1894); Hmp., F. I. Moths IV 468, f. 249 (1896) [redescr.]; Jordan, Seitz Lep. Ind. II 52 (1908) [redescr.].

Type:–*variata,* Swinh.

variata, Swinhoe

Chalcosiopsis variata, Swinh., A. M. N. H. (6) XIV 442-443, fig. (1894)[1]; *id.,* T. E. S. 1895, 38 (1895)[2]; Hmp., F. I. Moths IV 468-469, f. 249 (1896) ♀.[3]; Jordan, Seitz Lep. Ind. II 52, t. 8 f. L. (1908).4

Khasis.[3, 4] Shillong.[1, 2]

Of doubtful position.

The following species, originally described as belonging to the Zygæninæ, is perhaps not a Zygænid. It is omitted by Jordan in his revision of Indian species in Seitz Lep. Indo-austr. *Callartona purpurascens,* with which it is compared, "belongs to Tineidæ" according to Hampson (B. J. XIII 225; 1900) :–

microsticta, Hampson

Callartona microsticta, Hmp., B. J. XI 284 (1897) ♂.[1]

Khasis.[1]]

PROCRINÆ CHALCOSIANÆ)

Chalcosianæ : Hmpsn., F. I. Moths I 246-285 (1892) [key to genera]; Jordan, Seitz Lep. Ind. II 8 (1908) [characters]; *id.*, Lep. Pal. II 5-6 (1910); Hering, Archiv f. Naturgeschichte LXXXVIII, Abt. A, Heft 11, pp. 1-93, t. 1 (1922) [revis. of Oriental spp., key to genera, pp. 4-11].

Chalcosiidæ : Kirby, Handb. Lep. III 71 (1897).

18. Procris, Fabricius (1807).

Procris, Fabr., Illiger's Mag. VI 289 (1807); Jordan, Seitz Lep. Pal. II 6 (1910) [redescr.]; Hering Arch. f. Naturg. LXXXVIII (A. 11), p. 14 (1922). Type:–*statices*, Linn. [Europe].

elegans, Poujade

Thyrina elegans, Pouj., Ann. S. E. Fr. 1886, p. exliii (1886).[1] Moupin.[1, 2]

Illiberis elegans, Leech, T. E. S. 1898, 335 (1898).[2] W. China.[3]

Procris elegans, Jord., Seitz Lep. Pal. II 7, t. 1 f. D (1910).[3] Kwei-chow-Prov.[2] Amur.[2]

19. Zygænoprocris, Hampson (1900).

Zygaenoprocris, Hmp., B. J. XIII 225, fig. (1900). Type:–*chalcochlora*, Hmp.

chalcochlora, Hampson

Zygaenoprocris chalcochlora, Hmp., B. J. XIII 225, fig. (1900). ♂.[1] Chitral.[1]

Procris chalcochlora, Jordan, Seitz Lep. Pal. II 9 (1910).[2]

20. Callizygæna, Moore (1882).

Callizygaena, Feld., Reise Novara, Lep. IV, Erkl. t. 83 f. 4 (1874) [non-descr.]; Moore, Lep. Ceylon II 38 (1882); Hmp., F. I. Moths I 248, f. 165 (1892) [redescr.]; Snellen, Tijds. Ent. XLV 225-226 (1902) [redescr.]; Jordan, Seitz Lep. Ind. II 8 (1908) [redescr.]; Hering, Arch. f. Naurg. LXXXVIII, A. 11, pp. 11-12 (1922) [key to spp.]. Type:–*nivimacula*, Feld.

aurata, Cramer (Two subspecies)

(1) aurata aurata, Cramer

Sphinx aurata, Cram., Pap. Exot. III, t. 264 f. A (1780).[1] South India.[5-7] Madras.

Callizygaena auricincta, Swinh., Cat. Het. Mus. Oxon. I 62 Nilgiris.[2, 3]

(1892)[2]; Hmp., F. I. Moths I 249 (1892) ♂.[3]

Callizygaena aurata, Hmp., B. I. XIII 225 (1900) [synon.].[4]

Callizygaena auratus, Jordan, Seitz Lep. Ind. II 8-9 (1908)[5]; Hering, Arch. f. Naturg. LXXXVIII (A. 11), p. 12 (1922).[6]

Callizygaena auratus auratus, Jord., Seitz Lep. Ind. II 9 (1908).[7]

ab. gana, Swinhoe

Callizygaena gana, Swinh., T. E. S. 1891, 133, t. 8 f. 12 (1891)[8]; Hmp., F. I. Moths I 249 (1892) ♂.[9] — Ganjam.[8, 9] Madras.

Callizygaena auratus auratus ab. *gana,* Jordan, Seitz Lep. Ind. II 9, t. 1 f. F (1908).[10]

Callizygaena auratus f. *gana,* Hering, Arch. f. Naturg. LXXXVIII (A. 11), p. 12 (1922).[11]

(2) aurata nivimacula, Felder

Callizygaena nivimacula, Feld., Reise Novara, Lep. IV, t. 83 f. 4 (1874)[12]; Mo., Lep. Ceylon II 38, t. 95 f. 5 (1882)[13]; Hmp., F. I. Moths I 248, f. 165 (1892) ♀.[14] — Ceylon.[12, 13, 16] Trincomali.[14, 15]

Callizygaena auratus nivimacula, Jordan, Seitz Lep. Ind. II 8-9 (1908)[15]; Hering, Arch. f. Naturg. LXXXVIII (A. 11), p. 12 (1922)[16].

aurifasciata, Hering

Callizygaena aurifasciata, Hering, Arch. f. Naturg. LXXXVIII (A. 11), p. 12 (1922) ♂[1] — "Ost-indien."[1]

21. Lamprochloe, Hampson (1900).

Lamprochloë, Hmp., B. J. XIII 226, fig. (1900). — Type:–*albipuncta,* Hmp.

albipuncta, Hampson

Lamprochloë albipuncta, Hmp., B. J. XIII 226, fig. (1900) ♀.[1] — Pundaluoya.[1]

Callizygaena albipuncta, Jordan, Seitz Lep. Ind. II 9, t. 1 f. G (1908).[2]

22. Procotes, Butler (1876).

Procotes, Butl., J. Linn. Soc. XII 355, t. 28 f. 1 (1876); Hmp., F. I. Moths I 239, f. 155 (1892) [redescr.]; Jordan, — Type:–*diminuta,* Wlk.

Seitz Lep. Ind. II 9 (1908) [redescr.]; Hering, Arch. f. Naturg. LXXXVIII (A. 11), p. 13 (1922).

diminuta, Walker

Euchromia diminuta, Wlk., Cat. I 230-231 (1854).[1]

Procotes diminuta, Mo., Lep. Ceylon II 37, t. 95 f. 3 (1882)[2]; Hmp., F. I. Moths I 239, f. 155 (1892) ♂[3]; Jordan, Seitz Lep. Ind. II 9, t. 1 f. H (1908)[4]; Hering, Arch. f. Naturg. LXXXVIII (A. 11), p. 13 (1922).[5]

Ceylon.[1, 5]
Trincomali.[3]

23. Heteropan, Walker (1854).

Heteropan, Wlk., Cat. II 440 (1854); Hmp., F. I. Moths I 256 (1892) [redescr.]; Piepers and Snellen, Tijds. Ent. XLV 216 (1902) [redescr.]; Jordan, Seitz Lep. Ind. II (1908) [redescr.]; Hering, Arch. f. Naturg. LXXXVIII (A. 11), pp. 14-16 (1922) [key to spp.].

Type:–*scintillans,* Wlk.

Birtina, Wlk., Cat. XXXI 125 (1864).

Type:–*lycaenoides,* Wlk. [Papua].

analis, Jordan

Heteropan analis, Jord., Seitz Lep. Ind. II 13 (1908) ♂.[1]

Hsipaw.[1]
[N. Shan States, Burma].

cupreatus, Hampson

Heteropan cupreatus, Hmp., F. I. Moths I 256 (1892) ♂[1]; *id.,* Ill. Het. IX 63, t. 157, f. 10, (1893)[2]; Jord., Seitz Lep. Ind. II 11, t. 2 f. A (1908).[3]

Pundaluoya.[1, 2]

scintillans, Walker

Heteropan scintillans, Wlk., Cat. II 441 (1854)[1]; Mo., Lep. Ceylon II 43, t. 96 f. 3 (1882)[2]; Hmp., F. I. Moths I 256, f. 170 (1892) ♂[2]; Piepers and Snellen, Tijds. Ent. XLV 216-217 (1902)♂ [an haec species ?][4]; Jordan, Seitz Lep. Ind. II 11, t. 2 f. A (1908).[5]

Ceylon.[1, 2, 3, 5]
S. India.[5]
Burma.[5], Nicobars.[5]

Java.[4]

24. Arbudas, Moore (1879).

Arbudas, Moore, Lep. Atk., pp. 19-20 (1879); Hmp., F. I. Moths I 257, f. 171 (1892) [redescr.]; Hering, Arch. f. Naturg. LXXXVIII (A. 11), p. 18 (1922) [key to spp.].

Type:–*bicolor,* Mo.

bicolor, Moore

Arbudas bicolor, Mo., Lep. Atk., p. 20, t. 2 f. 19 (1879)[1]; Hmp., F. I. Moths I 257, f. 171 (1892)♂ ♀[2]; Swinh.,

Darjiling.[1]
Sikkim.[2, 4]

T. E. S. 1895, 34 (1895)[3]; Jord., Seitz Lep. Ind. II 13, t. 2 f. B (1908).[4] — Bhutan.[4] Shillong.[3]

flavimacula, Hampson (Two subspecies)

(1) flavimacula flavimacula, Hampson

Heteropan flavimacula, Hmp., F. I. Moths I 256 (1892)♂ ♀.[1] — Momeit (Burma).[1]

Arbudas flavimacula flavimacula, Jordan, Seitz Lep. Ind. II 13 (1908).[2] — ————

(2) flavimacula leucas, Jordan

Arbudas flavimacula leucas, Jord., Seitz Lep. Ind. II 13, t. 2 f. C (1908).[3] — Tonkin.[3]

leis, Swinhoe

Heteropan leis, Swinh., A. M. N. H. (6) XIV 442 (1894)[1]; *id.,* T. E. S. 1895, 34 (1895)[2]; Hmp., F. I. Moths IV 469 (1896) ♂.[3] — Khasis.[3] Shillong.[1, 2] Cherrapunji.[2]

Arbudas leis, Jord., Seitz Lep. Ind. II 13, t. 2 f. B (1908).[4]

ochrea, Elwes

Cyclosia ? ochrea, Elwes, P. Z. S. 1890, 385, t. 33 f. 4 (1890)♂ ♀.[1] — Nagas.[1, 2]

Heteropan ochreus, Hmp., F. I. Moths I 256 (1892) [redescr.].[2]

Arbudas ochrea, Jord., Seitz Lep. Ind. II 13, t. 2 f. B (1908).[3]

25. Trypanophora, Kollar (1844).

Trypanophora, Kollar, Hügel's Kaschmir IV (ii) 457 (1844); Wlk., Cat. II 434 (1854); Kirby, Cat. Het. I 58 (1892); Hmp., F. I. Moths I 249, f. 166 (1892) [redescr.]; Jordan, Seitz Lep. Ind. II 14 (1908) [redescr.]; *id.,* Seitz Lep. Pal. II 9 (1910) [redescr.]; Hering, Arch. f. Naturg. LXXXVIII (A. 11), pp. 19-20 (1922) [key to spp.]. — Type:–*semihyalina,* Koll.

Tryphanophora, Snellen, Tijds. Ent. XLV 217-218 [redescr.], t. 18 f. 10 [larva] (1902).

atkinsoni, Moore

Trypanophora atkinsoni, Mo., Lep. Atk., p. 15 (1879)[1]; Hering, Arch. f. Naturg. LXXXVIII (A. 11) p. 20 (1922).[2] — Darjiling.[1]

australis, Jordan

Trypanophora australis, Jord., Seitz Lep. Ind. II 15, t. 2 f. D (1908) ♂.[1] — Mercara.[1]

flavalis, Hampson

Trypanophora, flavalis, Hmp., F. I. Moths I 250 (1892) ♀.[1]; Jordan, Seitz Lep. Ind. II 15, t. 2 f. D (1908) ♀.[2] — Matelei (Burma).[1]

hosemanni, Hering

Trypanophora hosemanni, Hering, Arch. f. Naturg. LXXXVIII (A. 11), pp. 20-21 (1922) ♂.[1] — "Probably Darjiling."

semihyalina, Kollar

Trypanophora semihyalina, Kollar, Hügel's Kaschmir IV (ii) 457, t. 19 ff. 1-4 (1844)[1]; Wlk., Cat. II 434 (1854)[2]; Mo., P. Z. S. 1859, 197, t. 60 ff. 1, 2 (1859)[3]; Mo., Cat. Lep. E. I. Co. Mus. II, t. 14. f. 6 [larva] (1859)[4]; Hmp., F. I. Moths I 249-250, f. 166 (1892) [larva descr.][5]; Swinh., T. E. S. 1895, 33 (1895)[6]; Dudgeon, B. J. XII 296 (1899)[7]; de Niceville, Ind. Mus. Notes V 107, t. 10 ff. 4 [♀] *a, b* [larva], *c* [cocoon] (1903)[8]; Jordan, Seitz Lep. Ind. II 14-15 (1908)[9]; *id.,* Lep. Pal. II 10, t. 2 f. A (1910)[10]; Hering, Arch. f. Naturg. LXXXVIII (A. 11), p. 20 (1922).[11] — Kashmir.[1, 5, 10] Kangra.[5] Sikkim.[5] Shillong.[6] Calcutta.[5, 8] Dacca. Ganjam. E. Pegu.[5]

Syntomis humeralis, Wlk., Cat. VII 1593 (1856) [Hindostan].[12]

Syntomis glaucopoides, Wlk., Cat. XXXI 69 (1864) [S. Hindostan].[13]

Larva descr.[4 5 8], on *Terminalia catappa*[8], on *Rhaphiolepis.*[3]

taprobanes, Walker

Trypanophora taprobanes, Wlk, Cat. II 435 (1854)[1]; Mo., Lep. Ceylon II 40 t. 95 ff. 8, 8*a* (1882)[2]; Hmp., F. I. Moths I 250-251 (1892) [redescr.][3]; Jordan, Seitz Lep. Ind. II 15, t. 2 f. D (1908).[4] — Ceylon.[1-4]

Larva descr.[2 3], on *Lagerstroemia.*[2]

26. Cyclosia, Hubner (1822).

Cyclosia, Hb., Verz., p. 177 (1822); Wlk., Cat. II 414 (1854); Hmp., F. I. Moths I 269, f. 178 (1892) [redescr.]; *id.,* B. J. XVI 193 (1905) [synon.]; Piepers and Snellen, — Type:–*panthona,* Cram.

Tijds. Ent. XLV 187-188 (1902) [redescr.]; Jordan, Seitz Lep. Ind. II 17 (1908) [redescr.]; Turner, T. E. S. 1918, 176, f. 36 [neuration of *C. panthona*]; Hering, Arch. f. Naturg. LXXXVIII (A. 11), pp. 22-26 (1922) [key to spp.].

Epyrgis, Herrich-Schäffer, Aussereur. Schmett. I 7, ff. 5-10 (1853); Kirby, Cat. Het. I 44 (1892). — Type:–*midamia*, H.-S.

Pintia, Walker, Cat. II 280-281 (1854); Hmp., F. I. Moths I 258, f. 173 (1892) [redescr.]. — Type:–[*macularta*, Guér. =] *metachloros*, Wlk.

Isbarta, Walker, Cat. VII 1672 (1856); Hmp., F. I. Moths I 276, f. 184 (1892) [redescr.]; Snellen, Tijds. Ent. XLV 181 (1902) [redescr.]. — Type:–[*pieridoides*, H.-S.=] *glauca*, Wlk.

Klaboana, Moore, P. Z. S. 1879, 393 (1879). — Type:–*macularia*, Guér.

Callamesia, Butler, A. M. N. H. (5) XVI 345 (1885); Hmp., F. I. Moths I 277, f. 185 (1892) [redescr.]. — Type:–*midamia*, H.-S.

binghami, Butler

Epyrgis binghami, Butl., A. M. N. H. (5) X 374 (1882).[1]

Isbarta binghami, Hmp., F. I. Moths I 276-277, f. 184 (1892) ♀.[2]

Isbarta cyanescens, Hmp., T. E. S. 1895, 284 (1895) ♂[3]; *id.*, F. I. Moths IV 470 (1896) ♂[4]; *id.*, B. J. XX 95 (1910) [♂ of *binghami*].[5]

Cyclosia pieridoides binghami, Jordan, Seitz Lep. Ind. II 17 (1908).[6]

Tenasserim.[1-4, 6]

cuprea, Swinhoe

Isbarta cuprea, Swinh., T. E. S. 1891, 475 (1891)[1]; Hmp., F. I. Moths I 277 (1892) [redescr.][2]; Swinh., T. E. S. 1895, 37 (1895).[3]

Cyclosia inornata sobria ab. *cuprea*, Jordan, Seitz Lep. Ind. II 18 (1908).[4]

Khasis.[1, 2]
Shillong.[3]
E. Pegu.[2]

imitans, Butler

Epyrgis imitans, Butl., Ill. Het. V 24, t. 84 f. 1 (1881).[1]

Isbarta imitans, Hmp., F. I. Moths I 277 (1892) [redescr.][2]; Dudgeon, B. J. XII 301 (1899)[3]; Swinh., T. E. S. 1895, 37 (1895)[4]; Lefroy, Ind. Ins. Life, p. 503, f. 337 [lower fig.] (1909).[5]

Sikkim [low elev.][3, 6]
Bhutan. [2, 3, 6]
Shillong[4]
Cherrapunji.[4]
Karen Hills.[2]

Cyclosia imitans, Jordan, Seitz Lep. Ind. II 18, t. 2 f. i (1908)[6]; Hering, Arch. f. Naturg. LXXXVIII (A. 11), p. 32 (1922).[7]

macularia, Guerin (Five subspecies)

Gynautocera macularia, Guér., Delessert Souv. Ind., p. 83 t. 25 f. (1843).[1] — Mergui.[10]

Cyclosia macularia Jordan Seitz Lep. Ind. II 19-20 (1908)[2]; Hering, Arch. f. Naturg. LXXXVIII (A. 11) p. 30 (1922).[3]

(1) macularia macularia, Guerin

Gyclosia macularia macularia, Jord., Seitz Lep. Ind. II 19, t. 3 f. B (1908).[4]

Malacca.[3, 4]
Sumatra.[3, 4, 9]
Java.[5, 8, 11]
Nias.[12]
Makassar.[14]
Borneo.[3, 4]

(2) macularia metachloros, Walker

Pintiahmetac loros., Wlk. Cat. II 281 (1854)[5]; Mo., Cat. Lep. E.I. Co. Mus. II 321, t. 8^{a}, ff. 4, 4^{a} [larva, pupa] (1859).[6]

Cyclosia metachloros, Snellen, Iris VIII 132 (1895)[7]; Piepers and Snellen, Tijds. Ent. XLV 192-193, t. 1 f. 5 [larva] (1902) [synon.].[8]

Pintia cyan a, Butler, A. M. N. H. (5) XII 160 (1883) ♂[9]; Hmp., F. I. Moths I 259 (1892) [redescr.].[10]

Cyclosia macularia metachloros, Jord., Seitz Lep. Ind. II 19, t. 3 f. C (1908).[11]

(3) macularia purpura, Jordan

Cyclosia macularia purpurea, Jord., Seitz Lep. Ind. II 19 (1908).[12]

(4) macularia castigata, Jordan

Cyclosia macularia castigata, Jord., Seitz Lep. Ind. II 20, t. 3 f. B (1908).[13]

(5) macularia transita, Hering

Cyclosia macularia transita, Hering, Arch. f. Naturg. LXXXVIII (A. 11), pp. 30-31 (1922).[14]

Larva descr.[6, 8], on *Pierardia racemosa.*[8]

midamia, Herrich-Schaffer (Six subspecies)

Epyrgis midamia, Herr.-Schäff., Aussereur. Schmett. I, t. 2 f. 7 (1853) ♂.[1]

Sikkim.[4, 9]
Bhutan.[3, 4, 6, 9]

Cyclosia venusta, Wlk., Cat. II 416 (1854).[2]

Amesia stelligera, Butl., Ill. Het. V 23, t. 83 ff. 8, 9 (1881).[3]

Callamesia midama, Hmp., F. I. Moths I 278, f. 185 (1892) [redescr.][4]; Swinh., T. E. S. 1895, 37 (1895)[5]; Dudgeon, B. J. XII 302 (1899).[6]

Cyclosia midamia, Jordan, Seitz Lep. Ind. II 20-21 (1908)[7]; Hering, Arch. f. Naturg. LXXXVIII (A. 11), p. 34 (1922) [subspp.].[8]

Assam.[4, 9] Nagas.[4], Sylthet.[2, 4] Khasis.[5], Burma.[4] Tenasserim.[12] Ceylone.[9]

(1) midamia midamia, Herrich-Schaffer

Epyrgis midamia, H.-S:, *l. c.*[1]

Cyclosia midamia midamia, Jord., Seitz Lep. Ind. II 20, t. 3 f. E (1908).[9]

Epyrgis hormenia, H.-S., Aussereur. Schmett. I, t. 2 f. 8 (1853) ♀.[10]

Cyclosia midamia midamia form. foem. *hormenia*, Jordan, Seitz Lep. Ind. II 21, t. 3 f. E (1908).[11]

———————

Siam.[4]

Malacca.[14, 16] Singapore.[13]

Sumatra.[14, 16, 19] Java.[20], Nias.[17, 18] Berneo.[14, 16, 15] Tonkin.[9]

(2) midamia alcathoe

Cyclosia midamia alcathœ, Jord., Seitz Lep. Ind. II 21 (1908).[12]

(3) midamia submaculans, Walker

Cyclosia submaculans, Wlk., Journ. Linn. Soc. III 185 (1859).[13]

Cyclosia midamia submaculans, Jord., Seitz Lep. Ind. II 21 (1908).[14]

Amesia striata, Druce, A. M. N. H. (6) VII 142 (1891).[15]

Cyclosia midamia submaculans form *striata*, Druce; Jordan, Seitz Lep. Ind. II 21, t. 3 f. E (1908).[16]

(4) midamia trepsichrois, Butler

Amesia trepsichrois, Butl., Ent. Mo., Mag. XX 57 (1883) ♂.[17]

Cyclosia midamia trepsichrois, Jord., Seitz Lep. Ind. II 21, t. 3 f. E (1908).[18]

(5) midamia padangana

Cyclosia midamia padangana, Jordan, Seitz Lep. Ind. II 21 (1908).[19]

(6) midamia dolosa

Cyclosia midamia dolosa, Jordan, Seitz Lep. Ind. II 21 (1908).[20]

ni, Swinhoe

Cyclosia ni, Swinh., A. M. N. H. (8) XVI 177-178 (1915).[1]

Maymyo.[1]
Tenasserim.[1]

panthona, Cramer

Geometra panthona, Gram, Pap. Exot. IV, t. 322 f. C (1781 ?).[1]

Cyclosia panthona, Mo., Lep. Ceylon II 44, t. 26 ff. 1, 1*a*, 1*b* (1882)[2]; Hmp., F. I. Moths I 270 (1892) [redescr.][3]; Swinh., T. E. S. 1895, 36 (1895)[4]; Jordan, Seitz Lep. Ind. II 20, t. 3 ff. *C* [♂ ♀] (1908)[5]; Turner, T. E. S. 1918, 176, f. 36 [neuration].[6]

Larva descr.[2, 3], on *A. Lindleyana.*[2]

Nagrispur (Sikkim). Sikkim.[3]
Cachar.[3], Khasis.[3]
Shillong.[4]
Mergui.[3, 5]
Ceylon.[2, 3, 5]

Malacca.[3], Siam.[3]
Hong Kong.[3, 5]

papilionaris, Drury (Seven subspecies).

Noctua papilionaris, Drury, Ill. Exot. Ent. II 4, t. 2 f. 4 (1773) ♀ [China].[1]

Attacus papilionaris, Cram, Exot. Ins. I 45, t. 29 f. A (1779) ♀.[2]

Cyclosia enodis, Swinh., Cat. Het. Mus. Oxon. I 69, t. 2 f. 2 (1892) ♀.[3]

Cyclosia papilionaris, Hmp., F. I. Moths I 269-270, f. 178 (1892) ♂[4]; Swinh., T. E. S. 1895, 36 (1895)[5]; Dudgeon, B. J. XII 300 (1899)[6]; Piepers and Snellen, Tijds. Ent. XLV 189-191, t. 13 f. 4 [larva] (1902) [synon][7]; Hmp., B. J. XVI 193 (1905) [*ferrea* is ♂][8]; Jordan, Seitz Lep. Ind. II 18-19 (1908)[9]; Hering, Arch. f. Naturg. LXXXVIII (A. 11), pp. 33-34 (1922).[10]

Eterusia ferrea, Wlk., Cat. II 431 (1854) ♂.[11]

Sikkim.[4, 13, 24]
Bhutan.[24]
Sukna.
Cachar.[13]
Sylhet.[13]
Khasis.[4, 5]
Cherrapunji.[14]
Burma.[4, 13, 27, 32]
Shan States.[25]
Tenasserim.[25]
Mergui.[4]
Nicobars.[13, 29, 30, 32]
And Mans.[13,27,29,32]
mSouth India.[18]
Nilgiris.[16, 17, 19, 20, 22]
Travancore.[17, 19]
Coorg.

Pintia ferrea, Butl., Ill. Het. V 23 t. 83 f. 7 (1881) ♂[12]; Hmp., F. I. Moths I 258-259, f. 17 (1892) ♂[12]; Swinh., T. E. S. 1895, 35 (1895).[14]

Sumatra.[7, 15]
Java.[4, 7, 15]
Malacca.[15]
Hong Kong.[11, 13, 15]
China.[1, 4], Hainan.[15]

(1) papilionaris papilionaris, Drury

Noctua papilionaris Drury, *l. c.*[1]

Cyclosia pápilionâris pilionaris, Jordan, Seitz Lep. Ind. II 19, t. 3 f. A (1908).[15]

(2) papilionaris australinda, Hampson

Cyclosia australinda, Hmp., Ill. Het VIII 45, t. 139 f. 23 (1891) ♀.[16]; *id.,* F. I. Moths I 270 (1892) ♀.[17]

Cyclosia papilionaris australinda, Jordan, Seitz Lep. Ind. II 19 (1908).[18]

Pintia latipennis, Hmp., Ill. Het. VIII 45, t. 139 f. 22 (1891) ♂[19]; *id.,* F. I. Moths I 259 (1892) ♂[20]; Piepers and Snellen, Tijds. Ent. XLV 191 (1902)[21]; Hmp., B. J. XVI 193 (1905) [synon.][22]

(3) papilionaris venaria, Fabricius

Phalœna venaria, Fabr., Ent., Syst. III (ii) 156 (1793).[23]

Cyclosia papilionaris venaria, Jordan, Seitz Lep. Ind. II 19 (1908).[24]

(4) papilionaris adusta

Cyclosia papilionaris adjusta, Jordan, Seitz Lep. Ind. II 19 (1908).[25]

(5) papilionaris nigrescens, Moore

Cyclosia nigrescens, Mo., P. Z. S. 1877, 600 (1877)[26]; Hmp., F. I. Moths I 269 (1892) [redescr.][27]; *id.,* B. J. XVI 193 (1905) [var. of *papilionaris*].[28]

Cyclosia papilionaris nigrescens, Jordan, Seitz Lep. Ind. II 19 (1908).[29]

(6) papilionaris nicobarensis, Hering

Cyclosia papilionaris nicobarensis Hering, Arch. f. Naturg. XXXVIII (A. 11), pp. 33-34 (1922).[30]

(7) papilionaris parvula, Butler

[an subsp. vera aut synon. subsp. alicujus ?].

Cyclosia par vla, Butl., A. M. N. H. (5) XII 160[31]; Hmp., F. I. Moths I 269 (1892) [redescr.][32]; Hmp., B. J. XVI 193 (1905) [var. of *papilionaris*].[33]

Larva descr.[7], on *Solanum ardisioides.*[7]

rehfousi, Oberthur

Milionia rehfoúsi, Obth., Bull. S. E. Fr. 1910, 315 (1910).[1]

Psaphis rehfousi, Obth., Et. Lep. Comp. V. 330. t. 77. f. 703 (1911).[2]

Siac-lou.[1, 2]
[E. Tibmet border.]

27. Gynautocera, Guerin (1830)

Gynautocera, Guér., Mag. Zool. 1830, p. 12 (1830); Wlk., Cat. II 411 (1854); Hmp., F. I. Moths I 278-279, f. 186 (1892) [redescr.]; Snellen, Tijds. Ent. XLV 17-171 (1902) [redescr.]; Jordan, Seitz Lep. Ind. II 21 (1908) [redescr.]; Hering, Arch. f. Naturg. LXXXVIII (A. 11), pp. 34-35 (1922) [key to spp.].

Type:–*papilionaria,* Guér.

Gymnautocera, Westwood, Arcana Entom. I 20 (1841).

Type:–*papilionaria,* Guér.

pap lionaria, Guerin

Gynautocera papilionaria, Guér., Mag. Zool. 1830, p. 12 (1830)[1]; Wlk., Cat. II 411 (1854)[2]; Hmp., F. I. Moths I 279, f. 186 (1892) [redescr.][3]; Swinh., T. E. S. 1895, 37 (1895)[4]; Dudgeon, B. J. XII 302 (1899)[5]; Jordan, Seitz Lep. Ind. II 21, t. 3 f. F (1908).[6]

Gymnautocera papilionaria, Westw., Arc. Entom. I 20 (1841) [redescr.].[7]

Chalcosia selene, Kollar, Hügel's Kaschmir, IV, (ii), p. 463 (1844).[8]

Histia fraterna, Moore, P. Z. S., 1883, 15 (1893).[9]

Gynautocera fraterna, Butl., Ill. Het. VII 27, t. 121 f. 7 (1889).[10]

Gynautocera zara, Swinh., T. E S. 1891, 476 (1891)[11]; *id.,* T. E. S. 1895, 37 (1895) [*rara*-laps.].[12]

Kashmir.[8]
Jeolikote (Kumaon)
Himalaya.[3, 6]
Sikkim.[5]
Kalimpong.[5]
Bhutan.[5], Assam.[3, 6]
Nagas.[3]
Sylhet.[3]
Khasis.[4, 11]
Shillong.[12]
Calcutta [? error loci].[3, 6, 9, 10]
Bassein.[3]
Tenasserim.[3]

Annam.[6]
Tonkin.[6]
Hainan.[6]

28. Histia, Hubner (1822).

Histia, Hb., Verz., p. 198 (1822); Wlk., Cat. II 411 (1854); Kirby, Cat. Het. I 55 (1892); Hmp., F. I. Moths I 279, f. 187 (1892) [redescr.]; Snellen, Tijds. Ent. XLV 168-169 (1902) [redescr.]; Jordan, Seitz Lep. Ind. II 21 (1908) [redescr.]; *id.*, Lep. Pal. II 13 (1910) [redescr.]; larva of Javanese *libelluloides* descr.]; Hering, Arch. f. Naturg. LXXXVIII (A. 11), pp. 36-38 (1922) [key to spp.]. — Type:–*rhodpe*, Cram.

rhodope, Cramer (Six subspecies).

Bombyx rhodope, Cram., Pap. Exot. I, t. 30 f. F (1775).[1] — Darjiling., Sikkim.[8, 11]

Histia rhodope, Kirby, Cat. Hét. I 55 (1892)[2]; Snellen, Tijds. Ent. XLV 168 (1902).[3] — Bhutan.[11], Assam.[8] gauhati.

Zygaena flabellicornis, Fabr., Syst. Entom., p. 831 (1775).[4] — Khasis.[8, 9], Nagas.[8]

Histia flabellicornis, Hb., Verz., p. 198 (1822)[5]; Wlk., Cat. II 412 (1854)[6]; Preiss, Abbild. Nachtschmett., p. 3, t. 1 f. 6 (1888)[7]; Hmp., F. I. Moths I 280, f. 187 (1892) [redescr.][8], Swinh., T. E. S. 1895, 38 (1895)[9]; Leech, T. E. S. 1898, 348 (1898)[10]; Dudgeon, B. J. XII 302 (1899).[11] — Burma.[8] ---------- Hongkong.[8]

(1) rhodope rhodope, Cramer

Bombyx rhodope, Cram., *l. c.*[1]

Histia rhodope rhodope Jordan, Seitz Lep. Ind. II 22, t. 3 f. G (1908).[12]

ab. albimacula, Hampson

Histia albimacula, Hmp., F. I. Moths I 280 (1892)♂ *nec.* ♀.[13]; Swinh., T. E. S. 1895, 38 (1895)[14]; Roths., Novit. Zool. VI 441 (1899).[15] — Momeit (Burma).[13] Shan States.[14] Cherrapunji.[14] [*nec.* Java-*error loci.*]

Histia rhodope rhodope ab. *albimacula*, Jordan, Seitz Lep. Ind. II 22 (1908).[16]

(2) rhodope nilgira, Moore

Histia nilgira, Mo., Lep. Atk., pp. 18-19 (1879)[17]; Hmp., F. I. Moths I 280 (1892) [redescr.].[18] — Nilgiris.[17-19]

Histia rhodope nilgira, Jordan, Seitz Lep. Ind. II 22 (1908).[19]

(3) rhodope cometaris, Butler.

Histia cometaris, Butl., A. M. N. H. (5) X 374 (1882)[20]; Hmp., F. I. Moths I 280 (1892) ♂.[21]

Histia rhodope cometaris, Jord., Seitz Lep. Ind. II 22 (1908).[22]

Tenasserim.[20-22]
E. Pegu.[21]
Shan States.[22]

(4) rhodope nigrinus, Jordan

Histia rhodope nigrinus, Jord. Seitz Lep. Pal. II 13, t. 3 f. A (1910).[23]

(5) rhodope sumbawana, Hering

Histia rhodope sumbawana, Hering, Arch. f. Naturg. LXXXVIII (A. 11), p. 38 (1922) ♀.[24]

(6) rhodope ultima, Hering

Histia rhodope ultima, Hering, Arch. f. Naturg. XXXVIII (A. 11), p. 28 (1922) ♀.[25]

Larva descr.[11]

29. Pompelon, Walker (1854).

Pompeton, Walker, Cat. II 413 (1854); Hmp., F. I. Moths I 271, f. 179 (1892) [redescr.]; Snellen, Tijds. Ent. XLV 184-185 (1902) [redescr.]; Jordan, Seitz Lep. Ind. II 23 (1908) [redescr.]; Hering, Arch. f. Naturg. LXXXVIII (A. 11), pp. 39-40 (1922) [key to forms.].

Type: *marginatum,* Guèrin

Marginatum, Guerin

Gynautocera marginatum, Guér., Deless., Souv. Ind. p. 84, t. 25 f. 1 (184).[1]

Pompelon marginata, Wlk., C t. II 414 (1854).[2]

Pompelon marginata, Snellen, Tijds., Ent. XXII 74 (1879)[3]; *id.,* Iris. VIII 133 (1895)[4]; Piepers and Snellen, Tijds. Ent. XLV 185-187, t. 13 f. 3 (larav] (1902)[5]; Jordan, Seitz Lep. Ind. II 23 (1908) [variation][6]; Hering. Archiv. f. Naturg. LXXXVIII (A. 11.), pp. 39-40 (1922) [key to subspp.].[7]

Heterusia atrocyanea, Herrich-Schäffer, Aussereur. Schmett. I, ff. 157, 158 (1854).[8]

Pompelon ampliatum, Butl., P. Z. S. 1878, 387 (1878).[9]

Burma.[11]
E. Pegu.[15]
Mergui.[17]
Penang.[2, 17, 12]
Singapore.[15]
Sumatra.[5] Java.[2, 5]
Palawan.[5]
Clelebes.[5]
Philippines.[5, 13]
Boreno.[5, 15, 17]

Pompelon valentula, Swinh., P. Z. S. 1889, 401 (1890)[10]; Hmp., F. I. Moths I 271 (1892) ♀.[11]

Pompelon anethusssa, Grose-Smith, A. M. N. H. (6) VII 141 (1891).[12]

Pompelon philippensis, Grose-Smith, A. M. N. H. (6) VII 141 (1891).[13]

Pompelon subcyanea, Swinh., Cat. Het. Mus. Oxon. I 72 (1892)[14]; Hmp., F. I. Moths I 271, f. 179 (1892) [redescr.][15]

Pompelon rotundata, Swinh., Cat. Het. Mus. Oxon. I 72 (1892).[16]

Pompelon cyanosura, Druce, A. M. N . H. (6) XIV 24 (1894).[17]

Larva descr.,[5] on *Cinnamomum* sp.[5]

30. Hampsonia, Swinhoe (1894).

Hampsonia, Swinh., A. M. N. H. (6) XIV 443 (1894); Hmp., F. I. Moths IV 469, f. 250 (1896); Jodan, Seitz Lep. Ind. II 23 (1908) [redescr.]; Hering, Arch. f. Naturg. LXXXVIII (A. 11), p. 41 (1922).

Type: *pulcl errima,* Swinh.

pulcherrima, Swinhoe

Hampsonia pulcherrima, Swinh., A. M. N. H. (6) XIV 443 (1894)[1]; *id.,* T. E. S. 1895, 36, t. 1 ff. 9, 9*a* (1895)[2] Hmp., F. I. Moths IV 470, f. 250 (1896) ♀.[3]; Jordan, Seitz Lep. Ind. II 23, t. 4 f. C. (1908)♂ ♀.[4]; Hering, Arch. f. Naturg. LXXXVIII (A. 11), p. 41 (1922).[5]

Khasis.[3, 4]

Cherrapunji.[1, 2]

31. Campylotes, Westwood (1839).

Campylotes; Westw., Royle's Botany of Himalayas, Lep. p. liii (1839); *id.,* Arcana Entom. I 20 (1841) [redescr.]; Wlk., Cat. II 419 (1854); Kirby, Cat. Het. I 48 (1892); Hmp., F. I. Moths I 274, f. 182 (1892) [redescr.]; Jordan, Seitz Lep. Ind. II 23-24 (1908) [redescr.] *id.,* Lep. Pal. II 12 (1910) [redescr.]; Hering, Arch. f. Naturg. LXXXVIII (A. 11), p. 41 (1922) [key to spp.].

Type:– *histrionicus,* Westw.

atkinsoni, Moore

Campylotes atkinsoni, Mo., Lep. Atk. p. 17 (1879)[1]; Waterhouse, Aid Ident. Ins. II, t. 112 (1882)[2]; Hmp.,

Darjiling.[1]
Sikkim (high elev.).[5]
Sikkim.[3, 4]

F. I. Moths I 275 (1892) [redescr.][3]; Dudgeon, B. J. XII 301 (1899)[4]; Jordan, Seitz Lep. Ind. II 24, t. 4 f. C (1908).[5]

burmana, Hampson

Campylotes burmana, Hmp., Novit. Zool. XXVI 281 (1920).[1]

S. Chin Hills.[1]

desgodinsi, Oberthür (Two subspecies).

(1) desgodinsi desgodinsi, Oberthür.

Epyrgis desgodinsi, Obth., Et. Ent. IX 18, t. 2 f. 10 (1884).[1]

Campylotes desgodinsi, Hmp., F. I. Moths I 275 (1892) ♀.[2]; Leech, T. E. S. 1898, 347 (1898)[3]; Jord., Lep. Pal. II 13, t. 2 f. H. (1910).[4]

Tibet.[1, 2]
Moupin.[3]

W. China.[3, 4]

(2) desgodinsi splendida, Elwes

Campylotes desgodinsi, var. *splendida,* Elwes, P. Z. S. 1890, 384, 385, t. 33 f. 3 (1890)[5]; Hmp., F. I. Moths I 275 (1892).[6]

Campylotes desgodinsi splendida, Jordan, Seitz Lep. Ind. II 24, t. 4 f. C (1908).[7]

Nagas.[5-7]

histrionicus, Westwood

Campylotes histrionicus, Westw., Royle's Bot. Him., Lep. p. 53, t. 10 f. 1 (1839)[1]; *id.,* Arcana Entom. I 20 (1841) [redescr.][2]; Wlk., Cat. II 420 (1854)[3]; Preiss, Abbild. Nachtschmett., p. 3, t. 1 f. 4 (1888)[4]; Elwes, P. Z. S. 1890, 383 (1890)[5]; Hmp., F. I. Moths I 274- 275, f. 182 (1892) [redescr.][6]; Swinh., T. E. S. 1895, 37 (1895)[7]; Leech, T. E. S. 1898, 346 (1898)[8]; Dudgeon, B. J. XII 301 (1899)[9]; Hmp., P. Z. S. 1906, 493 (1906) [*histrionica*][10]; Jordan, Seitz Lep. Ind. II 24 (1908)[11]; *id.,* Lep. Pal. II 12, t. f. G (1910).[12]

Afghanistan.[12]
N.-W. India.[12]
Jeolikote (Kumaon).
Throughout Himalaya.[6]
Kashmir.[11]
Masuir., Sikkim.[9]
Lebong (Sikkim).
Sikkim (9,000-10,000ft).[11, 12]
Raitdong (Sikkim).[10]
Khasis.[6, 7, 11]
Shillong.
Cherrapunji.

Campylotes histrionicus var. *altissima,* Elwes, P. Z. S. 1890, 384, t. 33 f. 1 (1890).[13]

W. China.[8, 11, 12]

Campylotes excelsa, Obth., Et. Ent. XX 54, t. 10 f. 178.[14]

minima, Oberthür

Campylotes minima, Obth., Et. Ent. XIX 25, t. 6 f. 54 (1894)[1]; Leech, T. E. S. 1898, 347 (1898)[2]; Jordan, Seitz Lep. Pal. II 13, t. 2 f. (1910).[3] — Ta-tsien-lu.2, 3 Omei-shan.2, 3 [E. Tibet border.]

philomena, Oberthür

Campylotes philomena, Obth., Ét. Lep. Comp. XX 205-206, fig. (1923).[1] — Ta-tsien-lu.1

pratti, Leech

Campylotes pratti, Leech, Entom. XXIII 109 (1890)[1]; *id.,* T. E. S. 1898, 347 (1898)[2]; Jordan, Seitz Lep. Pal. II 13, t. 2 f. H (1910).[3] — Putsu-fong.2 [E. Tibet border]

Chang-yang.[1] [Centr. China]

romanovi, Leech

Campylotes romanovi, Leech, T. E. S. 1898, 346-347 (1898)♂ ♀.[1]; Jordan, Seitz Lep. Fal. II 12, t. 2 f. H (1910).[2] — Moupin.[1, 2] Wa-shan.[1, 2]

sikkimensis, Elwes

Campylotes sikkimensis, Elwes, P. Z. S. 1890, 384, t. 33 f. 2 (1890)[1]; Jordan, Seitz Lep. Ind. II 24, t. 4 f. C. (1908).[2] — Sikkim (high elev.)[1, 2] Assam (high elev.).[2]

Campylotes histrionicus sikkimensis, Hmp., F. I. Moths I 275 (1892).[3]

32. Erasmia, Hope (1840).

Erasmia, Hope, Trans. Linn. Soc. XVIII 446 (1840); Westw., Arcana Entom. I 19 (1841); Wlk., Cat. II 418 (1854); Hmp., Z. I. Moths I 273, f. 181 (1892); Kirby, Handbk. Lep. III 72 (1897); Jordan, Seitz Lep. Ind. II 24 (1908) [redescr.]; *id.,* Lep. Fal. II 12 (1910) [redescr.]; Hering, rch. f. Naturg. LXXXVIII (A. 11), pp. 41-42 (1922) [key to spp.]. — Type:–*pulchella,* Hope.

Amesia, Westwood, Duncan's Nat. Libr., Exot. Moths, p. 93 (1841); Hmp., F. I. Moths I 272, f. 180 (1892) [redescr.]; Kirby, Handb. Lep. III 73 (1897) [redescr.]; Snellen, Tijds. Ent. XLV 175 (1902) [redescr.]; Hering, Arch. f. Naturg. LXXXVIII (A. 11), p. 13 (1922) [key to spp.]. — Type:– *sanguiflua,* Drury.

aliris, Doubleday (Two subspecies).

(1) aliris aliris, Doubleday

Gynautocera aliris, Doubl., Ann. Nat. Hist. XIX 74 (1847).[1] — N. India.[5] Sikkim.[4]

Cyclosia aliris, Wlk., Cat. II 415-416 (1854).[2] — Darjiling. Assam.[4]

Amesia aliris, Butl. Ill. Het. V 24, t 83 f 10 (188)[3]; Hmp., F. I. Moths I 272-273 (1892)♂ ♀.[4] — Aagas.[4] Sylhet.[2, 4] Karea Hills.[4]

Erasmia aliris aliris, Jordan, Seitz Lep. Ind. II 25, t. 4 f. D (1908).[5]

(2) aliris analis

Erasmia aliris analis, Jord., Seitz Lep. Ind. II 25 (1908).[6] — Tenasserim.[6]

namouna, Doubleday

Gynautocera namouna, Doubl., Ann. Nat. His. XIX 74 (1874).[1] — Darjiling.[3] [?error loci]

———————

Epyrgis euplaeoides, Herr.-Sch., Aussereur, Schmett. I, t. 2 ff. 9, 10 (1853).[2] — Malacca.[8] Sumatra.[8]

Amesia hyala, Druce, P. Z. S. 1885, 518 (1885)[3]; Waterhouse, Aid Ident. Ins. II, t. 172 f. 3 (1886)[4]; Hmp., F. I. Moths I 273 (1892) [redescr.].[5] — Java.[7] Borneo.[5, 7]

Amesia euploeoides, Snellen, Iris VIII 133 (1895)[6]; *id.,* Tijds. nt. XLV 177-179 (1902) [redescr.].[7]

Erasmia namouna, Jordan, Seitz Lep. Ind. II 25, t. 4 f. E (1908) [variation; doubtless incorrectly recorded from Sikkim].[8]

pulchella, Hope (Two subspecies).

(1) pulchella, pulchella, Hope

Erasmia pulchella, Hope, Trans. Linn. Soc. XVIII 446, t. 31 f. 5 (1840)[1] Westw., Arcana Entom. I 19 (1841) [redescr.][2]; Westw., Nat. Libr., Exot. Moths, p. 91, t. 3 f. 2 (1841)[3]; Wlk., Cat. II 419 (1854)[4]; Freiss, Abbild. Nachtschmett., p. 3, t. 1 f. 3 (1888)[5]; Hmp., F. I. Moths I 273-274, f. 181 (1892)♂ ♀.[6]; Swinh., T. E. S. 1895, 37 (1895)[7]; Kirby, Handb. Lep. III 72-73, t. 78 f. 2 (1897)[8]; Dudgeon, B. J. XII 300 (1899).[9] — Khatmandu. Sikkim.[6, 10] Darjiling. Assam.[2, 6, 10] Nagas.[6] Sylhet.[1, 6] Khasis.[7] Burma [mont.]

Erasmia pulchella pulchella, Jordan, Seitz Lep. Ind. II 24, t. 4 f. B (1908).[10]

(2) pulchella chinensis, Jordan

Erasmia pulchella, Leech, T. E. S. 1898, 346 (1898).[11]

Erasmia pulchella chinensis, Jordan, Seitz Lep. Ind. II 24 (1908)[12]; *id.,* Lep. Pal. II 12, t. 2 f. G (1910)[13]; Hering, Arch. f. Naturg. LXXXVIII (A. 11), p. 42 (1922).[14]

Tibet.
Burma (mont.)[12]
W. China[11,13]
S. China.[14]

Larva on "wild coffee."[9]

sanguiflua, Drury (Four subspecies).

(1) sanguiflua sanguiflua, Drury

Noctua sanguiflua, Drury, Ill. Exot. Ent. II 55, t. 20 ff. 1, 2 (1773).[1]

Amesia sanguiflua, Westw., Dunc. Nat. Libr., Exot. Moths, p. 93, t. 3 f. 3 (1841)[3]; Wlk., Cat. II 415 (1854) [*Cyclosia*][3]; Hmp., F. I. Moths I 272, f. 180 (1892) [redescr.][4]; Swinh., T. E. S. 1895, 36 (1895)[5]; Kirby, Handb. Lep. III 73-75, t. 78 f. 3 (1897).[6]

Erasmia sanguiflua sanguiflua, Jordan, Seitz Lep. Ind. II 25, t. 4 f. D (1908).[7]

Sikkim.[4]
Khasis.[4, 5]
Shillong.[4]
Sylhet.[3, 4]
Burma.[4, 7]
————————
Surinam [error loci].[1]

(2) sanguiflua lugens, Dohrn

Erasmia sanguiflua lugens, Jord., Seitz Lep. Ind. II 25 (1908).[9]

Sumatra.[2]

(3) sanguiflua gedeana, Fruhstorfer

Erasmia sanguiflua gedeana, Jord., Seitz Lep. Ind. II 25 (1908).[11]

Amesia sanguiflua, Snellen, Tijds. Ent. XLV 176-177 (1902) [redescr.].[12]

Java.[11, 12]

(4) sanguiflua sauteri, Hering

Amesia sanguiflua sauteri, Hering, Arch. f. Naturg. LXXXVIII (A. 11), pp. 43-44 (1922)♂ ♀.[13]

Formosa.[13]

33. Agalope, Walker (1854).

Agalope, Walker, Cat. II 437 (1854); Hmp., F. I. Moths I 282, f. 190 (1892) [redescr.]; Jordan, Seitz Lep. Ind. II 25-26 (1908) [redescr.]; *id.,* Lep. Pal. II 13 (1910)

Type:– *hyalina,* Kollar.

[redescr.]; Hering, Arch. f. Naturg. LXXXVIII (A. 11), pp. 44-45 (1922) [key to spp. and forms].

Chelura [præocc.], Hope, Trans. Linn. Soc. XVIII 444 (1840); Westw., Arcana Entom. I 20 (1841) [redescr.]; Wlk., Cat. II 436 (1854); Hmp., F. I. Moths I 283, f. 191 (1892) [redescr.]. — Type:–*bifasciata*, Hope.

Achelura, Kirby, Cat. Het. I 56 (1892); Hmp., F. I. Moths IV 470 (1896); Snellen, Tijds. Ent. XLV 171-172 (1902) [redescr.]; Turner, T. E. S. 1918, 176 (1918) [neuration]. — Type:–*bifasciata*, Hope.

aurelia, Oberthür

Agalope aurelia, Obth., Ét. Lep. comp. XX 206-207, fig. (1923).[1] — Yunnan.[1]

basiflava, Moore

Chelura basiflava, Mo., P. Z. S. 1879, 391 (1879) ♂ [1]; Waterhouse, Aid Ident. Ins. II, t. 131 f. 3 (1883)[2]; Hmp., F. I. Moths I 284 (1892) [redescr.].[3]

Agalope basiflava, Jordan, Seitz Lep. Ind. II 26 (1908) ♂.[4]

Darjiling.[1]
Sikkim.[3, 4]

bieti, Oberthür

Chalcosia bieti, Obth., Et. Ent. XI 29, t. 6 f. 40 (1886).[1]

Agalope bieti, Jordan, Seitz Lep. Pal. II 13, t. 3 f. B (1910).[2]

Ta-tsien-lu.[1, 2]
[E. Tibet border].

bifasciata, Hope

Chelura bifasciata, Hope, Trans. Linn. Soc. XVIII 444 (1840)[1]; Westw., Arcana Entom. I 20 (184) [redescr.][2]; Hmp., F. I. Moths I 283-284, f. 191 ((1892) [redescr.][3]; Swinh., T. E. S. 1895 38 (1895).[4]

Achelura bifasciata, Dudgeon, B. J. XII 303 (1899) [habits][5]; Turner, T. E. S. 1918, 176 (1918) [neuration].[6]

Agalope bifasciata, Jordan, Seitz Lep. Ind. II 26, t. 4 f. F (1908).[7]

N.W. India.[7]
Naini Tal.[3]
Nepal.[1, 3, 7]
Khatmandu.
Sikkim.[3, 5, 7]
Bhutan.[5]
Assam.[1, 3, 7]
Shillong.[4], Burma.[7]
Maymyo.

davidi, Oberthür

Chalcosia davidi, Obth., Et. nt. IX 19, t. 1 f. 2 (1884).[1]

Agalope davidi, Leech, T. E. S. 1898, 348 (1898)[2]; Jordan, Seitz Lep. Fal. II 13, t. 3 f. A (1910).[3]

Tibet.[1]
Moupin.[2]
Omei-shan.[2]

W. China.[2, 3]
Centr. China.[2, 3]

dejeani, Oberthür

Agalope dejeani, Obth., Et. Ent. XVIII 20, t. 2 f. 24 (1893)[1]; Leech, T. E. S. 1898, 349 (1898) [*Chelura*][2]; Jordan, Seitz Lep. Pal. II 13, t. 3 f. A (1910).[3] — Ta-tsien-lu.[2, 3] [E. Tibet border.]

eronioides, Moore (Two subspecies).

(1) eronoides eronioides, Moore

Chelura eronioides, Mo., Lep. Atk., p. 15 (1879)[1]; Hmp., F. I. Moths I 284 (1892) [redescr.][2]; Swinh., T. E. S. 1895, 38 (1895)[3]; Elwes, B. J. XII 303 (1899).[4]

Agalope eronioides, Jordan, Seitz Lep. Ind. II 26, t. 4 f. F (1809)[5]; Jordan, Seitz Lep. Pal. II 13 (1910).[6]

Darjiling. Sikkim.[2, 4, 5] Assam.[5] Shillong.[3] Cherrapunji.[3]

(2) eronioides diluta, Jordan

Chelura eronioides, Leech, T. E. S. 1898, 349 (1898).[7] — Moupin.[7, 8]

Agalope eronioides diluta, Jordan, Seitz Lep. Pal. II 13, t. 3 f. B (1910).[8] — Omei-shan.[7, 8]

glacialis, Moore (Three subspecies).

(1) glacialis glacialis, Moore

Chelura glacialis, Mo., P. Z. S. 1872, 570 (1872)[1]; Hemp., F.I. Moths I 284 (1892)♂ ♀.[2]

Achelura glacialis, Dudgeon, B. J. XII 303 (1899).[3]

Agalope glacialis glacialis, Jordan, Seitz Lep. Ind. II 26 (1908).[4]

Agalope glacialis, Hering, Arch. f. Naturg. LXXXVIII (A. 11), p. 46 (1922).[5]

Darjiling.[1] Sikkim.[2, 4] Burma.[2, 4]

(2) glacialis parthenie, Jordan

Agalope glacialis parthenie, Jord., Seitz Lep. Ind. II 26, t. 4 f. F (1908).6 — Java.[6]

(3) glacialis angustifasciata, Hering

Agalope glacialis angustifasciata, Hering, Arch. f. Naturg. LXXXVIII (A. 11), p. 46 (1922) ♂.[7] — "Kwaitschou (China)."[7]

hyalina, Kollar

Chalcosia hyalina, Kollar, Hügel's Kaschmir IV (ii), p. 462 (1844).[1] — Afghanistan.[5] Yusimarg (Kashmir).

Agalope hyalina, Hmp., F. I. Moths I 282-283, f. 190 (1892) ♂ ♀.[2]; Dudgeon, B. J. XII 302-303 (1899)[3]; Jordan, Seitz Lep. Ind. II 26 (1908)[4]; *id.,* Lep. Pal. II 13, t. 3 f. B (1910).[5]	Dungagali (Hazara; 8,000 ft.). Simla to Skiim.[2] Masuri.[1] Sikkim.[3, 4]
Agalope basalis, Wlk., Cat. II 438 (1854) [N. India][6]; Mo., Cat. Lep. E. I. Co. Mus. II 330, t. 8*a* f. 5 (1859)[7]; Butl., Ill. Het. V 26, t. 84 f. 5 (1881).[8]	Darjiling.[9, 11] Bhutan.[3] Muktesar (Kumaon; 7,000 ft.).
Agalope primularis, Butl., P. Z. S. 1875, 392 (1875)[9]; *id.,* Ill. Het. V 26, t. 84 f. 7 (1881).[10]	Manipur.[2] Burma.[2, 5]
Agalope glacialis [*nec* Mo.], Butl., Ill. Het. V 26, t. 84 f. 6 (1881).[11]	

immaculata, Leech

Agalope immaculata, Leech, T. E. S. 1898, 348 (1898) ♂[1]; Jordan, Seitz Lep. Pal. II 13, t. 3 f. B (1910).[2]	Ta-tsien-lu.[1, 2] [E. Tibet border.]

lucia, Oberthür

Agalope lucia, Obth., Et. Lep. comp. XX 207, fig. (1923).[1]	Ta-tsien-lu.[1] [E. Tibet border.]

34. Elcysma, Butler (1881).

Elcysma, Butl., T. E. S. 1881, 4 (1881); Hmp., F. I. Moths I 284, f. 192 (1892) [redescr.]; Jordan, Seitz Lep. Ind. II 26 (1908) [redescr.]; *id.,* Lep. Pal. II 13 (1910) [redescr.; larva descr.; only one sp., *westwoodi,* divided into three forms]; Hering, Arch. f. Naturg. LXXXVIII (A. 11), p. 47 (1922) [key to subspp.].	Type:–*westwoodii,* Voll.

westwoodii, Vollenhoven (Three subspecies).

(1) westwoodii westwoodii, Vollenhoven.

Agalope westwoodii, Voll., Tijds. Ent. VI 136-137, t. 9 f. 3 (1863) ♂[1]	Japan.[1, 3, 4] Korea.[4]
Elcysma translucida, Butl., T. E. S. 1881, 4 (1881) ♂.[2]	China.[4]
Elcysma westwoodi, Elwes P. Z. S. 1890, 386-387, t. 34 f. 5 (1890).[3]	
Elcysma westwoodii, Leech, T. E. S. 1898, 349-350 (1898).[4]	

(2) westwoodii, dohertyi, Elwes

Elcysma dohertyi, Elwes, P. Z. S. 1890, 386-387, t. 34 f. 4 (1890) ♂[1]; Hmp., F. I. Moths I 285, f. 192 (1892) ♂[6]	Nagas.[5-7]

Elcysma westwoodi dohertyi, Jordan, Seitz Lep. Ind. II 26, t. 5 f. C (1908) ♂[7]; Hering, Arch. f. Naturg. LXXXVIII (A. 11), p. 47 (1922).[6]

(3) westwoodii caudata, Bremer

Chalcosia caudata, Brem., Lep. Ost. -Sib., p. 97, t. 8 f. 8 (1864)[9]; Elwes, P. Z. S. 1890, 387, t. 34 f. 6 (1890).[10] — Amur.[9] Askold.[10]

Elcysma westwoodii caudata, Staudinger, Rom. Mém. Lép. VI, 248.[11]

35. Boradia, Moore (1879).

Boradia, Mo., P. Z. S. 1879, 391-392 (1879); Hmp., F. I. Moths I 286, f. 193 (1892) [redescr.]; Jordan, Seitz Lep. Ind. II 26-27 (1908) [redescr.]; Hering, Arch. f. Naturg. LXXXVIII (A. 11), p. 47 (1922). — Type:–*carneola,* Mo.

carneola, Moore

Boradia carneola, Mo., P. Z. S. 1879, 392 (1879)♂ ♀.[1]; Butl., Ill. Het. VII 27, t. 121 f. 8 (1889)[2]; Hmp., F. I. Moths I 286, f. 193 (1892) ♂[8]; Jordan, Seitz Lep. Ind. II 27, t. 5 f. A (1908)[4]; Hering, Arch. f. Naturg. LXXXVIII (A. 11), p. 47 (1922).[5] — Kangra.[3] Dharmsala.[1, 4] Simla.[4]

36. Philopator, Moore (1865).

Philopator, Mo., P. Z. S. 1865, 800 (1865); Hmp., F. I. Moths I 281, f. 189 (1892) [redescr.]; Jordan Seitz Lep. Ind. II 27 (1908) [redescr.]; Hering, Arch. f. Naturg. LXXXVIII (A. 11), p. 47 (1922) [key to spp.]. — Type:–*basimaculata,* Mo.

basimaculata, Moore

Philopator basimaculata, Mo., P. Z. S. 1865, 800, t. 42 f. 6 (1865)[1]; Hmp., F. I. Moths I 282, f. 189 (1892)♂ ♀[2]; Swinh., T. E. S. 1895, 38 (1895)[3]; Dudgeon, B. J. XII 302 (1899) ♀.[4]; Hmp., P. Z. S. 1906, 493 (1906) [*basimacula*][5]; Jordan, Seitz Lep. Ind. II 27, t. 5 f. A (1908).[6] — Darjiling.[1] Sikkim.[2, 4, 6] Raitong (Sikkim).[5] Nagas.[2] Cherrapunji.[3] Manipur.[2]

rotunda, Hampson

Philopator rotunda, Hmp., F. I. Moths IV 470 (1896)♂ ♀.[1]; Dudgeon and Elwes, B. J. XII 302 (1899)[2]; Jordan, Seitz Lep. Ind. II 27, t. 5 f. A (1908).[3] — Sikkim.[1, 3], Khasis.[1, 2] Cherrapunji. Nagas.[1, 2] Manipur.[1, 3]

37. Cadphises, Moore (1865).

Cadphises, Mo., P. Z. S. 1865, 800 (1865); Hmp., F. I. Moths I 275, f. 183 (1892) [redescr.]; Jodan, Seitz Lep. Ind. II 27, (1908) [redescr.]; Hering, Archiv. f. Naturg. LXXXVIII (A. 11), p. 48 (1972) [key to spp.]. — Type:–*maculata,* Mo.

azim, Doubleday

Gynautocera azim, Doubl., Ann. Nat. Hist. XIX 74 (1847).[1] — N. India.[2]

Cadphises azim, Kirby, Cat. Het. I 42 (1892)[2]; Jordan, Seitz Lep. Ind. II 27 (1908) [Doubledays description of *azim* fits both *maculata* and *moorei;* bis type appears to be no longer in existence].[2]

maculata, Moore

Cadphises maculata, Mo., P. Z. S. 1865, 801, t. 42 f. 7 (1865)[1]; Hmp., F. I. Moths I 275-276, f. 183♀ (1892) [redescr.][2]; Dudgeon, B. J. XII 301 (1899)[3]; Jordan, Seitz Lep. Ind. II 27, t. 5 f. E (1908).[4] — Darjiling.[1] Sikkim.[2, 4] Khasis.[2]

moorei, Butler (Two subspecies).

(1) moorei moorei, Butler

Cadphises moorei, Butl., P. Z. S. 1875, 392 (1875)[1]; Hmp., F. I. Moths I 276 (1892) ♀.[2]; Swinh., T. E. S. 1895, 37 (1895)[3]; Dudgeon, B. J. XII 301 (1899) ♂[1]; Jordan, Seitz Lep. Ind. II 27, t. 5 f. E (1908).[5] — Darjiling.[1] Sikkim.[2, 5] Bhutan.[4, 5] Khasis.[2, 3] Nagas.[2]

---

(2) moorei javanensis, Hering

Cadphises moorei javanensis, Hering, Arch. f. Naturg. LXXXVIII (A. 11), p. 48 (1922)♂ ♀.[6] — Java.[6]

38. Corma, Walker (1864).

Corma, Wlk., XXXI 124 (1864); Hmp., F. I. Moths I 268, f. 177 (1892) [redescr.]; Jordan, Seitz Lep. Ind. II 27 (1908) [redescr.]; *id.,* Lep. Pal. II 11 (1910) [redescr.]; Hering, Arch. f. Naturg. LXXXVIII (A. 11), p. 48 (1992) [key to spp.]. — Type:–*obscurata,* W. [Java; Sumatra, Borneo.]

Codane, Mo., Lep. Atk., p. 17 (1879); Snellen, Tijds. XLV 208 (1902) [redescr.]. — Type:–*zenotia,* Doubl.

maculata, Hampson (Two subspecies).

(1) maculata maculata, Hampson

Corma maculata, Hmp., F. I. Moths I 268 (1892) ♀[1]; Swinh., T. E. S. 1895, 36 (1895).[2]

Corma maculata maculata, Jordan, Seitz Lep. Ind. II 27, t. 5 f. A (1908).[3]

Chin Hills.[1]
Shillong.

(2) maculata laranda, Druce

Corma laranda, Druce, A. M. N. H. (6) XVIII 235 (1896)[4]; Jordan, Seitz Lep. Pal. II 11, t. 2 f. E (1910).[5]

Corma maculata laranda, Jordan, Seitz Lep. Ind. II 27-28 (1908)[6]; Hering, Arch. f. Naturg. LXXXVIII (A. 11), p. 48 (1922).[7]

———
Hunan.[4]
W. China.[5, 6]
S. China.[5, 6]
Centr. China.[5]

zelica, Doubleday

Gynautocera zelica, Doubl., Ann. Nat. Hist. XIX 76, t. 7 f. 3 (1847).[1]

Pidorus zelica, Wlk., Cat. II 425 (1854).[2]

Codone zelica, Butl., Ill. Het. V 25, t. 84 f. 2 (1886).[3]

Codane leucomelas, Mo., J. A. S. B. LV 97 (1886).[4]

Codane neoterica, Swinch., T. E. S. 1890, 177 (1890).[5]

Corma zelica, Hmp., F. I. Moths I 268-269 (1892)♂ ♀.[6]; Jordan, Seitz. Lep. Ind. II 28, t. 5 f. B (1908).[7]

Sylhet.[6]
E. Pegu.[6]
Moulmein.[6]
Rangoon.[5]
———
Siam.[4]

zenotia, Doubleday

Gynautocera zenotia, Doubl., Ann. Nat. Hist. XIX 77, t. 7 f. 2 (1847).[1]

Pidorus zenotea, Wlk., Cat. II 425 (1854).[2]

Corma spoliata, Wlk., Cat. XXXI 125 (1864).[3]

Corma zenotea, Hmp., F. I. Moths I 268, f. 177 ♀ (1892)♂ ♀.[4]

Corma zenotia, Jordan, Seitz Lep. Ind. II 28, t. 5 f. B (1908).[5]

Khasis.[4]
Karen Hills.[4]
Moulmein.[2, 4]
———
Siam.[3, 4, 5]
Tonkin.[5]

39. Pseudonyctemera, Snellen (1903).

Pseudonyctemera, Snell., Tijds. Ent. XLV 210-211 (1903); Jordan, Seitz Lep Ind. II 28 (1908) [redescr.];

Type:– [*dissimulata,* Wlk.=] *ficta,* Wlk.

Hering, Arch. f. Naturg. LXXXVIII (A. 11), pp. 48-49 (1922). [key to forms].

dissimulata, Walker (Two subspecies).

(1) dissimulata dissimulata, Walker

Cyclosia dissimulata, Wlk., J. Linn. Soc. VI 96 (1862).[1]	Malacca.[3, 5]
Milleria pontioides, Butl., A. M. N. H. (5) VII 35 (1881).[2]	Singapore.[5] Java.[5]
Milleria lyra, Druce, A. M. N. H. (6) VII 143-144 (1891)♂ ♀.[3]	Billiton.[5] Borneo.[1, 2, 5, 4, 4a]
Cyclosia ficta, Wlk., J. Linn. Soc. VI 97 (1862)[4]; Swinh., Cat. Het. Mus. Oxon. I 70, t. 2 f. 4 (1892) [*Milleria*].[4a]	
Pseudonyctemera dissimulata dissimulata, Jordan, Seiz Lep. Ind. II 28, t. 5 f. A (1908).[5]	————

(2) dissimulata pura, Jordan

Pseudonyctemera dissimulata pura, Jord., Seitz Lep. Ind. II 28, t. 5 f. A (1908).[6] — Assam.[5]

invaria, Walker (Two subspecies).

(1) invaria invaria, Walker

Didina invaria, Wlk., Cat. XXXI 123-124 (1864)[1]; Swinh., Cat. Het. Mus. Oxon. I 72 (1892).[2] — Sumatra.[1, 2]

(2) invaria griseifusa, Hampson

Chalcosia griseifusa, Hmp., F. I. Moths I 264 (1892) ♀.[3]	E. Pegu.[3, 4] ————
Pseudonyctemera invaria griseifusa, Jordan, Seitz Lep. Ind. II 28, t. 5 f. B (1908)[4]; Hering, Arch. f. Naturg. LXXXVII (A. 11), p. 49 (1922) ♂[5]	Malaccca.[3, 5] Singapore.[3, 4]

40. HERPA. WALKER (1854).

Herpa, Wlk., Cat. II 441-442 (1854); Hmp., F. I. Moths I 257, f. 172 (1892) [redescr.]; Jordan, Seitz Lep. Ind. II 29 (1908) [redescr.]; *id.,* Lep. Pal. II 11 (1910) [redescr.]; Hering, Arch. f. Naturg. LXXXVIII (A. 11), p. 51 (1922) [key to spp.]. — Type:–*venosa,*Wlk.

basiflava, Oberthür

Herpa basiflava, Obth., Et. Ent. XV 21, t. 3 f. 25 (1891)[1]; Leech, T. E. S. 1898, 341 (1898)[2]; Jordan, Seitz Lep. — Ta-tsien-lu.[2] Wa-ssu-know.[2]

Ind. II 29 (1908) [probably a separate genus][3]; *id.*, Lep. Pal. II 11, t 2 f. D (1910).[4] — Chi-tou.[2] m[E.Tibet border.]

eleonora, Oberthur

Herpa eleonora, Obth., Et. Lep. comp. XX 208, fig. (1923).[1] — Ta-tsien-lu.[1]

luteola, Leech

Herpa luteola, Leech, T. E. S. 1898, 341 (1898) ♀.[1]; Jordan, Seitz Lep. Pal. II 11, t. 2 f D (1910) ♀.[2] — Wa-shan.[1, 2] [E. Tibet border.]

ochracea, Leech

Herpa ochracea, Leech, T. E. S. 1898, 340 (1898) ♂[2]; Jord., Seitz Lep. Pal. II 11, t. 2 f. D (1910) ♂.[2] — Wa-shan.[1, 2]

primulina, Elwes

Herpa subhyalina var. *primulina,* Elwes, P. Z. S. 1890, 382-383 (1890).[1] — Nagas.[1-3]

Herpa primulina, Hmp., F. I. Moths I 258 (1892) [redescr.][2]; Jordan, Seitz Lep. Ind. II 39 (1908).[3]

subhyalina, Moore

Herpa subhyalina, Mo., Lep. Atk., p. 18 (1879) ♂[1]; Hmp., F. I. Moths I 258 (1892) [redscr.][2]; Jordan, Seitz. Lep. Ind. II 29, t. 5 f. C (1908).[3] — Sikkim.[2, 3] Lachung Valley.[1] Assam.[3] Tse-kou.[4, 6]

Herpa elongatissima, Obth., Butl. S. E. Fr. 1910, 314 (1910)[4]; Hering, Arch. f. Naturg. LXXXVIII (A. 11), p. 51 (1922) [syn. of *subhyalina*][5]; Obth., Et. Lep. comp. V 330, t. 77 f. 699 (1911).[6]

venosa, Walker (Two subspecies).

(1) venosa venosa, Walker

Herpa venosa, Wlk., Cat. II 442 (1854)[1]; Mo., Cat. Lep. E. I. Co. Mus. II 330, t. 8*a* f. 6 (1859)[2]; Hmp., F. I. Moths I 257, f. 172 (1892) [redescr.][3]; Swinh., T. E. S. 1895, 35 (1895)[4]; Jordan, Seitz Lep. Ind. II 29, t. 5 f. C (1908).[5] — Sylhet.[1] Khasis.[3, 4] Shillong. Cherrapunji.

(2) venosa sinica, Oberthür

Herpa sinica, Obth., Et. Ent. XV 21 (1891).[6] — Kwei-tcheu.[6] Ta-tsien-lu.[7, 9]

Herpa venosa, Leech, T. E. S. 1898, 340-341 (1898).[7]

Herpa venosa sinica, Jordan, Seitz Lep. Ind. II 29 (1908)[8]; *id.,* Lep. Pal. II 11, t. 2 f. D (1910)[9]; Hering, Arch. f. Naturg. LXXXVIII (A. 11), p. 51 (1922).[10]

41. Psaphis, Walker (1854).

Psaphis, Wlk., Cat. II 433 (1854); Jordan, Seitz Lep. Ind. II 29 (1908) [redescr.]; Hering, Arch. f. Naturg. LXXXVIII (A. 11), p. 51 (1922) [kep to supp.]. Type:–*camadeva,* Doubl. [Penang].

Canerkes, Mo., P. Z. S. 1865, 802 (1865); Snellen, Tijds. Ent. XLV 173-174 (1902) [redescr.]. Type:– euschemoides, Mo.

Canerces, Hmp., F. I. Moths I 281 (1892). Type:– *euschemoides,* Mo,

euschemoides, Moore (Three subspecies).

(1) euschemoides euschemoides, Moore

Canerkes euschemoides, Mo., P. Z. S. 1865, 802, t. 42 f. 8 (1865).[1] Sikkim.[2-4] Sylhet.[1, 2] mE. Pegu.[2]

Canerces euschemoides, Hmp., F. I. Moths I 281, f. 188 (1892) ♂[2]; Dudgeon, B. J. XII 302 (1899).[3]

Psaphis euschemoides euschemoides, Jordan, Seitz Lep. Ind. II 30, t. 5 f. D (1908).[4]

(2) euschemoides javanicus, Rothschild

Canerces javanicus, Roths., Novit. Zool. III 56 (1896).[5] Java.[5-8]

Cancerces javana, Röber, Entom. Nachr. 1897, 7 (1897).[6]

Canerkes euschemoides, Snellen, Tijds. Ent. XLV 174-175 (1902) ♀.[7]

Psaphis euschemoides javanicus, Jord., Seitz Lep. Ind. II 30 (1908).[8]

(3) euschemoides scotais, Jordan

Psaphis euschemoides, scotais, Jord., Seitz Lep. Ind. II 30, t. 5 f. D (1908).[9] Borneo.[9]

42. Eucorma, Jordan (1908).

Eucorma, Jordan, Seitz Lep. Ind. II 30 (1908); Hering, Arch. f. Naturg. LXXXVIII (A. 11), p. 52 (1922) [key to spp.]. Type:–*intercisa,* Mo. [Java]

obliquaria, Fabricius (Two subspecies).

(1) obliquaria obliquaria, Fabricius

Phalœna obliquaria, Fb., Mantissa II 194 (1787)[1]; *id.*, Ent. Syst. III (ii) 156-157 (1794).[2]	Mergui.[3-5]
Soritia obliquaria, Hmp., F. I. Moths I 254 (1892) [redescr.].[3]	Malacca.[3, 4]
Milleria obliquaria, Piepers and Snellen, Tijds. Ent. XLV 201-202 (1902)♂ ♀.[4]	Siam.[1, 2, 5, 6] Sumatra.[5] Java.[4, 5] Billiton.[4]
Eucorma obliquaria, Jordan, Seitz Lep. Ind. II 30, t. 5 ff. E. F (1908).[5]	Celebes.[4][?] Borneo.[3-5]
Eterusia transversa, Wlk., Cat. XXXI 118 (1864).[6]	
Erasmia eusemoides, Feld., Reise Novara, Lep. IV, t. 83 f. 10 (1874).[7]	
Erasmia laja, Pag., Jahrb. Nass. Ver. Naturk. XXXVIII 11, t. 1 f. 5 (1885).[8]	

(2) obliquaria interrupta, Hering

Eucorma obliquaria interrupta, Hering, Arch. f. Naturg. LXXXVIII (A. 11), p. 52 (1922)♂ ♀.[9]	Nias.[9]

43. Eterusia, Hope (1840).

Eterusia, Hope, Trans. Linn. Soc. XVIII 445 (1840); Westwood, Arcana Entom. I 19 (1841) [redescr.]; Wlk., Cat. II 427 (1854); Kirby, Cat. Het. I 49 (1892); Jordan, Seitz Lep. Ind. II 31 (1908) [redescr.]; *id.*, Lep. Pal. II 10 (1910) [redescr.].	Type:–*tricolor,* Hope.
Heterusia, Doubleday, Zool. II 468 (1844); Hmp., F. I. Moths I 259, f. 174 (1982) [redescr.]; Hering, Arch. f. Naturg. LXXXVIII (A. 11), pp. 6062 (1922) [key to spp.].	Type:–*tricolor,* Hope.
Soritia, Wlk., Cat. II 435 (1854); Hmp., F. I. Moths I 251-252, f. 168 (1892) [redescr.]; Snellen, Tijds, Ent. XLV 203 (1902) [redescr.]; Hering, Arch. f. Naturg. LXXXVIII (A 11), pp. 55-58 (1922) [key to forms].	Type:–*pulchella,* Koillar.
Sephisa, Mo., Lep. Ceylon II 41 (1882).	Type:–*[aedea]cingala,* Mo.
Devanica, Mo., T. E. S. 1884, 355 (1884); Kirby, Handb. Lep. III 75 (1897) [*Heterusia* and *Sephisa* are *præoccupied*].	Type:–[*aedea*]cingala, Mo.
Prosopandrophila, Hering, Arch. f. Naturg. LXXXVIII (A. 11), pp. 60-62 (1922) [key to spp.].	Type:–*distincta,* Guér.[1]

ædea, Clerck (Five subspecies).

(1) ædea ædea, Clerck

Papilio ædea, Clerck, Icones, t. 4 f. 2 (1759)[1]; Linn., Amœn. Acad. VI 403 (1764).[2]

Heterusia aedea, Doubl., Zool. II 469, fig. (1844)[3]; Leech, T. E. S. 1898, 342 (1898)[4]; Wileman, T. E.S. 1911, 350-351 (1911).[5]

Eterusia aedea, Wlk., Cat. II 428 (1854)[6]; Jordan, Seitz Lep. Pal. II 10, t. 2 f. C (1910).[7]

Eterusia aedea aedea, Jord., Seitz Lep. Ind. II 34 (1908).[8]

China.[8]
Centr. China.[4, 5, 7]
W. China.[4, 5, 7]
Japan.[5] [?]

(2) ædea edocla, Doubleday

Heterusia edocla, Doubl., Zoologist III 469 (1847)[9]; Hmp., F. I. Moths I 261 (1892) [redescr.][10]; Dudgeon, B. J. XII 298 (1899) [larva on tea, etc.][11]; Fletcher, Rept. Third Entom. Meeting I 140 (1910).[12]

Eterusia edocla, Wlk., Cat. II 428 (1854)[13]; Swinh., T. E. S. 1895, 35 (1895).[14]

Heterusia signata, Möschl., Stett. ent. Zeit. XXXIII 341 (1872).[15]

Heterusia dulcis, Butl., Ill. Het. V 21, t. 83 f. 3 (1881).[16]

Heterusia aedea, Hmp., F. I. Moths I 262 (1892) [*partim*].[17]

Eterusia aedea edocla, Jordan, Seitz Lep. Ind. II 34, t. 6 f. E (1908) [subsp. from N. India and Burma].[18]

Nepal.[10,] Sikkim.[10, 12]
Darjiling.[16]
Cachar.[10, 12]
Nagas.[10], Shillong.[14]
Gauhati., Burma.[18]
Maymymo.

ab. 1, magnifica, Butler

Eterusia magnifica, Butl., T. E. S. 1879, 5 (1879)[19]; *id.,* Ill. Het. V 20, t. 83 f. 2 (1881)[20]; Swinh., T. E. S. 1895, 35 (1895)[21]; Jordan, Seitz Lep. Pal. II 10, t. 2 f. B (1910).[22]

Heterusia magnifica, Hmp., F. I. Moths I 261 (1892) [redescr.][23]; Leech, T. E. S. 1898, 342 (1898) [? ab. *lepcha*][24]; Antram, B. J. XVIII 431, tab. [larva a pest of tea][25]; Fletcher, Rept. Second Ent. Meeting, pp. 19-20 (1917)[26]; *id.,* Rept. Third Ent. Meeting I 139 (1919).[27]

Khatmandu
(Nepal).
Sikkim.[23]
Darjiling.[28]
Duars.[27,]
Nagas.[23]
Khasis.[21,]
Sylhet.[23]
Cachar.[19, 23]
Gauhati.
Jorhat.
W. China [?].[22, 24]

Heterusia cingala [*nec* Mo.], Barlow, Ind. Mus. Notes V 16-17 (1900).[28]

Eterusia aedea edocla ab. *magnifica,* Jordan, Seitz Lep. Ind. II 34 (1908).[29]

ab. 2, lepcha, Jordan

Eterusia aedea edocla ab. *lepcha,* Jordan, Seitz Lep. Ind. II 34 (1908).[30]

(3) ædea cingala, Moore

Heterusia cingala, Mo., A. M. N. H. (4) XX 343 (1877)[31]; Mo., Lep. Ceylon II 41, t. 96 f. 1 (1882) [*Sephisa*][32]; Hmp., F. I. Moths I 262 (1892) [redescr.; larva descr.][33]; Green. Ind. Mus. Notes IV, v, p. 279, t. 18 ff. 2*a* ♂, *b* ♀, *c* larva, *d* cocoon (1900) [occ. pest of tea in Ceylon; parasit, by *Exorista heterusia,* Coq.][34]; Rutherford, Trop. Agric. XLIII 128-129 (Aug. 1914) [as tea-pest].[35]

Ceylon.[31-36] Pusselawa.[34] Madulsima Permadeniya.

Eterusia aedea cingala, Jordan, Seitz Lep. Ind. II 34 (1908).[36]

(4) ædea, virescens, Butler

Heterusia virescens, Butl. Ill. Het. V 21, t. 83 f. 9 (1881)[37]; Hmp., F. I. Moths I 262 (1892) [redescr.][38]; Fletcher, South Ind. Ins., pp. 448-449, f. 326 (1914) [occ. local pest of tea][39]; Anstead, Planters' Chronicle X 195 (10 Apl. 1915)[40]; Fletcher, Rept. Second Ent. Meeting, pp. 19-20 (1917)[41]; *id.,* Rept. Third Ent. Meeting I 140 (1919).[42]

Khandala.[38] Nilgiris.[38, 39, 41, 42] Kotagiri. Wynaad.[39, 42] Shembaganur (Palnis).

Eterusia aedea virescens, Jordan, Seitz Lep. Ind. II 34, t. 6 f. E ((1908).[48]

(5) ædea formosana, Jordan

Eterusia aedea formosana, Jordan, Seitz Lep. Ind. II 34 (1908).[44]

Formosa.[44] ? Japan [see[5]].

Larva descr. [25, 33, 34, 32], on tea[11, 25-28, 34-35, 39, 40-42], on *Lagerstroemia,* etc.[32]

bicolor, Moore (Two subspecies).

(1) bicolor bicolor, Moore

Devanica bicolor, Mo., T. E. S. 1884, 355 (1884)[1]; Mo., J. A. S. B. LIII. 235 (1885).[2]

Cachar.[1, 7], Nagas.[8] Khasis.[7]

Soritia viridivena, Hmp., F. I. Moths I 253 (1882) ♀.[3]; Swinh., T. E. S. 1895, 34 (1895).[4]	Cherrapunji.[4] Sylhet.[2]
Soritia bicolor, Hmp., F. I. Moths IV 469 (1896) [synon.][5]; Hering, Arch. f. Naturg. LXXXVIII (A. 11), p. 60 (1922).[6]	————
Eterusia bicolor bicolor, Jordan, Seitz Lep. Ind. II 32, t. 6 f. B (1908).[7]	

(2) bicolor lata, Jordan

Eterusia bicolor lata, Jord., Seitz Lep. Ind. II 32, t. 6 f. B (1908).[8]	C. Tonkin.[8]

cecilia, Oberthür

Soritia cecilia, Obth., Et. Lep. comp. XX 210-211, fig. (1923).[1]	Moupin.[1,] Siao-lou.[1] Ya-tchéou.[1] [All on E. Tibet border.]

circinata, Herrich-Schäffer

Heterusia circinata, H.-S., Aussereur. Schmett., p. 57, ff. 156, 157 (1854).[1]	Sikkim.[3-5] Assam.[1-3, 5]
Eterusia fasciata, Wlk., Char. Undescr. Het., p. 5 (1869).[2]	Shillong.[4]
Soritia circinata, Hmp., F. I. Moths I 253 (1892) ♂[3]; Swinh., T. E. S. 1895, 34 (1895).[4]	
Eterusia circinata, Jordan, Seitz Lep. Ind. II 33, t. 6 f. C (1908).[5]	

culoti, Oberthür (Two subspecies).

(1) culoti culoti, Oberthür

Eterusia culoti, Obth., Bull. S. E. Fr. 1910, 314-315 (1910)[1]; Jordan, Seitz Lep. Pal. II 441, t. 56 f. K (1912)[2]; Obth., Et. Lep. comp. V 330, t. 77 f. 702 (1911).[3]	Moupin.[1] Tien-tsuen.[1] Tay-tou-ho.[1]

(2) culoti dubernardi, Oberthür

Eterusia dubernardi, Obth., Bull. S. E. Fr. 1910, 315 (1910)[4]; *id.,* Et. Lep. comp. V 330, t. 77 f. 701 (1911).[5]	Tse-kou.[4, 5] [E. Tibmet border.]
Heterusia culoti dubernardi, Hering, Arch. f. Naturg. LXXXVIII (A. 11), p. 64 (1922).[6]	

dichroa, Jordan

Eterusia dichroa, Jord., Seitz Lep. Ind. II 33 (1908) ♀.[1]	Khasis.[1]

distincta, Guérin (Five subspecies).

(1) distincta distincta, Guérin

Gynautocera distincta, Guér., Delessert, Souv. Ind. II 85, t. 24 f. 3 (1843) ♀.[1]

Chalcosia distincta, Wlk., Cat. II 423 (1854)[2]; Snellen, Midd. Sumatra, Lepid. p. 131 (1880)[3]; Pagenst. Jahrb. Nass. Ver. Naturk. XXXVIII 11, t. 1 f. 9 (1884)[4]; Hmp. F. I. Moths I 267 (1892) ♀[5]; Snellen, Iris VIII 132 (1895)[6]; Piepers and Snellen, Tijds. Ent. XLV 197-198 (1902)[7]; Hmp., B. J. XX 95 (1910) [synon.].[8]

Enterusia distincta distincta, Jordan, Seitz Lep. Ind. II 34, t. 6 f. B (1908).[9]

Prosopandrophila distincta, Hering, Arch. f. Naturg. LXXXVIII (A. 11), p. 61 (1922).[10]

Eterusia osseata, Wlk., Cat. XXXI, 120-121 (1864).[11]

Sumatra.[3, 7, 9, 11]
Penang.[9]
Malacca.[9]
Cambodia.[9]
Mergui.[5][?]
Assam.[2][?]

(2) distincta drataraja, Moore

Eterusia drataraja, Mo., Cat. Lep. E. I. Co. Mus. II 321, t. 8*a* f. 3 [♂] (1859) ♂[12]; Hmp., F. I. Moths I 262 (1892)♂ [*Heterusia*].[13]

Eterusia distincta drataraja, Jordan, Seitz Lep. Ind. II 34, t. 6 f. C (1908).[14]

Mergui.[13] [?]

Java.[12, 14]

(3) distincta œnone, Butler

Chalcosia œnone, Butl., Ent. Mo., Mag. XX 57 (1883).[15]

Eterusia distincta œnone, Jordan, Seitz Lep. Ind. II 34 (1908).[16]

Nias.[15, 16]

(4) distincta indistincta, Swinhoe

Chalcosia indistincta, Swinh., Cat. Het. Mus. Oxon. I 74-75 (1892)♂ ♀.[17]

Eterusia distincta indistincta, Jordan, Seitz Lep. Ind. II 34 (1908).[18]

Borneo.[17, 18]

(5) distincta albina, Jordan

Eterusia distincta albina, Jord., Seitz Lep. Ind. II 34 (1908).[19]

Mergui.[19]

Larva descr.[7], on *Melastoma polyanthum*,[7] *Jambosa densiflora*.[7]

elizabetha, Walker

Eterusia elizabetha, Wlk., Cat. II 433 (1854) ♀.[1]; Leech, T. E. S. 1898, 337 (1898) [*Soritia*][2]; Jordan, Seitz Lep. Ind. II 32 (1908)[3]; *id.*, Lep. Pal. II 10, t. 2 f. B (1910).[4]

Eterusia dirupta, Wlk., Cat. XXXI 119 (1864) ♂.[5]

Heterusia microcephala, Feld., Reise Novara, Lep. IV, t. 83 f. 7 (1874) ♂[6]

Ta-tsien-lu.[2, 4]
———————
W. China.[3]
N. China.[1, 3, 5]
Shanghai.[4]
Hongkong.[5]

lacreuzei, Oberthür

Eterusia lacreuzei, Obth., Bull. S. E. Fr. 1910, p. 315 (1910)[1]; Jordan, Seitz Lep. Pal. II 441, t. 56 f. i (1912)[2]; Obth., Et. Lep. comp. V 330, t. 77 f. 704 (1911).[3]

Tse-Kou.[1, 3]
[E. Tibet Border.]

lativitta, Moore

Heterusia fasciata [nec Wlk.], Möschler, Stett. ent. Zeit. XXXIII 340 (1872).[1]

Eterusia lativitta, Mo., Lep. Atk., p. 15 (1879)[2]; Hmp., F. I. Moths I 260 (1892) [redescr.; *Heterusia*][3]; Swinh., T. E. S. 1895 35 (1895)[4]; Jordan, Seitz Lep. Ind. II 33-34, t. 6 f. E (1908).[5]

Darjiling.[1, 2]
Sikkim.[3, 5]
Khasis.[3]
Shilong.[4]
Sylhet.[3] Nagas.[3]
Tenasserim.[5]

lydia, Oberthür

Soritia lydia, Obth., Et. Lep. comp. XX 209-210, fig. (1923).[1]

Siao-lou.[1]
[E. Tibet border.]

mœrens, Oberthür

Eterusia moerens, Obth., Bull S. E. Fr. 1910, p. 315 (1910)[1]; Jordan, Seitz Lep. Pal. II 441, t. 56 f. i (1912)[2]; Obth., Et. Lep. comp. V 330, t. 77, f. 698 (1911).[3]

Tse-kou.[1, 2, 3]
[E. Tibet border.]

nigribasalis, Hampson

Soritia nigribasalis, Hmp., F. I. Moths I 253 (1892) ♀.[1]; Swinh., T. E. S. 1895, 34 (1895).[2]

Eterusia nigribasalis, Jordan, Seitz Lep. Ind. II 32 (1908) ♀.[3]

Sikkim.[1, 3]
Cherrapunji.[2]

pulchella, Kollar (Three subspecies).

(1) pulchella pulchella, Kollar

Chalcosia pulchella, Koll., Hügel's Kaschmir IV (ii) 461 (1844).[1]♀

Afghanistan.[15]
Kashmir.[15]

Chalcosia leptalina, Koll., Hügel's Kaschmir IV (ii) 462 (1844).[2] ♂

Gynautocera sexpunctata, Doubl., Ann. Nat. Hist. XIX 77 (1847) [ab.].[3]

Eterusia sexpunctata, Wlk., Cat. II 432 (1854).[4]

Eterusia triliturata, Wlk., Cat. XXXI 119-120 (1864) ♂.[5]

Heterusia flavomaculata, Möschler, Stett. ent. Zeit. XXXIII 343 (1872).[6]

Heterusia octopunctata, Möschl., Stett. ent. Zeit. XXXIII 344 (1872).[7]

Beterusia cicada, Feld., Reise Novara, Lep. IV, t. 83 f. 8 (1874).[8]

Soritia olivascens, Mo., Lep. Atk., p. 16 (1879).[9]

Soritia leptalina, Hmp., F. I. Moths I 252-253, f. 168 (1892) [redescr.][10]; Swinh., T. E. S. 1895, 34 (1895)[11]; Leech, T. E. S. 1898, 336-337 (1898)[12]; Dudgeon, B. J. XII 296 (1899)[13]; Fletcher, Rept. Third Ent. Meeting I 140 (1919).[14]

Eterusia, leptalina, Jordan, Seitz Lep. Pal. II 10, t. 2 f. A (1910).[15]

Eterusia pulchella, Jordan, Seitz Lep. Ind. II 31-32 (1908).[16]

Soritia pulchella, Hering, Arch. f. Naturg. LXXXVIII (A. 11), p. 50 (1922).[17]

Dungagali.[14]
Masuri.[1, 2, 14, 5]
Kurnaon Hills.
Darjiling.[14]
Khasis.[9, 11, 14]
Shillong.[14]
Sylhet.[3, 7, 10, 4]
Cherrapunji.
Maymyo.[14]
Bernardmyo.[10]
Kalaw (Burma).

W. China.[12, 15]
Centr. China.[12, 15]

(2) pulchella major, Jordan

Eterusia pulchella major, Jord., Seitz Lep. Ind. II 32 (1908).[18]

Tonkin.[18]

(3) pulchella xanthophlebia, Hering

[? a berr. nec susp.].

Soritia pulchella xanthophlebia, Hering, Arch. f. Naturg. LXXXVIII (A. 11), pp. 58-59 (1922)♀.[19]

Khasis.[19]

Larva on apple,[14] rose,[13] tea,[13] etc.

raja, Moore

Eterusia raja, Mo., Cat. Lep. E. I. Co. Mus. II 320, t. 8[a] f. 2 (1859).[1]

Darjiling.[1]
Sikkim.[2, 4]

Heterusia raja, Hmp., F. I. Moths I 259-260 (1892) ♀.[2]; Dudgeon, B. J. XII 298 (1899).[3]

Eterusia rajah, Jordan, Seitz Lep. Ind. II 33, t. f. D (1908).[4]

Tukvar.[3]
Assam.[4]

repleta, Walker

Eterusia repleta, Wlk., Cat. XXXI 118 (1864) ♂.[1]

Eterusia alompra, Mo., Lep. Atk., p. 16 (1879) [ab.].[2]

Eterusia urania, Schaus, Ent. Amer., p. 39 (1890) [ab.].[3]

Eterusia submarginalis, Swinh., Cat. Het. Mus. Oxon I 81 (1892) ♀.[4]

Eterusia repleta, Swinh., Cat. Het. Mus. Oxon II 587 (1900).[5]

Heterusia alompra, Hmp., F. I. Moths I 260 (1892) [redescr.][6]; Elwes, B. J. XII 298 (1899).[7]

Eterusia repleta, Jord., Seitz Lep. Ind. II 33 (1908)[8].

Eterusia repleta ab. *urania,* Jord., Seitz Lep. Ind. II 33, t. 6 f. D (1908).[9]

Heterusia repleta, Hering, Arch. f. Naturg. LXXXVIII (A. 11), p. 65 (1922).[10]

Sikkim.[6]
Bhutan.[6, 7]
Sibsagar., Nagas.[6]
Burma.[6]
Ruby Mines distr.[2]

Cambodia.[1, 6, 4]

shahama, Moore

Eterusia shahama, Mo., P. Z. S. 1865, 801 (1865)[1]; Waterhouse, Aid Ident. Ins. II, t. 127 f. 3 (1883)[2]; Hmp., F. I. Moths I 253 (1892) [redescr.; *Soritia*][3]; Jordan, Seitz Lep. Ind. II 32, t. 6 f. F (1908).[4]

Darjiling.[4]
Sikkim.[3]

sublutea, Walker

Eterusia sublutea, Wlk., Cat. II 430 (1854).[1]

Heterusia scintillans, Herrich-Schäffer, Aussereur. Schumett. I, ff. 154, 155 (1854).[2]

Heterusia tricolor, Hmp., F. I. Moths I 260-261 (1892) [*partim*].[3]

Eterusia sublutea, Jordan, Seitz Lep. Ind. II 33, t. 6 f. D (1908).[4]

Sikkim.[4], Sylhet.[1]
Assam.[4]

Tricolor, Hope

Eterusia tricolor, Hope, Trans. Linn. Soc. XVIII 445, t. 31 f. 8 (1840)[1]; Westw., Dunc, Nat. Libr., Exotic Moths, p. 39, t. 3 f. 1 (1841)[2]; *id.,* Arcana Entom. I 19 (1841) [redescr.][3]; Wlk., Cat. II 428 (1854)[4]; Swinh., T. E. S. 1895, 35 (1895).[5]

Sikkim.[6, 8, 10]
Bhutan.[8, 10]
Nepal.[6], Assam.[1, 3, 10]
Sylhet.[6], Khasis.[5]
Nagas.[6]

Heterusia tricolor, Hmp., F. I. Moths I 260-261, f. 174 (1892) [*partim*][6]; Leech, T. E. S. 1898, 341-342 (1898)[7]; Dudgeon, B. J. XII 298 (1899).[8] — E. Pegu.[6] Omei-shan.[7] [E. Tibet border.]

Devanica tricolor, Kirby, Handb. Lep. III 76, t. 78 f. 1 (1897).[9] — W. China.[11]

Eterusia tricolor, Jordan, Seitz Lep. Ind. II 33 (1908)[16]; *id.,* Lep. Pal. II 10, t, 2 f. B (1910).[11]

Heterusia trimacula, Möschler, Stett. ent. Zeit. XXXIII 342 (1872).[12]

Larva descr.[8]

viridibasalis, Dudgeon

Soritia viridibasalis, Dudgeon, B. J. XVI 399 (1905) ♂[1]; Hmp., B. J. XX (1910)[2]; Jordan, Seitz Lep. Ind. II 32 (1908) [*Enterusia*].[3] — Tenasserim.[1, 3]

44. Pidorus, Walker (1854).

Pidorus, Wlk., Cat. II 424 (1854); Hmp., F. I. Moths I 254, f. 169 (1892) [redescr.]; Piepers and Snellen, Tijds. Ent. XLV 203 (1902) [not considered distinct from *Soritia*]; Jordan, Seitz Let. Ind. II 35 (1908) [redescr.]; *id.,* Lep. Pal. II 10 (1910) [redescr.]; Hering, Arch. f. Naturg. LXXXVIII (A. 11), pp. 65-66 (1922) [key to spp.]. — Type:–*gloucopis,* Drury.

Laurion, Wlk., Cat. II 426 (1854). — Type:–*circe,* Herr.-Schäff.

albifascia, Moore (Two subspecies).

(1) albifascia albifascia, Moore

Milleria albifascia, Mo., Lep. Atk., p. 19 (1879).[1] — Nagas.[2], Khasis.[3]

Pidorus albifascia, Hmp., F. I. Moths I 255 (1892) ♂[2]; Swinh., T. E. S. 1895, 34 (1895).[3] — Sylhet.[2] Cherrapunji.[1] Momeit (Burma).[2]

Pidorus albifascia albifascia, Jordan, Seitz Lep. Ind. II 35, t 6 f. G (1908).[4]

(2) albifascia steleus, Jordan

Pidorus albifascia steleus, Jord., Seitz Lep. Ind. II 35 (1908)[5]; Hering, Arch. f. Naturg. LXXXVIII (A. 11), p. 66 (1922).[6] — Tonkin.[5, 6]

circe, Herrich-Schäffer

Chalcosia circe, H.-S., Aussereur. Schmett., f. 2 (1853).[1]

Laurion metallica, Wlk., Cat. II 426 (1854).[2]

Pidorus circe, Hmp. F. I. Moths I 255, f. 169 (1892) [redescr.][3]; Swinh., T. E. S. 1895, 34 (1895)[4]; Jordan, Seitz Lep. Ind. II 35, t. 6 f. F (1908).[5]

Sikkim.[5]
Khasis.[3, 4]
Shillong.
Nagas.[3], Cachar.[3]
Sylhet.[2]

gemina, Walker

Laurion gemina, Wlk., Cat. II 427 (1854).[1]

Pidorus geminus, Hmp., F. I. Moths I 254 (1892) [redescr.][2]; Swinch., T. E. S. 1895, 34 (1895)[3]; Leech, T. E. S. 1898, 339 (1898)[4]; Dudgeon, B. J. XII 297 (1899)[5]; Hmp., B. J. XVI 193 (1905).[6]

Pidorus gemina, Jordan, Seitz Lep. Ind. II 35 (1908)[7]; *id.,* Lep. Pal. II 11, t. 2 f. B (1910).[6]

Sikkim.[2, 5]
Bhutan.[5]
Sylhet.[1, 2], Khasis.[3]
Maymyo (Burma)
Moulmein.[2]
Padaung (Burma).
Omei-shan.[4]

W. China.[4, 8]
Cambodia.[2]
Hongkong.[1, 2]
Borneo.[2], Hainan.[7]

glaucopis, Drury (Two subspecies).

(1) glaucopis glaucopis, Drury

Bombyx glaucopis, Drury, Ill. Exot. Ent. II, t. 6 f. 4 (1773)[1]; Stoll, Pap. Exot. IV, t. 322 f. D (1781 ?).[2]

Bombyx rubrocollaris, Göze, Ent. Beytr. III (iii) 66 (1781).[3]

Pidorus glaucapis, Wlk., Cat. II 424 (1854)[4]; Hmp., F. I. Moths I 255 (1892) [redescr.] [*partim*][5]; Swinh., T. E. S. 1895, 34 (1895)[6]; Hering, Arch. f. Naturg. LXXXVIII (A. 11), p. 66 (1922).[7]

Pidorus glaucopis glaucopis, Jordan, Seitz Lep. Ind. II 35, t. 6 f. F (1908).[8]

Bengal.[1]
Darjiling distr.
Sikkim.[5]
Bhutan.[5], Cachar.[5]
Nagas.[3], Sylhet.[4, 5]
Khasis.[6],
Shillong.
Burma.[8]

Tonkin.[8]

(2) glaucopis atratus, Butler

Pidorus atratus, Butl., A. M. N. H. (4) XX 401 (1877)[9]; *id.,* Ill. Het. II 9, t. 23 f. 9 (1878).[10]

Pidorus glaucopis, Hmp., F. I. Moths I 255 (1892) [*partim*][11]; Leech, T. E. S. 1898, 338 (1898).[12]

W. China.[12, 13]
Korea.[12, 13]
Manchuria.[13]
Japan.[9-12]

Pidorus glaucopis atratus, Jordan, Seitz Lep. Pal. II 11, t. 2 f C. (1910).[13]

eechi, Jordan

Pidorus fasciatus, Leech, T. E. S. 1898, 339 (1898)♂ ♀.[1] Moupin.[1,2]

Pidorus leechi, Jordan, Seitz Lep. Pal. II 11, t. 2 f. D (1910).[2] Omei-shan.[1,2]

leno, Swinhoe

Pidorus leno, Swinh., A. M. N. H. (7) VI 305 (1900)♂ ♀[1]; Jordan, Seitz Lep. Ind. II 35, t. 6 f. F (1908).[2] Assam.[2] Jantia Hills.[1] Shillong. ——— Hainan.

miles, Butler

Laurion miles, Butl., Ill. Het. V 25, t. 34 f. 3 (1881).[1] Darjiling.[1] Sikkim.[2,3]

Pidorus miles, Hmp., F. I. Moths I 255 (1892) ♂[2]; Jordan, Seitz Lep. Ind. II 36, t. 7 f. A (1908).[3] Bhutan.[2,3]

truncatus, Jordan

Pidorus truncatus, Jord., Seitz. Lep. Ind. II 36, t. 6 f. G (1908)♂ ♀.[1] Sikkim.[1]

45. Chalcosia, Hübner (1822).

Chalcosia, Hb., Verz., p. 173 (1822); Wlk., Cat. II 420 (1854); Kirby, Cat. Het. I 44 (1892); Hmp., F. I. Moths I 264, f. 176 (1892) [redescr.]; Snellen, Tijds. Ent. XLV 195 (1902) [redescr.; *Eterusia* is included]; Jordan, Seitz Lep. Ind. II 36 (1908) [redescr.]; *id.,* Lep. Pal. II 10 (1910) [redescr.]; Turner, T. E. S. 1918, 176, f. 37 (1818) [neuration of *C. affinis*]; Hering, Arch. f. Naturg. LXXXVIII (A. 11), pp. 69-71 (1922) [key to spp.]. Type:–*pectinicornis,* Linn.

Milleria, Herrich-Schäffer, Aussereur. Schmett. I 78 (1856); Kirby, Cat. Het. I 43 (1892); Hmp., F. I. Moths I 262, f. 175 (1892) [redescr.]; Snellen, Tijds. Ent. XLV 199 (1902)[redescr.]; Hering, Arch. f. Naturg. LXXXVIII (A. 11), pp. 66-67 (1922) [key to spp.; considered to be distinct from *Chalcosia*]. Type:–*adalifa,* Double. *(virginalsi,* H.-S).[1]

adalifa, Doubleday (Three subspecies).

(1) adalifa adalifa, Doubleday

Gynautocera adalifa, Doubl., Ann. Nat. Hist. XIX 76 (1847)[1]; Butl., Ill. Het. V 22, t. 83 f. 5 (1891).[2] Sikkim.[5,10] Darjiling., Bhutan.[10]

Milleria adalifa, Kirby, Cat. Het. I 43 (1892) [*partim*].[3]

Chalcosia adalifa, Hmp., F. I. Moths I 265 (1892) ♀.[4]

Chalcosia adalifa adalifa, Jordan, Seitz Lep. Ind. II 36, t. 7 f. A (1908).[5]

Chalcosia virginalis, Herr.-Sch., Aussereur. Schmett. I f. 4 (1853).[6]

Milleria virginalis Hmp., F. I. Moths I 263, f. 175 (1892) [redescr.][7]; Swinh., T. E. S. 1895, 35 (1895)[8]; Leech, T. E. S. 1898, 343 (1898)[9]; Dudgeon, B. J. XII 299 (1899).[10]

Cyclosia subcyanescens, Wlk., Cat. II 417 (1854).[11]

Sylhet.[7, 11]
Khasis.[4]
Shillong.[8]
Burma.[5, 7]
Omei-shan.[9]

W. China.[9]
Tonkin.[5]

ab. 1, fuliginosa, Walker

Cyclosia fuliginosa, Wlk. Cat. II 418 (1854).[12]

Milleria fuliginosa, Butl., Ill. Het. V 22, t. 83 f. 6 (1881).[13]

Milleria virginalis, var. *fuliginosa,* Hmp., F. I. Moths I 263 (1892).[14]

Chalcosia adalifa ab. *fuliginosa,* Jordan, Seitz Lep. Ind. II 36, t. 7 f. A (1908).[15]

Sylhet.[12]

ab. 2, cardinalis, Moore

Cyclosia cardinalis, Mo., Lep. Atk., p. 18 (1879).[16]

Milleria cardinalis, Hmp., F. I. Moths I 264 (1892) ♂.[17]

Milleria adalifa cardinalis, Hering, Arch. f. Naturg. LXXXVIII (A. 11), p. 67 (1922) [redescr.][18].

Sikkim.[17]
Darjiling.[16, 18]

ab. 3. cyanivena, Hampson

Milleria cyanivena, Hmp., F. I. Moths I 263 (1892) ♂.[19]

Chalcosia adalifa ab. *cyanivena,* Jordan, Seitz. Lep. Ind. II 36 (1908).[20]

Bernardmyo.[19]

(2) adalifa candida, Vollenhoven

Chalcosia candida, Voll., Tijds., Ent. VI 135-136 (1863).[21]

.........*forbesi,* Druce.

Chalcosia adalifa candida, Jordan, Seitz Lep. Ind. II 36-37, t. 7 f. A (1908).[23]

Chalcosia adalifa candida ♂ f. *nitida,* Jordan, t. c., p. 37 (1908).[24]

Sumatra.[21, 23, 24]
Java.[23]

(3) adalifa fuhoshonis, Strand

Milleria adalifa fuhoshonis, Hering, Arch. f. Naturg. LXXXVIII (A. 11), p. 67 (1922).[25] — Formosa.[25]

affinis, Gúerin

Gynautocera affinis, Gúer., Delessert, Souv. Ind. II 86, t. 24 f. 2 (1843).[1]

Chalcosia affinis, Mo., Lep. Ceylon II 45, t. 98 f. 2 (1882)[2]; Hmp., F. I. Moths I 266 (1892)[3]; Jordan, Seitz Lep. Ind. II 37, t. 7 f. B (1908)[4]; Turner, T. E. S. 1918, 176, f. 37 [neuration] (1918).[5]

W. Ghats.[3] Nilgiris.[3] S. India.[4] Ceylon.[2, 3]

alpherakyi, Leech

Chalcosia alpherakyi, Leech, T. E. S. 1898, 345 (1898)♂ ♀[1]; Jordan, Seitz Lep. Pal. II 12, t. 2 f. F (1910).[2] — China-Tibet frontier.[1, 2] (10,000 feet).

hamiltoni, Swinhoe

Milleria hamiltoni, Swinh., T. E. S. 1891, 475-476 (1891) ♂[1]; Hmp., F. I. Moths I 264 (1892) ♂[2]; Swinh., T. E. S. 1895, 35 (1895).[3]

Chalcosia hamiltoni, Jordan, Seitz Lep. Ind. II 37, t. 7 f. B (1908).[4]

Khasis.[1, 2, 4] Shillong.[3]

pectinicornis, Linnæus (Six subspecies).

(1) pectinicornis pectinicornis, Linnæus — China.[1, 4, 5, 6, 7, 10, 15]

Sphinx pectinicornis, Linn., Syst. Nat. (ed. X) I 495 (1758)[1]; Edw., Gleanings t. 226 (1758).[2]

Gynautocera pectinicornis, Gúer., Deless. Souv. Ind. p. 88, t. 24 f. 4 (1843)[3] [? haec subsp.].

Chalcosia pectinicornis, Jordan, Seitz Lep. Pal. II 11, t. 2 f. E (1910).[4]

Papilio thallo, Linn., Syst. Nat. (ed. XII) I (ii) 656 (1767).[5]

Sphinx thallo, Donovan, Ins. China, t. 41 f. 1 (1798).[6]

Chalcosia thallo, Hmp., F. I. Moths I 266-267 (1892) [*partim*].[7]

Sphinx auxo, Linn., Syst. Nat. (ed. XII), p. 805 (1767).[8]

Chalcosia auxa, Jordan, Seitz Lep. Ind. II 37 (1908)[9]; Hering, Arch. f. Naturg. LXXXVIII (A. 11), p. 72 (1922).[10]

Chalcosia auxo thallo, Hering. *l. c.*[11]

Bombyx tiberina, Cramer. Pap. Exot. I, t. 32 ff. C, D (1775).[12]

Chalcosia tiberina, Hb., Samml. Exot. Schmett. ii (1816-1824) [♀. *nec.* ♂][13]; Wlk., Cat. II 422 (1854).[14]

Chalcosia guerini, Kirby, Cat. Het. I 45 (1892).[15]

(2) pectinicornis idæoides, Herrich-Schäffer

Chalcosia tiberina, [*nec* Cram.], Hb., Samml. Exot. Schmett. ii (1816-1824) [♂, *nec* ♀.].[16]

Chalcosia idaeoides, H.-S., Aussereur. Schmett. I, f. 6 (1853)[17]; Hmp., F. I. Moths I 265 (1892) [*partim*] [redescr.][18]; Swinh., T. E. S. 1895, 35 (1895).[19]

Chalcosia auxo idaeoides, Jordan, Seitz Lep. Ind. II 37, t. 7 f. B (1908)[20]; Hering., Arch. f. Naturg. LXXXVIII (A. 11), p. 72 (1922).[21]

Chalcosia pectinicornis, [*nec* Linn], Wlk, Cat. II 420 (1854).[22]

Khasis.[19, 20]
Shillonmg.

(3) petinicornis diana, Butler

Chalcosia diana, Butl., P. Z. S. 1877, 815 (1877).[23]

Chalcosia auxo diana, Jordan, Seitz Lep. Ind. II 37, t. 7 f. C (1908)[24]; Hering, Arch. f. Naturg. LXXXVIII (A. 11), p. 72 (1922).[25]

Formosa.[23, 24]

(4) pectinicornis albata, Moore

Chalcosia albata, Mo., P. Z. S. 18 1879, 390 (1879).[26]

Chalcosia idaeoides, Hmp, F. I. Moths I 265 (1892) [*partim*].[27]

Chalcosia auxo albata, Jordan, Seitz Lep. Ind. II 37 (1908)[28]; Hering, Arch. f. Naturg. LXXXVIII (A. 11), p. 72 (1922).[29]

Dharmsala.[23]
N.-W. Himalaya.[27-29]
Jeolikote (Kumaon).
Nepal.[27-20]
Khatmandu.

(5) pectinicornis argentata, Moore

Chalcosia argentata, Mo., Lep. Atk., pp. 17-18 (1879)[30]; Hmp. F. I. Moths I 265 (1892) [redescr.] [*partim*].[31]

Chalcosia auxoargentata, Jordan, Seitz Lep. Ind. II 37, t. 7 f. C. (1908)[32]; Hering, Arch. f. Naturg. LXXXVIII (A. 11), p. 72 (1922).[33]

"Hazar D." (Hazara Dist.?).[30]
Sikkim.[32]
N.-E. Assam.[32]

Chalcosia adalifa [*nec* Doubl.], Wlk., Cat. II 421 (1854).[34]

(6) pectinicornis campa, Jordan

Chalcosia auxo campa, Jord., Seitz Lep. Ind. II 37 (1908)[35]; Hering, Arch. f. Naturg. LXXXVIII (A. 11), p. 72 (1922).[36] — Tonkin.[35]

phalænaria, Gúerin

Gynautocera phalaenaria, Gúer., Delessert, Souv. Ind. II 84, t. 24 f. 1 (1843).[1]

Heterusia pulchella, H.-S., Aussereur. Schmett., ff. 160, 161 (1854).[2]

Chalcosia phalaenaria, Jordan, Seitz. Lep. Ind. II 38 (1908) [Seven subspp.:–1, *birmanica,* Jord., L. Burma; 2, *annamitica,* Jord., Annam; 3. *latifasciata,* Butl., Malay Penins.; 4, *coliadoides,* Wlk., Borneo, Sumatra; 5, *obtusa,* Jord., Palawan; 6, *enganica,* Jord., Engano; 7, *phalaenaria,* Gúer., Java].[3]

Chalcosia phalaenaria, Hering, Arch. f. Naturg. LXXXVIII (A. 11), pp. 72-73 (1922) [Three new subspp.:– *intermedia,* Sumatra; *glaucoplaga,* Sumatra; *albocellularis,* Luzon].[4]

Lower Burma.[5]
-----------. Annam.[3]
Bhilippines.[4]
Borneo.[3], Java.[3]
Sumatra.[3, 4]

phalænaria birmanica, Jordan

Chalcosia phalaenaria birmanica, Jordan, Seitz Lep. Ind. II 38 (1908)♂ ♀.[5] — Lower Burma.[5]

pretiosa, Walker

Chalcosia pretiosa, Wlk., Cat. XXXI 116 (1864)[1]; Mo., Lep. Ceylon II 44, t. 96 f. 4 (1884)[2]; Hmp. F. I. Moths I 267 (1892) [redescr.][3]; Jordan, Seitz Lep. Ind. II 37 (1908).[4]

Ceylone.[1-3]
Haldummulla.[4]
Pundaluoya.[4]

ab. 1, albina, Hampson

Chalcosia pretiosa var. *albina,* Hmp., F. I. Moths I 267 (1892) ♂[5]; *id.,* Ill. Het. IX, t. 157 f. 5 (1893).[6]

Chalcosia pretiosa albina, Jordan, Seitz Lep. Ind. II 37 (1908).[7]

ab. 2, eximia, Jordan

Chalcosia pretiosa eximia, Jord., Seitz Lep. Ind. II 37, t. 7 f. B (1908).[8] — Horton Plaing.[3]

remota, Walker

Eterusia remota, Wlk., Cat. II 431 (1854).[1]	Tibet (?).
Laurion remota, Butl., Ill. Het. II 9, t. 23 f. 10 (1878).[2]	W. China.[3]
Chalcosia remota, Jord., Seitz Lep. Pal. II 12, t. 2 f. F (1910).[3]	N. China.[1, 2] Korea.[3], Japan.[3]

reticularis, Leech

Chalcosia reticularis, Leech, T. E. S. 1898, 344 (1898) ♀.[1]; Jord., Seitz Lep. Pal. II 12, t. 2 f. E (1910) ♀.[2]	Ta-tsien-lu.[1, 2]

suffusa, Leech

Chalcosia suffusa, Leech, T. E. S. 1898, 343-344 (1898)♂ ♀[1]; Jord., Seitz Lep. Pal. II 11-12, t. 2 f. E (1910)[2]; Hering, Arch. f. Naturg. LXXXVIII (A. 11) pp. 73-74 (1922) [redescr.].[3]	Omei-shan.[1] Chia-ting-fu.[1] W. China.[2]

syfanica, Oberthür

Chalcosia syfanica, Obth., Et. Ent. XIX 33, t 6 f. 43 (1894)[1]; Jord., Seitz Lep. Pal. II 12, t. 2 f F. (1910).[2]	Ta--tsien-lu.[1, 2]

thibetana, Oberthür

Arbudas thibetana, Obth., Et. Ent. XIX 32, t. 6 f. 44 (1894).[1]	Ta-tsien-lu.[2]
Chalcosia thibetana, Jord., Seitz Lep. Pal. II 12, t. 2 f. F (1910).[2]	

venosa, Walker

Chalcosia venosa, Wlk., Cat. II 422-423 (1854)[1]; Mo., Lep. Ceylon II 45, t. 97 ff. 2, 2*a* (1882)[2]; Jordan, Seitz Lep. Ind. II 37, t. 7 f. B (1908)[3]; Hering, Arch. f. Naturg. LXXXVIII (A. 11), p. 72 (1922)[4].	Ceylon.[1-6, 9] Pundaluoya.[7] Kandy.
Chalcosia quadrifasciata, Mo., Lep. Ceylon II. 45, t. 96 f. 3 (1882) ♂.[5]	
Chalcosia similata, Mo., Lep. Ceylon, II 45, t. 97 ff. 3, 3*a* (1882)♂ ♀.[6]	
Chalcosia myrrhina, Hmp., F. I. Moths I 265 (1892) ♂[7]; *id.,* Ill, Het. IX, t. 157 f. 14 (1893).[8]	
Chalcosia thallo, Hmp., F. I. Moths I 266-267 (1892) [*partim*].[9]	

46. Chalcophædra, Jordan (1908).

Chalcophaedra, Jord., Seitz Lep. Ind. II 39 (1908); Hering Arch. f. Naturg. LXXXVIII (A. 11), p. 74 (1922).	Type:–*zuleika,* Doubl.

zuleika, Doubleday

Gynautocera zuleika, Doubl., Ann. Nat. Hist. XIX 76, t. 7 f. 4 (1877).[4]	Sikkim.[3, 6], Bhutan.[3]
Chalcosia zuleika, Wlk., Cat. II 423 (1854).[2]	Sylhet.[2], Khasis.[5]
Chalcophaedra zuleika, Jordan, Seitz Lep. Ind. II 39, t. 7 f. D (1908).[3]	Cherrapunji.[6] Shillong.
Chalcosia corrusca, H.-S., Aussereur. Schmett. I, f. 1 (1853).[4]	Burma.[3]
Chalcosia corrusca, Hmp., F. I. Moths I 266 (1892) [redescr.][5]; Swinh., T. E. S. 1895, 36 (1895).[6]	

47. Rhodopsona, Jordan (1910) (? 1908).*

Rhodopsona, Jordan, Seitz Lep. Pal. II 10 (1910); *id.,* Lep. Ind. II 39 (1908) [apparently published after Lep. Pal. !] Hering, Arch. f. Naturg. LXXXVIII (A. 11), p. 74 (1922) [key to spp.].	Type:–*costata,* Wlk.

bocki, Swinhoe

Soritia bocki, Swinh., A. M. N. H. (7) XVI 145 (1905)♂ ♀.[1]	Hsipaw.[1] (Burma)
Rhodopsona bocki, ed., Seitz Lep. II 39, t. 7 f. E (1908).[2]	Mandalay.

costata, Walker

Retina costata, Wek., Cat. II 439 (1854)[1]; Butl., Ill. Het. III 9, t. 43 f. 7 (1879)[2]; Leech, T. E. S. 1898, 337 (1898).[3]	Omei-Shan.[3]
Rhodopsona costata, Jordan, Seitz Lep. Pal. II 10, t. 2 f. c. (1910).[4]	W. China.[3, 4] Centr. China.[3, 4] N. China.[1, 3, 4]

jordani, Oberthür

Rhodopsona jordani, Obth., Bull. S. E. Fr. 1910, 314 (1910)[1]; *id.,* Et. Lep. comp. V. 330, t. 77, f. 696 (1911)[2]; Obth., Seitz Lep. Pal. II 441 (1912).[2]	Moupin.[1] E. Tibet border.[2] Siao-lu.[3]

reverdini, Oberthür

Rhodopsona reverdini, Obth., Bull. S. E. Fr. 1910, 314 (1910); *id.,* Et. Lep. Comp. V 330, t. 77, f. 697 (1911).[2]	Siao-lu.[1, 2]

rubiginosa, Leech

Retina rubiginosa, Leech, T. E. S. 1898, 338 (1898) ♂.[1]	Tibet(?).
Rhodopsona rubiginosa, Jordan, Seitz Lep. Pal. II 10, t. 2, f. C (1910).[2]	W. China.[2] kwei-chow Prov.[1]

*1907 according to the Zoological Record.

48. Retina, Walker (1854).

Retina, Wlk., Cat. II 438-439 (1854); Jordan, Seitz Lep. Ind. II 39 (1908) [redescr.]; Hering, Arch. f. Naturg. LXXXVIII (A. 11), p. 75 (1922) [key to spp.]. — Type:–*rubrivitta,* Wlk.

rubrivitta, Walker

Retina rubrivitta, Wlk., Cat. II 439 (1854)[1]; Butl., Ill. Het. V 25, t. 81, f. 4 (1881)[2]; Jordan, Seitz Lep. Ind. II 39, t. 7 f. E (1905).[3]

Soritia rubrivitta, Hmp., F. I. Moths I 252 (1892) ♀.[4]; Swinh., T. E. S. 1895, 34 (1895).[5]

Retina rubrivittata, Hering, Arch. f. Naturg. LXXXVIII (A. 11), p. 75 (1922).[6]

Sikkim.[3,4] Sibsagar.[4] Sylhet.[1,4] Cherrapunji.[5]

49. Phlebohecta, Hampson (1892).

Phlebohecta, Hmp., F. I. Moths I 251, f. 167 (1892); Jordan, Seitz Lep. Ind. II 39 (1908) [redescr.]; Hering, Arch. f. Naturg. LXXXVIII (A. 11), pp. 75-76 (1922) [key to spp.]. — Type:–*fuscescens,* Ma.

Paraphlebia, Felder, Reise Novara, Lep. IV 7 (1874). — Type:–*lithosina,* Felder.

eupoma, Swinhoe

Herpa eupoma, Swinh., A. M. N. H. (6) XIX 166 (1897) ♂[1]; Hmp., B. J. XI 284-285 (1897).[2]

Phlebohecta ? eupomia, Jordan, Seitz Lep. Ind. II 40, t. 7 t. F. (1908) [probably represents a different genus.][3]

Khasis.[2] Jaintias.[1,2]

fuscescens, Moore

Soritia fuscescens, Mo., Lep. Atk., p. 16 (1879).[1]

Retina ? fuscescens, Elwes, P. Z. S. 1890. 386, t. 32 f. 12♀ (1890).[2]

Phlebohecta fuscescens, Hmp., F. I. Moths I 251 f. 167 (1892)♂ ♀.[3] Swinh., T. E. S. 1895. 33 (1895)[4]; Jordan, Seitz Lep. Ind. II 39, t. 7 f. E (1908).[5]

Sikkim.[2,3,5] Darjiling.[1] Bhutan.[5] Shillong.[4]

lithosina, Felder (Two subspecies).

(1) lithosina lithosina, Felder

Paraphlebia lithosina, Feld., Reise Novara, Lep. IV, t. 83, f. 6 (1874).[1] — Borneo.[1,2]

Phlebohecta lithosina lithosina, Jordan, Seitz Lep. Ind. II 39, t. 7 f. F (1908).[2] — Burma.[2]

(2) lithosina flavicosta, Elwes

Retina ? flavicosta, Elwes, P. Z. S. 1890, 386, t. 32 f. 1♀ (1890).[3] — Sikkim.[3, 4, 6] Burma.[4]

Phlebohecta flavicosta, Hmp., F. I. Moths I 251 (1892) ♀[4]; Hmp., B. J. XIII 226 (1900) [synon.].[5]

Phlebohecta lithosina flavicosta, Jord., Seitz Lep. Ind. II 39 (1908).[6]

Soritia lithosia, Obth., Et. Ent. XIX 26, t. 5 f. 25 (1894).[7]

50. Pseudoscaptesyle, Hering (1922).

Pseudoscaptesyle, Hering, Arch. f. Naturg. LXXXVIII (A. 11), pp. 76-77 (1922). — Type:–*circumdata*, Wlk.

circumdata, Walker (Three subspecies).

(1) circumdata circumdata, Walker

Eterusia circumdata, Wlk., Cat. XXXI 121 (1864) [Hindostan].[1] — Sikkim.[2, 3]

Pidorus circumdatus, Hmp., F. I. Moths I 254 (1892) [*partim*].[2]

Eterusia circumdata circumdata, Jordan, Seitz Lep. Ind. II 32-33 (1908).[3]

Pseudoscaptesyle circumdata, Hering, Arch. f. Naturg. LXXXVIII (A. 11), p. 77 (1922).[4]

(2) circumdata purpuralis, Jordan

Pidorus circumdatus, Hmp., F. I. Moths I 254 (1892) [*partim*][5]; Swinh., T. E. S. 1895, 34 (1895).[6] — Khasis.[7] Shillong.[6] Cherrapunji.[6]

Eterusia circumdata purpuralis, Jordan, Seitz. Lep. Ind. II 33, t. 6 f. B (1908).[7]

(3) circumdata citrana, Hering

Pseudoscaptesyle circumdata citrana, Hering, Arch. f. Naturg. LXXXVIII (A. 11), p. 77 (1922) ♂.[8] — Sumatra.[8]

Phaudinae

Phaudinae : Hmp., F. I. Moths I 285-286 (1892) [key to genera]; Jordon, Seitz Lep. Ind. II 6 (1908) [redescr.]; *id.,* Lep. Pal. II 4 (1910) [redescr.].

51. Phauda, Walker (1854)

Phauda, Wlk., Cat. I 256 (1854); Kirby Cat. Het. I 111 (1892); Hmp., F. I. Moths I 287, f. 196 (1892) [redescr.]; Snellen, Tijds., Ent. XLV 238-239 (1902) [redescr.]; Jordan, Seitz Lep. Ind. II 6-7 (1908) [redescr. ; larva descr.]; *id.,* Lep. Pal. II 5 (1910) [redescr.].

Type:–*flammans,* Wlk.

Xenares, Herrich-Schäffer, Aussereur. Schmett I 7, f. 223 (1854).

Type:–*triadum* Wlk. [N. China]

dichroa, Jordan

Phauda dichroa, Jord., Seitz Lep. Ind. II 7, t. 1 f. E (1908) ♀.[1]

Port Blair.[1] (Andamans)

erythra, Jordan

Phauda triadum erythra, Jordan, Seitz Lep. Ind. II 7, t. 1 f. E (1908) ♀.[1]

Belgaum.[4]

Larva descr.[1]

flammans, Walker

Euchromia [*Phauda*] *flammans,* Wlk., Cat. I 257 (1854) [N. India].[1]

Simla.[3, 8]
Lahore.[8]

Phauda, flammans, Butl., Ill. Het. I 20, t. 9 f. 2 (1877)[2]; Hmp., F. I. Moths I 287, f. 196 (1892) [redescr.][3]; Swinh., T. E. S. 1895, 38 (1895)[4]; Dudgeon, B. J. XII 303 (1899)[5]; Piepers and Snellen, Tijds. Ent. XLV 241, t. 13 f. 13 [larva] (1902)[6]; Hmp., P. Z. S. 1906, 493 (1906)[7]; Jordan, Seitz. Lep. Ind. II 7, t. 1 f. E (1908)[8].

Sikkim.[3, 5, 8]
Raitdong (Sikkim).[7], Pusa.
Cherrapunji.[4]
Burma.[3, 8]

Java.[6]

Larva descr.[6], on *Ficus* sp.[6]

fuscalis, Swinhoe

Phauda fuscalis, Swinh., Cat. Het. Mus. Oxon. I 60 (1892) ♀.[1]; Hmp., F. I. Moths I 288 (1892) ♀.[2]; Dudgeon, B. J. XII 303 (1899)[3]; Jordan, Seitz Lep. Ind. II 7, t. 1 f. E (1908) ♀.[4]

Sikkim.[3]
Assam.[1, 2, 4]

pratti, Leech

Phauda pratti, Leech, Entom, XXIII 81 (1890)[1]; *id.,* T. E. S. 1898. 350 (1898)[2]; Jordan, Seitz Lep. Pal. II 5, t. 1 f. A (1910).[3] — Moupin.[2, 3] Wa-shan.[2, 3] [E. Tibet borde] ———— Ichang [C. China].[1]

52. Alophogaster, Hampson (1892)

Alophogaster, Hmp., F. I. Moths I 287, f. 195 (1892); Jordan, Seitz Lep. Ind. II 7 (1908) [redescr.]. — Type:–*rubribasis,* Hmp.

rubribasis, Hampson

Alophogaster rubribasis, Hmp., F. I. Moths I 287, f. 195 (1892); ♂[1] Jordan, Seitz Lep. Ind. II 7, t. 1 f. F (1908).[2] — Nagas.[1, 2] Khasis.[2, 3]

Alophogaster serraticornis, Hmp., F. I. Moths IV 471 (1896) [♀. "♂ " ex. err.][3].

53. Phaudopsis, Hampson (1900)

Phaudopsis, Hmp., B. J. XIII 226, fig. (1900); Jordan, Seitz Lep. Ind. II 8 (1908) [redescr.]. — Type:–*igneola,* Hmp.

igneola, Hampson

Phaudopsis igneola, Hmp., B. J. XIII 226, fig. (1900)♂ ♀.[1]; Jordan, Seitz Lep. Ind. II 8, t. 1 f. F (1908).[2] — Khasis.[1, 2]

54. Akesina, Moore (1888)

Akesina, Moore, P. Z. S. 1888. 396 (1888); Jordan, Seitz Lep. Ind. II 8 (1908) [redescr.]. — Type:–*basalis,* Mo.

Anomœotes, Felder, Reise Novara, Lep. IV, t. 100 f. 5 (1874) [non-descr.]; Butl., Ill. Het. VII 28 (1889); Kirby, Cat. Het. I 59 (1892); Hmp., F. I. Moths I 286, f. 194 (1892) [redescr.]. — Type:–*basalis,* Mo.

basalis, Moore

Anomœotes basalis, Mo., P. Z. S., 1888. 396 (1888) [*Akesina*][1]; Butl., Ill. Het. VII 28, t. 121 f. 9 (1889)[2]; Hmp. F. I. Moths I 286, f. 194 (1892) ♂[3]; Jordan, Seitz Lep. Ind. II 8 (1908) [*Akesina*].[4] — Kangra.[1-4]

55. Pseudopsyche, Edwards (1882)

Pseudopsyche, H. Edwards, Papilio II 124 (1882); Jordan, Seitz Lep. Pal. II 5 (1910) [redescr.]. — Type:–*exigua,* Edw.[Arizona]

yarka, Oberthür Ta-tsien-lu.[1]

Pseudopsyche ? yarka, Obth., Et. Ent. XVIII 42, t. 4 f. 49 (1894)[1]; Jordan, Seitz Lep. Pal. II 5, t. 1 f. B (1910).[2]

HIMANTOPTERINÆ

Himantopterinæ : Jordan, Seitz Lep. Ind. II 5 (1908) [characters].

Thymaridæ : Kirby, Cat. Het. I 60 (1892).

56. Himantopterus, Wesmael (1836)

Himantopterus, Wesmael, Bull. Acad. Brux. III 162 (1836); Kirby, Cat. Het. I 60 (1892); Hmp., F. I. Moths I 288, f. 197 (1892) [redescr.]; Snellen, Tijds. Ent. XLV 166-167, t. 14 f. 1 [larva] (1902) [redescr.]; Jordan, Seitz Lep. Ind. II 6 (1908) [redescr.]. Type:– *fuscinervis,* Wesm. [Java]

Thymara, Doubleday, Zoologist I 197 (1848); Wlk., Cat. II 383 (1854); Kirby, Cat. Het. I 60 (1892); Kirby, Handbk. Lep. III 83-84 (1897). Type:–*zaida,* Doubl.

caudatus, Moore

Thymara caudata, Mo., P. Z. S. 1879, 394, t. 32 f. 3 (1879)[1]; Elwes, T. E. S. 1890, 333, t. 10 ff. 9, 11, 12 (1890).[2]

Himantopterus caudatus, Hmp., F. I. Moths I 288 (1892) ♂[3]; Ramakrishna Ayyar, B. J. XXIV 378 (1916)[4]; Jordan, Seitz Lep. Ind. II 6, t. 1 f. D (1908).[5]

Nilgiris.[2, 3, 5] Wynaad.[3] Bababudin Hills.[4] S. India.[5] N. India.[5] [?] Burma.[1, 3, 5] [?] Pankabari (Skiikim).[2][?]

dohertyi, Elwes (Two subspecies)

(1) dohertyi dohertyi, Elwes

Himantopterus dohertyi, Elwes, T. E. S. 1890, 332, t. 10 ff. 1-7 (1890)[1]; Hmp., F. I. Moths I 288 (1892) [redescr.][2]

Himantopterus dohertyi dohertyi, Jordan, Seitz Lep. Ind. II 6 (1908).[3]

Naga Hills.[1, 3]

(2) dohertyi elwesi, Jordan

Himantopterus dohertyi, Swinh., T. E. S. 1895. 38 (1895).[4]

Himantopterus doherty elwesi, Jordan, Seitz Lep. Ind. II 6, t. 1 f. D (1908).[5]

Khasis.[5] Cherrapunji.[4] Shillong.

Larva on *Quercus griffithii* at Shillong.

zaida, Doubleday (Two subspecies)

(1) zaida zaida, Doubleday

Thymara zaida, Doubl., Zoologist I 197 (1843)[1]; Wlk., Cat. II 383 (1854)[2]; Elwes, T. E. S. 1890, 333, t. 10 ff. 8, 10, (1890).[3]

Khasis.[4] Shillong.[5] Cherrapunji.[2]

Himantopterus zaida, Hmp., F. I. Moths I 289, f. 197 (1892) ♂[4]; Swinh., T. E. S. 1895. 38 (1895).[5]

Himantopterus zaida zaida, Jordan, Seitz Lep. Ind. II 6, t. 1 f. D (1908).[6]

(2) zaida flavescens, Jordan

Himantopterus zaida flavescens, Jord., Seitz Lep. Ind. II 6 (1908).[7]

Dawna Hills.[7]

225 species.

www.ingramcontent.com/pod-product-compliance
Ingram Content Group UK Ltd.
Pitfield, Milton Keynes, MK11 3LW, UK
UKHW020232290926
15470UKWH00002B/493

9 789386 071064